GPS Waypoints:
Idaho

This handy reference provides over 12,000 Idaho waypoints that are ready to enter into your GPS receiver. In a matter of seconds you can find the precise coordinates for any of Idaho's mountains, lakes, cities, towns, parks, campgrounds, airports, hospitals, highway intersections, and much more. A few keystrokes later the coordinates you need can be in your GPS receiver, ready to guide you to the location of your choice. Enjoy!

compiled by

Michael Ferguson

Glassford Publishing
Boise, Idaho

Library of Congress Catalog Card Number: 98-89844

Ferguson, Michael H., 1950-
GPS Waypoints: Idaho
1. Global Positioning System. 2. Outdoor Recreation 3. Travel Guides

ISBN 1-892182-16-5

Printed in the United States of America
10 9 8 7 6 5 4 3 2 1

Disclaimer: The information provided in this book is accurate to the best knowledge of the author and publisher. However, much of it is taken from materials supplied by the U.S. Government. There is no warranty, therefore, express or implied, as to the accuracy of the data contained herein.

Contents

[1] Arches, Beaches, Caves, Craters, Falls, Lava Flows, Swamps, Woods

Important Information About This Waypoint Guide

The features (and their coordinates) listed in this waypoint guide are primarily derived from U.S. Government databases. The Highway Intersection data is obtained from the U.S. Transportation Department's "National Transportation Atlas Database." Most other data is obtained from the U.S. Department of the Interior's "Geographic Names Information System."

Every effort has been made to assure that these waypoints have been reported as accurately as possible. However, it is impossible to independently verify the accuracy of each feature that is listed. Consequently, these coordinates must not be used as a sole means of navigation. DO NOT RELY ON THESE COORDINATES IN SITUATIONS WHERE YOUR HEALTH OR SAFETY IS AT RISK.

The airport coordinates listed in this book are intended for ground use. DO NOT RELY ON THESE AIRPORT COORDINATES FOR AVIATION PURPOSES.

The marine features listed in this book provide approximate position information. DO NOT RELY ON THEM TO STEER OR OTHERWISE NAVIGATE A WATERCRAFT.

The highway intersection coordinates listed in this book are based on the North American Datum of 1983 (NAD83). All other coordinates are based on the North American Datum of 1927 (NAD27). Be sure your GPS receiver is set to the correct datum prior to entering coordinates taken from this guide. Also note that all coordinates are listed as ***degrees, minute, seconds***. You will need to set your receiver to that coordinate format prior to entering waypoint coordinates taken from this guide. After these coordinates have been entered into your receiver you may change your receiver's datum and/or coordinate system format to suit your needs.

Some of the coordinates listed represent area features (Basins, Bays, Flats, Lakes, Parks, etc). Waypoints that represent area-type features are usually based on the approximate geographic center of the feature. An exception to this rule occurs in the case of populated places, i.e. cities and towns. In this category the coordinates may represent the center of the area's population distribution, the location of the central Post Office, City Hall, etc. In most cases you will be well within the "area" of an area-type feature before your receiver indicates you are there. In other words, for a typical area feature you will probably reach its perimeter before you reach the waypoint that represents it.

Many of the feature categories represent point features. Arches, Caves, Mountains, Pillars and Summits are examples of point feature categories. Be careful, however, since some features you might think of as a point are actually areas. Highway Intersections are a good example. In most cases they are indeed point in nature, the point being where the two routes' centerlines cross. However, in some cases (urban freeways, for example) the intersection is fairly complex and spread-out over a relatively large geographic area.

The accuracy standard for the point features included in this guide is within ±5 seconds, or about 175 meters. The effect of Selective Availability adds 100 meters of error (at 95% probability) when applicable. This means that any error in the listed coordinate data will be compounded by the amount of receiver error that occurs when navigating in the field. Since the coordinate data is considered accurate to within 175 meters, Selective Availability adds another 100 meters, for a total potential error of ±275 meters. Additional error may arise due to poor satellite geometry, multipath interference, typographic errors, or other error sources. See ***GPS Land Navigation***, also by Michael Ferguson, for more information on the sources of GPS receiver errors.

How This Book Is Organized

The waypoints listed in this book are grouped into 22 major feature categories, plus a miscellaneous category that contains 8 minor feature categories that each have a small number of items - arches, caves, falls, swamps, etc. Each category's features are sorted alphabetically by feature name. If identical feature names occur, they are further sorted alphabetically by county. If identical feature names (within a category) occur within a county, the items are further sorted by latitude, from north to south. This sorting system is intended to make it easier to find a specific feature, especially in cases where duplicate names occur. In the case of the miscellaneous category, the first sort is done on the feature type.

In the case of Highway Intersections the "feature name" is a composite of the two routes that intersect. Each intersection can be named in two ways: "Route 1 Route 2," and "Route 2 Route 1." Only the higher ordered variant is listed - in this case, "Route 1 Route 2." This means you always need to look up the lower number in a pair of routes that intersect. Also, be aware that regardless of their numbers, Interstate Highways are ordered above U.S. Highways, and U.S. Highways are ordered above State Highways. This means that all Interstate Highways are listed first, followed by U.S. Highways. State Highways only appear as second elements of Interstate Highway and U.S. Highway "feature names." Intersections of two State Highways are not included in this listing.

Be aware that a feature might not be listed in the category you expect. For example, campgrounds can be listed in either the Camps & Campgrounds category, or the Parks & Reserves category. A few features named "Mountain" appear in the Ridges & Slopes category. Features named "Rock" usually appear in the Rocks & Pillars category, but a few are listed in the Mountains, Peaks, Summits category. In short, don't be afraid to look in several categories to find an "elusive" feature!

The information provided for each feature includes the county it is located in. Among other things, this can help differentiate features with identical names. Since only one county is listed, there will be cases where features that involve multiple counties will be limited to showing only one of two or more counties. Examples of this situation include summits that are on the border of two or more counties, and lakes that lie in two or more counties.

How To Get The Most From This Book And These Waypoints

This book of waypoints has one primary purpose: to help you get the most benefit and fun from your GPS receiver. While it is targeted at GPS receiver users, this book is also useful to any topographic map user, including those who are not GPS owners. That's because virtually all the features listed on these pages correspond exactly with the features shown on United States Geological Survey topographic maps. In a other words, this book can also serve as the index (for the feature categories included) to the entire set of 1,704 7.5-minute USGS topographic maps for the state of Idaho!

Before getting into the ways to use the coordinates provided in this book, let's look at a few ways to make storing and using this book easier. First, if you plan to take this book with you (i.e., in your pack or glove box), it may be worthwhile to place it in a ziplock plastic bag to protect it from the elements and/or its surroundings. A gallon size bag works well.

Second, you may want to keep a short (3 1/2" to 5 1/2") bookmark in the book. This isn't to keep your place, but rather to serve as a straight-edge when you are looking up features. You can even use a medium-sized sheet of Post-It note paper, with the added benefit that your "straight-edge" will stick to the page and not slip!

Finally, if you prefer a book that lays flat, you can take this book to a suitable quick print shop (Kinko's, etc.) and have the spine removed and a spiral binding put in its place. This will cost a few dollars, and you'll lose the spine, but your book will lay perfectly flat. Incidently, producing this book with a spiral binding is more expensive, so you would have paid several dollars more anyway. Now let's look at two of the most basic ways to use the waypoint coordinates contained in this book:

Finding A Location In The Real World

Perhaps the most basic use for the coordinates listed in this book is entering them in your GPS receiver. Say you want to "find" Borah Peak. You look under the heading for Mountains and see it is in Custer county at 44°08'14"N, 113°46'46"W. Enter these coordinates into your receiver and the next time you have a satellite lock you will be able to see the exact distance and direction from where you are to Idaho's highest peak (as an added bonus, the listing also provides Borah Peak's elevation of 12,662 feet). All you need to do is "GoTo" the waypoint you created using the coordinates from this book!

You can use this procedure with any coordinates from this book. Say you're at a scenic overlook and you want to know the distance and direction to a peak you think is in view. Look it up in this book, enter its coordinates into your receiver, "GoTo" the peak and your receiver will tell you its exact distance and direction. If you have an accurate compass handy, you'll even be able to pick it out of a "lineup" of many peaks.

You can even use this book effectively in routine everyday situations. Say you're going to drive from Boise, Idaho to Oakley, Idaho - for the first time. You can easily get the coordinates for Oakley from the Cities, Towns, Historic Places category. You can also see on your highway map that you need to drive east on Interstate 84 then turn south on State Route 27. You obtain the exit coordinates by looking up the intersection named "I 84 SR 27." With just these two waypoints in your receiver you'll have the ability to know the exact distance (as the crow flies) to either your freeway exit or your destination. No more wondering how far, and no more worrying about daydreaming right past your exit.

Even if you have a topographic map that includes the feature you're interested in, it can be much quicker and simpler to obtain the coordinates you want by looking them up in this book. That way you don't need to fuss with a Latitude/Longitude ruler, UTM grid reader, or dividers. However, if the waypoint is very important ("mission critical," so to speak) you will want to use those tools to double-check the coordinates.

Finding A Location On A Map

Another basic use for this guide is finding items on a particular map. Say you are working with a USGS 1:100,000 map of southwestern Idaho and you want to find Basque Spring. You think it is somewhere on your map, but there are dozens of springs and hundreds of named features on that map. Rather than scanning the map to try and find what may be a needle in a haystack, just look up Basque Spring in the Springs section of this book and note its latitude and longitude. Then, using the latitude and longitude grid on your map, eyeball its approximate location. You should be able to pinpoint Basque Spring (if it's on your map) in a matter of seconds.

This method is very much like using the letters and numbers that serve as the look-up grid on street maps and phone book maps. Just remember there are 60 seconds in a minute, and 60 minutes in a degree. The only difference between this guide and the index to a street map is you will need to interpolate using these angular units to sight in your "crosshairs."

You can even apply this method with highway maps, as long as the map shows latitude and longitude. Highway intersections, towns, and other major features should match, but be careful with feature names. The names in this book are identical to those used by the U.S. Geological Survey, but they may not always match the names used on commercial maps.

When you look-up a feature by its name, you may find there are more than one feature (of a particular category) that share the same name. If you know the county that can help narrow your choices, since the feature's county is included in the listing. If you only know the feature's approximate location within the state, the map on page 12 can be used with either the county information or the latitude/longitude information associated with the feature to narrow the choices.

If you want to estimate the distance between features, convert latitude and/or longitude to their respective ground distances. Each minute of latitude is 1.15 miles in length, and each minute of longitude is approximately .8 miles in length. Each second of latitude is 101 feet in length, and each second of longitude is approximately 70 feet in length. Please note the conversion for latitude is constant, but the conversion for longitude depends on the latitude. The conversion figures above are for a latitude of 45°, the approximate center of Idaho.

UTM Coordinates – Why They're Not Used In This Book

Many readers will wonder why this book uses latitude and longitude, rather than Universal Transverse Mercator (UTM) for its coordinate system. The answer is very simple: convenience.

Very few maps (other than U.S. Geological Survey topographic maps) show the UTM coordinate system. On the other hand, most maps (including USGS topographic maps) show the latitude/longitude coordinate system. Listing the waypoints in this book using latitude/longitude gives **you** the most flexibility when it comes to using this book with a wide variety of maps. And, listing these coordinates as latitude/longitude doesn't prevent you from using UTM. Just load the desired coordinates into your GPS receiver using the latitude/longitude coordinate system, then switch your receiver to UTM. It will convert those saved coordinates to UTM automatically!

While it's true that UTM is far easier to use when working with coordinates on a USGS topographic map, it's also true that latitude/longitude coordinates are just as easy as UTM when it comes to entering coordinates into your receiver. It's a common misconception that you must pick one coordinate system and stick with it. In reality, you have as much flexibility as your receiver supplies. So, by using latitude and longitude, the coordinates listed in this book are compatible with the largest number of maps.

Idaho's Counties & Major Cities

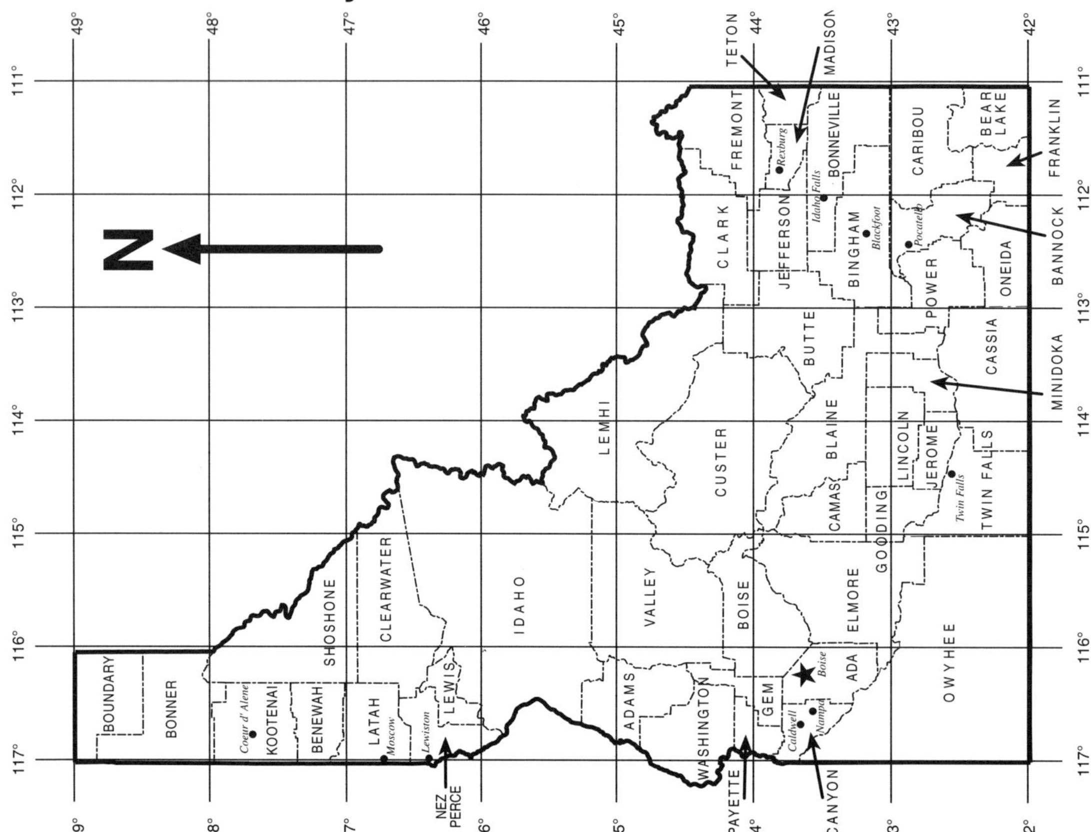

Airport, Heliport, or Landing Strip	County	Latitude	Longitude	Elev.
Aberdeen Municipal Airport	Bingham	42°55'16"N	112°52'49"W	4470
Albion Municipal Airport	Cassia	42°23'52"N	113°33'24"W	4777
Allen H Tigert Airport	Caribou	42°38'54"N	111°34'52"W	5839
American Falls Airport	Power	42°47'56"N	112°49'20"W	4419
Anderson-Plummer Airport	Latah	46°55'28"N	116°57'27"W	2580
Antelope Valley Airport	Butte	43°40'38"N	113°36'07"W	6180
Arco Airport	Butte	43°36'19"N	113°19'51"W	5328
Arco Municipal Airport	Butte	43°36'18"N	113°19'44"W	5320
Athol Airport	Bonner	47°57'27"N	116°40'27"W	2410
Atlanta Airport	Elmore	43°48'49"N	115°08'03"W	5500
Bancroft Municipal Airport	Caribou	42°43'17"N	111°51'44"W	5435
Bannock Regional Medical Center Heliport	Bannock	42°52'04"N	112°25'49"W	4550
Bear Lake County Airport	Bear Lake	42°14'58"N	115°20'22"W	5928
Bear Trap Airport	Blaine	42°58'30"N	113°20'59"W	4716
Benewah Community Hospital Heliport	Benewah	47°18'51"N	116°33'59"W	2235
Bennett Airport	Elmore	43°19'12"N	115°27'00"W	4360
Bernard USFS Airport	Valley	44°58'47"N	114°44'02"W	3626
Big Creek Airport	Valley	45°08'00"N	115°19'10"W	5743
Big Creek Heliport	Valley	45°08'00"N	115°19'05"W	5750
Big Island Airport	Clearwater	46°41'51"N	115°58'58"W	2150
Big Southern Butte Airport	Butte	43°25'58"N	113°03'17"W	5061
Bird Number Two Airport	Bonner	48°14'07"N	116°23'32"W	2192
Boise Air Terminal	Ada	43°33'42"N	116°14'20"W	2858
Boise Cascade Heliport	Ada	43°39'12"N	116°12'18"W	2890
Bottle Bay Seaplane Base	Bonner	48°28'45"N	116°26'40"W	2063
Boundary County Airport	Boundary	48°43'35"N	116°17'40"W	2331
Bowman Field	Fremont	43°59'35"N	111°32'59"W	5100
Bradley Field	Ada	43°38'42"N	116°15'59"W	2626
Bredding Ranch Landing Area	Power	42°48'22"N	113°02'12"W	4551
Brooks Seaplane Base	Kootenai	47°40'20"N	116°47'06"W	2125
Bruce Meadows Airport	Valley	44°24'56"N	115°18'57"W	6370
Buhl Municipal Airport	Twin Falls	42°35'47"N	114°47'46"W	3660
Buhl Municipal Airport	Twin Falls	42°35'47"N	114°47'44"W	3645
Burley Municipal Airport	Cassia	42°32'33"N	113°46'14"W	4150
Cabin Creek USFS Airport	Valley	45°08'37"N	114°55'41"W	4289
Caldwell Industrial Airport	Canyon	43°38'31"N	116°38'04"W	2429
Caldwell Municipal Airport	Canyon	43°39'59"N	116°42'27"W	2349
Camas County Airport	Camas	43°20'31"N	115°47'51"W	5058
Cambell Ferry Airstrip	Idaho	45°29'13"N	115°19'45"W	2644
Carey Airport	Blaine	43°18'30"N	113°56'00"W	4783
Cascade Airport	Valley	44°29'38"N	116°00'55"W	4742
Cavanaugh Bay Airport	Bonner	48°31'08"N	116°49'16"W	2484
Cayuse Creek Airport	Clearwater	46°39'58"N	115°04'21"W	3500
Cedar Mountain Sky Ranch	Kootenai	47°53'55"N	116°39'30"W	2445
Challis Airport	Custer	44°31'23"N	114°13'00"W	5072
Chamberlain Landing Field	Idaho	45°22'25"N	115°11'52"W	5651
Chamberlain USFS Airport	Idaho	45°22'25"N	115°11'52"W	5765
Cliffs Landing Strip	Owyhee	42°38'24"N	116°59'05"W	5011

Airport, Heliport, or Landing Strip	County	Latitude	Longitude	Elev.
Coeur D'Alene Air Terminal	Kootenai	47°46'27"N	116°49'06"W	2318
Coeur D'Alene Resort Heliport	Kootenai	47°40'21"N	116°46'58"W	2130
Cold Meadow USFS Airport	Idaho	45°17'37"N	114°56'42"W	7030
Concord Landing Strip	Idaho	45°35'06"N	115°40'50"W	7655
Copper Basin Airport	Custer	43°48'07"N	113°49'50"W	7920
Copper Basin Landing Strip	Custer	43°48'08"N	113°49'50"W	
Cottonwood Municipal Airport	Idaho	46°02'18"N	116°19'52"W	3474
Council Municipal Airport	Adams	44°45'04"N	116°26'41"W	2963
Council Municipal Airport	Adams	44°43'45"N	116°26'35"W	2926
Coxs Well Airport	Blaine	43°13'05"N	113°13'35"W	5030
Craigmont Municipal Airport	Lewis	46°14'52"N	116°28'41"W	3785
Cumming Triangle Landing Area	Lemhi	45°26'33"N	113°59'39"W	3950
CX Ranch Airport	Bonner	48°08'17"N	116°11'48"W	2071
Davison Ranch Airport	Canyon	43°39'30"N	116°45'01"W	2370
Deadwood Dam Airstrip	Valley	44°17'52"N	115°38'26"W	5400
Dietrich Landing Area	Lincoln	42°55'15"N	114°15'44"W	4200
Dixie USFS Airport	Idaho	45°31'09"N	115°30'52"W	5148
Donnelly Airport	Valley	44°43'42"N	116°05'18"W	4860
Downey Airport	Bannock	42°25'35"N	112°06'30"W	4906
Driggs Teton Peaks Municipal Airport	Teton	43°44'22"N	111°06'01"W	6195
Dubois Municipal Airport	Clark	44°10'00"N	112°13'30"W	5123
Dworshak Heliport	Clearwater	46°30'44"N	116°17'32"W	1615
E I Regional Medical Center Heliport	Bonneville	43°28'10"N	111°59'29"W	4713
Eckhart International Airport	Boundary	48°59'35"N	116°30'05"W	1756
Elk City Airport	Idaho	45°49'18"N	115°26'18"W	4097
Elk River Airport	Clearwater	46°47'14"N	116°10'32"W	2820
Emmett Municipal Airport	Gem	43°51'06"N	116°32'12"W	2350
Falconberry Landing Strip	Custer	44°41'56"N	114°45'52"W	4808
Fanning Field	Bonneville	43°30'47"N	112°04'01"W	4728
Fish Lake Airport	Idaho	46°19'49"N	115°03'44"W	5646
Flat Top Airstrip	Blaine	43°29'38"N	113°55'14"W	5841
Floating Feather Airport	Ada	43°42'06"N	116°18'40"W	2680
Flying B Ranch Landing Strip	Valley	44°58'05"N	114°43'55"W	3647
Flying Y Ranch Airport	Adams	44°47'40"N	116°31'54"W	3180
Foster Ranch Landing Strip	Custer	44°39'24"N	114°38'44"W	5533
Fountains Airport	Latah	46°42'48"N	116°59'49"W	2550
Frank Field	Canyon	43°40'24"N	116°46'16"W	2365
Friedman Memorial Airport	Blaine	43°30'24"N	114°17'49"W	5315
Garden Valley Airport	Boise	44°04'02"N	115°55'51"W	3177
Garden Valley Heliport	Boise	44°04'02"N	115°55'51"W	3150
Gimlet Airport	Blaine	43°37'19"N	114°20'37"W	5600
Glenns Ferry Airport	Elmore	42°56'38"N	115°19'43"W	2536
Glenns Ferry Municipal Airport	Elmore	42°56'45"N	115°19'45"W	2536
Glider Airport	Ada	43°42'53"N	116°17'45"W	2815
Golden Age Mine Heliport	Boise	44°00'10"N	115°48'40"W	4950
Gooding Municipal Airport	Gooding	42°55'03"N	114°45'41"W	3729
Gooding Municipal Airport	Gooding	42°54'54"N	114°45'56"W	3729
Grace Airport	Caribou	42°36'30"N	111°43'40"W	5598

Airports, Heliports, Landing Strips

Airport, Heliport, or Landing Strip	County	Latitude	Longitude	Elev.
Graham USFS Airport	Elmore	43°57'19"N	115°16'18"W	5726
Granite Airport	Bonner	47°59'05"N	116°40'55"W	2260
Grasmere Airport	Owyhee	42°22'23"N	115°52'41"W	5134
Grindstone Ag Airport	Elmore	42°52'11"N	115°22'35"W	3125
Hackney Airpark	Kootenai	47°57'25"N	116°40'35"W	2445
Halls Hilltop Landing Strip	Latah	46°51'08"N	116°22'52"W	3040
Happy Hollow Ranch Airport	Adams	44°56'34"N	116°40'56"W	4652
Hawk Haven Airport	Kootenai	47°45'20"N	116°51'30"W	2333
Hazelton Municipal Airport	Jerome	42°34'36"N	114°08'03"W	4172
Hazelton Municipal Airport	Jerome	42°34'34"N	114°08'09"W	
Henley Airstrip	Bonner	47°54'32"N	116°42'17"W	2345
Henrys Lake Airport	Fremont	44°38'07"N	111°20'40"W	6555
Hickman Airport	Kootenai	47°23'33"N	116°52'19"W	2500
Hidden Lakes Airport	Valley	44°13'30"N	116°10'43"W	4845
High Valley Bills Airport	Valley	44°14'22"N	116°08'37"W	4883
Hollow Top Airport	Blaine	43°19'26"N	113°35'22"W	5369
Homedale Municipal Airport	Owyhee	43°36'56"N	116°55'16"W	2210
Hooks Airport	Kootenai	47°55'44"N	116°42'36"W	2350
Howe Airport	Butte	43°50'21"N	113°02'45"W	4930
Hubler Field	Canyon	43°41'42"N	116°38'12"W	2385
Hubofs Heliport	Kootenai	47°43'59"N	116°59'56"W	2000
Idaho City USFS Airport	Boise	43°49'28"N	115°50'35"W	3920
Idaho County Airport	Idaho	45°56'31"N	116°07'26"W	3309
Idaho State Emergency Airstrip	Idaho	45°25'27"N	116°18'30"W	1773
Indian Creek USFS Airport	Valley	44°45'41"N	115°06'23"W	4701
Interstate Airport	Cassia	42°16'55"N	113°18'22"W	4580
Jerome County Airport	Jerome	42°43'33"N	114°27'15"W	4038
Jerome County Airport	Jerome	42°43'32"N	114°27'16"W	4015
Johnson Creek Airport	Valley	44°54'44"N	115°29'05"W	4933
Jump West Parachute Center Airport	Canyon	43°43'44"N	116°31'19"W	2550
Kamiah Municipal Airport	Idaho	46°13'10"N	116°00'45"W	1194
Klosterman Landing Area	Minidoka	42°40'27"N	113°46'34"W	4205
Kooskia Municipal Airport	Idaho	46°07'58"N	115°58'40"W	1263
Kootenai Medical Center Heliport	Kootenai	47°41'46"N	116°47'38"W	2230
Krassel USFS Airport	Valley	44°58'28"N	115°43'44"W	3982
Laidlaw Corrals Airport	Lincoln	43°02'14"N	113°43'58"W	4439
Lake Pend Oreille Seaplane Base	Bonner	48°13'00"N	116°21'35"W	2063
Landmark United States Forest Service Airport	Valley	44°38'34"N	115°31'55"W	6662
Landmark USFS Airport	Valley	44°38'33"N	115°31'57"W	6662
Lanham Field	Gem	43°52'40"N	116°32'10"W	2343
Larkin Airport	Ada	43°27'46"N	116°21'29"W	2750
Leadore Airport	Lemhi	44°40'28"N	113°21'10"W	6018
Lee Williams Memorial Airport	Washington	44°27'42"N	116°45'24"W	2617
Lemhi County Airport	Lemhi	45°07'18"N	113°52'51"W	4045
Lewiston Airport	Nez Perce	46°22'28"N	117°00'51"W	1438
Loon Creek Landing Strip	Custer	44°35'27"N	114°49'23"W	5420
Mackay Airport	Custer	43°54'33"N	113°36'00"W	5891
Mackay Bar Airport	Idaho	45°22'48"N	115°30'17"W	2172
Magee Airport	Shoshone	47°50'30"N	116°15'04"W	3002
Magic Reservoir Airport	Camas	43°16'57"N	114°23'40"W	4844
Magic Valley Regional Medical Center Heliport	Twin Falls	42°33'53"N	114°29'40"W	3675
Mahoney Creek Airstrip	Valley	44°44'39"N	114°55'14"W	4618
Mahoney Creek USFS Airport	Valley	44°44'41"N	114°55'14"W	4618
Malad City Airport	Oneida	42°10'15"N	112°17'18"W	4503
Malta Airport	Cassia	42°18'41"N	113°22'39"W	4564
May Airport	Lemhi	44°36'38"N	113°53'38"W	5324
McCall Airport	Valley	44°53'40"N	116°05'58"W	5023
McCall Memorial Hospital Heliport	Valley	44°54'33"N	116°06'34"W	5018
McCarley Field	Bingham	43°12'36"N	112°20'54"W	4488
Mercy Heliport	Canyon	43°33'16"N	116°34'00"W	2535
Midway Airport	Bingham	43°27'19"N	112°48'16"W	5005
Mile Hi Landing Strip	Valley	45°09'04"N	114°59'54"W	5880
Minidoka Memorial Hospital Heliport	Minidoka	42°37'16"N	113°41'07"W	4156
Moose Creek Airport	Idaho	46°07'28"N	114°55'20"W	2393
Morgan Ranch Airport	Valley	44°33'12"N	115°18'15"W	5634
Mountain Home Municipal Airport	Elmore	43°07'53"N	115°43'45"W	3164
Mountain Home Municipal Airport	Elmore	43°07'31"N	115°43'36"W	
Mud Lake Airport	Jefferson	43°50'54"N	112°29'54"W	4787
Mullins Airstrip	Idaho	45°42'53"N	115°21'42"W	4299
Murphy Airport	Owyhee	43°13'00"N	116°32'53"W	2855
Murphy Hot Springs Airport	Owyhee	42°01'27"N	115°20'04"W	5820
Murray Airport	Bonner	48°21'22"N	116°30'15"W	2150
Nampa Municipal Airport	Canyon	43°35'05"N	116°31'38"W	2530
Nampa Valley Heliport	Ada	43°33'03"N	116°15'15"W	2795
New Meadows Airport	Adams	44°58'42"N	116°17'00"W	3908
Nez Perce Municipal Airport	Lewis	46°14'15"N	116°14'30"W	3201
Nichols Ranch Airport	Kootenai	47°40'41"N	117°01'26"W	2430
Oakley Municipal Airport	Cassia	42°14'03"N	113°52'36"W	4650
Olmstead Sky Ranch	Bonner	48°21'03"N	116°33'11"W	2140
Orofino Airport	Clearwater	46°29'28"N	116°16'22"W	1002
Orofino Municipal Airport	Clearwater	46°29'29"N	116°16'25"W	1000
Orogrande Landing Strip	Idaho	45°43'48"N	115°31'35"W	4405
Otterson Ranch Airport	Kootenai	47°44'25"N	116°59'50"W	2150
Owen Ranches Inc Airport	Owyhee	42°47'45"N	115°43'59"W	2620
Parma Airport	Canyon	43°46'44"N	116°56'15"W	2225
Payette Municipal Airport	Payette	44°05'30"N	116°54'00"W	2228
Picabo Airport	Blaine	43°18'30"N	114°03'45"W	4828
Pine Airport	Elmore	43°27'59"N	115°18'33"W	4232
Pistol Creek Ranch Landing Strip	Valley	44°44'07"N	115°08'37"W	4800
Pocatello Regional Airport	Power	42°54'47"N	112°35'39"W	4449
Preston Airport	Franklin	42°06'27"N	111°54'44"W	4726
Priest Lake USFS Airport	Bonner	48°34'28"N	116°57'27"W	2611
Priest River Municipal Airport	Bonner	48°11'25"N	116°54'30"W	2187
Q B One Airport	Jefferson	43°36'05"N	112°14'33"W	4875
Quaking Aspen Butte Landing Strip	Butte	43°25'37"N	113°10'05"W	5205
Rainbow Ranch Airport	Bonneville	43°24'27"N	111°58'45"W	4750

Airport, Heliport, or Landing Strip	County	Latitude	Longitude	Elev.
Ranch Aero Airport	Kootenai	47°49'44"N	116°46'57"W	2315
Reed Ranch Airport	Valley	44°53'30"N	115°42'44"W	4153
Rehn Ranch Landing Area	Cassia	42°29'43"N	113°07'43"W	4930
Rexburg Airport	Madison	43°49'57"N	111°48'18"W	4858
Reynolds Airport	Lincoln	43°02'38"N	114°09'51"W	4268
Riddle Airport	Owyhee	42°11'03"N	116°06'52"W	5356
Rigby Airport	Jefferson	43°38'40"N	111°55'40"W	4845
Riggins Flight Strip	Idaho	45°25'27"N	116°18'33"W	1773
Rockford Municipal Airport	Bingham	43°11'25"N	112°31'52"W	4465
Root Ranch Landing Strip	Idaho	45°18'48"N	115°01'29"W	5594
Round Mountain Aerodrome	Kootenai	47°54'55"N	116°47'59"W	2440
Running Creek Ranch Airport	Idaho	45°54'51"N	114°50'05"W	2969
Russell W Anderson Strip	Bingham	43°11'04"N	112°27'56"W	4450
Saint Alphonsus Helistop	Ada	43°36'48"N	116°15'19"W	2800
Saint Josephs Regional Medical Center Heliport	Nez Perce	44°54'33"N	116°06'34"W	
Saint Lukes Heliport	Ada	43°36'48"N	116°11'35"W	2740
Saint Maries Municipal Airport	Benewah	47°19'40"N	116°34'35"W	2127
Sandpoint Airport	Bonner	48°17'48"N	116°33'42"W	2126
Savage Ranch Landing Area	Cassia	42°28'25"N	113°51'30"W	4254
Seven Devils Airport	Adams	45°00'40"N	116°41'16"W	4487
Seven Devils Ranch Landing Strip	Adams	45°00'40"N	116°41'15"W	4487
Shearer Airport	Idaho	45°59'30"N	114°50'24"W	2634
Shoshone BLM Heliport	Lincoln	42°55'55"N	114°24'44"W	3980
Shoshone County Airport	Shoshone	47°32'50"N	116°11'18"W	2223
Shoshone Landing Area	Lincoln	42°56'37"N	114°23'20"W	3995
Silva Ranch Airport	Custer	44°15'53"N	115°01'33"W	6400
Silverwood Airport	Kootenai	47°54'30"N	116°42'30"W	2350
Simplot Airstrip	Lemhi	44°48'34"N	114°48'31"W	4040
Simplot Airstrip	Lemhi	44°48'34"N	114°48'31"W	4040
Sky Island Ranch Airport	Benewah	47°19'20"N	116°38'20"W	2880
Slate Creek Airport	Idaho	45°40'22"N	116°18'16"W	1660
Sluder Airstrip	Blaine	43°24'13"N	114°16'18"W	5017
Smiley Creek Airport	Blaine	43°54'55"N	114°47'47"W	7160
Smith Prairie Airport	Elmore	43°29'53"N	115°32'50"W	4958
Smith Ranch Airport	Kootenai	47°45'23"N	117°01'22"W	2370
Snake River Seaplane Base	Nez Perce	46°24'00"N	117°03'00"W	
Soldier Bar USFS Airport	Valley	45°06'18"N	114°47'50"W	4190
Spencer Ranch Landing Strip	Idaho	45°51'10"N	116°40'02"W	4284
Stanford Field	Fremont	43°57'02"N	111°41'00"W	4966
Stanley Airport	Custer	44°12'31"N	114°56'01"W	6403
Star 'S' Ranch Airport	Custer	43°58'55"N	114°02'34"W	6660
Stibnite Airport	Valley	44°54'01"N	115°19'56"W	6526
Stocking Meadows Airport	Shoshone	46°56'10"N	115°51'50"W	3850
Strawberry Glen Airport	Ada	43°39'54"N	116°17'25"W	2602
Sunrise Skypark	Owyhee	43°25'01"N	116°42'17"W	2240
Symms Airport	Canyon	43°34'10"N	116°46'33"W	2680
Taylor Ranch Landing Area	Valley	45°06'15"N	114°51'16"W	3835
Teton Peaks Airport	Teton	43°44'15"N	111°06'15"W	6202

Airport, Heliport, or Landing Strip	County	Latitude	Longitude	Elev.
Thomas Creek Airport	Valley	44°43'26"N	115°00'11"W	4400
Timber Basin Ranch Airport	Bonner	48°13'25"N	116°26'15"W	2220
Tracy Ranch Airport	Camas	43°17'12"N	115°04'41"W	5071
Treeport Airport	Bonner	47°58'41"N	116°47'30"W	2500
Triangle Airstrip	Owyhee	42°47'16"N	116°38'10"W	5090
Twin Bridges Airport	Custer	43°56'38"N	114°06'34"W	6893
Twin Falls Airport	Twin Falls	42°28'54"N	114°29'12"W	4151
University of Idaho Heliport	Latah	46°43'38"N	117°01'22"W	2604
Upper Loon Creek USFS Airport	Custer	44°33'07"N	114°50'53"W	5500
Valenov Ranch Airport	Bonner	48°17'48"N	117°00'51"W	2425
Valley County Hospital Heliport	Valley	44°31'08"N	116°02'50"W	4800
Vines Landing Strip	Valley	45°07'57"N	114°59'54"W	4110
Wallace Ranger Station Heliport	Shoshone	47°29'37"N	115°57'32"W	2720
Warm Springs Creek Airport	Boise	44°08'32"N	115°18'48"W	4831
Warren Airport	Idaho	45°16'06"N	115°40'57"W	5896
Weatherby Landing Area	Elmore	43°46'12"N	115°31'09"W	4400
Weatherby USFS Airport	Elmore	43°49'30"N	115°19'51"W	4494
Weiser Municipal Airport	Washington	44°12'25"N	116°57'41"W	2112
Whelans Heliport	Canyon	43°35'50"N	116°46'52"W	2565
Wood Brothers Ranch Airport	Nez Perce	46°22'56"N	116°30'42"W	3400
Young Heliport	Ada	43°43'00"N	116°23'17"W	2625
Young Landing Area	Owyhee	42°57'17"N	115°49'00"W	2711

Bars, Bends, Reefs, Shallows

Bar, Bend, Reef, or Shallow	County	Latitude	Longitude	Elev.
American Bar	Idaho	45°54'49"N	116°25'35"W	
Badger Creek Bar	Butte	44°04'47"N	113°12'16"W	
Bailey Bar	Idaho	45°34'05"N	115°12'07"W	
Bargamin Bar	Idaho	45°34'04"N	115°11'27"W	
Bear Bar	Idaho	45°31'15"N	115°05'15"W	
Bear Bar	Nez Perce	46°00'01"N	116°55'02"W	
Big Bar	Adams	45°08'06"N	116°44'18"W	
Big Bar	Idaho	45°32'15"N	116°31'40"W	
Big Bend	Teton	43°39'12"N	111°10'29"W	
Big Bend Of Sheep Creek	Owyhee	42°27'21"N	115°42'45"W	
Big Foot Bar	Elmore	43°06'07"N	116°14'01"W	
Big Hole	Valley	44°27'00"N	115°13'51"W	
Big Holes	Owyhee	42°26'22"N	115°50'13"W	
Big Mallard Creek Bar	Idaho	45°32'10"N	115°16'19"W	
Blackhawk Bar	Idaho	45°37'35"N	116°18'08"W	
Blue Bird Bar	Idaho	45°24'03"N	115°28'37"W	
Bobcat Bar	Nez Perce	46°08'01"N	116°56'09"W	
Boise Bar	Idaho	45°26'28"N	115°26'34"W	
Bonanza Bar	Power	42°37'58"N	113°07'49"W	
Buckhorn Bar	Valley	44°56'05"N	115°44'11"W	
Bull Creek Bar	Idaho	45°26'37"N	115°43'31"W	2063
Butcher Bar	Idaho	45°33'30"N	116°17'58"W	
Cache Bar	Lemhi	45°19'08"N	114°38'09"W	
California Bar	Lemhi	45°11'18"N	114°08'34"W	
Campbell Flat	Idaho	45°41'25"N	116°18'48"W	
Cedar Creek Bar	Custer	44°00'27"N	113°45'12"W	
Chimney Bar	Idaho	45°17'32"N	116°40'03"W	1840
Chimney Bar	Idaho	45°17'23"N	116°40'13"W	1700
Clarks Hole	Idaho	45°28'02"N	116°30'45"W	2500
Cooper Bar	Idaho	45°54'15"N	116°23'49"W	
Cooper Bar	Idaho	45°43'55"N	116°18'26"W	
Corey Bar	Idaho	45°29'38"N	114°58'58"W	
Corn Creek Bar	Lemhi	45°22'14"N	114°41'15"W	2960
Cottonwood Point	Benewah	47°19'52"N	116°37'02"W	2236
Cougar Rapids Bar	Nez Perce	45°57'37"N	116°53'13"W	
Coxey Creek Bar	Valley	45°08'20"N	115°01'59"W	
Coyote Hole	Owyhee	42°06'12"N	116°47'15"W	
Cunningham Bar	Idaho	45°22'01"N	114°41'08"W	
Deadman Bar	Valley	44°57'47"N	115°39'13"W	
Deer Creek Bar	Butte	44°01'45"N	113°15'19"W	
Deer Park Bar	Idaho	45°31'51"N	115°06'41"W	
Devils Elbow	Shoshone	47°44'55"N	116°01'41"W	2540
Devils Elbow	Shoshone	47°37'14"N	116°12'23"W	
Disappointment Bar	Idaho	45°25'22"N	114°52'50"W	
Dry Bar	Idaho	46°05'44"N	115°06'08"W	
Eagle Bar	Adams	45°13'09"N	116°42'27"W	
Ebenezer Bar	Lemhi	45°18'18"N	114°30'51"W	
Elbow Bend	Idaho	46°12'05"N	114°43'20"W	

Bar, Bend, Reef, or Shallow	County	Latitude	Longitude	Elev.
Elkhorn Bar	Idaho	45°30'57"N	115°19'36"W	
Elkhorn Bar	Idaho	45°26'56"N	114°55'17"W	
Fall Creek Bar	Idaho	45°28'32"N	115°21'09"W	
Farewell Bend	Washington	44°17'54"N	117°13'05"W	2080
Fawn Creek Bar	Idaho	45°24'59"N	114°51'16"W	
Fisher Bottom	Bonneville	43°35'40"N	111°28'22"W	
Fivemile Bar	Idaho	45°24'44"N	115°28'04"W	
Gaines Bar	Idaho	45°28'30"N	115°23'27"W	
Geneva Bar	Nez Perce	45°53'39"N	116°50'36"W	
Goddard Bar	Idaho	46°06'02"N	115°32'50"W	
Groundhog Bar	Idaho	45°28'47"N	115°20'58"W	
Haney Bar	Idaho	45°24'31"N	115°28'37"W	
Hard Boil Bar	Valley	45°08'35"N	115°03'45"W	
Harpers Bend	Nez Perce	46°29'23"N	116°27'06"W	
Hole in Rocks	Fremont	44°10'50"N	111°17'08"W	
Horse Bend	Idaho	45°40'06"N	116°17'07"W	
Horseshoe Bend	Idaho	45°40'08"N	116°16'54"W	1490
Hospital Bar	Valley	44°50'03"N	114°47'20"W	
Indian Creek Bar	Idaho	45°23'48"N	115°36'11"W	2600
Indian Point	Elmore	43°21'08"N	115°31'57"W	
Island Bar	Idaho	45°25'02"N	116°15'32"W	1740
Jackson Bar	Idaho	45°23'46"N	115°29'06"W	2320
John Day Bar	Idaho	45°34'58"N	116°17'48"W	
Johnson Bar	Idaho	46°06'29"N	115°33'32"W	1960
Johnson Bar	Idaho	45°34'31"N	116°18'14"W	
Johnson Bar	Idaho	45°27'28"N	116°33'56"W	
Johnson Bar	Lemhi	44°14'19"N	112°59'04"W	
Jones Bar	Camas	43°37'24"N	114°54'33"W	
Kenmitzer Bar	Idaho	45°32'06"N	115°17'45"W	
Kirby Bar	Idaho	45°34'57"N	116°28'47"W	1220
Kirkwood Bar	Idaho	45°34'08"N	116°29'47"W	
Lantz Bar	Idaho	45°25'10"N	114°52'01"W	
Large Bar	Idaho	45°40'12"N	116°17'00"W	1520
Lemhi Bar	Idaho	45°28'33"N	115°23'05"W	
Lightfoot Bar	Camas	43°36'32"N	114°57'01"W	
Little Elkhorn Bar	Idaho	45°30'20"N	115°19'45"W	
Long Tom Bar	Idaho	45°27'35"N	115°52'50"W	
Lucile Bar	Idaho	45°31'16"N	116°18'17"W	1700
Lucky Creek Bar	Idaho	45°23'26"N	114°47'31"W	
Ludwig Bar	Idaho	45°23'33"N	115°29'20"W	
Lufkin Bottom	Bonneville	43°34'52"N	111°27'55"W	
Lyons Bar	Idaho	45°48'18"N	116°18'49"W	
Maggie Bend	Idaho	46°09'05"N	115°55'59"W	
Monumental Bar	Valley	45°09'40"N	115°07'47"W	
Mormon Bend	Custer	44°15'43"N	114°50'50"W	
Moscow Bar	Clearwater	46°47'05"N	115°28'21"W	
Mulky Bar	Butte	44°10'41"N	113°21'31"W	
Music Bar	Idaho	45°24'48"N	116°10'57"W	

Bar, Bend, Reef, or Shallow	County	Latitude	Longitude	Elev.
Nixon Bar	Idaho	45°30'37"N	115°03'43"W	
Over Easy Bar	Valley	45°08'32"N	115°03'19"W	
Oxbow	Idaho	46°02'00"N	116°37'37"W	
Oxbow Bend	Boise	44°04'15"N	115°39'48"W	
Oxbow, The	Adams	44°59'04"N	116°49'37"W	1688
Painter Bar	Idaho	45°25'03"N	115°27'52"W	
Pine Bar	Idaho	45°29'54"N	116°33'05"W	1400
Pittsburg Bar	Idaho	45°38'04"N	116°28'53"W	
Ragtown Bar	Idaho	45°45'14"N	116°33'06"W	
Rainbow Bend	Teton	43°42'20"N	111°10'07"W	
Rattlesnake Bar	Idaho	46°04'12"N	114°51'44"W	
Rattlesnake Bar	Idaho	45°32'54"N	115°09'04"W	
Reed Bar	Idaho	45°48'14"N	115°41'54"W	
Regan Bend	Gem	43°56'43"N	116°21'16"W	
Richardson Bar	Idaho	45°32'18"N	115°15'22"W	
Rocky Bar	Blaine	43°31'00"N	113°56'35"W	
Rocky Bar	Idaho	45°23'26"N	115°29'37"W	
Rotten Bottom	Teton	43°36'07"N	111°11'52"W	
Russel Bar	Idaho	45°34'31"N	116°29'01"W	1200
Russell Bar	Idaho	45°40'02"N	116°18'44"W	
Salzer Bar	Lemhi	45°32'06"N	114°03'12"W	
Sheep Creek Bar	Idaho	45°28'10"N	115°48'10"W	
Sherwin Bar	Idaho	45°36'07"N	116°16'29"W	
Shorts Bar	Idaho	45°51'25"N	116°17'43"W	
Slicker Bar	Idaho	45°38'15"N	116°16'55"W	
Soft Boil Bar	Valley	45°08'26"N	115°02'47"W	
Soldier Bar	Valley	45°06'19"N	114°47'53"W	
Spindle Creek Bar	Idaho	45°23'30"N	114°46'51"W	
Spring Bar	Idaho	45°25'33"N	116°09'05"W	
Squaw Bar	Idaho	45°30'38"N	116°17'52"W	
Sunny Bar	Butte	44°00'22"N	113°09'20"W	
Sunny Bar	Idaho	45°30'49"N	115°02'34"W	
Swartz Bar	Idaho	45°51'02"N	116°17'52"W	
Taylor Bar	Idaho	45°41'03"N	116°18'35"W	
Tie Bend	Jefferson	43°38'30"N	111°41'20"W	4995
Twentyfive Mile Bar	Idaho	46°04'24"N	115°22'27"W	
Twentymile Bar	Idaho	46°05'13"N	115°27'58"W	
Twilegar Bar	Idaho	45°32'32"N	116°18'13"W	1800
Upper Butcher Bar	Idaho	45°32'56"N	116°18'22"W	
Upper Sherwin Bar	Idaho	45°35'32"N	116°17'10"W	
Waldo Bar	Idaho	45°53'52"N	114°46'42"W	
Warm Springs Bar	Idaho	45°51'02"N	114°55'24"W	
Warm Springs Bar	Idaho	45°18'03"N	116°40'15"W	1435
Weavers Hole	Camas	43°18'00"N	114°56'51"W	5045
White House Bar	Idaho	45°58'19"N	116°31'57"W	
Whitley Bottom	Payette	43°58'08"N	116°56'49"W	
Widow Bar	Idaho	45°28'57"N	115°20'55"W	
Wilson Bar	Idaho	45°39'46"N	116°17'06"W	1540
Yellow Pine Bar	Idaho	45°32'38"N	115°15'03"W	

Basins, Flats, Meadows, Plains

Basin, Flat, Meadow, or Plain	County	Latitude	Longitude	Elev.
Aikers Field	Owyhee	42°18'13"N	115°10'17"W	5129
Alder Creek Flats	Benewah	47°13'01"N	116°41'04"W	
Alexander Flats	Elmore	43°46'39"N	115°31'29"W	
Alkali Flat	Bingham	43°01'38"N	112°00'21"W	
Anchor Meadow	Idaho	45°35'23"N	115°51'30"W	
Ant Basin	Adams	45°07'56"N	116°27'28"W	
Antelope Basin	Owyhee	42°29'18"N	116°36'00"W	
Antelope Basin	Owyhee	42°08'32"N	115°53'29"W	
Antelope Flat	Custer	44°18'40"N	114°03'45"W	
Antelope Flat	Fremont	44°16'58"N	111°35'27"W	
Antelope Flat	Owyhee	43°34'19"N	111°31'25"W	5005
Antelope Flat	Owyhee	43°32'24"N	111°28'06"W	
Antelope Park	Fremont	44°23'15"N	111°24'48"W	
Antelope Pocket	Twin Falls	42°19'55"N	114°47'49"W	
Antelope Swale	Gem	44°21'48"N	116°11'34"W	
Ants Basin	Custer	44°03'30"N	114°38'17"W	
Aquinaldo Flat	Valley	44°46'17"N	115°04'38"W	
Arbuckle Basin	Adams	44°40'27"N	116°19'13"W	4980
Argora	Clark	44°22'38"N	112°36'28"W	6280
Armstrong Meadows	Bonner	48°45'02"N	116°52'26"W	
Arnett Meadows	Clearwater	46°34'52"N	115°41'46"W	
Aspen Flat	Valley	44°50'59"N	115°47'20"W	
Auxor Basin	Bonner	48°16'39"N	116°13'53"W	
Avondale Basin	Owyhee	43°03'26"N	116°44'12"W	
Ax Park	Lemhi	45°33'43"N	114°04'59"W	
Ayers Meadows	Valley	44°26'50"N	115°18'44"W	
Babs Flat	Elmore	43°07'46"N	115°17'54"W	
Badger Basin	Lemhi	45°19'22"N	113°49'35"W	
Badger Basin	Lemhi	44°55'58"N	114°02'36"W	
Badger Meadows	Clearwater	46°52'33"N	116°15'22"W	
Baldy Basin	Lemhi	44°58'15"N	113°43'52"W	
Ballinger Flat	Elmore	43°27'54"N	115°39'35"W	
Ballys Hole	Bonneville	43°28'01"N	111°33'47"W	
Bannister Basin	Payette	44°04'08"N	116°35'57"W	
Barber Flat	Adams	44°57'02"N	116°46'11"W	
Barber Flat	Elmore	43°48'36"N	115°32'01"W	
Barton Flats	Custer	44°01'27"N	113°50'57"W	
Basin, The	Bingham	43°14'45"N	111°42'51"W	6530
Basin, The	Idaho	45°53'48"N	116°22'30"W	
Bathtub Meadows	Shoshone	47°04'45"N	115°34'46"W	
Baxter Basin	Owyhee	43°01'34"N	116°59'24"W	
Bear Basin	Adams	44°56'57"N	116°08'24"W	
Bear Flat	Cassia	42°13'16"N	114°10'18"W	
Bear Meadow	Fremont	44°06'54"N	111°11'14"W	
Bear Park	Blaine	43°10'32"N	113°22'03"W	
Bear Wallow	Adams	45°09'03"N	116°22'33"W	
Bear Wallow	Custer	44°14'30"N	114°11'26"W	
Bearskin Meadows	Fremont	44°22'42"N	115°29'56"W	

Basin, Flat, Meadow, or Plain	County	Latitude	Longitude	Elev.
Beaver Flat	Idaho	46°18'31"N	115°22'08"W	
Beaver Flat	Owyhee	42°03'41"N	115°12'58"W	5430
Beaver Meadows	Idaho	46°33'02"N	114°29'53"W	6856
Beaver Meadows	Twin Falls	42°04'17"N	114°54'31"W	
Bedrock Flat	Adams	44°28'29"N	116°28'39"W	
Ben Mills Flat	Owyhee	42°24'19"N	116°40'37"W	
Bennett Creek Basin	Elmore	43°17'12"N	115°27'54"W	
Benton Meadows	Nez Perce	46°07'18"N	116°48'27"W	
Bernards Bedground	Owyhee	42°01'44"N	115°40'35"W	
Big Basin	Blaine	43°43'33"N	114°07'06"W	
Big Basin	Caribou	42°36'31"N	111°21'04"W	
Big Creek Meadows	Adams	44°52'02"N	116°13'25"W	
Big Creek Meadows	Idaho	45°36'14"N	115°30'37"W	
Big Flat	Bannock	42°47'59"N	112°28'46"W	
Big Flat	Gem	44°06'32"N	116°24'15"W	3800
Big Flat	Lemhi	45°16'38"N	113°53'19"W	
Big Flat	Owyhee	42°02'47"N	115°19'18"W	5698
Big Flat	Washington	44°42'57"N	116°43'52"W	
Big Grassy	Fremont	44°16'49"N	111°09'21"W	
Big Hank Meadow	Shoshone	47°49'22"N	116°05'47"W	
Big Lost River Sinks	Butte	43°46'30"N	112°52'02"W	
Big Meadows	Boise	44°05'34"N	115°08'57"W	
Big Meadows	Bonner	48°26'34"N	117°00'32"W	
Big Meadows	Shoshone	47°58'51"N	116°13'15"W	
Big Meadows	Shoshone	47°49'46"N	116°15'29"W	
Big Meadows	Valley	44°17'23"N	115°29'19"W	
Bighorn Basin	Custer	44°06'49"N	114°38'40"W	
Bimerick Meadows	Idaho	46°16'20"N	115°26'37"W	
Birch Basin	Butte	44°10'38"N	113°12'44"W	
Birch Creek Sinks	Butte	43°51'45"N	112°43'13"W	
Birch Flat	Boise	44°04'20"N	115°44'37"W	
Bismark Basin	Idaho	45°12'30"N	115°07'08"W	
Bismark Meadows	Bonner	48°37'18"N	116°57'51"W	
Black Grove	Bonneville	43°32'38"N	111°10'15"W	
Black Rock Pocket	Owyhee	42°06'34"N	115°39'33"W	
Blackstone Desert	Owyhee	42°28'37"N	115°42'23"W	
Blizzard Basin	Butte	43°30'05"N	113°39'40"W	
Bloom Meadows	Clearwater	46°51'27"N	116°17'33"W	
Bogus Basin	Boise	43°45'51"N	116°06'06"W	
Boise Basin	Boise	43°54'21"N	115°56'00"W	
Bouffard Flat	Idaho	45°41'59"N	116°14'19"W	
Boulder Basin	Blaine	43°50'37"N	114°30'27"W	
Boulder Basin	Idaho	45°49'14"N	116°19'44"W	
Boulder Basin	Lemhi	45°23'40"N	114°17'56"W	
Boulder Flat	Idaho	46°20'18"N	115°18'47"W	
Boulder Flat	Lemhi	44°45'37"N	113°42'48"W	
Boulder Meadows	Bonner	48°45'41"N	117°00'14"W	
Bowman Flat	Camas	43°11'10"N	114°51'00"W	

Basin, Flat, Meadow, or Plain	County	Latitude	Longitude	Elev.
Box Flat	Idaho	45°52'26"N	116°24'30"W	
Brace Flat	Owyhee	42°20'45"N	116°40'29"W	
Bradbury Flat	Custer	44°25'22"N	114°08'52"W	
Bradshaw Basin	Custer	44°18'12"N	114°11'45"W	
Brebner Flat	Shoshone	47°11'54"N	115°48'28"W	
Broken Wagon Flat	Owyhee	42°38'35"N	115°47'27"W	
Bronson Meadow	Latah	46°51'31"N	116°27'42"W	
Bronson Meadows	Shoshone	47°23'18"N	116°04'59"W	
Browns Basin	Owyhee	42°00'03"N	115°53'50"W	
Browns Basin	Owyhee	42°00'01"N	115°53'38"W	
Browns Basin	Valley	45°07'28"N	114°54'30"W	
Browns Bench	Twin Falls	42°03'49"N	114°47'25"W	
Browns Meadow	Latah	46°50'34"N	116°45'22"W	
Bruce Meadows	Valley	44°24'37"N	115°19'19"W	
Bruneau Desert	Owyhee	42°34'05"N	115°11'25"W	
Buck Flat	Owyhee	42°12'45"N	115°13'34"W	5030
Buck Meadows	Idaho	45°59'28"N	115°25'53"W	
Buck Meadows	Idaho	45°45'22"N	115°56'48"W	
Buck Pasture	Owyhee	42°02'26"N	115°58'55"W	
Buckbrush Flats	Cassia	42°18'02"N	114°16'49"W	
Bucks Basin	Valley	44°42'19"N	116°15'37"W	
Buckskin Basin	Bingham	43°03'44"N	112°18'09"W	
Buckskin Basin	Caribou	42°53'44"N	112°08'28"W	
Bull Basin	Owyhee	42°35'35"N	116°37'10"W	
Bull Basin	Owyhee	42°27'51"N	116°16'55"W	
Bull Basin	Owyhee	42°20'37"N	116°53'18"W	
Bull Lake	Owyhee	42°16'10"N	117°00'14"W	5339
Bull Meadow	Owyhee	43°05'01"N	116°44'13"W	
Bull Pasture	Oneida	42°09'41"N	112°31'39"W	5144
Bull Run Cove	Idaho	45°45'52"N	116°11'21"W	
Bullhead Basin	Owyhee	42°35'01"N	116°37'11"W	
Bumblebee Meadow	Shoshone	47°37'59"N	116°16'53"W	
Buncel Basin	Owyhee	42°22'20"N	116°06'03"W	
Burgess Flats	Blaine	43°24'35"N	114°00'30"W	5095
Burnam Flat	Owyhee	42°59'17"N	116°38'58"W	
Burnt Basin	Cassia	42°02'14"N	113°02'40"W	
Burnt Basin	Valley	44°35'12"N	116°15'47"W	
Burnt Flats	Idaho	45°50'24"N	116°12'18"W	
Burnt Wagon Basin	Adams	44°37'30"N	116°10'33"W	
Cache Meadows	Valley	44°20'17"N	115°25'24"W	
Calamity Meadows	Washington	44°44'18"N	116°39'02"W	
Camas Flats	Blaine	43°23'35"N	114°27'19"W	
Camas Meadows	Clark	44°23'53"N	111°54'38"W	
Camas Meadows	Lemhi	44°42'15"N	114°22'45"W	
Camas Prairie	Camas	43°19'14"N	114°48'30"W	
Camp Ten Meadow	Clearwater	46°43'15"N	115°44'26"W	
Campbell Flats	Blaine	43°26'50"N	114°01'40"W	5310
Canteen Meadows	Idaho	46°08'49"N	115°11'57"W	

Basin, Flat, Meadow, or Plain	County	Latitude	Longitude	Elev.
Carey Borrow Pits	Bonner	48°08'16"N	116°50'31"W	2121
Caribou Basin	Bonneville	43°09'42"N	111°17'58"W	
Casey Meadow	Clearwater	46°39'44"N	115°51'49"W	
Caton Meadow	Valley	44°52'27"N	115°34'14"W	
Cayuse Meadows	Idaho	45°48'12"N	116°04'06"W	
Cedar Field	Power	42°40'09"N	113°00'42"W	
Cedar Flats	Idaho	46°06'54"N	114°59'52"W	
Centennial Flat	Custer	44°21'43"N	114°16'57"W	
Center Basin	Teton	43°37'50"N	111°14'35"W	
Chain Meadows	Idaho	46°22'14"N	114°53'07"W	
Chalk Flat	Elmore	42°58'21"N	115°32'53"W	
Chamberlain Basin	Custer	44°01'43"N	114°35'58"W	
Chamberlain Basin	Idaho	45°21'45"N	115°10'58"W	
Chamberlain Meadows	Shoshone	46°56'45"N	115°13'32"W	
Chase Meadow	Idaho	45°50'01"N	116°09'15"W	
Chattin Flat	Elmore	43°01'18"N	116°05'58"W	
Cheatbeck Basin	Bear Lake	42°29'42"N	111°36'48"W	
Cherryville Flat	Franklin	42°01'41"N	111°44'34"W	5748
Chicken Flat	Owyhee	42°13'57"N	116°00'22"W	
China Basin	Elmore	43°50'56"N	115°08'53"W	
China Flat	Owyhee	43°11'04"N	116°48'02"W	
Chipmunk Meadow	Owyhee	43°10'43"N	116°51'28"W	
Chokecherry Flat	Adams	45°10'31"N	116°24'14"W	
Christianson Meadows	Clearwater	46°47'03"N	116°08'53"W	
Chubb Flat	Caribou	42°59'15"N	111°31'58"W	6245
Clark Valley	Caribou	42°57'36"N	111°28'49"W	
Clifton Basin	Franklin	42°10'29"N	112°03'22"W	
Clover Flat	Owyhee	42°27'17"N	115°22'34"W	4390
Clover Hollow	Elmore	43°02'46"N	115°39'52"W	
Club Meadows	Idaho	45°15'52"N	115°06'06"W	
Clyde Flat	Cassia	42°15'57"N	113°38'41"W	
Coal Mine Basin	Owyhee	43°12'35"N	117°00'55"W	
Coffee Flat	Washington	44°37'27"N	116°58'08"W	6585
Cold Meadows	Idaho	45°17'15"N	114°56'41"W	
Colson Basin	Power	42°21'32"N	112°48'03"W	
Commissary Basin	Bingham	43°11'46"N	111°50'33"W	6543
Con Shea Basin	Owyhee	43°15'59"N	116°30'19"W	
Connet Flat	Gooding	43°10'52"N	114°46'24"W	5448
Connor Flat	Cassia	42°18'56"N	113°35'53"W	7850
Coonrod Basin	Gem	44°09'01"N	116°23'05"W	
Cooper Flat	Idaho	45°53'13"N	114°36'17"W	
Copeland Flats	Valley	44°53'13"N	116°08'15"W	
Copenhagen Basin	Bear Lake	42°19'18"N	111°34'26"W	
Copper Basin	Custer	43°48'25"N	113°52'43"W	
Copper Basin	Lemhi	44°54'35"N	113°50'34"W	
Copper Basin Flat	Custer	43°48'32"N	113°51'23"W	
Corduroy Meadows	Idaho	45°47'21"N	116°01'56"W	
Corduroy Meadows	Valley	44°27'39"N	115°27'11"W	

Basins, Flats, Meadows, Plains

Basin, Flat, Meadow, or Plain	County	Latitude	Longitude	Elev.
Corral Basin	Custer	44°14'26"N	114°06'19"W	
Corral Flat	Lemhi	45°17'18"N	114°27'08"W	
Cottonwood Basin	Twin Falls	42°15'39"N	114°25'25"W	
Cottonwood Basin	Valley	44°42'51"N	116°17'42"W	
Cottonwood Flats	Idaho	46°17'23"N	115°54'18"W	
Cougar Basin	Adams	44°35'29"N	116°14'37"W	
Cougar Basin	Idaho	45°26'12"N	116°31'09"W	3800
Cougar Basin	Valley	45°02'53"N	115°16'44"W	
Cougar Flat	Elmore	43°27'18"N	115°37'29"W	
Cougar Meadows	Latah	46°55'15"N	116°26'44"W	
Cow Basin	Lemhi	44°55'35"N	114°46'15"W	
Cow Flats	Idaho	45°26'08"N	116°01'19"W	
Cow Meadow	Owyhee	42°21'06"N	116°53'42"W	
Cowboy Basin	Owyhee	42°26'47"N	116°37'16"W	
Coyote Flat	Owyhee	42°09'37"N	116°46'45"W	
Crab Creek Basin	Owyhee	42°22'41"N	116°02'31"W	
Crane Basin	Custer	44°23'16"N	114°00'22"W	
Crane Meadow	Valley	44°27'02"N	115°30'44"W	
Crane Meadow	Valley	44°26'51"N	115°29'15"W	
Crane Meadows	Idaho	45°17'09"N	115°17'39"W	
Cranes Flat	Bingham	43°06'53"N	111°38'34"W	
Crater Meadows	Clearwater	46°33'47"N	115°17'07"W	
Crescent Meadow	Valley	45°11'09"N	115°02'21"W	
Crooked Creek Flat	Bingham	43°02'23"N	111°36'22"W	
Crooks Corral Basin	Idaho	45°34'26"N	116°24'53"W	
Crows Nest	Owyhee	43°10'28"N	116°50'37"W	
Crows Nest Basin	Idaho	45°49'32"N	116°29'52"W	
Crows Nest Flat	Owyhee	42°35'13"N	115°15'26"W	3900
Crystal Pit	Butte	43°26'21"N	113°33'20"W	
Cub Basin	Franklin	42°06'34"N	111°36'44"W	
Curlew National Grassland	Oneida	42°12'30"N	112°44'59"W	
Curly Jack Flat	Oneida	42°19'03"N	112°11'06"W	
D Bar Basin	Owyhee	42°25'28"N	116°27'25"W	
Daggett Meadow	Bonner	48°21'54"N	116°59'58"W	
Danielson Basin	Bingham	43°02'37"N	112°08'49"W	
Danish Flat	Bear Lake	42°17'09"N	111°30'10"W	
Darlington Sinks	Custer	43°50'47"N	113°25'58"W	
Daveggio Meadows	Shoshone	47°09'43"N	115°59'38"W	
Davis Basin	Franklin	42°11'54"N	112°03'35"W	
Davis Flat	Camas	43°16'26"N	114°51'40"W	
Deadman Flat	Elmore	43°37'43"N	113°08'04"W	
Deadman Flat	Elmore	42°52'47"N	115°20'36"W	
Deadwood Basin	Valley	44°18'48"N	115°40'05"W	5093
Debbitt Basin	Shoshone	47°08'24"N	115°57'24"W	
Decker Flat	Custer	44°03'31"N	114°51'58"W	
Deer Flat	Bonneville	43°22'30"N	111°11'03"W	
Deer Flat	Canyon	43°42'55"N	117°00'57"W	
Deer Flats	Butte	44°14'23"N	113°18'33"W	
Deer Park	Boundary	48°52'00"N	116°20'41"W	
Deer Parks	Jefferson	43°46'46"N	112°02'34"W	
Della Basin	Bonneville	43°08'38"N	111°22'41"W	
Dempsey Flat	Idaho	45°19'12"N	116°18'29"W	
Dempsey Meadows	Gooding	43°11'14"N	115°04'48"W	
Devils Basin	Lemhi	44°33'02"N	113°35'12"W	
Devils Hole	Elmore	43°28'36"N	115°41'06"W	
Devils Hole	Idaho	45°18'54"N	116°15'01"W	
Devils Washbasin (historical)	Gooding	42°51'47"N	114°54'06"W	2740
Diamond A Desert	Owyhee	42°11'15"N	115°35'27"W	
Diamond Basin	Owyhee	43°06'37"N	116°37'05"W	
Diamond Boulder Flat	Caribou	42°51'09"N	111°13'04"W	
Diamond Flat	Caribou	42°51'23"N	111°13'08"W	7330
Dillinger Meadows	Idaho	45°29'33"N	115°11'38"W	
Dirty Head	Oneida	41°59'39"N	112°05'16"W	
Dishpan	Owyhee	42°05'24"N	115°25'40"W	5515
Ditto Flat	Camas	43°30'55"N	114°40'45"W	
Dodge Basin	Custer	44°33'47"N	114°04'25"W	
Doe Flat	Owyhee	42°13'29"N	115°13'30"W	5030
Dollar Creek Meadow	Valley	44°42'22"N	115°45'50"W	
Donahue Basin	Twin Falls	42°16'27"N	114°22'09"W	
Doumecq Plains	Idaho	45°50'30"N	116°21'27"W	
Dove Spring	Owyhee	42°45'47"N	115°19'31"W	
Dry Basin	Caribou	42°38'55"N	111°18'04"W	
Dry Basin	Franklin	42°16'14"N	111°36'11"W	7892
Dry Basin	Idaho	45°24'37"N	116°35'49"W	1600
Dry Buck Meadow	Valley	44°09'26"N	116°08'53"W	
Dry Meadow	Idaho	45°27'56"N	115°14'34"W	
Duckworth Meadows	Bonneville	43°12'27"N	111°30'23"W	6230
Dunn Basin	Bingham	43°05'44"N	111°57'02"W	5830
Durfee Meadow	Cassia	42°11'47"N	114°06'00"W	
Dutch Flat	Clark	44°05'37"N	112°15'07"W	
Dutch Flat	Latah	46°44'49"N	116°48'51"W	
Dutch Oven	Owyhee	42°30'31"N	116°45'37"W	
Dutchler Basin	Lemhi	45°25'38"N	114°14'20"W	
Earthquake Meadows	Idaho	45°51'46"N	115°57'33"W	
East Boulder Meadow	Lemhi	45°20'56"N	114°08'34"W	
Ecks Flat	Adams	44°56'19"N	116°11'55"W	
Eddy Basin	Custer	44°39'21"N	114°22'01"W	
Egan Basin	Franklin	42°04'04"N	111°34'38"W	
Eightmile Prairie	Kootenai	47°54'20"N	116°43'24"W	
Elba Basin	Cassia	42°15'12"N	113°33'41"W	
Elk Basin	Twin Falls	42°11'59"N	114°17'34"W	
Elk Flat	Idaho	45°46'52"N	116°11'36"W	
Elk Flat	Teton	43°40'06"N	111°18'26"W	
Elk Meadow	Custer	44°07'10"N	114°53'10"W	
Elk Meadow	Custer	44°15'55"N	115°06'24"W	
Elk Meadows	Bannock	42°44'46"N	112°28'15"W	

Basin, Flat, Meadow, or Plain	County	Latitude	Longitude	Elev.
Elk Meadows	Idaho	46°39'27"N	114°21'58"W	
Elk Meadows	Idaho	45°48'17"N	116°05'22"W	5200
Elk Meadows	Idaho	45°16'24"N	116°04'59"W	6701
Emmett Bench	Gem	43°54'13"N	116°33'49"W	
Enos Meadow	Valley	45°08'57"N	115°49'10"W	
Ephraim Valley	Caribou	42°30'31"N	111°08'04"W	
Erickson Meadow	Latah	46°50'51"N	116°27'33"W	
Euchre Valley	Bonneville	43°27'06"N	111°53'04"W	
Eureka Meadows	Clearwater	46°37'54"N	115°51'33"W	3148
Fall Creek Basin	Bonneville	43°20'32"N	111°28'42"W	
Faust Meadows	Idaho	45°15'46"N	115°35'54"W	
Fawn Meadow	Valley	45°08'54"N	115°23'28"W	
Ferguson Basin	Adams	44°42'41"N	116°36'14"W	4720
Ferguson Flat	Gooding	43°06'18"N	114°53'22"W	
Firbox Meadows	Lemhi	44°24'11"N	113°21'18"W	
Firebox Meadows	Lemhi	44°25'08"N	113°21'12"W	
First Basin	Lemhi	44°52'36"N	113°51'13"W	
Fish Creek Basin	Bannock	42°33'51"N	111°54'51"W	
Fish Creek Meadows	Idaho	46°20'23"N	115°33'35"W	4240
Fivemile Meadows	Caribou	42°43'59"N	111°36'30"W	
Flatiron, The	Adams	45°08'09"N	116°43'32"W	
Florence Basin	Idaho	45°31'42"N	116°00'13"W	
Floyd Meadows	Idaho	45°18'52"N	115°45'48"W	
Foolhen Meadows	Valley	45°02'56"N	115°55'18"W	
Fork Meadow	Owyhee	42°06'30"N	115°58'25"W	
Fort Hall Bottoms	Blaine	43°03'23"N	112°32'51"W	
Fortune Meadows	Shoshone	47°05'30"N	115°52'23"W	
Fortynine Meadows	Shoshone	47°05'40"N	115°52'40"W	
Four Lakes Basin	Custer	44°03'06"N	114°36'19"W	
Fox Flat	Bear Lake	42°23'42"N	111°13'33"W	
Fox Park	Idaho	46°01'32"N	114°56'17"W	
Franklin Basin	Franklin	42°04'15"N	111°36'39"W	7861
Fritzer Flat	Lemhi	45°12'07"N	114°18'58"W	
Frog Meadows	Lemhi	45°04'22"N	114°32'22"W	
Frosty Meadows	Idaho	45°15'12"N	115°59'14"W	
Frying Pan Basin	Owyhee	42°24'50"N	116°24'53"W	
Frying Pan, The	Power	42°58'30"N	113°12'15"W	
Frymier Field	Power	42°38'02"N	113°06'13"W	
Fuller Pasture	Cassia	42°14'36"N	114°09'58"W	
Gallagher Pasture	Boise	44°04'33"N	115°46'52"W	3400
Garrison Flat	Idaho	45°27'41"N	116°20'25"W	
Germania Basin	Custer	43°59'35"N	114°39'10"W	
Germer Basin	Custer	44°22'28"N	114°14'10"W	
Gertrudes Meadow	Owyhee	42°37'31"N	116°58'38"W	
Ghoul Basin	Lemhi	44°59'45"N	113°31'10"W	
Gibson Basin	Franklin	42°01'57"N	111°33'34"W	
Gilman Flat	Blaine	43°28'02"N	114°22'53"W	
Gleason Meadow	Bonner	48°24'20"N	116°58'59"W	

Basin, Flat, Meadow, or Plain	County	Latitude	Longitude	Elev.
Goat Basin	Valley	45°06'47"N	114°48'20"W	
Goe Flat	Bonneville	43°35'52"N	111°35'06"W	
Gold Dollar Flat	Clearwater	46°27'41"N	115°46'03"W	3641
Gold Fork Meadow	Valley	44°39'16"N	115°50'15"W	
Gold Meadows	Idaho	46°22'28"N	115°06'23"W	
Goodman Flat	Elmore	43°08'42"N	115°17'34"W	
Goodwin Meadows	Idaho	45°47'05"N	116°03'42"W	
Goose Pasture	Twin Falls	42°10'15"N	114°55'48"W	
Granite Basin	Adams	44°43'24"N	116°16'04"W	
Granite Basin	Boise	44°09'29"N	115°49'31"W	
Grassy Flat	Owyhee	42°40'01"N	116°24'07"W	
Grassy Flat	Valley	44°23'39"N	116°04'56"W	
Grave Meadow	Idaho	45°57'33"N	115°05'02"W	
Graves Meadow	Latah	46°57'58"N	116°31'43"W	
Greasewood Flat	Adams	44°44'26"N	116°37'26"W	5710
Green Basin	Bear Lake	42°15'14"N	111°34'26"W	7953
Greenfield Flat	Gem	44°29'23"N	116°09'38"W	
Grizzly Basin	Washington	44°42'30"N	116°40'49"W	
Halogeton Flat	Owyhee	42°07'33"N	116°52'56"W	
Hand Meadows	Idaho	45°17'24"N	115°15'54"W	
Hanna Flats	Bonner	48°34'28"N	116°58'00"W	
Hansen Basin	Fremont	44°00'30"N	111°53'37"W	
Hanson Meadows	Clearwater	46°41'50"N	114°55'45"W	
Hard Creek Basin	Idaho	45°09'07"N	116°09'02"W	
Hard Creek Meadow	Idaho	45°08'37"N	116°10'47"W	
Hard Creek meadows	Idaho	45°08'23"N	116°15'23"W	4000
Hardscrabble Pasture (historical)	Boise	44°14'02"N	115°52'57"W	4520
Harlan Meadows	Idaho	45°29'04"N	115°13'54"W	
Harrison Flats	Kootenai	47°25'20"N	116°42'12"W	
Harscrabble Pasture	Valley	44°14'12"N	115°54'01"W	3420
Hartly Meadows	Adams	45°01'44"N	116°06'56"W	
Haskins Flat	Latah	46°44'11"N	116°56'52"W	2610
Hawes Pasture	Owyhee	42°31'29"N	116°37'09"W	
Hayden Basin	Lemhi	44°49'50"N	113°44'09"W	
Haypress Meadows	Idaho	45°17'43"N	115°10'02"W	
Henderson Flats	Ada	43°06'45"N	116°17'37"W	
Henley Basin	Owyhee	42°32'15"N	116°34'44"W	
Henley Basin	Washington	44°24'05"N	117°08'50"W	
Hennessey Meadow	Valley	44°53'15"N	115°27'31"W	
Henry Meadow	Fremont	43°59'51"N	111°17'11"W	
Henrys Lake Flat	Fremont	44°35'47"N	111°20'21"W	
Hidden Valley	Canyon	43°27'42"N	116°40'20"W	
Hidden Valley	Kootenai	47°48'11"N	116°56'54"W	
High Basin	Bingham	43°10'39"N	111°52'48"W	
High Prairie	Elmore	43°19'43"N	115°13'47"W	
Highland Flats	Boundary	48°35'15"N	116°26'34"W	
Hill Pasture	Owyhee	42°32'14"N	116°03'39"W	
Hiltsley Flat	Idaho	45°25'57"N	116°22'33"W	3840

Basins, Flats, Meadows, Plains

Basin, Flat, Meadow, or Plain	County	Latitude	Longitude	Elev.
Hirschi Flat	Clark	44°29'50"N	111°55'39"W	
Hog Meadow	Latah	46°49'21"N	116°27'09"W	
Hole, The	Bear Lake	42°33'00"N	111°20'35"W	
Hoodoo Meadows	Lemhi	45°03'23"N	114°33'30"W	
Horse Basin	Custer	44°11'31"N	114°03'16"W	
Horse Basin	Franklin	42°12'49"N	111°39'05"W	
Horse Basin	Owyhee	42°02'25"N	116°34'20"W	
Horse Basin	Twin Falls	42°17'42"N	114°26'14"W	
Horse Flat	Adams	44°46'32"N	116°33'10"W	
Horse Flat	Bear Lake	42°09'53"N	111°31'53"W	
Horse Flat	Idaho	45°35'43"N	115°27'01"W	
Horse Flat	Owyhee	42°27'08"N	116°58'35"W	
Horse Heaven	Teton	43°39'34"N	111°15'01"W	
Horse Heaven Meadows	Idaho	44°47'11"N	114°37'33"W	
Horse Pasture Basin	Adams	45°10'54"N	116°34'42"W	
Horse Wallow	Custer	43°53'56"N	113°50'05"W	
Horsefly Meadows	Idaho	46°08'09"N	114°44'46"W	
Horsehoe Basin	Bear Lake	42°14'31"N	111°36'11"W	7967
Horsethief Basin	Valley	44°30'59"N	115°54'46"W	
Hot Springs Meadow	Idaho	45°28'23"N	115°06'31"W	
Howard Flat	Power	42°34'25"N	112°43'47"W	
Hubbard Basin	Adams	44°39'48"N	116°10'57"W	
Huckleberry Basin	Caribou	42°38'55"N	111°22'23"W	
Huckleberry Flat	Boise	43°53'46"N	115°59'50"W	
Huddles Hole	Butte	43°27'12"N	113°22'39"W	5601
Huddles Hole	Butte	43°26'56"N	113°22'08"W	
Hughes Meadows	Boundary	48°52'23"N	117°00'05"W	
Hurdy Flat	Boise	43°58'15"N	115°35'30"W	5160
Icicle Flat	Idaho	46°10'37"N	116°25'41"W	
Indian Flat	Franklin	42°03'54"N	111°42'02"W	
Indian Hay Meadows	Owyhee	42°03'00"N	115°59'43"W	
Indian Meadows	Idaho	46°24'17"N	115°04'20"W	
Indian Meadows	Owyhee	42°41'49"N	116°46'24"W	
Inside Desert	Owyhee	42°18'30"N	115°33'15"W	
Iron Basin	Custer	44°07'50"N	114°39'52"W	
Jack Pine Flats	Bonner	48°25'21"N	116°50'50"W	
Jackass Flat	Custer	44°43'38"N	114°58'25"W	
Jacks Creek Meadows	Owyhee	42°25'43"N	116°08'28"W	
Jarvis Pasture	Owyhee	42°10'26"N	116°27'29"W	
Jasper Flats	Blaine	43°22'11"N	114°06'53"W	
Jeffs Flats	Custer	44°32'55"N	114°20'10"W	
Jensen Meadow	Bonneville	43°12'08"N	111°11'29"W	
Joe Jump Basin	Custer	44°11'44"N	114°18'44"W	
Joes Basin	Cassia	42°04'01"N	114°07'34"W	
John Boyd Flat	Twin Falls	42°14'24"N	114°58'35"W	
Joiner Flat	Owyhee	42°05'51"N	115°11'47"W	5370
Jolley Flat	Elmore	42°58'39"N	115°08'30"W	
Joseph Plains	Idaho	45°55'21"N	116°32'56"W	
Josper Flats	Blaine	43°23'03"N	114°07'45"W	
J-P Desert	Owyhee	42°08'25"N	115°46'03"W	
Jug Meadows	Valley	44°50'10"N	115°58'41"W	
Juliette Basin	Custer	44°23'09"N	114°22'37"W	
Jumbo Basin	Valley	44°56'51"N	115°58'58"W	
Junction Valley	Cassia	41°56'23"N	113°44'37"W	
Juniper Basin	Owyhee	42°04'06"N	116°26'46"W	
Justice Park	Bannock	42°41'32"N	112°22'40"W	6680
Kelly Meadows	Idaho	45°18'53"N	115°49'06"W	
Kettles, The	Custer	44°06'08"N	114°37'17"W	
Keystone Meadows	Idaho	45°13'36"N	115°41'41"W	
Kidder Flat	Idaho	45°50'51"N	116°25'06"W	
Kimbal Basin	Owyhee	42°12'43"N	116°32'09"W	
Kings Bowl	Power	42°56'58"N	113°12'56"W	
Kit Price Prairie	Shoshone	47°44'34"N	116°00'30"W	
Kiwah Meadow	Valley	44°50'38"N	115°16'24"W	
Kooskooskia Meadows	Idaho	46°22'52"N	114°41'02"W	
Kootenai Orchard	Boundary	48°32'54"N	116°27'16"W	
Krutschmer Field	Owyhee	42°02'30"N	115°54'00"W	6344
Kruze Meadows	Nez Perce	46°08'33"N	116°46'38"W	
Lacey Flat	Idaho	45°49'50"N	116°24'04"W	
Lake Basin	Adams	44°43'15"N	116°15'09"W	6920
Lake Basin	Custer	44°01'50"N	114°10'29"W	
Larabee Meadows	Nez Perce	46°07'10"N	116°44'26"W	
Larkspur Park	Minidoka	42°59'32"N	113°31'37"W	4484
Lava Flat	Adams	44°29'01"N	116°24'27"W	
Lee Flat	Idaho	45°50'13"N	116°17'13"W	
Lehman Basin	Custer	43°56'01"N	113°51'32"W	
Lew Hand Meadow	Clearwater	46°25'38"N	115°50'01"W	
Linehan Flat	Owyhee	43°01'08"N	116°42'47"W	
Litte Park	Blaine	43°17'44"N	113°41'18"W	
Little Antelope Flat	Custer	44°23'16"N	114°04'21"W	
Little Basin	Blaine	43°42'16"N	114°08'19"W	
Little Basin	Butte	44°01'48"N	113°24'05"W	
Little Basin	Custer	44°19'01"N	114°56'57"W	
Little Bradshaw Basin	Custer	44°17'36"N	114°13'16"W	
Little Cabin Meadows	Clearwater	46°33'54"N	115°12'22"W	
Little Camas Prairie	Elmore	43°22'04"N	115°23'20"W	
Little Fiddler Flat	Elmore	43°29'22"N	115°41'32"W	
Little Flat	Camas	43°13'10"N	114°55'39"W	
Little Flat	Caribou	42°48'19"N	111°49'21"W	5750
Little Horse Basin	Owyhee	42°05'48"N	116°39'09"W	
Little Jacks Creek Basin	Owyhee	42°35'10"N	116°13'55"W	
Little Laidlaw Park	Blaine	43°19'01"N	113°31'10"W	
Little Lodgepole Meadow	Idaho	45°20'44"N	115°11'23"W	
Little Lost River Sinks	Butte	43°46'08"N	112°58'58"W	
Little Mallard Meadows	Idaho	45°35'05"N	115°25'13"W	
Little Pony Meadows	Idaho	45°12'09"N	115°41'54"W	

Basin, Flat, Meadow, or Plain	County	Latitude	Longitude	Elev.
Little Prairie	Butte	43°23'15"N	113°27'30"W	
Little Sage Hen Basin	Valley	44°19'07"N	116°09'23"W	
Little Sage Hen Flat	Elmore	43°15'49"N	115°32'54"W	
Little Valley	Bonneville	43°04'40"N	111°31'40"W	
Little Willow Flat	Washington	44°12'55"N	116°31'14"W	
Litz Basin	Franklin	42°07'12"N	111°39'45"W	
Lobauer Basin	Valley	45°06'46"N	114°53'26"W	
Long Meadow	Idaho	45°36'12"N	115°48'30"W	
Long Meadow	Latah	46°44'38"N	116°22'10"W	
Long Meadows	Fremont	44°05'28"N	111°06'49"W	
Lost Basin	Adams	45°08'26"N	116°34'20"W	
Lost Basin	Adams	44°50'31"N	116°46'58"W	
Lost Basin	Valley	44°32'09"N	115°51'03"W	
Lost Knife Meadows	Idaho	46°22'10"N	114°55'23"W	
Lost Meadows	Idaho	45°48'00"N	116°03'41"W	
Lost Meadows	Valley	44°18'10"N	115°27'07"W	
Lost Packer Meadows	Idaho	45°27'47"N	114°47'45"W	
Lost Park Meadows	Idaho	46°37'59"N	114°29'25"W	
Lower Bench	Elmore	43°30'11"N	115°41'06"W	3800
Lower Deer Flat	Canyon	43°36'54"N	116°46'12"W	
Lower Lodgepole Meadow	Idaho	45°19'06"N	115°12'00"W	
Lower Meadow	Idaho	45°18'56"N	114°52'52"W	
Lower Ramey Meadows	Idaho	45°16'09"N	115°11'31"W	
Lower Trapper Flat	Valley	44°47'42"N	115°29'11"W	
Lower Trout Creek Meadows	Idaho	45°24'59"N	115°18'03"W	
Lower Twentymile Meadows	Idaho	45°45'37"N	115°44'54"W	
Lykow Flat	Boise	43°51'26"N	115°48'08"W	
Lynx Meadows	Idaho	45°48'21"N	114°58'50"W	
Mace Meadows	Valley	44°19'17"N	115°27'07"W	
Macon Flat	Camas	43°18'52"N	114°27'19"W	
Macumber Meadows	Latah	46°59'13"N	116°41'38"W	
Magpie Basin	Cassia	42°15'37"N	114°15'26"W	
Mahogany Basin	Bear Lake	42°18'35"N	111°31'09"W	
Mahoney Flat	Blaine	43°19'31"N	114°20'05"W	
Manada Flat	Owyhee	42°27'26"N	116°56'29"W	
Marion Meadows	Idaho	46°25'59"N	114°40'07"W	
Marshall Meadow	Idaho	45°21'20"N	115°51'16"W	
Marten Meadows	Idaho	45°56'54"N	115°22'54"W	
Mason Meadows	Latah	46°38'51"N	116°21'19"W	
McCalla Basin	Idaho	45°29'15"N	116°28'13"W	4800
McComas Meadows	Idaho	45°54'12"N	115°54'57"W	
McCormack Meadows	Bonner	48°17'57"N	116°44'10"W	
McGaffee Basin	Idaho	45°24'05"N	116°34'55"W	3400
McGowan Basin	Lemhi	44°58'23"N	114°20'45"W	
McMullen Basin	Twin Falls	42°15'24"N	114°24'04"W	
Meade Basin	Caribou	42°30'04"N	111°14'37"W	
Meadow of Doubt	Idaho	45°27'39"N	115°10'22"W	
Meadows, The	Custer	44°07'56"N	114°43'22"W	
Meadows, The	Oneida	42°11'27"N	112°45'40"W	
Melon Valley	Twin Falls	42°38'38"N	114°47'26"W	
Meyers Cove	Lemhi	44°49'50"N	114°30'10"W	
Mica Flats	Kootenai	47°37'48"N	116°51'13"W	
Mica Meadows	Shoshone	47°11'44"N	116°13'10"W	
Michaud Flats	Power	42°55'32"N	112°35'16"W	
Middle Meadow	Idaho	45°28'17"N	115°10'42"W	
Middle Meadows	Idaho	45°17'36"N	114°53'23"W	
Mile Flat	Valley	45°08'47"N	115°05'37"W	
Mission Flats	Kootenai	47°33'20"N	116°22'51"W	
Missouri Flat	Latah	46°45'54"N	117°01'48"W	
Montour Valley	Gem	43°55'53"N	116°19'11"W	
Moody Meadow	Madison	43°40'54"N	111°27'42"W	
Moonlight Meadow	Bear Lake	42°23'44"N	111°35'42"W	
Moorehead Flat	Adams	44°57'05"N	116°11'04"W	
Moores Flat	Elmore	43°22'56"N	115°13'37"W	
Moose Creek Plateau	Fremont	44°29'54"N	111°07'03"W	
Moose Jaw Meadow	Idaho	45°20'40"N	115°15'11"W	
Moose Meadows	Idaho	46°07'08"N	115°08'52"W	
Moose Meadows	Idaho	45°18'40"N	115°07'02"W	
Moose Meadows	Latah	46°52'44"N	116°26'59"W	
Morgan Meadow	Bonneville	43°02'28"N	111°18'12"W	
Morgans Pasture	Bingham	43°22'55"N	112°21'30"W	
Moss Flat	Owyhee	43°02'25"N	116°55'41"W	
Mountain Meadow	Bonner	48°05'52"N	116°53'12"W	
Mountain Meadow	Valley	44°29'21"N	115°22'16"W	
Mountain Meadows	Idaho	45°41'57"N	115°13'40"W	
Moyer Basin	Lemhi	44°57'26"N	114°18'30"W	
Mud Flat	Boise	43°55'53"N	115°57'39"W	
Mud Flat	Idaho	45°15'11"N	116°17'21"W	
Mud Flat	Owyhee	42°36'12"N	116°33'10"W	
Mud Flat	Owyhee	42°10'32"N	116°16'21"W	
Mud Flat	Owyhee	42°05'01"N	115°07'28"W	5610
Mud Flat	Twin Falls	42°06'42"N	114°50'27"W	6910
Mud Flats	Owyhee	42°05'37"N	115°59'53"W	
Mud Lake	Twin Falls	42°23'45"N	114°38'03"W	
Mule Meadows	Clark	44°30'47"N	112°00'57"W	
Murphy Flat	Owyhee	43°10'35"N	116°27'41"W	
Musselshell Meadows	Clearwater	46°21'07"N	115°44'56"W	3182
Never Again Flats	Clearwater	46°39'26"N	115°05'23"W	3540
Nichol Flat	Owyhee	42°21'19"N	116°13'52"W	
Nickel Creek Pocket	Owyhee	42°28'11"N	116°42'10"W	
Noble Basin	Owyhee	42°57'21"N	116°58'01"W	
North Ant Basin	Caribou	42°26'20"N	111°35'17"W	
North Antelope Flat	Fremont	44°18'18"N	111°36'16"W	
North Basin	Lemhi	44°54'23"N	113°49'54"W	
North Bench	Boundary	48°45'47"N	116°18'55"W	
North Center Lake	Lincoln	42°52'51"N	113°53'13"W	

Basins, Flats, Meadows, Plains

Basin, Flat, Meadow, or Plain	County	Latitude	Longitude	Elev.
North Sinker Flat	Owyhee	43°06'41"N	116°42'54"W	
North Star Meadows	Idaho	45°32'00"N	115°26'24"W	
North Worm Creek Basin	Franklin	42°10'54"N	111°40'29"W	
North-South Ski Bowl	Benewah	47°03'57"N	116°39'24"W	3802
No-see-um Meadows	Idaho	46°25'09"N	115°18'11"W	
Nugent Meadows	Owyhee	43°03'59"N	116°42'11"W	
Nut Basin	Idaho	45°31'58"N	116°10'25"W	
Nutmeg Flat	Washington	44°15'26"N	116°41'01"W	
Old Canyon Meadow	Oneida	42°17'04"N	112°15'41"W	
Old Maids Clearing, The	Bonner	47°59'44"N	116°33'37"W	
Old Man Meadows	Idaho	46°13'50"N	115°13'18"W	
Olive Meadows	Washington	44°46'59"N	116°39'27"W	6420
O'Neil Basin	Twin Falls	41°42'15"N	115°09'12"W	
Oregon Lake	Owyhee	42°18'35"N	117°01'30"W	5204
Otts Basin	Bonner	48°11'56"N	116°39'01"W	
Oviatt Meadows	Clearwater	46°45'26"N	116°16'24"W	
Owyhee Desert	Owyhee	41°47'31"N	116°57'53"W	5321
Oxford Basin	Franklin	42°15'51"N	112°04'54"W	7043
Oxford Meadows	Clearwater	46°36'00"N	115°38'01"W	
Pack River Flats	Bonner	48°17'51"N	116°22'46"W	
Pack Saddle Basin	Bingham	43°07'32"N	111°42'00"W	6490
Packer Meadows	Bonner	48°43'02"N	117°00'07"W	
Packer Meadows	Idaho	46°38'16"N	114°33'08"W	
Packsaddle Basin	Teton	43°43'38"N	111°21'17"W	
Paddelford Flat	Blaine	43°19'42"N	113°46'59"W	4890
Paddy Flat	Valley	44°46'54"N	115°57'14"W	
Paint Pot, The	Custer	44°13'05"N	114°11'44"W	
Paradise Basin	Bonneville	43°27'13"N	111°11'08"W	
Paradise Flat	Adams	45°13'00"N	116°30'43"W	
Paradise Flat	Adams	44°56'48"N	116°42'16"W	
Paris Flat	Bear Lake	42°13'12"N	111°35'00"W	7866
Pasadena Valley	Elmore	42°56'18"N	115°11'39"W	
Pasture, The	Elmore	42°58'02"N	115°10'19"W	
Pealy Flat	Idaho	45°51'36"N	116°17'03"W	
Peel Tree Basin	Lemhi	44°54'33"N	114°03'00"W	5380
Peer Field	Owyhee	42°31'12"N	116°53'14"W	
Pen Basin	Valley	44°38'00"N	115°32'00"W	
Peterson Flat	Idaho	45°26'30"N	115°55'30"W	4480
Phantom Meadow	Idaho	45°22'13"N	114°51'42"W	
Phoebe Meadow	Valley	44°56'03"N	115°40'07"W	
Pidgeon Flat	Boise	44°06'39"N	115°39'48"W	
Pimble Point	Idaho	45°51'14"N	116°29'23"W	
Pine Basin	Twin Falls	42°16'16"N	114°18'23"W	
Pine Creek Bench	Bonneville	43°29'04"N	111°22'59"W	
Pine Creek Meadow	Valley	44°17'33"N	116°09'47"W	
Pine Flat	Bear Lake	42°13'34"N	111°28'15"W	
Pine Flat	Boise	44°03'59"N	115°40'44"W	
Pine Flat	Shoshone	47°48'24"N	115°57'10"W	

Basin, Flat, Meadow, or Plain	County	Latitude	Longitude	Elev.
Pine Grass Flat	Washington	44°41'23"N	116°57'02"W	
Piute Basin	Owyhee	42°11'58"N	116°36'55"W	
Pixley Basin	Owyhee	42°48'32"N	116°25'58"W	
Placer Basin	Adams	45°06'46"N	116°36'48"W	
Placer Flat	Blaine	43°37'15"N	114°35'33"W	
Pocatello Bench	Bannock	42°56'36"N	112°25'51"W	
Pocatello Valley	Oneida	42°03'08"N	112°28'44"W	
Poison Flat	Blaine	43°34'44"N	114°33'35"W	
Poison Flat	Madison	43°37'50"N	111°35'19"W	
Poker Meadows	Valley	44°26'08"N	115°20'09"W	
Pony Flats	Clearwater	46°39'07"N	115°06'45"W	
Pony Meadows	Valley	45°11'23"N	115°41'24"W	
Porcupine Flat	Valley	44°36'56"N	115°49'48"W	
Porcupine Meadow	Idaho	45°32'01"N	115°51'12"W	
Pot Hole Basin	Twin Falls	42°02'48"N	114°18'00"W	
Povert Flat	Custer	44°18'25"N	114°21'56"W	
Poverty Flat	Blaine	43°24'35"N	114°17'02"W	
Poverty Flat	Lemhi	45°18'41"N	114°29'28"W	
Poverty Flats	Franklin	42°14'09"N	111°54'40"W	5013
Powell Pasture	Idaho	46°31'09"N	114°41'35"W	
Pury Flat	Elmore	43°35'37"N	115°49'47"W	
Putney Meadows	Fremont	44°06'00"N	111°05'12"W	
Quaker Flat	Bonneville	43°18'40"N	111°05'16"W	
Queen Creek Meadow	Idaho	45°25'44"N	115°05'51"W	
Racetrack Meadow	Lemhi	45°17'03"N	114°01'59"W	
Rammage Meadows	Gem	44°24'50"N	116°12'17"W	
Randall Flat	Latah	46°45'23"N	116°49'27"W	
Rathdrum Prairie	Kootenai	47°46'07"N	116°50'26"W	
Rattlesnake Basin	Bear Lake	42°34'37"N	111°21'41"W	
Rattlesnake Basin	Idaho	45°17'23"N	116°24'17"W	
Rattlesnake Basin	Owyhee	42°18'05"N	115°59'35"W	
Red Basin	Owyhee	42°19'52"N	116°48'41"W	
Red Sinks	Franklin	42°01'04"N	111°30'16"W	
Red Top Meadows	Idaho	45°21'33"N	115°16'49"W	
Reed Point	Idaho	45°51'20"N	116°26'45"W	
Rentile Field	Owyhee	42°23'38"N	117°01'27"W	5230
Republican Flats	Idaho	45°21'39"N	115°42'06"W	
Reservation Basin	Owyhee	42°51'16"N	116°24'50"W	
Reservoir Flat	Valley	45°04'27"N	115°38'21"W	
Robideaux Meadows	Clearwater	46°45'13"N	116°10'00"W	
Rock Creek Basin	Fremont	44°34'04"N	111°28'45"W	
Rock Flat	Adams	44°55'58"N	116°10'13"W	
Rock Flat	Owyhee	42°22'19"N	116°58'48"W	
Rocky Comfort Flat	Adams	44°57'58"N	116°41'15"W	
Rooker Basin	Lemhi	44°56'15"N	114°20'12"W	
Rooney Basin	Idaho	45°26'13"N	116°11'25"W	
Ross Fork Basin	Camas	43°47'53"N	114°59'14"W	
Round Bottom Meadows	Idaho	45°36'55"N	115°59'57"W	

Basin, Flat, Meadow, or Plain	County	Latitude	Longitude	Elev.
Round Meadow	Valley	45°09'20"N	115°26'08"W	
Round Meadow	Valley	44°48'45"N	115°33'34"W	
Round Meadows	Clearwater	46°44'52"N	116°19'14"W	
Round Prairie	Boundary	48°57'08"N	116°16'16"W	
Round Valley	Adams	45°07'18"N	116°19'32"W	
Rowlands Park	Latah	46°45'37"N	116°53'03"W	2850
Ruby Meadows	Idaho	45°13'53"N	115°52'52"W	
Ryan Pasture	Owyhee	42°15'51"N	116°47'41"W	
Rye Flat	Boise	43°48'54"N	115°57'26"W	
Rye Grass Flat	Blaine	43°15'54"N	114°18'34"W	
Rye Patch	Idaho	45°21'17"N	115°45'23"W	
Rye Patch	Owyhee	43°06'41"N	116°23'11"W	
Ryegrass Flat	Butte	43°03'19"N	112°52'13"W	
Saddle Horse Basin	Oneida	42°12'22"N	112°59'23"W	
Sage Hen Basin	Bingham	43°06'44"N	112°11'13"W	
Sage Hen Basin	Gem	44°19'36"N	116°10'41"W	
Sage Hen Flat	Elmore	43°15'33"N	115°31'53"W	
Sagebrush Basin	Owyhee	42°31'47"N	116°12'56"W	
Sagehen Flats	Twin Falls	42°18'52"N	114°18'14"W	
Sands Basin	Owyhee	43°25'09"N	116°58'49"W	
Sater Meadows	Valley	45°04'40"N	116°06'04"W	
Saw Log Basin	Bingham	43°17'17"N	111°57'32"W	5860
Schweitzer Basin	Bonner	48°22'03"N	116°37'42"W	
Scott Valley	Valley	44°33'53"N	115°54'26"W	
Secesh Meadows	Idaho	45°14'11"N	115°48'33"W	
Service Flats	Idaho	45°49'07"N	116°09'58"W	
Shake Meadow	Clearwater	46°26'17"N	115°59'20"W	
Shares Basin	Owyhee	43°21'52"N	116°50'53"W	
Shea Meadows	Latah	46°51'42"N	116°29'11"W	
Sheep Pen Basin	Custer	44°22'51"N	113°54'06"W	
Shillam Flat	Idaho	45°48'01"N	116°24'00"W	
Shingle Flat	Adams	44°46'59"N	116°21'45"W	
Shoshone Basin	Twin Falls	42°03'00"N	114°25'52"W	
Sinks, The	Franklin	42°19'56"N	111°40'05"W	
Sisters Basin	Shoshone	47°09'07"N	115°44'35"W	4010
Six Lake Basin	Adams	45°11'43"N	116°35'40"W	7457
Sky Ranch Flat	Blaine	43°25'08"N	114°19'47"W	
Slade Flat	Elmore	43°04'26"N	115°38'27"W	
Slate Basin	Clark	44°15'48"N	112°43'33"W	
Slaters Flat	Ada	43°26'34"N	115°59'07"W	
Sly Meadows	Benewah	47°16'37"N	116°23'17"W	
Smiley Meadows	Custer	43°41'43"N	113°45'58"W	
Smith Meadow	Owyhee	42°28'29"N	116°47'13"W	
Smith Meadows	Latah	46°59'42"N	116°40'59"W	
Smith Meadows	Latah	46°50'48"N	116°31'19"W	
Smith Prairie	Elmore	43°29'20"N	115°35'39"W	
Snake River Plain	Bonneville	43°27'47"N	111°57'36"W	
Snake River Plains	Lincoln	43°02'31"N	113°52'31"W	

Basin, Flat, Meadow, or Plain	County	Latitude	Longitude	Elev.
Sneakfoot Meadows	Idaho	46°26'59"N	114°39'01"W	
Snow Valley	Bonner	48°16'47"N	116°59'40"W	
Snowbird Meadows	Shoshone	47°54'25"N	116°18'11"W	
Soard Flat	Idaho	45°24'10"N	116°22'01"W	
Soldier Meadows	Idaho	46°26'15"N	115°34'47"W	
Sommercamp Basin	Owyhee	43°06'11"N	116°54'32"W	
Sommercamp Flat (historical)	Owyhee	43°00'21"N	116°49'49"W	
Sonners Flat	Blaine	43°15'41"N	114°15'22"W	
South Ant Basin	Caribou	42°25'40"N	111°35'29"W	
South Cheatbeck Basin	Caribou	42°28'01"N	111°35'56"W	
South Worm Creek Basin	Franklin	42°10'23"N	111°40'12"W	
Spaulding Basin	Bonneville	43°17'44"N	111°04'16"W	
Spokane Meadows	Shoshone	47°05'17"N	115°46'33"W	
Sponge Meadows	Idaho	46°20'16"N	115°06'13"W	
Spring Basin	Custer	44°22'09"N	114°28'23"W	
Spring Basin	Custer	44°08'03"N	114°03'04"W	
Spring Basin	Custer	44°04'54"N	114°32'36"W	
Spring Creek Basin	Owyhee	42°11'26"N	116°54'27"W	
Sproule Flat	Elmore	43°20'24"N	115°31'09"W	
Spud Basin	Custer	44°29'37"N	114°02'44"W	
Spud Patch Flat	Blaine	43°17'02"N	114°16'09"W	
Squaw Flat	Adams	44°46'05"N	116°15'06"W	
Squaw Flat	Camas	43°15'49"N	114°46'43"W	
Squaw Flat	Owyhee	42°09'53"N	116°03'40"W	
Squaw Meadow	Idaho	45°36'18"N	115°44'03"W	
Squaw Meadows	Owyhee	42°21'32"N	116°28'11"W	
Squawboard Meadow	Lemhi	44°58'03"N	114°10'17"W	
SS Field	Owyhee	42°03'56"N	115°57'38"W	6386
Stamp Meadows	Fremont	44°30'41"N	111°21'55"W	
Stampede Park	Caribou	42°40'15"N	111°37'43"W	5750
Stanley Basin	Custer	44°12'08"N	114°57'41"W	
Star Meadows	Idaho	46°30'58"N	114°40'50"W	
State Line Meadows	Owyhee	41°59'51"N	115°59'33"W	
State Line Meadows	Owyhee	42°00'01"N	115°59'27"W	
Stauffer Flat	Owyhee	42°26'35"N	116°52'28"W	
Steer Basin	Cassia	42°16'35"N	114°15'52"W	
Stewart Flat	Caribou	42°42'18"N	111°12'07"W	
Stewarts Meadow	Valley	44°29'59"N	116°07'23"W	
Stickney Basin	Idaho	45°27'08"N	116°33'23"W	2320
Stinking Water Basin	Washington	44°14'47"N	116°43'13"W	
Stocking Meadows	Clearwater	46°56'22"N	115°52'03"W	
Stolle Meadows	Valley	44°35'02"N	115°40'41"W	
Stony Creek Meadows	Clearwater	46°53'15"N	116°04'53"W	4440
Stony Meadow	Valley	44°33'18"N	115°49'08"W	
Storm Creek Flat	Idaho	45°34'49"N	114°37'22"W	
Stover Meadow	Valley	44°47'08"N	116°00'32"W	
Stratton Basin	Valley	44°27'16"N	115°38'16"W	
Stratton Meadow	Idaho	45°14'43"N	115°45'12"W	

Basins, Flats, Meadows, Plains

Basin, Flat, Meadow, or Plain	County	Latitude	Longitude	Elev.
Strawberry Basin	Custer	44°04'33"N	114°39'43"W	
Striker Basin	Owyhee	43°13'20"N	116°32'18"W	
Strode Basin	Owyhee	43°30'27"N	117°00'39"W	
Sucker Flat	Twin Falls	42°36'41"N	114°35'11"W	
Summit Flat	Boise	43°59'27"N	115°42'00"W	
Summit Flat	Idaho	45°44'24"N	115°35'00"W	
Summit Flat	Owyhee	42°42'14"N	116°23'34"W	
Summit Flat	Owyhee	42°15'34"N	116°01'10"W	
Summit Flat	Owyhee	42°14'51"N	116°01'24"W	
Sunflower Flat	Custer	44°43'55"N	115°01'02"W	
Sunflower Flat	Valley	44°50'04"N	116°10'48"W	
Swan Basin	Lemhi	44°34'56"N	113°25'37"W	
Swan Flat	Bear Lake	41°59'54"N	111°29'26"W	
Swartz Meadow	Idaho	45°46'50"N	116°06'47"W	4960
Swensen Basin	Custer	44°02'08"N	114°00'28"W	
Table Meadows	Idaho	45°55'58"N	115°30'43"W	
Tamarack Flat	Gem	44°23'33"N	116°14'05"W	
Tamarack Flat	Washington	44°35'16"N	116°58'25"W	
Teapot Basin	Elmore	43°10'55"N	115°30'26"W	
Tee Meadow	Latah	46°50'08"N	116°29'53"W	
Telegraph Flat	Bear Lake	42°09'27"N	111°34'58"W	
Tenlake Basin	Boise	43°57'59"N	115°01'18"W	
Tenmile Flat	Idaho	45°56'55"N	115°47'47"W	
Tenmile Meadows	Idaho	45°40'52"N	115°42'19"W	
Tepee Flats	Idaho	45°32'15"N	115°40'25"W	
Teton Basin	Teton	43°40'00"N	111°10'00"W	
Teton Basin	Teton	43°42'00"N	111°10'30"W	
Texas Basin	Owyhee	43°13'06"N	116°58'39"W	
The Settlement	Bonner	48°10'09"N	116°51'25"W	
Thomas Flats	Owyhee	43°09'33"N	116°20'14"W	
Thompson Flat	Idaho	45°35'50"N	114°43'03"W	
Thoroughbred Flat	Twin Falls	42°09'35"N	114°19'40"W	
Three Blaze Meadow	Idaho	45°22'14"N	115°16'19"W	
Three Forks Meadow	Owyhee	42°05'24"N	116°02'07"W	
Three Links Meadows	Idaho	46°11'43"N	115°09'05"W	
Timothy Flats	Idaho	46°04'34"N	115°15'59"W	
Tobacco Meadow	Owyhee	42°35'55"N	116°46'33"W	
Towhead Basin	Valley	45°08'45"N	114°59'25"W	
Toy Flat	Owyhee	42°52'40"N	116°34'24"W	
Tractor Flat	Bingham	43°37'32"N	112°32'25"W	
Tranquil Basin	Valley	44°21'38"N	115°42'03"W	6338
Trapper Flat	Lemhi	45°09'43"N	114°17'00"W	
Trapper Flat	Valley	44°47'44"N	115°27'23"W	
Trappers Flat	Boise	44°02'33"N	115°23'26"W	6590
Trench Mortar Flat	Butte	43°24'45"N	113°31'15"W	
Triangle Flat	Owyhee	42°47'28"N	116°43'54"W	
Tripod Meadow	Valley	44°18'21"N	116°07'55"W	
Turk Meadow	Owyhee	42°06'20"N	115°58'27"W	6140

Basin, Flat, Meadow, or Plain	County	Latitude	Longitude	Elev.
Twelvemile Meadows	Lemhi	44°56'12"N	113°52'08"W	
Twentymile Flat	Owyhee	42°55'51"N	116°06'14"W	
Twogood Flat	Idaho	45°48'26"N	116°27'50"W	
Tygee Creek Basin	Fremont	44°38'24"N	111°15'17"W	7550
Tygee Creek Basin	Fremont	44°38'13"N	111°14'22"W	
Tyndall Meadows	Valley	44°33'48"N	115°33'00"W	
Ucon Basin	Valley	44°22'12"N	115°43'52"W	
Upper Basin	Clearwater	46°53'22"N	116°08'50"W	
Upper Deer Flat	Canyon	43°30'32"N	116°35'49"W	
Upper Deer Flat	Canyon	43°29'14"N	116°35'59"W	
Upper Lodgepole Meadow	Idaho	45°18'24"N	115°12'47"W	
Upper Ramey Meadows	Idaho	45°17'13"N	115°13'16"W	
Upper Red Top Meadows	Idaho	45°20'46"N	115°18'50"W	
Upper Trapper Flat	Valley	44°47'48"N	115°24'51"W	
Upper Trout Creek Meadows	Idaho	45°24'00"N	115°18'42"W	
Vassar Meadows	Latah	46°50'55"N	116°32'42"W	
Wag Meadows	Idaho	46°23'07"N	114°53'03"W	
Wagner Field	Owyhee	42°00'01"N	115°54'47"W	
Wagon Box Basin	Owyhee	42°31'00"N	116°33'50"W	
Walcot Basin	Owyhee	42°03'09"N	116°52'12"W	
Wapiti Meadows	Idaho	45°19'21"N	115°04'13"W	
Wardenhoff Meadows	Valley	44°49'36"N	115°33'12"W	
Warner Flat	Power	42°33'38"N	112°43'49"W	
Warren Meadow	Idaho	45°17'48"N	115°42'23"W	
Washington Basin	Custer	44°00'16"N	114°38'55"W	
Waters Flat	Clark	44°28'27"N	111°55'00"W	6580
Weippe Prairie	Clearwater	46°21'29"N	115°54'56"W	
Weiser Cove	Washington	44°13'51"N	116°49'46"W	
Weitas Meadows	Idaho	46°25'59"N	115°28'55"W	
Wells Bench	Clearwater	46°31'24"N	116°11'50"W	
West Fork Basin	Idaho	45°14'19"N	115°05'51"W	
West Weiser Flat	Washington	44°16'09"N	117°02'48"W	
Wet Creek Basin	Butte	44°00'28"N	113°24'16"W	
Wet Meadows	Idaho	45°28'17"N	115°15'58"W	
Wet Meadows	Latah	46°51'50"N	116°31'44"W	
Wet Meadows	Valley	44°24'20"N	115°30'10"W	
Whiskey Dick Flat	Kootenai	47°46'54"N	116°59'45"W	
Whiskey Flat	Bear Lake	42°26'06"N	111°09'58"W	
White Bird Meadow	Idaho	45°30'57"N	115°11'24"W	
White Cow Basin	Owyhee	42°18'06"N	116°38'24"W	
White Flat	Elmore	43°27'36"N	115°22'58"W	
White Horse Basin	Lemhi	45°09'22"N	114°13'00"W	
Whitehawk Basin	Valley	44°17'06"N	115°34'23"W	
Wids Meadow	Owyhee	42°04'42"N	115°58'34"W	6435
Wilkins Island	Owyhee	41°57'47"N	115°22'35"W	
Willow Flat	Franklin	42°08'22"N	111°37'17"W	
Willow Flat	Idaho	45°37'45"N	116°02'44"W	
Willow Meadow	Lemhi	45°23'21"N	114°28'06"W	

Basin, Flat, Meadow, or Plain	County	Latitude	Longitude	Elev.
Willow Patch	Custer	44°25'50"N	114°29'24"W	
Wilson Flat	Bingham	43°01'47"N	111°41'17"W	6180
Wilson Flat	Elmore	43°21'58"N	115°27'24"W	
Wilson Meadows	Gem	44°25'56"N	116°09'09"W	
Wilson Pasture	Owyhee	42°30'56"N	116°49'36"W	
Wilson Pasture	Owyhee	42°17'07"N	116°55'06"W	
Wind River Meadows	Idaho	45°35'38"N	115°55'54"W	
Windmill Flat	Caribou	42°42'56"N	111°41'02"W	6268
Wino Basin	Custer	44°24'41"N	113°58'32"W	
Wolf Flat	Bonneville	43°35'56"N	111°36'36"W	
Woodruff Flat	Idaho	45°50'44"N	116°16'47"W	3000
Woods Basin	Custer	44°12'45"N	114°20'05"W	
Wounded Doe Licks	Idaho	46°17'49"N	115°03'36"W	
Y P Desert	Owyhee	42°03'35"N	116°41'11"W	
Ziegler Basin	Custer	44°10'19"N	114°18'52"W	

Bay, Cove, or Harbor	County	Latitude	Longitude	Elev.
Aberdeen Lodge Bay	Kootenai	47°27'45"N	116°51'54"W	
Albeni Cove	Bonner	48°10'32"N	117°07'15"W	
Bear Creek Bay	Bonner	48°38'13"N	116°51'03"W	
Beauty Bay	Kootenai	47°37'00"N	116°40'58"W	
Bell Bay	Kootenai	47°28'17"N	116°50'24"W	
Bennett Bay	Kootenai	47°38'45"N	116°43'11"W	
Berven Bay	Kootenai	47°46'05"N	116°44'56"W	
Big Bay	Twin Falls	42°06'35"N	114°44'57"W	
Big Hole	Bingham	42°57'07"N	112°46'08"W	
Big Water Hole	Elmore	43°07'05"N	115°08'18"W	
Black Bay	Kootenai	47°42'10"N	116°55'34"W	
Black Bay	Kootenai	47°31'31"N	116°48'01"W	
Black Rock Bay	Kootenai	47°31'18"N	116°50'48"W	
Bloomsburg Bay	Kootenai	47°25'53"N	116°46'56"W	
Blue Creek Bay	Kootenai	47°37'58"N	116°40'04"W	
Bottle Bay	Bonner	48°42'09"N	116°51'47"W	
Bottle Bay	Bonner	48°14'41"N	116°27'01"W	
Bozanta Tavern Bay	Kootenai	47°46'01"N	116°45'09"W	
Browns Bay	Kootenai	47°25'13"N	116°46'22"W	
Bruces Eddy	Clearwater	46°31'32"N	116°18'07"W	1300
Bruneau Arm	Owyhee	42°55'00"N	115°52'14"W	2455
Camp Bay	Bonner	48°11'54"N	116°22'24"W	
Carey Bay	Kootenai	47°24'42"N	116°45'50"W	
Carlin Bay	Kootenai	47°32'26"N	116°46'39"W	
Casco Bay	Kootenai	47°39'36"N	116°48'47"W	
Cavanaugh Bay	Bonner	48°31'50"N	116°49'40"W	
Cave Bay	Kootenai	47°28'03"N	116°52'29"W	
Cleland Bay	Kootenai	47°26'29"N	116°47'55"W	
Coopers Bay	Kootenai	47°45'08"N	116°44'21"W	
Cottonwood Bay	Kootenai	47°26'51"N	116°50'02"W	
Cougar Bay	Kootenai	47°40'17"N	116°48'54"W	
Cramps Bay	Kootenai	47°46'16"N	116°43'34"W	
Crescent Bay	Kootenai	47°32'44"N	116°49'24"W	
Deadmans Eddy	Shoshone	47°37'25"N	116°11'48"W	
Distillery Bay	Bonner	48°40'56"N	116°52'24"W	
Driftwood Bay	Kootenai	47°35'33"N	116°47'44"W	2140
Duck Bay	Valley	44°56'52"N	116°04'45"W	
Echo Bay	Kootenai	47°57'10"N	116°31'14"W	
Echo Bay	Kootenai	47°36'46"N	116°46'11"W	
Elliot Bay	Bonner	48°12'23"N	116°21'34"W	
Ellisport Bay	Bonner	48°14'02"N	116°17'27"W	
Everwell Bay	Kootenai	47°36'33"N	116°48'58"W	2155
Fullers Bay	Kootenai	47°26'05"N	116°47'21"W	
Gand Bay	Kootenai	47°33'43"N	116°47'00"W	2140
Garfield Bay	Bonner	48°10'50"N	116°25'57"W	
Gotham Bay	Kootenai	47°35'44"N	116°47'04"W	
Green Bay	Bonner	48°10'37"N	116°24'29"W	
Half Round Bay	Kootenai	47°30'52"N	116°48'53"W	

Bays, Coves, Harbors

Bay, Cove, or Harbor	County	Latitude	Longitude	Elev.
Happy Cove	Kootenai	47°34'31"N	116°49'06"W	
Hole, The	Elmore	43°56'49"N	115°07'40"W	8397
Honeysuckle Bay	Kootenai	47°45'08"N	116°45'16"W	
Huckleberry Bay	Bonner	48°40'56"N	116°49'55"W	
Huckleberry Bay	Valley	44°57'17"N	116°04'19"W	4990
Idlewilde Bay	Kootenai	47°57'30"N	116°32'53"W	
Indian Creek Bay	Bonner	48°36'29"N	116°50'12"W	
Kalispell Bay	Bonner	48°33'50"N	116°55'28"W	
Kid Island Bay	Kootenai	47°38'31"N	116°47'54"W	
Kilroy Bay	Bonner	48°07'00"N	116°24'40"W	
Kootenai Bay	Bonner	48°18'16"N	116°30'14"W	
Little Bay	Clearwater	46°34'42"N	116°15'16"W	1600
Little Cottonwood Bay	Kootenai	47°27'14"N	116°50'28"W	
Little Hole	Bingham	42°55'17"N	112°47'22"W	
Loffs Bay	Kootenai	47°33'28"N	116°49'14"W	
Lowmeister Bay	Kootenai	47°26'17"N	116°47'45"W	
Luby Bay	Bonner	48°32'32"N	116°55'02"W	
Mallard Bay	Bonner	48°13'13"N	116°41'32"W	2062
Martin Bay	Bonner	48°13'42"N	116°22'10"W	
Martin Bay	Kootenai	47°31'57"N	116°47'40"W	
McLeans Bay	Kootenai	47°48'01"N	116°41'16"W	
Mica Bay	Kootenai	47°36'14"N	116°50'33"W	
Mokins Bay	Kootenai	47°46'53"N	116°40'18"W	
Moscow Bay	Kootenai	47°37'21"N	116°42'28"W	
Mosquito Bay	Bonner	48°44'31"N	116°50'10"W	
Murphy Bay	Bonner	48°14'05"N	116°33'49"W	
Neeley Cove	Bonneville	43°25'35"N	111°04'09"W	
Nortons Bay	Owyhee	42°06'51"N	114°44'16"W	
Oden Bay	Bonner	48°17'56"N	116°27'06"W	
O'Gara Bay	Kootenai	47°24'11"N	116°44'28"W	
O'Rourke Bay	Kootenai	47°45'29"N	116°41'07"W	
Outlet Bay	Bonner	48°29'46"N	116°53'02"W	
Owens Bay	Bonner	48°12'58"N	116°17'46"W	
Pilgrim Cove	Valley	44°55'56"N	116°03'50"W	
Pine Cove	Bonner	48°06'28"N	116°25'30"W	
Powderhorn Bay	Kootenai	47°29'53"N	116°49'07"W	
Reeder Bay	Bonner	48°37'21"N	116°53'08"W	
Rockford Bay	Kootenai	47°30'14"N	116°53'14"W	
Scenic Bay	Kootenai	47°58'43"N	116°32'32"W	
Seagull Bay	Power	42°49'29"N	112°48'13"W	
Shingle Bay	Kootenai	47°24'51"N	116°44'29"W	
Shoshone Bay	Bonner	48°31'48"N	116°53'18"W	
Sixteen-to-One Bay	Kootenai	47°28'19"N	116°53'12"W	
Skinner Bay	Kootenai	47°47'18"N	116°41'23"W	
Squaw Bay	Bonner	48°43'42"N	116°49'48"W	
Squaw Bay	Kootenai	47°37'25"N	116°44'47"W	
Steamboat Bay	Bonner	48°31'17"N	116°50'50"W	
Sun Up Bay	Kootenai	47°29'06"N	116°54'39"W	
Sunrise Bay	Bonner	48°14'39"N	116°22'19"W	
Swede Bay	Kootenai	47°36'42"N	116°48'13"W	
Teachers Bay	Bonner	48°42'27"N	116°51'25"W	
Tule Bay	Bonner	48°44'01"N	116°51'19"W	
Turner Bay	Kootenai	47°34'22"N	116°47'09"W	
Wagon Wheel Bay	Valley	44°56'58"N	116°05'53"W	
Walts Bay	Twin Falls	42°04'41"N	114°45'53"W	
Whiskey Rock Bay	Bonner	48°03'16"N	116°27'32"W	
Windy Bay	Kootenai	47°44'59"N	116°42'20"W	
Windy Bay	Kootenai	47°28'50"N	116°54'20"W	
Wolf Lodge Bay	Kootenai	47°37'22"N	116°40'21"W	

Camp or Campground	County	Latitude	Longitude	Elev.
Abbot Campground	Elmore	43°36'29"N	115°13'01"W	
Albeni Cove Campground	Bonner	48°10'34"N	116°59'53"W	2120
Albeni Falls Campground	Bonner	48°10'35"N	116°59'50"W	2080
Albert Moser Campground	Franklin	42°08'17"N	111°41'38"W	
Alice Pittinger Girl Scout Camp	Valley	44°56'17"N	116°04'09"W	5005
Allison Creek Campground	Idaho	45°25'08"N	116°09'44"W	
Alpine Campground	Bonneville	43°11'48"N	111°02'35"W	
Alturas Inlet Campground	Blaine	43°54'24"N	114°52'39"W	
Amanita Campground	Valley	44°42'06"N	116°07'48"W	4860
Antelope Campground	Gem	44°20'07"N	116°11'06"W	4990
Antimony Camp	Valley	44°55'42"N	115°28'45"W	
Apgar Campground	Idaho	46°12'50"N	115°32'09"W	
Aquarius Campground	Clearwater	46°50'28"N	115°36'58"W	
Arrow Campsite	Idaho	45°35'34"N	114°49'01"W	
Bad Bear Campground	Boise	43°54'03"N	115°42'27"W	
Baker Creek Campground	Blaine	43°47'06"N	114°33'51"W	6675
Baker Creek Campground	Blaine	43°44'43"N	114°33'51"W	
Bald Mountain Campground	Boise	43°44'57"N	115°44'37"W	
Banks Campground	Boise	44°04'30"N	116°07'07"W	2800
Bargamin Bar Campsite	Idaho	45°33'59"N	115°11'36"W	
Baritoe Campground	Bonner	48°32'30"N	116°52'41"W	
Bark Camp	Clearwater	46°49'10"N	116°01'59"W	
Barneys Park Campground	Valley	44°19'34"N	115°38'49"W	5321
Base Camp Campground	Idaho	45°37'22"N	114°50'40"W	7724
Basin Creek Campground	Custer	44°15'48"N	114°49'07"W	
Baumgartner Campground	Camas	43°36'21"N	115°04'31"W	
Bayhorse Campground	Custer	44°24'41"N	114°24'02"W	
Bear Camp	Caribou	42°49'27"N	112°06'10"W	
Bear Campground	Adams	45°04'37"N	116°37'34"W	
Bear Creek Campground	Custer	43°59'15"N	113°27'28"W	
Bear Creek Campground	Idaho	45°51'50"N	115°36'58"W	4050
Bear Gulch Campground	Fremont	44°09'09"N	111°17'03"W	
Bear Gulch Campground	Twin Falls	42°13'41"N	114°22'46"W	
Bear Packer Camp	Idaho	46°02'40"N	115°15'02"W	1790
Bear Valley Campground	Valley	44°24'40"N	115°22'08"W	6396
Bears Oil and Roots Camp	Idaho	46°34'32"N	114°54'34"W	6360
Beauty Creek Campground	Kootenai	47°36'24"N	116°40'04"W	2150
Beaver Creek Campground	Clark	44°24'54"N	112°11'34"W	
Beaver Creek Campground	Custer	44°24'51"N	115°08'45"W	
Beaver Creek Campground	Franklin	42°01'15"N	111°31'42"W	
Beaver Creek Campground	Shoshone	47°04'55"N	115°21'14"W	3618
Beaver Creek Youth Camp	Boise	43°57'57"N	115°36'27"W	5135
Bell Bay Campground	Kootenai	47°28'36"N	116°50'26"W	2561
Belle Creek Campground	Idaho	46°27'01"N	115°38'17"W	
Bentz Cow Camp	Idaho	45°38'47"N	115°58'17"W	
Big Bend Campsite	Custer	44°36'41"N	115°16'41"W	6794
Big Creek Camp (historical)	Lemhi	44°26'46"N	113°35'53"W	
Big Creek Campground	Lemhi	44°26'32"N	113°35'53"W	6560
Big Creek Campground	Shoshone	47°18'13"N	116°07'09"W	2380
Big Eddy Campground	Clearwater	46°31'31"N	116°18'20"W	1680
Big Eddy Campground	Valley	44°13'14"N	116°06'21"W	4120
Big Eightmile Campground	Lemhi	44°36'22"N	113°34'16"W	7600
Big Elk Creek Campground	Bonneville	43°19'33"N	111°07'02"W	
Big Hank Campground	Shoshone	47°49'27"N	116°06'00"W	2707
Big Indian Camp	Lemhi	45°27'12"N	114°37'02"W	
Big Mallard Creek Campground	Idaho	45°32'13"N	115°16'11"W	2800
Big Roaring River Lake Campground	Elmore	43°37'08"N	115°26'34"W	8100
Big Smoky Campground	Camas	43°36'29"N	114°54'01"W	
Big Snag Campsite	Valley	44°41'40"N	115°08'52"W	4830
Big Spring Campground	Caribou	42°45'53"N	112°05'45"W	
Big Trinity lake Campground	Elmore	43°37'26"N	115°25'29"W	7840
Bighorn Campground	Camas	43°36'00"N	115°03'41"W	
Birch Creek Campground	Boise	43°39'46"N	115°43'59"W	
Bird Creek Campground	Elmore	43°37'14"N	115°10'29"W	
Bitch Creek Cow Camp	Fremont	43°59'05"N	111°09'31"W	
Black Lake Campground	Adams	45°11'22"N	116°33'40"W	7240
Black Lee Campground	Valley	44°59'09"N	115°57'14"W	
Black Pine Campground	Nez Perce	46°09'32"N	116°47'11"W	
Black Rock Campground	Boise	43°47'43"N	115°35'14"W	
Blind Creek Campground	Custer	44°16'51"N	114°43'55"W	5980
Blowfly Campground	Custer	44°42'33"N	114°19'37"W	6480
Blue Point Campground	Valley	44°30'52"N	115°34'38"W	6040
Boiling Springs Campground	Valley	44°21'35"N	115°51'30"W	4040
Bostetter Campground	Cassia	42°09'57"N	114°10'05"W	
Boundary Campground	Blaine	43°43'18"N	114°19'29"W	
Boundary Campground	Idaho	45°53'21"N	116°02'02"W	2080
Bowery Creek Trailhead Camp	Custer	44°01'39"N	114°27'50"W	6440
Bowns Campground	Camas	43°36'25"N	114°52'50"W	
Boy Scout Camp (historical)	Owyhee	43°26'12"N	116°49'59"W	
Boy Scout Camp Campground	Valley	44°36'03"N	115°17'18"W	7000
Boyd Creek Campground	Idaho	46°04'52"N	115°26'30"W	
Brackenbury Campground	Cassia	42°19'32"N	113°38'45"W	
Bradley Memorial Camp	Custer	44°23'53"N	115°08'50"W	
Breakwater Campground	Kootenai	47°39'14"N	116°21'35"W	
Bridge Campground	Blaine	43°31'34"N	114°28'48"W	
Bridge Creek Campground	Idaho	45°46'58"N	115°12'44"W	
Brig Campground	Kootenai	47°57'55"N	116°34'42"W	2300
Browns Camp	Latah	46°40'04"N	116°20'04"W	
Browns Camp	Valley	45°02'40"N	115°44'48"W	
Bruce Meadows Campground	Valley	44°25'36"N	115°19'51"W	6365
Bruces Eddy Campground	Clearwater	46°30'37"N	116°17'16"W	1600
Bruin Creek Campground	Idaho	45°31'02"N	115°04'20"W	
Brush Lake Campground	Boundary	48°53'15"N	116°19'43"W	3000
Buck Camp	Owyhee	42°05'41"N	115°47'38"W	
Buck Mountain Campground	Valley	44°40'55"N	115°32'20"W	6601
Buckhorn Bar Campground	Valley	44°56'10"N	115°44'20"W	

Camps & Campgrounds

Camp or Campground	County	Latitude	Longitude	Elev.
Buckhorn Campground	Valley	44°56'19"N	115°44'14"W	
Buffalo Campground	Fremont	44°25'28"N	111°22'06"W	
Bull Basin Camp	Owyhee	42°20'27"N	116°54'15"W	
Bull Trout Lake Campground	Boise	44°18'11"N	115°15'20"W	6940
Bull Trout Lake Campground	Boise	44°17'49"N	115°15'17"W	6970
Bumblebee Campground	Shoshone	47°38'01"N	116°16'43"W	2240
Bungalow Cow Camp	Idaho	45°31'46"N	116°05'12"W	
Burgdorf Campground	Idaho	45°16'30"N	115°54'26"W	6200
Buttermilk Campground	Fremont	44°26'01"N	111°25'27"W	
Button Hook Group Camp	Kootenai	47°57'22"N	116°34'25"W	2260
Cabin Creek Campground	Adams	44°39'28"N	116°16'20"W	4128
Cable Hole Campsite	Valley	44°33'02"N	115°17'46"W	5634
Cache Bar Campground	Lemhi	45°19'07"N	114°38'10"W	2980
Cache Mountain Camp	Idaho	46°22'47"N	115°24'26"W	4760
Caldwell Labor Camp	Canyon	43°42'14"N	116°42'33"W	
California Creek Bar Campground	Idaho	45°27'00"N	115°45'37"W	2070
Camas Cow Camp	Camas	43°30'03"N	114°44'08"W	
Cameron Campsite	Valley	44°43'53"N	114°56'29"W	4284
Camp	Clearwater	46°45'15"N	115°40'36"W	
Camp Codge	Bonner	48°43'03"N	116°50'15"W	
Camp Creek Access Area	Lemhi	44°56'14"N	113°57'44"W	4300
Camp Easterseal	Kootenai	47°26'39"N	116°50'19"W	
Camp Easton	Kootenai	47°36'01"N	116°46'28"W	
Camp Ee Da How	Boise	43°56'39"N	115°38'40"W	
Camp Eight	Boundary	48°56'15"N	116°24'45"W	
Camp Fifty-Seven	Clearwater	46°35'08"N	115°42'53"W	
Camp Five	Boundary	48°51'55"N	116°22'32"W	
Camp Forty	Shoshone	46°56'59"N	116°04'15"W	
Camp Forty-Four	Shoshone	47°07'18"N	115°52'47"W	
Camp Forty-Three	Clearwater	46°45'59"N	116°07'15"W	
Camp Grizzly	Latah	46°56'33"N	116°39'23"W	
Camp Henry Historic Site	Fremont	43°59'42"N	111°17'06"W	
Camp Holly Lake	Minidoka	42°39'54"N	113°41'47"W	
Camp Ho-Nok	Bear Lake	42°32'02"N	111°34'56"W	6360
Camp Howard Cow Camp	Idaho	45°43'43"N	116°24'25"W	
Camp Little Lemhi	Bonneville	43°21'48"N	111°14'26"W	
Camp Luthy	Blaine	43°55'46"N	114°51'25"W	
Camp Man Apu	Blaine	43°50'13"N	114°25'02"W	
Camp Martin	Idaho	46°26'32"N	115°36'16"W	
Camp Mivoden	Kootenai	47°46'53"N	116°40'18"W	
Camp Moosehorn	Idaho	46°18'07"N	115°37'05"W	
Camp Nine	Boundary	48°50'46"N	116°16'19"W	
Camp Perkins	Blaine	43°55'48"N	114°50'14"W	
Camp Sixty	Clearwater	46°41'54"N	115°37'54"W	
Camp Stidwell	Bonner	48°09'32"N	116°29'35"W	2390
Camp Ta-Man-a-Wis	Bonneville	43°21'44"N	111°19'30"W	
Camp Tendoy	Bannock	42°41'21"N	112°22'00"W	
Camp Thomas	Idaho	45°54'18"N	116°39'13"W	

Camp or Campground	County	Latitude	Longitude	Elev.
Camp Three	Shoshone	47°07'48"N	116°06'05"W	
Camp Wilderness	Franklin	42°02'26"N	111°38'03"W	
Campbell Creek Recreation Site	Valley	44°29'58"N	116°05'03"W	4940
Campbell Ranch	Power	42°36'59"N	113°00'04"W	
Campbells Cow Camp	Adams	45°12'59"N	116°23'46"W	
Campbells Ferry	Idaho	45°29'21"N	115°19'53"W	
Campbells Pond Access Area	Clearwater	46°33'43"N	115°52'27"W	4000
Canoe Camp	Bonner	48°28'56"N	116°52'32"W	
Canyon Campground	Camas	43°37'41"N	114°51'29"W	
Canyon Campground	Valley	44°11'17"N	116°06'52"W	3920
Carrie Creek Campground	Camas	43°33'12"N	114°45'34"W	6100
Cascade Lake 4H Camp	Valley	44°43'10"N	116°06'00"W	4830
Castle Butte Camp	Idaho	46°24'16"N	115°11'23"W	2430
Castle Creek Campground	Idaho	45°49'41"N	115°58'03"W	
Castle Rock Campground	Elmore	43°24'41"N	115°23'38"W	4210
Cave Campsite	Lemhi	44°50'05"N	114°47'31"W	4080
CCC Campground	Shoshone	47°17'15"N	116°07'15"W	2298
Cedar Creek Campground	Shoshone	47°03'04"N	116°17'17"W	2760
Cedar Flats Campground	Idaho	46°06'55"N	114°59'48"W	2220
Cedar Grove Camp	Latah	47°01'23"N	116°41'57"W	
Center Camp Corrals	Boise	44°10'29"N	115°44'16"W	7500
Central Idaho Camp	Blaine	43°48'31"N	114°36'43"W	
Chaparral Campground	Elmore	43°36'52"N	115°12'09"W	
Cherry Creek Campground	Bannock	42°18'16"N	112°08'04"W	
Cherry Springs Campground	Bannock	42°45'16"N	112°23'40"W	
Cherry Springs Campground Picnic Area	Bannock	42°44'54"N	112°23'34"W	
Chesterfield Cow Camp	Caribou	42°55'38"N	111°46'07"W	6105
Chinook Bay Campground	Custer	44°09'32"N	114°54'43"W	
Chinook Campground	Idaho	45°12'45"N	115°48'32"W	
Cloverleaf Campground	Franklin	42°05'40"N	111°31'51"W	
Cluvewr Creek Campsite	Custer	44°46'45"N	114°52'07"W	4140
Cold Spring Campground	Bear Lake	42°30'39"N	111°34'56"W	6375
Cold Spring Campground	Elmore	43°35'53"N	115°26'42"W	
Cold Springs Camp	Clearwater	46°47'14"N	115°18'10"W	6731
Cold Springs Campground	Adams	44°56'58"N	116°26'24"W	4833
Cold Springs Campground	Valley	44°11'37"N	116°06'44"W	3923
Cold Springs Campsite	Valley	44°34'08"N	115°18'07"W	5634
Cold Springs Cow Camp	Idaho	45°30'25"N	116°27'52"W	
Cold Water Camp	Power	42°37'14"N	113°06'59"W	
Colt Creek Campground	Idaho	46°25'24"N	114°37'22"W	
Conrad Crossing Campground	Shoshone	47°09'31"N	115°24'57"W	3630
Copper Basin Cow Camp	Custer	43°49'03"N	113°54'45"W	
Copper Camp	Valley	45°10'16"N	115°11'55"W	
Copper Creek Camp	Blaine	43°36'27"N	113°55'49"W	
Copper Creek Campground	Boundary	48°59'07"N	116°09'07"W	2630
Corey Bar Campground	Idaho	45°29'40"N	114°59'10"W	2800
Corral Creek Cow Camp	Custer	43°51'07"N	113°51'25"W	
Cotter Campground	Idaho	45°52'57"N	116°01'19"W	

Camp or Campground	County	Latitude	Longitude	Elev.
Cottonwood Campground	Blaine	43°39'27"N	114°27'28"W	
Cottonwood Campground	Boise	43°37'57"N	115°49'27"W	3360
Cottonwood Campground	Owyhee	42°54'38"N	115°52'56"W	2465
Cove Camp	Idaho	46°09'19"N	115°13'55"W	6440
Cove Camp	Idaho	45°50'38"N	116°03'04"W	
Cow Camp	Clark	44°27'46"N	112°45'59"W	
Cow Camp	Custer	44°23'02"N	114°28'25"W	
Cow Creek Campsite	Lemhi	44°48'53"N	114°48'24"W	4020
Cow Creek Forest Camp	Elmore	43°22'00"N	115°33'10"W	
Cozy Cove Campground	Valley	44°17'28"N	115°39'04"W	5335
Crags Campground	Lemhi	45°06'10"N	114°31'21"W	
Craig Camp	Owyhee	42°24'00"N	116°49'08"W	
Crooked River Number Five Campground	Idaho	45°47'37"N	115°31'48"W	4100
Crooked River Number Three Campground	Idaho	45°47'48"N	115°31'44"W	4170
Curlew Campground	Oneida	42°04'19"N	112°41'26"W	
Custer Number 1 Campground	Custer	44°23'58"N	114°39'46"W	
Dagger Campsite	Valley	44°31'43"N	115°17'03"W	5630
Deadman Campground	Valley	44°57'47"N	115°39'38"W	
Deadwater Campground	Lemhi	45°23'15"N	114°03'00"W	3600
Deadwood Campground	Boise	44°04'52"N	115°39'28"W	
Dean Camp	Camas	43°13'41"N	114°53'49"W	
Deep Creek Campground	Boundary	48°35'22"N	116°23'52"W	
Deep Creek Campground	Idaho	45°42'51"N	114°42'30"W	
Deer Creek Campground	Blaine	43°31'44"N	114°30'07"W	6020
Deer Flat Campground	Valley	44°24'32"N	115°33'10"W	6315
Deer Park Campground	Elmore	43°54'51"N	115°23'50"W	4850
Delyle Campground (historical)	Bonner	48°02'11"N	116°09'55"W	
Devils Elbow Campground	Shoshone	47°46'16"N	116°01'53"W	2603
Diamond Match Camp	Clearwater	46°48'00"N	116°05'01"W	
Dickensheet Campground	Bonner	48°27'06"N	116°53'58"W	
Dirt Oven Campground	Boundary	48°55'06"N	116°40'30"W	
Dismal Camp	Idaho	45°21'04"N	114°56'59"W	
Ditch Creek Campground	Idaho	45°44'50"N	115°17'47"W	4528
Dixon Cow Camp	Idaho	45°36'15"N	116°25'00"W	
Dog Creek Campground	Elmore	43°31'39"N	115°18'21"W	
Dolly Lake Campsite	Custer	44°44'14"N	115°08'56"W	4825
Dome House Campsite	Valley	44°39'03"N	115°09'55"W	4860
Dry Camp	Idaho	46°26'21"N	115°16'09"W	5998
Dry Creek Campground	Oneida	42°03'29"N	112°08'33"W	
Dutchman Flat Campground	Custer	44°15'57"N	114°43'13"W	
East Camos Campground	Clark	44°28'38"N	111°55'15"W	6620
East Fork Public Campground (historical)	Bonner	48°14'37"N	116°06'36"W	
Easterseal Camp	Kootenai	47°26'39"N	116°50'19"W	2300
Eastside Campground	Gem	44°19'56"N	116°10'25"W	4950
Ebenezer Bar Campground	Lemhi	45°18'18"N	114°30'53"W	3134
EC Rettig Campground	Clearwater	46°36'56"N	115°48'15"W	3280
Edna Creek Campground	Boise	43°57'45"N	115°37'17"W	
Eightmile Campground	Bear Lake	42°29'16"N	111°35'02"W	

Camp or Campground	County	Latitude	Longitude	Elev.
Eightmile Campground	Custer	44°25'35"N	114°37'13"W	6840
Elk Bar Campground	Valley	45°07'14"N	114°43'19"W	3550
Elk Creek Campground	Valley	44°24'55"N	115°28'09"W	
Elk Spring Campground	Elmore	43°42'28"N	115°22'11"W	
Elkhorn Bar Campground	Idaho	45°26'55"N	114°55'11"W	2800
Emerald Creek Campground	Latah	47°00'22"N	116°19'35"W	3000
Emigration Campground	Bear Lake	42°22'11"N	111°33'22"W	
Evans Creek Campground	Elmore	43°24'01"N	115°24'48"W	4210
Evergreen Campground	Adams	44°53'36"N	116°23'17"W	3850
Fall Creek Cow Camp	Bonneville	43°21'56"N	111°30'33"W	5800
Falls Campground	Bonneville	43°26'13"N	111°21'41"W	
Father and Sons Campground	Cassia	42°09'47"N	114°11'05"W	
Federal Gulch Campground	Blaine	43°40'08"N	114°09'08"W	
Fir Creek Campground	Valley	44°25'42"N	115°17'03"W	
Fir Springs Campground	Valley	44°21'46"N	115°37'58"W	5380
Fire Island Campsite	Valley	44°38'13"N	115°11'24"W	4860
Fish Creek Meadow Campground	Idaho	45°51'28"N	116°04'53"W	
Five Points Campground	Camas	43°32'32"N	114°49'03"W	
Fivemile Campground	Idaho	45°43'02"N	115°32'22"W	4560
Flat Rock Campground	Custer	44°17'27"N	114°43'01"W	
Floodwood Camp	Clearwater	46°55'37"N	115°56'54"W	2240
Fly Camp (historical)	Clearwater	46°51'20"N	115°14'22"W	6300
Fly Creek Campsite	Lemhi	44°40'17"N	114°33'15"W	8080
Fly Flat Campground	Shoshone	47°06'46"N	115°23'24"W	3491
Fog Mountain Campground	Idaho	46°03'23"N	115°18'33"W	1900
Footrot Corrals Campground	Idaho	46°28'13"N	115°33'14"W	5080
Forest Camp	Elmore	43°21'01"N	115°32'02"W	
Forks Of Big Pine Creek Campground	Boise	44°05'59"N	115°45'29"W	
Four Echoes, Camp	Kootenai	47°28'09"N	116°55'54"W	
Fourmile Campground	Elmore	43°53'38"N	115°28'30"W	
Fourmile Campground	Valley	44°55'42"N	115°44'29"W	3920
Fourmile Campground	Valley	44°51'47"N	115°41'31"W	4191
Fourteen Mile Tree Campsite	Idaho	45°41'20"N	115°10'03"W	
Franklin Labor Camp	Canyon	43°38'41"N	116°33'05"W	
Freeman Creek Campground	Clearwater	46°34'32"N	116°16'33"W	1600
Freezeout Camp	Shoshone	46°59'36"N	116°01'22"W	
Full Stomach Camp	Idaho	46°19'20"N	115°37'14"W	4500
Gardells Campsite	Custer	44°34'00"N	115°17'56"W	5634
Gedney Creek Campground	Idaho	46°03'26"N	115°18'44"W	1698
George, Camp	Clearwater	46°37'36"N	115°20'51"W	4666
Gilmore Camp	Kootenai	47°57'35"N	116°34'26"W	2310
Glacier View Campground	Custer	44°08'52"N	114°54'40"W	6530
Glade Creek Camp	Idaho	46°37'22"N	114°34'25"W	5165
Glade Creek Campground	Idaho	46°13'12"N	115°31'35"W	
Glover Campground	Idaho	46°04'09"N	115°21'44"W	
Gold Creek Campground	Shoshone	47°08'16"N	115°24'28"W	3280
Golden Gate Campground	Valley	44°56'08"N	115°29'04"W	
Gooding Cow Camp	Camas	43°30'57"N	114°43'37"W	

Camps & Campgrounds

Camp or Campground	County	Latitude	Longitude	Elev.
Graham Bridge Campground	Boise	43°57'49"N	115°16'25"W	
Grandad Campsite	Idaho	45°22'02"N	114°51'53"W	
Grandjean Campground	Custer	44°08'55"N	115°09'04"W	5160
Granite Creek Campground	Bonner	48°41'22"N	116°59'47"W	
Granite Creek Cow Camp	Fremont	44°01'32"N	111°04'12"W	
Granite Spring Campground	Elmore	43°23'40"N	115°32'39"W	
Grave Creek Campground	Bingham	43°02'23"N	111°51'17"W	5880
Gravel Creek Campground	Caribou	42°56'12"N	111°22'43"W	
Gravel Creek Campground	Caribou	42°56'12"N	111°22'43"W	6620
Grayback Gulch Campground	Boise	43°48'26"N	115°51'39"W	3983
Green Flat Campground	Idaho	46°23'39"N	115°12'50"W	
Greensward Camp	Idaho	46°27'12"N	115°14'50"W	6300
Greyhound Creek Campsite	Custer	44°38'56"N	115°09'56"W	4850
Greylock Campground	Custer	44°24'43"N	114°38'26"W	
Grouse Campground	Adams	45°04'06"N	116°10'01"W	
Guard Station Campground	Valley	44°45'25"N	115°06'47"W	4701
Halfway House Campground	Idaho	45°30'34"N	115°31'36"W	
Hand Camp	Idaho	45°13'53"N	115°18'04"W	
Hawkseye Camp	Shoshone	47°46'23"N	115°57'47"W	
Hayfork Campground	Boise	43°54'31"N	115°41'49"W	
Hazard Campground	Idaho	45°12'06"N	116°08'31"W	7070
Helende Campground	Boise	44°05'52"N	115°28'11"W	4181
Heyburn Mountain Campground	Custer	44°07'54"N	114°55'19"W	
Heyburn Youth Camp	Benewah	47°21'43"N	116°46'31"W	
Hibbs Cow Camp	Idaho	45°20'49"N	116°34'43"W	
Hidden Creek Campground	Clearwater	46°49'56"N	115°10'43"W	
Hill Campground	Kootenai	47°58'16"N	116°34'34"W	2319
Hoffman Campground	Bonneville	43°10'07"N	111°04'53"W	
Hollywood Point Campground	Gem	44°19'36"N	116°10'39"W	5470
Holman Creek Campground	Custer	44°14'57"N	114°31'44"W	
Home Canyon Campground	Bear Lake	42°20'13"N	111°13'21"W	
Homers Campground	Valley	44°19'21"N	115°38'50"W	5318
Honeysuckle Campground	Kootenai	47°44'23"N	116°28'27"W	
Hood Ranch Campground	Valley	44°43'50"N	114°59'37"W	4353
Hoodoo Lake Campground	Idaho	46°19'19"N	114°39'04"W	
Horse Camp	Caribou	42°57'54"N	111°55'46"W	6601
Horse Camp	Clearwater	46°34'42"N	115°22'33"W	
Horse Camp	Idaho	46°19'15"N	115°11'45"W	
Horse Creek Campsite	Lemhi	45°23'50"N	114°43'55"W	3000
Horse Creek Hot Springs Campground	Lemhi	45°30'15"N	114°27'34"W	7555
Horse Sweat Pass Camp	Idaho	46°25'05"N	115°21'56"W	4737
Horsesteak Meadow Camp	Idaho	46°22'34"N	115°32'20"W	4410
Horsetail Campsite	Lemhi	44°50'17"N	114°47'00"W	4022
Hospital Bar Campground	Valley	44°50'07"N	114°47'11"W	4040
Hot Spring Campground	Boise	44°03'42"N	115°55'04"W	3104
Hot Springs Campground	Boise	44°03'42"N	115°55'04"W	
Howard Camp	Idaho	46°31'54"N	115°04'01"W	
Huckleberry Campground	Adams	45°05'02"N	116°36'51"W	
Huckleberry Flat Campground	Shoshone	47°16'07"N	116°05'13"W	2230
Hudlow Camp	Kootenai	47°49'11"N	116°31'15"W	
Hungery Creek Camp	Idaho	46°21'59"N	115°25'11"W	3100
Hyndman Campground	Blaine	43°39'40"N	114°13'33"W	
Ice Hole Campground	Valley	44°53'17"N	115°29'55"W	
Ice Springs Campground	Elmore	43°28'59"N	115°23'46"W	4960
Idaho Outdoor Assn Campground	Elmore	43°46'22"N	115°32'04"W	
Independence Campground	Shoshone	47°53'12"N	116°13'24"W	
Indian Creek Campground	Bonner	48°36'41"N	116°49'49"W	
Indian Creek Campground	Idaho	45°47'21"N	114°45'45"W	
Indian Grave Camp	Idaho	46°29'55"N	115°08'47"W	
Indian Spring Campsite	Valley	44°46'09"N	115°05'26"W	4635
Inlet Campground	Custer	44°14'42"N	115°03'50"W	
Iron Creek Campground	Custer	44°11'56"N	115°00'32"W	
Iron Lake Campground	Lemhi	44°54'20"N	114°11'35"W	8810
Ivydale Organization Camp	Boise	43°48'19"N	115°50'43"W	4160
Jackson Landing Campground	Fremont	44°26'46"N	111°26'11"W	6325
Jacksons Cow Camp	Idaho	46°21'44"N	115°31'16"W	
Jerry Johnson Campground	Idaho	46°28'34"N	114°54'30"W	
Jerusalem Artichokes Campsite	Idaho	46°22'32"N	115°31'26"W	4394
Johnson Camp	Clearwater	46°51'21"N	116°01'47"W	
Johnson Creek Campground	Elmore	43°56'23"N	115°16'59"W	
Johnstons Camp	Owyhee	42°01'25"N	115°53'40"W	
Jons Camp Campsite	Custer	44°38'43"N	115°10'26"W	4840
Jordan Creek Campground	Shoshone	47°56'45"N	116°05'51"W	
Jumbo Camp	Idaho	45°33'33"N	115°41'19"W	
Juniper Spring Campground (historical)	Cassia	42°17'19"N	114°15'36"W	
Justrite Campground	Washington	44°32'27"N	116°57'08"W	4200
Kennally Creek Campground	Valley	44°46'57"N	115°52'29"W	5800
Kestrel Campground	Kootenai	47°57'26"N	116°35'33"W	2395
Kiawanis Campground	Washington	44°30'46"N	116°57'08"W	3780
Kirkham Hot Springs Campground	Boise	44°04'21"N	115°32'31"W	3970
Kirkman Campground	Twin Falls	42°12'51"N	114°18'44"W	
Kirkwood Cow Camp	Idaho	45°33'23"N	116°24'56"W	
Kit Price Campground	Shoshone	47°44'27"N	116°00'20"W	2560
Knife Edge Campground	Idaho	46°13'38"N	115°28'24"W	
Krassel Creek Campground	Valley	44°58'51"N	115°43'37"W	
Kum Ba Yah Camp	Twin Falls	42°10'30"N	114°17'10"W	
Lafferty Campground	Adams	44°56'24"N	116°39'15"W	4340
Lake Creek Campsite	Custer	44°38'33"N	115°10'45"W	4840
Lake Fork Campground	Valley	44°55'23"N	115°56'42"W	5340
Lake View Campground	Custer	44°14'51"N	115°03'35"W	
Lantz Bar Campsite	Idaho	45°25'15"N	114°51'59"W	2832
Larios Camp	Owyhee	42°09'32"N	115°33'08"W	
Last Chance Campground	Adams	44°59'22"N	116°11'23"W	
Lightning Strike Campground	Valley	45°11'21"N	114°42'45"W	3600
Lime Creek Campground	Elmore	43°26'40"N	115°07'52"W	5320
Lintz Campground (historical)	Shoshone	47°09'31"N	115°25'13"W	

Camp or Campground	County	Latitude	Longitude	Elev.
Little Boulder Campground	Latah	46°46'19"N	116°27'25"W	2640
Little Boulder Transfer Camp	Custer	44°05'24"N	114°26'33"W	6155
Little Creek Campsite	Custer	44°43'37"N	114°59'39"W	4400
Little North Fork Campground	Shoshone	47°03'59"N	115°50'57"W	
Little Roaring River Lake Campground	Elmore	43°37'45"N	115°26'35"W	7880
Little Smoky Campground	Camas	43°32'50"N	114°49'03"W	
Little Soldier Campground	Valley	44°44'13"N	115°01'53"W	4540
Little Trinity Lake Campground	Elmore	43°37'46"N	115°25'48"W	7800
Little West Fork Campground	Custer	44°42'08"N	114°18'54"W	6360
Little Wilson Creek Campground	Elmore	43°22'40"N	115°26'03"W	4200
Lodgepole Campground	Valley	44°51'58"N	115°41'44"W	
Lola Creek Campground	Custer	44°24'28"N	115°10'47"W	
Lolo Creek Campground	Idaho	46°17'36"N	115°45'03"W	
Lonesome Cove Camp	Idaho	46°32'41"N	114°59'41"W	6420
Long Camp and ASA Smith Mission	Lewis	46°14'47"N	116°02'27"W	1160
Long Tom Campground	Lemhi	45°18'01"N	114°34'47"W	
Lost Spring Campground	Lemhi	44°50'41"N	114°27'50"W	
Lower Coffee Pot Campground	Fremont	44°29'26"N	111°23'50"W	
Lower Grouse Campground	Lemhi	44°52'23"N	114°46'00"W	3820
Lower Jackass Flat Campground	Custer	44°43'23"N	114°57'45"W	4280
Lower O'Brien Campground	Custer	44°15'27"N	114°41'39"W	
Lower Penstemon Campground	Twin Falls	42°11'48"N	114°16'57"W	
Luby Bay Campground	Bonner	48°32'59"N	116°55'25"W	
Lyons Mining Camp	Idaho	45°48'17"N	116°18'40"W	1800
Lytle Camp	Idaho	45°58'54"N	115°41'47"W	
Mackay Bar Campground	Idaho	45°23'33"N	115°29'35"W	2280
Magary Camp	Clearwater	46°33'19"N	115°21'48"W	
Magpie Creek Campsite	Idaho	45°32'55"N	115°09'16"W	
Magruder Campground	Idaho	45°44'11"N	114°45'27"W	
Mahoney Campsite	Custer	44°45'02"N	114°54'55"W	4240
Mahoney Campsite	Lemhi	44°39'37"N	114°32'31"W	8420
Mammouth Springs Campground	Shoshone	47°06'17"N	115°37'41"W	
Marble Creek Campground	Custer	44°44'35"N	115°01'00"W	4460
Marble Creek Campground	Shoshone	47°11'53"N	116°04'01"W	2960
Marble Creek Campground	Valley	44°44'40"N	115°01'18"W	4460
Marble Creek Number Two Campsite	Custer	44°44'35"N	115°01'00"W	4460
Marsh Creek Transfer Camp	Custer	44°24'38"N	115°11'03"W	6440
Marshall Creek Campground	Custer	44°15'15"N	114°41'05"W	
McCoy Creek Campground	Bonneville	43°11'02"N	111°06'00"W	
McCrea Bridge Campground	Fremont	44°27'45"N	111°23'57"W	
McMullen Cow Camp	Twin Falls	42°15'11"N	114°23'56"W	
Meadow Creek Campground	Boundary	48°49'10"N	116°08'47"W	
Meadow Creek Campground	Idaho	45°49'46"N	115°55'36"W	
Meadow Lake Campground	Lemhi	44°26'02"N	113°19'00"W	9195
Merrys Bay Campground	Clearwater	46°30'43"N	116°15'59"W	2000
Middle Fork Campground	Lemhi	45°17'58"N	114°35'44"W	3360
Middle Fork Peak Campground	Lemhi	44°57'45"N	114°38'36"W	
Mill Canyon Campground	Caribou	42°48'36"N	111°21'45"W	6760

Camp or Campground	County	Latitude	Longitude	Elev.
Mill Creek Campground	Custer	44°28'21"N	114°26'25"W	
Mill Creek Campground	Fremont	44°26'55"N	111°25'32"W	
Mokins Bay Campground	Kootenai	47°47'04"N	116°39'52"W	
Monday Camp	Valley	44°55'30"N	115°20'04"W	
Montpelier Canyon Campground	Bear Lake	42°19'53"N	111°13'49"W	
Mormon Bend Campground	Custer	44°15'42"N	114°50'28"W	
Mountain View Campground	Boise	44°04'44"N	115°36'11"W	
Mountain View Campground	Custer	44°09'25"N	114°54'40"W	
Mud Lake Cow Camp	Canyon	43°47'47"N	116°33'21"W	
Muellers Camp	Benewah	47°18'48"N	116°27'32"W	
Muleshoe Creek Camp	Idaho	46°20'45"N	114°38'37"W	
Murdock Campground	Blaine	43°48'13"N	114°25'10"W	
Mush Campground	Clearwater	46°45'26"N	115°22'07"W	
Narrows Campground	Camas	43°33'27"N	114°49'27"W	
Navigation Campground	Bonner	48°47'41"N	116°54'34"W	
Nighthawk Campground	Kootenai	47°57'36"N	116°35'23"W	2382
Ninemeyer Campground	Elmore	43°45'21"N	115°34'04"W	
Nineteenmile Camp	Clearwater	46°29'49"N	115°30'32"W	
Nordman Campground	Bonner	48°40'09"N	116°57'31"W	2760
North Cove Campground	Bonner	48°34'26"N	116°53'43"W	
North Fork Campground	Blaine	43°47'16"N	114°25'29"W	
North Fork Campground	Idaho	45°38'22"N	116°07'07"W	
North Meadow Camp	Idaho	45°55'00"N	115°53'43"W	
North Shore Campground	Blaine	43°55'08"N	114°51'57"W	7044
No-see-um Camp	Idaho	46°25'27"N	115°18'10"W	6120
Oakie Point Campsite	Valley	44°38'53"N	115°10'06"W	4850
O'Hara Campground	Idaho	46°05'09"N	115°30'31"W	
O'Hara Youth Camp	Idaho	46°05'21"N	115°31'04"W	
Old Camp	Clearwater	46°43'40"N	115°51'23"W	3200
Old Camp 10	Clearwater	46°43'29"N	115°44'16"W	3510
Old Camp 11	Clearwater	46°42'36"N	115°44'34"W	3643
Old Camp 6	Clearwater	46°43'08"N	115°47'48"W	3680
Old Camp J	Clearwater	46°49'38"N	115°40'05"W	
One Spoon Campground (historical)	Boise	44°13'37"N	115°54'37"W	
Orogrande Campground	Idaho	45°41'51"N	115°32'43"W	4760
Orogrande Summit Campground	Idaho	45°38'24"N	115°36'49"W	7250
Orogrande Summit Campground	Idaho	45°38'23"N	115°36'48"W	
Osprey Campground	Bonner	48°30'22"N	116°53'15"W	
Outlet Campground	Bonner	48°29'56"N	116°53'32"W	
Pack Bridge Campsite	Valley	44°45'37"N	115°06'15"W	4700
Packsaddle Campground	Boise	44°13'20"N	115°41'56"W	
Packsaddle Campground	Shoshone	47°13'55"N	115°43'42"W	
Paddy Flat Campground	Valley	44°46'34"N	115°56'33"W	
Palisades Creek Campground	Bonneville	43°23'53"N	111°13'12"W	
Papoose Campground	Idaho	45°24'17"N	116°23'45"W	3680
Papoose Creek Campground	Valley	45°10'32"N	114°43'15"W	3600
Paradise Campground	Idaho	45°51'40"N	114°44'17"W	
Paradise Campground	Washington	44°32'37"N	116°57'06"W	4260

Camps & Campgrounds

Camp or Campground	County	Latitude	Longitude	Elev.
Park Creek Campground	Boise	44°07'01"N	115°34'49"W	
Park Creek Campground	Custer	43°50'09"N	114°15'28"W	
Parrot Placer Campground	Lemhi	45°11'28"N	114°41'45"W	3540
Pebble Beach Campsite	Custer	44°47'02"N	114°50'57"W	4120
Pebble Cow Camp	Caribou	42°45'48"N	112°05'14"W	
Pen Basin Campground	Valley	44°37'54"N	115°31'24"W	6685
Penny Spring Campground	Valley	44°42'50"N	115°41'30"W	5080
Pete Forks Campground	Idaho	46°23'09"N	115°35'06"W	5260
Peterson Campground	Kootenai	47°58'13"N	116°35'09"W	2290
Pettit Campground	Twin Falls	42°10'58"N	114°16'53"W	
Phi Kappa Campground	Custer	43°51'31"N	114°13'05"W	
Phillips Creek Transfer Campground	Custer	44°36'07"N	114°48'58"W	5411
Picnic Point Campground	Valley	44°39'16"N	115°40'17"W	5294
Pine Bar Campground	Caribou	42°58'21"N	111°12'26"W	
Pine Creek Campground	Boise	44°05'59"N	115°45'29"W	
Pine Creek Campground	Teton	43°34'25"N	111°12'22"W	
Pine Creek Flat Campsite	Custer	44°45'44"N	114°53'37"W	4240
Pine Flats Campground	Boise	44°03'47"N	115°40'52"W	3700
Pioneer Forest Camp	Camas	43°29'26"N	114°49'52"W	
Pistol Creek Campground	Valley	44°43'25"N	115°08'56"W	4813
Piute Basin Camp	Owyhee	42°12'42"N	116°30'24"W	
Plowboy Campground	Bonner	48°46'11"N	116°52'47"W	
Poet Creek Campground	Idaho	45°43'23"N	115°01'59"W	
Point Campground	Custer	44°08'08"N	114°55'54"W	
Poison Creek Campground	Valley	44°40'11"N	116°06'36"W	4860
Poker Meadows Campground	Valley	44°25'36"N	115°19'50"W	
Polecamp Flat Campground	Custer	44°18'13"N	114°43'08"W	
Ponderosa Campground	Valley	45°03'44"N	115°45'31"W	
Porcupine Campground	Franklin	42°05'43"N	111°31'03"W	
Porcupine Lake Campground	Bonner	48°14'38"N	116°11'04"W	4820
Portable Soup Camp	Idaho	46°23'17"N	115°28'49"W	3700
Porters Camp	Clearwater	46°30'37"N	115°53'09"W	2740
Post Camp	Owyhee	42°28'53"N	116°58'10"W	
Potlatch Camp 58	Idaho	45°53'23"N	115°55'18"W	3212
Poverty Flat Campground	Valley	44°49'23"N	115°42'10"W	
Powell Campground	Idaho	46°30'44"N	114°43'18"W	
Powerplant Campground	Elmore	46°39'47"N	115°43'20"W	
Prairie Campground	Blaine	43°49'03"N	114°35'45"W	
Priest River Campground	Bonner	48°10'38"N	116°54'07"W	2070
Pungo Creek Campground	Valley	44°45'55"N	115°04'23"W	4580
Queens River Campground	Elmore	43°49'16"N	115°12'33"W	
Quick Stop Campsite	Custer	44°43'13"N	115°08'38"W	4813
Race Creek Campground	Idaho	46°02'39"N	115°16'59"W	
Race Track Campground	Idaho	46°02'47"N	115°17'38"W	1780
Rackliff Campground	Idaho	46°05'07"N	115°29'36"W	
Rainbow Point Campground	Valley	44°42'12"N	116°07'52"W	4850
Rainy Day Point Campground	Idaho	45°45'43"N	115°42'07"W	5948
Rainy Hill Campground	Kootenai	47°28'30"N	116°35'10"W	2140

Camp or Campground	County	Latitude	Longitude	Elev.
Rattlesnake Campground	Elmore	43°35'03"N	115°41'43"W	
Rattlesnake Campground	Valley	44°15'58"N	115°52'44"W	3630
Rattlesnake Creek Campground	Lemhi	45°02'54"N	114°42'13"W	
Raven Creek Campground	Idaho	45°45'43"N	114°46'57"W	
Red Horse Campground	Idaho	45°53'26"N	115°37'17"W	4080
Red Horse Campground	Idaho	45°53'26"N	115°37'16"W	4100
Red River Campground	Idaho	45°45'02"N	115°16'13"W	4640
Red River Campground	Idaho	45°44'55"N	115°16'16"W	4732
Redfish Inlet Transfer Camp	Custer	44°05'58"N	114°57'08"W	
Redfish Outlet Campground	Custer	44°08'21"N	114°55'03"W	
Reeder Bay Campground	Bonner	48°37'31"N	116°53'27"W	
Retreat Camp	Idaho	46°22'33"N	115°27'40"W	3420
Retreat Camp	Idaho	46°22'29"N	115°27'38"W	3484
Rhett Creek Campground	Idaho	45°28'22"N	115°23'37"W	
River Bend Campground	Camas	43°35'04"N	114°58'09"W	
Riverside Campground	Custer	44°16'01"N	114°50'57"W	
Riverside Campground	Elmore	43°48'32"N	115°07'47"W	
Riverside Campground	Fremont	44°15'47"N	111°27'14"W	
Riverside Campground	Valley	44°20'29"N	115°39'24"W	5330
Robert E Lee Campground	Boise	43°54'20"N	115°26'01"W	
Robinson Lake Campground	Boundary	48°58'11"N	116°12'57"W	2645
Rock Creek Camp	Fremont	44°06'29"N	111°10'25"W	
Rock Creek Campground	Lemhi	44°40'50"N	114°44'45"W	
Rock Island Campground	Valley	44°47'02"N	114°51'07"W	4120
Rocky Bluff Campground	Idaho	45°37'57"N	116°00'36"W	
Rocky Point Campground	Bonner	48°34'20"N	116°53'35"W	2500
Rocky Ridge Lake Campground	Idaho	46°26'29"N	115°29'29"W	5700
Rooks Creek Campground	Blaine	43°39'08"N	114°31'32"W	6420
Round Lake Campground	Bonner	48°09'57"N	116°38'00"W	2202
Ruby Creek Campground	Clearwater	46°43'54"N	115°04'40"W	
Sack Creek Campground	Valley	44°21'30"N	115°24'22"W	
Saddle Camp	Idaho	46°30'52"N	115°06'13"W	
Saddle Camp	Idaho	45°20'18"N	116°28'48"W	
Sage Hen Campground	Gem	44°20'06"N	116°10'39"W	4990
Sagehen Campground	Bingham	43°01'15"N	111°49'33"W	5900
Saint Charles Campground	Bear Lake	42°06'46"N	111°26'50"W	
Salmon Point Campground	Valley	44°57'06"N	115°44'08"W	
Salmon River Campground	Custer	44°15'02"N	114°52'07"W	
Salmon River Campground	Custer	44°14'55"N	114°52'09"W	
Salmon Trout Camp	Idaho	46°17'56"N	115°38'41"W	3575
Salyer Cow Camp	Oneida	42°14'06"N	112°47'38"W	
Sams Creek Campground	Idaho	45°32'10"N	115°29'44"W	
Savage Camp	Idaho	46°29'14"N	115°31'26"W	4821
Sawmill Campground	Blaine	43°40'00"N	114°09'45"W	
Sawtooth Camp	Blaine	43°48'27"N	114°25'19"W	
Schipper Campground	Cassia	42°19'22"N	114°16'02"W	
Schneider Campground	Bonner	48°33'41"N	116°54'11"W	
Scott Campground	Kootenai	47°57'58"N	116°35'41"W	2325

Camp or Campground	County	Latitude	Longitude	Elev.
Scout Mountain Campground	Bannock	42°41'37"N	112°21'29"W	
Selway Falls Campground	Idaho	46°02'23"N	115°17'39"W	
Senator Creek Campground	Shoshone	47°52'38"N	116°08'00"W	
Seven Devils Campground	Idaho	45°20'50"N	116°31'00"W	7620
Shadowy St Joe Campground	Benewah	47°19'33"N	116°23'44"W	2150
Sheep Trail Campground	Custer	44°18'22"N	115°03'19"W	
Sheepeater Hot Springs Campground	Valley	44°37'39"N	115°11'48"W	4870
Shelf Campsite	Custer	44°48'07"N	114°49'54"W	4080
Shiefer Campground	Valley	45°10'20"N	115°34'51"W	3200
Shoreline Campground	Valley	44°39'25"N	115°40'07"W	5245
Shoshone Camp Work Center	Shoshone	47°42'27"N	115°58'12"W	2490
Shoshone Forest Camp	Shoshone	47°42'23"N	115°58'20"W	
Silver Campground	Bonner	48°33'48"N	116°53'57"W	
Silver Creek Campground	Valley	44°21'14"N	115°48'48"W	4590
Silver Sage Girl Scout Camp	Bannock	42°51'45"N	112°20'49"W	
Simplot Campground	Lemhi	44°48'30"N	114°48'43"W	
Sing Lee Campground	Idaho	45°53'07"N	115°37'24"W	3980
Sissons Campground	Shoshone	47°44'03"N	116°00'30"W	
Sixmile Project Camp	Valley	44°17'59"N	115°52'07"W	3750
Skeleton Campground	Camas	43°35'27"N	115°01'04"W	
Skookumchuck Campground	Idaho	45°42'12"N	116°18'51"W	1721
Slaughter Gulch Campground	Adams	44°58'13"N	116°27'28"W	4828
Slide Creek Campground	Idaho	46°05'06"N	115°27'06"W	1620
Slims Campground	Idaho	46°01'48"N	115°17'21"W	
Small Prairie Camp	Idaho	46°17'10"N	115°39'36"W	3530
Smith Lake Campground	Boundary	48°46'37"N	116°15'46"W	3010
Smoky Bear Campground	Blaine	43°55'13"N	114°51'40"W	
Smoky Forest Camp	Adams	45°09'34"N	116°22'41"W	
Snowberry Campground	Kootenai	47°58'05"N	116°33'00"W	2260
Snowmobile Parking Campground	Valley	44°18'03"N	116°05'13"W	4500
Sockeye Campground	Custer	44°07'52"N	114°55'00"W	6570
Soldier Forest Camp	Camas	43°28'57"N	114°49'31"W	
Sourdough Saddle Campground	Idaho	45°43'18"N	115°48'20"W	6080
Sousie Creek Campground	Clearwater	46°48'00"N	115°39'18"W	2340
South Fork Campground	Idaho	45°49'30"N	115°57'44"W	2470
South Fork Campground	Lemhi	44°43'32"N	114°38'23"W	
South Fork Salmon River Campground	Valley	44°39'11"N	115°42'03"W	5160
Spencer Camp	Owyhee	42°18'20"N	116°37'06"W	
Spirit Pines Church Camp	Bonner	47°59'05"N	116°54'35"W	
Spirit Revival Ridge Camp	Idaho	46°24'57"N	115°19'18"W	6658
Spring Bar Campground	Idaho	45°25'37"N	116°09'04"W	1844
Spring Camp Ranch	Idaho	45°51'50"N	116°34'32"W	
Spring Creek Campground	Bonneville	43°27'00"N	111°23'58"W	
Spring Creek Campground	Washington	44°34'13"N	116°56'45"W	4790
Spring Creek Campground	Washington	44°34'13"N	116°56'45"W	4840
Spring Point Campground	Bonner	48°14'11"N	116°35'11"W	2104
Spruce Tree Campground	Shoshone	47°02'16"N	115°20'48"W	
Squaw Camp	Owyhee	42°00'08"N	115°41'14"W	

Camp or Campground	County	Latitude	Longitude	Elev.
Squaw Creek Campground	Shoshone	47°17'46"N	115°46'28"W	2700
Stanley Lake Campground	Custer	44°14'55"N	115°03'09"W	
Star Creek Camp	Adams	45°09'02"N	116°23'54"W	
Star Hope Camp	Custer	43°44'48"N	113°56'27"W	
State Campground Recreation Site	Idaho	45°20'30"N	116°21'07"W	2200
State Land Campground	Custer	44°43'13"N	115°00'06"W	4409
State Line Camp	Owyhee	41°59'54"N	116°36'05"W	5237
Steamboat Camp	Shoshone	47°44'21"N	116°11'19"W	
Steel Creek Campground	Clark	44°27'56"N	112°01'17"W	
Steer Basin Campground	Cassia	42°16'47"N	114°15'34"W	
Stoddard Creek Campground	Clark	44°25'05"N	112°12'58"W	
Sugar Factory Camp	Canyon	43°34'50"N	116°34'59"W	
Summit Camp	Bonner	48°03'30"N	116°13'19"W	
Summit Campground	Bannock	42°21'00"N	112°16'31"W	
Summit Lake Campground	Valley	44°38'42"N	115°35'08"W	7290
Summit View Campground	Bear Lake	42°33'31"N	111°17'43"W	
Sunflower Campsite	Custer	44°43'48"N	115°01'04"W	4425
Sunny Gulch Campground	Custer	44°10'35"N	114°54'33"W	6360
Surprise Creek Camp	Idaho	45°32'19"N	114°37'26"W	
Survey Creek Campground	Valley	45°03'19"N	114°43'27"W	3600
Swamp Camp	Lemhi	45°23'32"N	114°33'04"W	
Swinging Bridge Campground	Valley	44°10'18"N	116°07'12"W	3670
Table Meadows Campground	Idaho	45°56'10"N	115°30'45"W	4840
Table Meadows Campground	Idaho	45°56'08"N	115°30'43"W	4835
Talmaks Campground	Lewis	46°10'32"N	116°33'51"W	
Tappan Island Campsite	Lemhi	44°52'28"N	114°45'34"W	3893
Tawakani Camp	Twin Falls	42°10'16"N	114°17'06"W	
Teepee Creek Campground	Valley	44°56'25"N	115°44'18"W	3860
Telephone Camp	Idaho	45°29'08"N	115°13'02"W	
Tenmile Campground	Boise	43°53'52"N	115°42'32"W	
Tennessee Creek Campground	Valley	44°24'55"N	115°23'03"W	
Thatcher Creek Campground	Custer	44°22'04"N	115°08'40"W	
The Sinque Hole Camp	Idaho	46°30'13"N	115°08'26"W	5970
The Smoking Place Camp	Idaho	46°29'27"N	115°09'07"W	6400
Third Fork Project Camp	Gem	44°22'35"N	116°18'19"W	
Thompson Flat Campground	Cassia	42°19'30"N	113°37'23"W	
Thorn Creek Campground	Boise	43°44'45"N	115°53'00"W	3800
Thorn Creek Fire Camp	Adams	44°58'13"N	116°10'06"W	
Three Bear Camp	Clearwater	46°32'24"N	115°33'06"W	5051
Three Pines Campground	Bonner	48°33'52"N	116°53'37"W	2500
Tie Creek Campground	Boise	44°12'32"N	115°55'30"W	
Tin Can Flat Campground	Shoshone	47°13'48"N	115°37'12"W	
Tincup Campground	Caribou	43°00'17"N	111°06'07"W	
Tincup Campground	Custer	44°35'50"N	114°48'45"W	
Tipton Flat Campground	Elmore	43°35'23"N	115°35'49"W	
Tom Creek Campsite	Valley	44°49'28"N	114°48'06"W	4120
Trail Creek Campground	Valley	44°16'37"N	115°52'26"W	3695
Trail Creek Campground	Valley	44°16'37"N	115°52'25"W	3693

Camps & Campgrounds

Camp or Campground	County	Latitude	Longitude	Elev.
Trail Flat Campsite	Valley	44°36'29"N	115°15'54"W	6794
Transfer Camp	Custer	44°36'53"N	114°21'20"W	6600
Transfer Campground	Custer	44°36'08"N	114°48'21"W	
Trout Creek Campground	Valley	44°44'50"N	115°33'15"W	6330
Troutdale Campground	Elmore	43°42'59"N	115°37'27"W	3533
Turner Flat Campground	Shoshone	47°14'13"N	115°39'12"W	
Twentymile Bar Campground	Idaho	46°05'12"N	115°28'09"W	1960
Twentythree Mile Campground	Idaho	46°04'37"N	115°24'42"W	1750
Twin Bridges Campground	Valley	44°48'21"N	115°31'15"W	
Twin Creek Campground	Lemhi	45°36'30"N	113°58'08"W	
Twin Creek Campground	Shoshone	47°05'11"N	115°46'32"W	
United Brethren Camp	Clearwater	46°31'49"N	116°26'20"W	
United States Forest Service Camp 216	Boundary	48°55'53"N	116°39'28"W	3720
Upper Coffee Pot Campground	Fremont	44°29'28"N	111°21'54"W	
Upper Grouse Campground	Lemhi	44°52'15"N	114°46'04"W	
Upper Jackass Flat Campground	Custer	44°43'27"N	114°58'10"W	4360
Upper O'brien Campground	Custer	44°15'34"N	114°41'51"W	
Upper Payette Campground	Valley	45°07'32"N	116°01'32"W	
Upper Payette Lake Campground	Valley	45°07'22"N	116°01'41"W	5800
Upper Penstemon Campground	Twin Falls	42°11'40"N	114°17'06"W	
Van Creek Campground	Idaho	45°25'42"N	116°08'18"W	2040
Velvet Falls Campground	Valley	44°35'45"N	115°17'27"W	7200
Victory Cove Church Camp	Valley	44°56'07"N	116°04'54"W	5025
Wagonhammer Campground	Lemhi	45°23'06"N	113°57'38"W	3650
Waldron Campground	Kootenai	47°57'32"N	116°35'28"W	2390
Ward Campground	Kootenai	47°57'53"N	116°34'23"W	2310
Warm Lake Lodge Campground	Valley	44°39'08"N	115°39'20"W	5325
Warm River Campground	Fremont	44°07'14"N	111°18'39"W	
Warm Springs Campground	Blaine	43°40'22"N	114°25'50"W	
Warm Springs Campground	Boise	44°09'04"N	115°18'36"W	
Washington Creek Campground	Clearwater	46°42'16"N	115°33'18"W	2160
Wedge Camp	Clearwater	46°30'43"N	115°28'09"W	
Weitas Meadows Campground	Idaho	46°26'07"N	115°28'52"W	5400
Wendover Campground	Idaho	46°30'36"N	114°47'01"W	
West Fork Campground	Custer	44°22'16"N	114°45'03"W	
West Pine Creek Girls Camp	Bonneville	43°33'39"N	111°17'53"W	
Whaa-Laa Campground	Kootenai	47°22'27"N	116°52'55"W	2734
White Creek Campground	Custer	44°47'36"N	114°50'24"W	4040
White Sand Campground	Idaho	46°30'27"N	114°41'09"W	
Whitehouse Campground	Idaho	46°30'23"N	114°46'25"W	
Whitetail Campground	Kootenai	47°57'50"N	116°33'40"W	2290
Whitewater Campground	Idaho	45°31'57"N	115°17'57"W	2522
Whitey Cox Campground	Custer	44°46'58"N	114°51'21"W	4180
Whitman Hollow Campground	Bear Lake	42°20'40"N	111°11'40"W	
Wild Goose Campground	Idaho	46°08'09"N	115°37'31"W	
Wildhorse Campground	Custer	43°49'21"N	114°05'42"W	
Wildhorse Lake Campground	Idaho	45°39'21"N	115°39'00"W	
Willow Creek Campground	Boise	43°57'33"N	115°31'49"W	

Camp or Campground	County	Latitude	Longitude	Elev.
Willow Creek Campground	Boise	43°38'39"N	115°45'48"W	
Willow Creek Campground	Elmore	43°36'24"N	115°08'33"W	
Wilson Cow Camp	Idaho	45°36'19"N	115°55'02"W	6430
Wilson Cow Camp	Idaho	45°24'19"N	116°33'18"W	
Wilson Creek Campground	Lemhi	45°01'50"N	114°42'26"W	
Windy Ridge Camp	Idaho	46°31'18"N	115°11'29"W	
Windy Saddle Campground	Idaho	45°21'00"N	116°30'38"W	7590
Winter Camp	Owyhee	42°33'15"N	115°30'17"W	
Witter Ridge Campsite	Idaho	45°33'49"N	114°48'51"W	
Wolftone Campground	Blaine	43°31'58"N	114°27'57"W	
Wood River Campground	Blaine	43°47'35"N	114°27'31"W	
Wyant Camp	Idaho	45°16'15"N	116°25'22"W	
Yellow Pine Campground	Valley	44°57'17"N	115°29'44"W	
Yellowjacket Campground	Lemhi	45°04'00"N	114°33'00"W	8000

City, Town, or Historic Place	County	Latitude	Longitude	Elev.
Abbey (historical)	Twin Falls	42°37'54"N	114°45'19"W	3406
Aberdeen	Bingham	42°56'39"N	112°50'15"W	4400
Aberdeen Junction	Bingham	43°13'23"N	112°28'12"W	4462
Abstein Place	Valley	44°58'15"N	115°29'09"W	4960
Acequia	Minidoka	42°40'05"N	113°35'46"W	4165
Adair	Shoshone	47°20'32"N	115°36'40"W	3707
Addie	Boundary	48°57'13"N	116°09'55"W	2540
Adelaide	Minidoka	42°48'26"N	113°43'16"W	4340
Advent Hollow (historical)	Latah	46°50'15"N	116°59'00"W	2820
Agatha	Nez Perce	46°30'39"N	116°34'33"W	920
Ahsahka	Clearwater	46°30'08"N	116°19'23"W	990
Aiken	Bingham	43°12'33"N	112°24'21"W	4460
Aikman (historical)	Gem	43°49'00"N	116°24'15"W	2812
Alameda	Bannock	42°53'24"N	112°27'12"W	4480
Albion	Cassia	42°24'46"N	113°34'38"W	4730
Alder City (historical)	Custer	43°52'21"N	113°31'10"W	5740
Alder Creek	Benewah	47°13'22"N	116°36'24"W	2240
Alexander	Caribou	42°39'04"N	111°42'30"W	5734
Algoma	Bonner	48°11'49"N	116°34'16"W	2179
Allendale	Canyon	43°40'35"N	116°52'26"W	2426
Allens Spur	Boundary	48°44'26"N	116°21'55"W	1781
Almo	Cassia	42°06'01"N	113°37'58"W	5390
Alpha	Valley	44°23'28"N	116°00'25"W	4760
Alpine	Adams	44°35'37"N	116°29'38"W	2819
Alridge	Bingham	43°12'41"N	112°00'15"W	4977
Alton	Bear Lake	42°13'39"N	111°09'03"W	6080
Alturas City (historical)	Elmore	43°48'03"N	115°09'34"W	5190
Amalga	Minidoka	42°34'37"N	113°43'26"W	4150
American Falls (county seat)	Power	42°47'10"N	112°51'13"W	4415
Ammon	Bonneville	43°28'11"N	111°57'57"W	4720
Amsco	Canyon	43°45'53"N	116°38'54"W	2500
Amsterdam	Twin Falls	42°17'45"N	114°35'02"W	4681
Anderson	Franklin	42°01'14"N	111°57'22"W	4564
Anderson Place	Twin Falls	42°14'41"N	114°57'00"W	5465
Annis	Jefferson	43°43'42"N	111°56'07"W	4826
Antelope (historical)	Butte	43°40'20"N	113°38'28"W	6300
Apple Valley	Canyon	43°50'24"N	116°59'44"W	2205
Appleton	Gooding	42°44'53"N	114°38'04"W	3480
Arbon	Power	42°27'21"N	112°34'03"W	5165
Archabal	Valley	44°50'49"N	116°07'16"W	5005
Archer	Madison	43°42'49"N	111°46'55"W	4915
Arco (county seat)	Butte	43°38'12"N	113°17'58"W	5325
Arimo	Bannock	42°33'36"N	112°10'14"W	4750
Arling	Valley	44°37'31"N	116°02'57"W	4835
Arrow	Nez Perce	46°28'41"N	116°46'01"W	860
Artesian City	Twin Falls	42°25'02"N	114°09'37"W	4232
Asbestos Point	Idaho	45°49'05"N	115°56'23"W	3176
Ashton	Fremont	44°04'18"N	111°26'51"W	5260

City, Town, or Historic Place	County	Latitude	Longitude	Elev.
Aspendale	Latah	46°40'37"N	116°29'47"W	2820
Athol	Kootenai	47°56'53"N	116°42'25"W	2391
Atlanta	Elmore	43°48'06"N	115°07'33"W	5383
Atlas	Kootenai	47°42'02"N	116°49'43"W	2210
Atomic City	Bingham	43°26'42"N	112°48'43"W	5015
Austin (historical)	Twin Falls	42°39'23"N	114°50'54"W	3340
Avery	Shoshone	47°15'02"N	115°48'15"W	2500
Avon	Latah	46°49'34"N	116°36'46"W	2725
Bacon	Adams	44°38'10"N	116°32'49"W	2780
Baker	Lemhi	45°05'41"N	113°44'01"W	4356
Bancroft	Caribou	42°43'13"N	111°53'06"W	5420
Banida	Franklin	42°13'52"N	111°56'30"W	4795
Banks	Boise	44°04'50"N	116°07'23"W	2840
Banner	Boise	44°01'12"N	115°32'05"W	5810
Bannock	Power	42°51'41"N	112°43'19"W	4410
Barber	Ada	43°33'55"N	116°06'44"W	2780
Barite	Blaine	43°33'23"N	114°19'32"W	5440
Barlow	Jefferson	43°40'16"N	112°00'44"W	4797
Barrymore	Jerome	42°40'51"N	114°27'44"W	3845
Barton	Washington	44°23'04"N	116°47'56"W	2380
Basalt	Bingham	43°18'56"N	112°09'49"W	4587
Basin	Cassia	42°14'44"N	113°47'01"W	5003
Bassett	Jefferson	43°39'04"N	112°05'15"W	4768
Bates	Teton	43°41'30"N	111°12'00"W	
Bayhorse	Custer	44°23'52"N	114°18'39"W	6100
Bayview	Kootenai	47°58'49"N	116°33'33"W	2160
Beachs Corner	Bonneville	43°32'33"N	111°57'48"W	4783
Bear	Adams	45°01'28"N	116°40'16"W	4380
Bear Creek Summer Home Area	Boise	44°10'00"N	115°11'35"W	5280
Bear Lake Hot Springs	Bear Lake	42°06'31"N	111°15'52"W	5940
Bear Lake Sands	Bear Lake	42°03'06"N	111°15'10"W	5989
Beatty	Ada	43°36'30"N	116°20'24"W	2635
Bedstead Corner	Fremont	44°16'33"N	111°47'32"W	6315
Beeman	Nez Perce	46°24'20"N	116°31'27"W	3120
Beer Bottle Crossing	Adams	44°36'31"N	116°13'48"W	6020
Beetville	Cassia	42°29'19"N	113°49'50"W	4225
Bellevue	Blaine	43°27'49"N	114°15'35"W	5190
Bellgrove	Kootenai	47°32'02"N	116°54'50"W	2457
Belmont	Kootenai	47°55'34"N	116°38'19"W	2462
Belvidere	Valley	44°28'32"N	116°01'57"W	4745
Bench	Caribou	42°30'13"N	111°40'46"W	5485
Benewah	Benewah	47°13'40"N	116°47'00"W	2776
Bengoechea Place	Owyhee	42°18'17"N	115°50'49"W	5010
Bennington	Bear Lake	42°23'28"N	111°19'15"W	6054
Berenice	Butte	43°49'40"N	112°58'24"W	4834
Berger	Twin Falls	42°28'11"N	114°34'26"W	4239
Bergstrom Place	Twin Falls	42°11'40"N	114°56'09"W	5475
Bern	Bear Lake	42°20'23"N	111°23'07"W	5964

Cities, Towns, Historic Places

City, Town, or Historic Place	County	Latitude	Longitude	Elev.
Besslen	Lincoln	42°54'31"N	114°10'22"W	4145
Best Corner	Clearwater	46°33'43"N	116°08'59"W	3140
Bickel (historical)	Twin Falls	42°31'41"N	114°12'35"W	4161
Big Cedar	Idaho	46°04'52"N	115°48'48"W	2490
Big Creek	Valley	45°07'38"N	115°19'24"W	5760
Big Eddy	Boise	44°12'58"N	116°06'23"W	4100
Big George	Nez Perce	46°30'10"N	116°29'31"W	940
Big Springs	Fremont	44°29'54"N	111°15'18"W	6390
Bills	Twin Falls	42°31'56"N	114°17'08"W	4052
Bingo Creek Landing	Clearwater	46°45'26"N	115°44'09"W	3180
Black Bear	Shoshone	47°30'58"N	115°51'05"W	3380
Black Canyon	Gem	43°55'36"N	116°26'29"W	
Black Pine	Oneida	42°04'21"N	112°58'46"W	4812
Blackbird	Lemhi	45°06'48"N	114°13'08"W	5038
Blackcloud	Shoshone	47°30'17"N	115°54'13"W	3040
Blackfoot (county seat)	Bingham	43°11'26"N	112°20'39"W	4498
Blackrock	Bannock	42°47'49"N	112°19'34"W	4510
Blacks (historical)	Ada	43°27'50"N	116°00'30"W	3540
Blacks Creek	Ada	43°28'40"N	116°08'14"W	3249
Blacktail	Bonner	48°07'14"N	116°29'07"W	2067
Blackwell	Kootenai	47°41'52"N	116°48'25"W	2220
Blaine	Camas	43°20'33"N	114°35'41"W	5027
Blaine	Latah	46°39'26"N	116°56'52"W	2818
Blake	Clearwater	46°28'24"N	116°10'36"W	1340
Blanchard	Bonner	48°01'01"N	116°58'58"W	2290
Bliss	Gooding	42°55'37"N	114°56'55"W	3262
Bloomington	Bear Lake	42°11'31"N	111°24'02"W	5969
Blue Dome	Clark	44°09'37"N	112°54'38"W	6086
Boehls	Clearwater	46°52'29"N	115°54'16"W	1498
Bogle	Shoshone	47°21'11"N	115°44'12"W	3000
Boise (county seat)	Ada	43°36'49"N	116°12'09"W	2730
Boise Hills Village	Ada	43°37'57"N	116°11'20"W	2840
Boise Junction	Ada	43°36'18"N	116°15'21"W	2720
Boles	Idaho	45°54'38"N	116°30'11"W	4540
Bonanza	Custer	44°22'14"N	114°43'37"W	6400
Bone	Bonneville	43°18'45"N	111°47'40"W	6060
Bonners Ferry (county seat)	Boundary	48°41'29"N	116°18'55"W	1930
Borax (historical)	Shoshone	47°26'04"N	115°40'15"W	4000
Border	Bear Lake	42°10'59"N	111°02'48"W	6072
Boulder	Blaine	43°50'26"N	114°30'20"W	9140
Boulder City (historical)	Boundary	48°36'10"N	116°04'06"W	3225
Bovard (historical)	Latah	46°39'55"N	116°42'05"W	1800
Bovill	Latah	46°51'32"N	116°23'33"W	2870
Bowmont	Canyon	43°27'22"N	116°32'24"W	2650
Box Canyon	Fremont	44°22'44"N	111°24'17"W	6180
Boyer	Bonner	48°19'23"N	116°26'34"W	2106
Bradley	Shoshone	47°32'47"N	116°09'35"W	2200
Bramwell	Gem	43°50'18"N	116°34'55"W	2375

City, Town, or Historic Place	County	Latitude	Longitude	Elev.
Bridge	Cassia	42°07'46"N	113°20'30"W	4750
Broadford	Blaine	43°28'11"N	114°16'46"W	5182
Bronx	Bonner	48°21'07"N	116°32'47"W	2130
Broten	Bonner	48°13'16"N	116°24'59"W	2326
Broten PO (historical)	Bonner	48°13'16"N	116°24'59"W	2330
Brownlee	Boise	44°00'32"N	116°14'40"W	3818
Bruce Eddy	Clearwater	46°30'49"N	116°17'23"W	1270
Bruneau	Owyhee	42°52'50"N	115°47'47"W	2525
Brunning	Clearwater	46°29'03"N	116°08'34"W	2900
Buckingham	Payette	43°59'21"N	116°53'33"W	2242
Budge	Minidoka	42°36'32"N	113°49'25"W	4151
Bugtown (historical)	Canyon	43°39'34"N	116°40'26"W	2320
Buhl	Twin Falls	42°35'57"N	114°45'31"W	3740
Buist	Oneida	42°19'38"N	112°36'17"W	5203
Bullion	Blaine	43°28'22"N	114°22'33"W	5542
Bullion (historical)	Shoshone	47°25'56"N	115°40'12"W	4200
Buncel Place	Owyhee	42°33'01"N	116°01'25"W	4515
Bundy	Nez Perce	46°21'41"N	116°45'23"W	1253
Bunn	Shoshone	47°30'40"N	115°53'53"W	3157
Burgdorf	Idaho	45°16'38"N	115°54'43"W	6140
Burke	Shoshone	47°31'13"N	115°49'09"W	3761
Burley (county seat)	Cassia	42°32'09"N	113°47'31"W	4165
Burmah	Lincoln	43°07'46"N	114°15'34"W	4500
Burns	Boundary	48°37'34"N	116°23'38"W	1850
Burton	Madison	43°47'51"N	111°51'26"W	4832
Buswell (historical)	Latah	46°49'59"N	117°01'38"W	2620
Butte City	Butte	43°36'36"N	113°14'36"W	5315
Byrne	Madison	43°41'31"N	111°44'45"W	5054
Cabarton	Valley	44°26'08"N	116°02'21"W	4744
Cabinet	Bonner	48°05'04"N	116°04'21"W	2160
Cable Car Crossing	Idaho	45°27'14"N	115°56'31"W	1950
Cache	Teton	43°47'17"N	111°09'44"W	6050
Calder	Shoshone	47°16'40"N	116°11'25"W	2190
Caldwell (county seat)	Canyon	43°39'47"N	116°41'11"W	2385
Calendar	Idaho	45°37'27"N	115°40'08"W	6520
Camas	Jefferson	44°00'27"N	112°13'13"W	4814
Cambridge	Bannock	42°27'02"N	112°06'57"W	4910
Cambridge	Washington	44°34'22"N	116°40'30"W	2651
Cameron	Nez Perce	46°36'21"N	116°34'16"W	2600
Camp Creek Summer Home Area	Boise	44°10'26"N	115°13'17"W	5400
Canfield	Idaho	45°50'16"N	116°22'15"W	4220
Canyon	Kootenai	47°34'31"N	116°24'42"W	2244
Cape Horn (historical)	Custer	44°24'18"N	115°10'02"W	6520
Carbon Center	Shoshone	47°33'30"N	115°54'00"W	2960
Carbonate (historical)	Shoshone	47°27'19"N	115°45'50"W	3980
Cardiff Mill	Clearwater	46°28'15"N	115°47'37"W	3120
Cardiff Spur	Clearwater	46°34'28"N	115°48'41"W	3439
Cardwell	Benewah	47°12'33"N	116°33'45"W	2520

City, Town, or Historic Place	County	Latitude	Longitude	Elev.
Carey	Blaine	43°18'28"N	113°56'38"W	4790
Careywood	Bonner	48°02'05"N	116°38'32"W	2280
Caribel	Idaho	46°15'29"N	115°56'37"W	2820
Caribou City	Bonneville	43°06'11"N	111°15'50"W	6890
Carmen	Lemhi	45°14'33"N	113°53'33"W	3860
Carothers Place (historical)	Owyhee	42°53'01"N	116°25'55"W	3710
Carrietown	Camas	43°36'18"N	114°41'58"W	7420
Cascade (county seat)	Valley	44°30'59"N	116°02'27"W	4760
Castle Rocks	Elmore	43°19'42"N	115°18'49"W	5228
Castleford	Twin Falls	42°31'15"N	114°52'03"W	3866
Casto	Custer	44°34'05"N	114°50'52"W	5660
Cataldo	Shoshone	47°32'56"N	116°19'43"W	2150
Cathedral Pines	Blaine	43°46'39"N	114°31'33"W	6550
Cavendish	Nez Perce	46°33'37"N	116°25'59"W	2938
Cayuse Junction	Clearwater	46°35'57"N	114°51'11"W	6342
Cedar	Twin Falls	42°35'38"N	114°42'17"W	3775
Cedar Creek	Bonner	48°01'06"N	116°26'43"W	2100
Cedarhill	Oneida	42°16'01"N	112°43'50"W	5370
Cedron	Teton	43°37'20"N	111°11'25"W	6085
Centerville	Boise	43°54'46"N	115°53'29"W	4220
Central	Caribou	42°38'06"N	111°48'39"W	5550
Central Cove	Canyon	43°36'26"N	116°51'42"W	2380
Cerro Grande	Bingham	43°26'28"N	112°55'57"W	4985
Chalk Cut	Elmore	42°58'52"N	115°31'56"W	2900
Challis (county seat)	Custer	44°30'17"N	114°13'51"W	5288
Chambers (historical)	Latah	46°55'30"N	116°46'58"W	2560
Chapin	Teton	43°38'36"N	111°06'36"W	6146
Chatcolet	Benewah	47°22'20"N	116°45'45"W	2136
Chausse	Bear Lake	42°10'51"N	111°04'59"W	6054
Cherry Creek	Oneida	42°05'54"N	112°13'41"W	4463
Cherrylane	Nez Perce	46°30'58"N	116°40'50"W	857
Cherryville	Franklin	42°02'26"N	111°44'55"W	4940
Chesley	Nez Perce	46°20'57"N	116°33'07"W	3396
Chester	Fremont	43°59'58"N	111°34'09"W	5068
Chesterfield	Caribou	42°52'01"N	111°54'04"W	5445
Chilco	Kootenai	47°51'50"N	116°44'44"W	2309
Chilly	Custer	44°04'40"N	113°52'41"W	6311
China Hat	Owyhee	42°06'06"N	115°56'21"W	6183
China Hill	Caribou	42°33'10"N	111°49'36"W	5450
Chubbuck	Bannock	42°55'15"N	112°27'55"W	4470
Clagstone	Bonner	48°01'11"N	116°48'20"W	2170
Clarendon Hot Springs	Blaine	43°33'34"N	114°24'51"W	5640
Clark Fork	Bonner	48°08'43"N	116°10'29"W	2091
Clark Tree	Idaho	46°17'49"N	115°42'38"W	3220
Clarkia	Shoshone	47°00'39"N	116°15'07"W	2830
Clarkson	Bingham	43°12'42"N	112°25'04"W	4470
Clarksville	Kootenai	47°45'14"N	116°43'55"W	2260
Clawson	Teton	43°47'53"N	111°06'36"W	6186

City, Town, or Historic Place	County	Latitude	Longitude	Elev.
Clayton	Custer	44°15'34"N	114°24'03"W	5471
Claytonia	Owyhee	43°34'03"N	116°49'55"W	2260
Clearwater	Idaho	46°01'21"N	115°53'24"W	2544
Cleft	Elmore	43°14'22"N	115°50'59"W	3230
Clementsville	Teton	43°52'36"N	111°22'09"W	5985
Cleveland	Franklin	42°20'24"N	111°42'45"W	4930
Clicks	Lewis	46°16'49"N	116°32'02"W	3700
Cliff (historical)	Custer	43°52'33"N	113°40'38"W	7900
Cliffs	Owyhee	42°38'21"N	116°58'46"W	4990
Clifton	Franklin	42°11'24"N	112°00'26"W	4849
Clover	Twin Falls	42°30'51"N	114°41'12"W	4110
Clover (historical)	Valley	45°08'02"N	114°59'41"W	4180
Cloverdale	Ada	43°37'11"N	116°19'59"W	2643
Clyde	Butte	44°08'09"N	113°14'46"W	5890
Coats	Minidoka	UNKNOW		
Cobalt	Lemhi	45°05'35"N	114°13'51"W	5020
Cocolalla	Bonner	48°06'29"N	116°36'59"W	2220
Coeur d'Alene (county seat)	Kootenai	47°40'40"N	116°46'46"W	2187
Coeur d'Alene Junction	Kootenai	47°45'24"N	116°54'41"W	2225
Coffee Point	Bingham	43°09'23"N	112°58'02"W	4841
Colburn	Bonner	48°23'50"N	116°32'03"W	2180
Cold House	Gem	UNKNOW		
Coleman	Bonner	48°00'15"N	116°54'21"W	2298
Collier Place	Owyhee	42°34'45"N	116°14'22"W	5002
Collins	Bingham	43°12'08"N	112°22'30"W	4475
Collister (subdivision)	Ada	43°39'30"N	116°14'50"W	2660
Coltman	Bonneville	43°37'04"N	112°00'41"W	4778
Comical Turn	Owyhee	42°33'41"N	116°08'37"W	5600
Conant	Cassia	42°16'34"N	113°26'56"W	4770
Concord	Idaho	45°35'05"N	115°40'58"W	7655
Concrete	Washington	44°21'19"N	116°47'40"W	2338
Conda	Caribou	42°43'42"N	111°31'54"W	6190
Conkling Park	Kootenai	47°24'16"N	116°45'26"W	2160
Connor	Cassia	42°16'52"N	113°30'02"W	4923
Coolin	Bonner	48°28'47"N	116°50'54"W	2515
Cooper	Clearwater	46°26'48"N	116°03'14"W	3168
Copeland	Boundary	48°54'08"N	116°23'16"W	1782
Copperville	Idaho	45°45'11"N	116°19'24"W	1460
Cora	Latah	47°00'14"N	116°57'50"W	2880
Corbin Junction	Kootenai	47°53'28"N	116°43'11"W	2336
Cornwall	Latah	46°42'32"N	116°51'53"W	2600
Cornwall	Shoshone	47°31'07"N	115°50'12"W	3800
Cotterel	Cassia	42°31'02"N	113°27'30"W	4509
Cotton	Bonneville	43°25'35"N	112°05'15"W	4656
Cottonwood	Idaho	46°02'55"N	116°20'55"W	
Cottonwood (historical)	Owyhee	42°29'10"N	116°00'26"W	5190
Coulam	Franklin	42°13'39"N	111°58'57"W	4747
Council	Valley	44°43'48"N	116°26'14"W	2940

Cities, Towns, Historic Places

City, Town, or Historic Place	County	Latitude	Longitude	Elev.
Council (county seat)	Adams	44°43'45"N	116°26'15"W	2953
Cow Creek	Clearwater	46°29'45"N	115°55'55"W	2480
Cox	Bingham	43°22'25"N	112°03'10"W	4643
Craig Junction	Lewis	46°16'32"N	116°31'58"W	3780
Craigmont	Lewis	46°14'29"N	116°27'58"W	3760
Crane (historical)	Washington	44°23'07"N	116°26'35"W	3270
Cream Can Junction	Blaine	42°56'19"N	113°23'05"W	4636
Crescent	Latah	46°38'36"N	116°25'28"W	2840
Crill Place	Owyhee	42°49'33"N	116°27'10"W	4160
Cross Trails	Idaho	45°28'35"N	115°41'27"W	5844
Crossport	Boundary	48°42'04"N	116°13'13"W	1796
Crouch	Boise	44°06'55"N	115°58'12"W	3021
Crystal	Power	42°39'14"N	112°29'44"W	5280
Crystal	Washington	44°10'02"N	116°53'24"W	2115
Crystal (historical)	Custer	44°16'03"N	114°19'30"W	5377
Culdesac	Nez Perce	46°22'29"N	116°40'20"W	1680
Cully Moores	Clearwater	46°31'50"N	115°59'26"W	3260
Culver	Bonner	48°19'08"N	116°27'18"W	2111
Cuprum	Adams	45°05'12"N	116°41'18"W	4300
Curry	Twin Falls	42°33'50"N	114°33'16"W	3725
Custer	Custer	44°23'15"N	114°41'42"W	6456
Dairy Creek	Oneida	42°27'45"N	112°25'45"W	5662
Dale (historical)	Valley	44°50'18"N	116°29'00"W	3200
Dalton Gardens	Kootenai	47°43'47"N	116°46'09"W	2262
Daniels	Oneida	42°22'23"N	112°24'44"W	5400
Dans Place	Owyhee	42°16'21"N	115°49'53"W	5070
Darby	Teton	43°41'39"N	111°03'55"W	6319
Darlington	Butte	43°48'49"N	113°24'50"W	5607
Davidson	Idaho	45°50'35"N	116°35'40"W	4260
Dayton	Franklin	42°06'47"N	111°59'34"W	4818
De Lamar	Owyhee	43°01'28"N	116°49'50"W	5460
De Lamar	Owyhee	43°01'20"N	116°49'30"W	5460
De Smet	Benewah	47°08'46"N	116°54'53"W	2600
Deal	Canyon	43°30'24"N	116°32'26"W	2565
Deary	Latah	46°47'58"N	116°33'18"W	2860
Declo	Cassia	42°31'06"N	113°37'38"W	4215
Deep Creek	Boundary	48°37'24"N	116°23'32"W	1866
Deep Creek	Twin Falls	42°35'37"N	114°50'43"W	3629
Dehlin	Bonneville	43°23'04"N	111°42'32"W	6170
DelMonte	Gem	43°52'08"N	116°32'32"W	2345
Delta	Shoshone	47°36'28"N	115°56'25"W	2540
Democrat	Owyhee	43°05'47"N	116°46'28"W	5420
Dempsey (historical)	Bannock	42°37'25"N	111°59'38"W	5082
Dent	Clearwater	46°37'26"N	116°12'08"W	2140
Denton	Bonner	48°12'10"N	116°14'50"W	2100
Denver	Idaho	45°59'54"N	116°14'06"W	3146
DeVeny Place	Idaho	45°21'27"N	116°23'39"W	2310
Devils Ladder	Adams	45°10'42"N	116°26'04"W	7400
Dew Drop Place	Owyhee	42°01'28"N	115°56'38"W	6670
Dewey	Owyhee	43°02'25"N	116°45'41"W	5820
DeWoff	Blaine	42°42'24"N	113°15'35"W	4296
Diamond	Washington	44°24'51"N	116°46'30"W	2500
Dickens Place	Owyhee	42°08'59"N	115°58'17"W	5810
Dickensheet Junction	Bonner	48°27'07"N	116°54'25"W	2540
Dickey	Custer	44°08'04"N	113°54'16"W	6345
Dickshooter	Owyhee	42°23'30"N	116°30'03"W	5330
Dietrich	Lincoln	42°54'34"N	114°15'52"W	4130
Dingle	Bear Lake	42°13'10"N	111°16'02"W	5952
Dixie	Elmore	43°19'02"N	115°26'46"W	4750
Doles	Canyon	43°40'11"N	116°46'58"W	2320
Don	Power	42°54'34"N	112°31'45"W	4445
Doniphan	Blaine	43°24'39"N	114°29'01"W	6021
Donnelly	Valley	44°43'54"N	116°04'48"W	4864
Dorsey (historical)	Shoshone	47°27'46"N	115°44'50"W	3800
Dover	Bonner	48°15'06"N	116°36'36"W	2060
Downata Hot Springs	Bannock	42°23'19"N	112°05'18"W	4800
Downey	Bannock	42°25'43"N	112°07'25"W	4865
Doyle Place	Owyhee	42°50'43"N	116°19'59"W	4140
Driggs (county seat)	Teton	43°43'24"N	111°06'38"W	6116
Drummond	Fremont	43°59'55"N	111°20'40"W	5607
Dry Forty	Owyhee	42°27'04"N	116°24'45"W	5780
Dryden	Idaho	46°08'10"N	116°12'02"W	3088
Dubois (county seat)	Clark	44°10'35"N	112°13'48"W	5145
Dudley	Kootenai	47°32'31"N	116°25'24"W	2135
Dufort	Bonner	48°09'55"N	116°35'45"W	2211
Duncan Place	Twin Falls	42°02'08"N	114°48'32"W	5390
Eagle	Ada	43°41'44"N	116°21'11"W	2565
Eagle	Shoshone	47°38'37"N	115°55'01"W	2540
Eagle Nest	Valley	44°36'26"N	115°56'32"W	7646
Easley Hot Springs	Blaine	43°46'50"N	114°32'32"W	6610
East Greenacres	Kootenai	47°45'09"N	116°58'08"W	2160
East Hope	Bonner	48°14'31"N	116°17'40"W	2180
East Kamiah	Idaho	46°12'10"N	116°00'25"W	1248
Eastport	Boundary	48°59'58"N	116°10'49"W	2633
Eaton	Washington	44°16'33"N	117°04'49"W	2104
Eccles	Fremont	44°19'37"N	111°18'03"W	6261
Echo Beach	Kootenai	47°53'16"N	116°52'32"W	2340
Eddyville	Kootenai	47°36'58"N	116°43'50"W	2200
Eden	Jerome	42°36'21"N	114°12'36"W	3955
Edgemere	Bonner	48°04'33"N	116°48'51"W	2176
Edmonds	Madison	43°54'47"N	111°52'48"W	4858
Edwardsburg	Valley	45°07'08"N	115°19'30"W	5700
Egin	Fremont	43°56'12"N	111°50'12"W	4885
Egypt	Franklin	42°05'48"N	111°50'12"W	4739
Eiffie	Payette	44°01'32"N	116°55'02"W	2200
Eighteenmile	Clark	44°18'01"N	111°54'06"W	6240

City, Town, or Historic Place	County	Latitude	Longitude	Elev.
Eileen	Boundary	48°46'35"N	116°09'45"W	2380
Elba	Cassia	42°14'54"N	113°33'38"W	5162
Elizabeth Park	Shoshone	47°31'47"N	116°05'31"W	2380
Elk City	Idaho	45°49'37"N	115°26'09"W	3980
Elk Creek	Shoshone	47°32'02"N	116°03'25"W	2400
Elk River	Clearwater	46°47'01"N	116°10'44"W	2860
Elk Summit	Idaho	46°19'26"N	114°38'56"W	5790
Elkhorn Village	Blaine	43°40'46"N	114°19'54"W	6080
Ellis	Custer	44°41'31"N	114°02'51"W	4648
Elmira	Bonner	48°28'47"N	116°27'41"W	2149
Elo (historical)	Valley	44°53'00"N	116°03'34"W	5090
Emerald Creek	Shoshone	47°04'19"N	116°19'38"W	2730
Emida	Benewah	47°06'57"N	116°35'49"W	2850
Emigrant Crossing	Elmore	43°02'53"N	115°17'08"W	3496
Emmett (county seat)	Gem	43°52'25"N	116°29'54"W	2375
Empire City (historical)	Owyhee	42°58'42"N	116°42'37"W	7880
Enaville	Shoshone	47°33'45"N	116°14'57"W	2200
English point	Kootenai	47°47'12"N	116°42'32"W	2570
Enrose	Canyon	43°42'06"N	116°44'45"W	2325
Era	Butte	43°34'50"N	113°34'43"W	5940
Erlmo	Shoshone	47°15'35"N	116°02'57"W	2279
Estes	Latah	46°46'40"N	117°01'08"W	2629
Ethelton	Shoshone	47°15'18"N	115°54'17"W	2400
Eugene	Nez Perce	46°26'40"N	116°33'04"W	2940
Evans Landing	Bonner	48°04'32"N	116°31'50"W	2060
Evergreen	Adams	44°53'25"N	116°23'21"W	3816
Excelsior Beach	Kootenai	47°52'49"N	116°52'33"W	2340
Fairburn (historical)	Nez Perce	46°15'14"N	116°43'52"W	4200
Fairfield (county seat)	Camas	43°20'48"N	114°47'27"W	5065
Fairview	Franklin	42°00'48"N	111°52'33"W	4517
Fairview	Power	42°51'01"N	112°52'07"W	4394
Fairview	Twin Falls	42°32'06"N	114°47'21"W	3918
Fairylawn	Owyhee	42°34'11"N	116°59'18"W	4930
Falcon	Shoshone	47°20'56"N	115°40'26"W	3400
Fall Creek	Idaho	45°48'44"N	115°39'06"W	3510
Falls City	Jerome	42°40'50"N	114°25'24"W	3838
Farnum	Fremont	43°59'55"N	111°24'32"W	5500
Featherville	Elmore	43°36'36"N	115°15'26"W	4552
Felt	Teton	43°52'23"N	111°11'02"W	6045
Feltham	Washington	44°12'26"N	116°55'48"W	2120
Fenn	Idaho	45°57'48"N	116°15'19"W	3275
Ferdinand	Idaho	46°09'09"N	116°23'18"W	3730
Ferguson	Shoshone	47°33'36"N	115°55'24"W	
Fernan Lake Village	Kootenai	47°40'26"N	116°44'40"W	2153
Fernwood	Benewah	47°06'44"N	116°23'30"W	2740
Filer	Twin Falls	42°34'13"N	114°36'25"W	3765
Fingal	Bingham	42°58'05"N	112°48'32"W	4401
Fir Grove	Kootenai	47°53'40"N	116°54'33"W	2340

City, Town, or Historic Place	County	Latitude	Longitude	Elev.
Firth	Bingham	43°18'19"N	112°10'56"W	4555
Fischer	Canyon	43°37'10"N	116°34'58"W	2460
Fish Haven	Bear Lake	42°02'13"N	111°23'44"W	5951
Five Corners	Clearwater	46°38'24"N	116°10'20"W	3028
Flat Creek	Benewah	47°13'56"N	116°29'25"W	2998
Flat Rock	Fremont	44°30'02"N	111°20'08"W	6380
Fletcher	Lewis	46°17'22"N	116°24'36"W	3511
Flint	Owyhee	42°55'00"N	116°46'55"W	5220
Florence	Idaho	45°30'04"N	116°01'39"W	6080
Forebay	Nez Perce	46°25'25"N	116°57'33"W	780
Forest	Lewis	46°08'53"N	116°39'24"W	4514
Forney	Lemhi	44°59'59"N	114°20'05"W	5667
Fort Boise (historical)	Canyon	43°49'45"N	117°00'44"W	2193
Fort Hall	Bingham	43°02'00"N	112°26'15"W	4448
Fort Wilson	Payette	44°01'45"N	116°50'39"W	2188
Four Corners	Bonner	48°17'18"N	116°59'21"W	2393
Four Corners	Owyhee	42°06'19"N	116°36'25"W	5222
Fourway Junction	Idaho	45°19'58"N	115°21'44"W	7300
Fox Creek	Teton	43°39'02"N	111°06'36"W	6146
France	Fremont	43°58'21"N	111°16'28"W	5865
Franklin	Franklin	42°00'49"N	111°48'23"W	4504
Fraser	Clearwater	46°23'28"N	116°08'15"W	3100
Freedom	Caribou	42°58'58"N	111°02'41"W	5778
Freeze	Latah	46°57'43"N	116°56'30"W	2520
French (historical)	Payette	44°07'20"N	116°38'35"W	2486
French Corner	Payette	44°03'27"N	116°34'46"W	2475
French Creek	Idaho	45°25'24"N	116°01'35"W	1940
Frisco	Shoshone	47°30'44"N	115°51'30"W	3320
Fritser Ford	Valley	45°05'12"N	115°37'40"W	3400
Frost Place	Owyhee	42°11'33"N	115°56'24"W	5665
Fruitland	Payette	44°00'28"N	116°54'56"W	2225
Fruitvale	Adams	44°48'55"N	116°26'21"W	3083
Fuller	Gooding	42°55'46"N	114°49'21"W	3401
Fullmer	Bingham	43°18'31"N	112°40'08"W	4690
Gale	Madison	43°45'55"N	111°43'07"W	5268
Galena	Blaine	43°52'17"N	114°39'23"W	7300
Gannett	Blaine	43°21'36"N	114°10'30"W	4917
Garden City	Ada	43°37'20"N	116°14'14"W	2660
Garden City (historical)	Custer	44°30'27"N	114°13'26"W	5171
Garden Valley	Boise	44°05'24"N	115°57'04"W	3143
Gardena	Boise	43°58'33"N	116°11'24"W	2670
Garfield	Bonner	48°11'03"N	116°25'42"W	2160
Garfield	Jefferson	43°38'07"N	111°57'47"W	4818
Garrard Ranch	Cassia	42°19'00"N	114°00'03"W	4580
Garwood	Kootenai	47°49'55"N	116°46'35"W	2318
Gay	Bingham	43°02'54"N	112°07'03"W	5630
Gayway Corner	Payette	44°01'33"N	116°55'21"W	2195
Gem	Shoshone	47°30'30"N	115°52'01"W	3230

Cities, Towns, Historic Places

City, Town, or Historic Place	County	Latitude	Longitude	Elev.
Genesee	Latah	46°33'03"N	116°55'28"W	2675
Geneva	Bear Lake	42°21'31"N	111°03'52"W	6173
Gentry	Shoshone	47°28'27"N	115°53'39"W	2837
Georgetown	Bear Lake	42°28'56"N	111°22'12"W	5990
German Settlement	Nez Perce	46°25'07"N	116°30'03"W	3200
Gerrard	Bonneville	43°23'04"N	112°02'21"W	4655
Gerrit	Fremont	44°12'31"N	111°16'13"W	5961
Gibbonsville	Lemhi	45°33'20"N	113°55'20"W	4541
Gibbs	Kootenai	47°41'27"N	116°48'10"W	2130
Gibler	Idaho	45°44'17"N	115°23'19"W	4340
Gibson	Bingham	43°06'02"N	112°25'03"W	4465
Gifford	Nez Perce	46°26'36"N	116°33'20"W	2960
Gilmore	Lemhi	44°27'32"N	113°16'08"W	7160
Gimlet	Blaine	43°36'15"N	114°20'56"W	5575
Givens Hot Springs	Owyhee	43°25'20"N	116°42'50"W	2315
Giveout	Bear Lake	42°25'07"N	111°09'28"W	6847
Glencoe	Bear Lake	42°00'40"N	111°24'27"W	5930
Glendale	Adams	44°50'05"N	116°24'23"W	3360
Glendale	Franklin	42°07'48"N	111°42'15"W	5140
Glengary	Bonner	48°13'19"N	116°22'25"W	2080
Glenns Ferry	Elmore	42°57'18"N	115°18'00"W	2560
Glenwood	Idaho	46°14'39"N	115°49'55"W	3130
Godwin	Twin Falls	42°30'46"N	114°34'26"W	4035
Golconda	Shoshone	47°28'29"N	115°52'12"W	2940
Gold Creek	Shoshone	47°08'14"N	115°24'25"W	3375
Gold Point	Idaho	45°46'59"N	115°23'34"W	4270
Goldburg	Custer	44°23'08"N	113°38'42"W	6025
Golden	Idaho	45°48'44"N	115°40'44"W	3463
Good Grief	Boundary	48°56'58"N	116°10'35"W	2607
Gooding (county seat)	Gooding	42°56'20"N	114°42'44"W	3573
Goodrich	Adams	44°39'13"N	116°33'25"W	2751
Goshen	Bingham	43°18'35"N	112°04'57"W	
Goshen Junction	Bingham	43°18'30"N	112°10'43"W	4567
Grace	Caribou	42°34'34"N	111°43'47"W	5540
Graham (historical)	Boise	43°58'08"N	115°16'25"W	5800
Grainville	Fremont	44°01'38"N	111°22'02"W	5445
Grand Junction	Kootenai	47°43'45"N	116°58'22"W	2138
Grand View	Owyhee	42°59'23"N	116°05'33"W	2365
Grandjean	Custer	44°09'37"N	115°10'00"W	5180
Grandview	Bingham	43°03'11"N	112°47'15"W	4445
Grangeville (county seat)	Idaho	45°55'36"N	116°07'17"W	3670
Granite	Boise	43°56'27"N	115°58'00"W	4300
Granite	Bonner	48°05'16"N	116°25'29"W	2120
Granite	Bonner	48°00'38"N	116°40'10"W	2260
Grant	Jefferson	43°38'27"N	112°00'45"W	4785
Grasmere	Owyhee	42°22'36"N	115°52'54"W	5125
Gray	Bonneville	43°03'00"N	111°22'42"W	6402
Green	Idaho	45°27'08"N	115°56'45"W	2434

City, Town, or Historic Place	County	Latitude	Longitude	Elev.
Greencreek	Idaho	46°06'26"N	116°15'48"W	3183
Greenleaf	Canyon	43°40'17"N	116°48'55"W	2410
Greentimber (historical)	Fremont	44°05'07"N	111°16'28"W	5645
Greenwood	Jerome	42°34'36"N	114°02'55"W	4101
Greer	Clearwater	46°23'24"N	116°10'27"W	1100
Greys Landing	Twin Falls	42°07'58"N	114°43'42"W	5389
Grimes Pass	Boise	44°02'41"N	115°51'23"W	3270
Gross	Gem	44°18'48"N	116°18'24"W	3450
Grouse	Custer	43°41'18"N	113°36'42"W	6160
Groveland	Bingham	43°13'12"N	112°22'18"W	4490
Guffey	Owyhee	43°17'50"N	116°32'47"W	2270
Gurney	Nez Perce	46°25'35"N	116°57'03"W	780
Guyaz	Bonneville	43°28'54"N	111°38'45"W	5810
Gwenford	Oneida	42°08'11"N	112°19'45"W	4440
Haden (historical)	Teton	43°49'30"N	111°10'44"W	6005
Hagerman	Gooding	42°48'44"N	114°53'52"W	2959
Hahn	Lemhi	44°22'20"N	113°13'00"W	7140
Hailey (county seat)	Blaine	43°31'11"N	114°18'52"W	5329
Haley	Clearwater	46°29'43"N	115°55'28"W	2540
Hamer	Jefferson	43°55'38"N	112°12'19"W	4814
Hamilton Corner	Payette	43°57'10"N	116°46'17"W	2277
Hammett	Elmore	42°56'45"N	115°27'55"W	2531
Hampton	Latah	46°54'52"N	116°50'54"W	2540
Hand Place	Twin Falls	42°08'41"N	114°58'07"W	5910
Hansberg	Clearwater	46°32'17"N	116°08'24"W	2680
Hansen	Twin Falls	42°31'47"N	114°18'19"W	4025
Hardscrabble Campground	Boise	44°14'23"N	115°53'54"W	3220
Harer	Bear Lake	42°11'22"N	111°10'15"W	6000
Harlem	Bonner	48°02'21"N	116°48'41"W	2175
Harpster	Idaho	45°59'12"N	115°57'45"W	1600
Harrell Place	Twin Falls	42°04'39"N	114°47'07"W	5190
Harris Landing	Kootenai	47°47'06"N	116°42'06"W	2260
Harrisburg	Idaho	46°17'49"N	116°01'29"W	3100
Harrison	Kootenai	47°27'16"N	116°47'04"W	2200
Hart	Madison	43°53'03"N	111°44'18"W	4902
Harvard	Latah	46°55'03"N	116°43'43"W	2585
Harvey Place	Owyhee	42°30'46"N	116°04'39"W	5350
Hatch	Caribou	42°49'10"N	111°51'05"W	5500
Hatwai	Nez Perce	46°26'11"N	116°58'02"W	785
Hauser	Kootenai	47°46'23"N	117°01'37"W	2220
Havens	Bingham	43°16'14"N	112°35'24"W	4590
Hawgood	Jefferson	43°52'11"N	112°09'02"W	4820
Hawkins	Bannock	42°32'37"N	112°20'29"W	5278
Hawley	Blaine	42°44'03"N	113°23'42"W	4342
Hawleys Landing	Benewah	47°21'21"N	116°46'00"W	2140
Haycrop	Kootenai	47°45'59"N	116°53'34"W	2246
Hayden	Kootenai	47°45'58"N	116°47'08"W	2278
Hayden Lake	Kootenai	47°45'32"N	116°45'21"W	2300

City, Town, or Historic Place	County	Latitude	Longitude	Elev.
Haytown (historical)	Jerome	42°41'34"N	114°30'05"W	3771
Hazel (historical)	Cassia	42°23'28"N	113°45'55"W	4318
Hazelton	Jerome	42°35'47"N	114°08'07"W	3955
Headquarters	Clearwater	46°37'48"N	115°48'30"W	3140
Heath	Washington	44°45'22"N	116°52'52"W	2900
Heath (historical)	Washington	44°45'07"N	116°46'07"W	7460
Heglar	Cassia	42°28'25"N	113°08'47"W	4726
Heise	Jefferson	43°38'31"N	111°41'01"W	4998
Helena	Adams	45°10'13"N	116°39'05"W	6660
Helmer	Latah	46°48'03"N	116°28'09"W	2830
Heman	Fremont	43°57'23"N	111°47'48"W	4906
Henry	Caribou	42°54'25"N	111°31'48"W	6132
Herbert	Madison	43°42'23"N	111°39'33"W	5593
Herman	Bonneville	43°08'26"N	111°25'48"W	6430
Herrick	Shoshone	47°16'22"N	116°06'20"W	2223
Heyburn	Minidoka	42°33'31"N	113°45'47"W	4152
Hibbard	Madison	43°51'19"N	111°50'16"W	4850
High Valley	Valley	44°13'27"N	116°09'14"W	4859
Highbridge	Clark	44°17'00"N	112°12'42"W	5540
Highlands	Ada	43°38'46"N	116°11'34"W	2940
Hill City	Camas	43°18'02"N	115°03'01"W	5092
Hillcrest	Ada	43°33'49"N	116°10'55"W	2850
Hillsdale (historical)	Adams	44°30'38"N	116°25'27"W	3200
Hillsdale (historical)	Jerome	42°34'33"N	114°12'54"W	4050
Hinckley	Madison	43°51'54"N	111°51'27"W	4830
Hobson	Cassia	42°31'55"N	113°55'23"W	4145
Hoelzle Place	Twin Falls	42°12'02"N	114°57'08"W	5370
Holbrook	Oneida	42°09'43"N	112°39'11"W	4779
Holbrook Summit	Oneida	42°09'55"N	112°27'10"W	5812
Hollister	Twin Falls	42°21'12"N	114°34'27"W	4515
Hollywood	Clearwater	46°33'19"N	115°50'08"W	8295
Homedale	Owyhee	43°37'04"N	116°55'58"W	2237
Hoover	Adams	44°45'09"N	116°26'26"W	2962
Hope	Bonner	48°14'52"N	116°18'22"W	2230
Horsecamp	Shoshone	47°11'29"N	115°52'51"W	4200
Horseshoe Bend	Boise	43°54'53"N	116°11'49"W	2630
Hot Springs	Owyhee	42°47'31"N	115°43'01"W	2592
Houston	Custer	43°52'59"N	113°34'33"W	5840
Howard	Lewis	46°16'35"N	116°17'50"W	3120
Howe	Butte	43°47'01"N	113°00'14"W	4824
Howell	Latah	46°44'09"N	116°48'57"W	2740
Howell (historical)	Boise	43°47'34"N	116°15'30"W	3360
Howelltown	Kootenai	47°47'09"N	116°58'48"W	2403
Hoyt	Shoshone	47°15'11"N	115°55'02"W	2380
Huetter	Kootenai	47°42'04"N	116°51'01"W	2133
Humphrey	Clark	44°29'18"N	112°13'58"W	6500
Hunt	Jerome	42°40'42"N	114°14'57"W	3955
Huston	Canyon	43°36'37"N	116°46'56"W	2515

City, Town, or Historic Place	County	Latitude	Longitude	Elev.
Hydra	Jerome	42°41'58"N	114°31'05"W	3738
Hynes	Minidoka	42°36'44"N	113°53'27"W	4177
Idaho City (county seat)	Boise	43°49'43"N	115°50'01"W	3906
Idaho Falls (county seat)	Bonneville	43°28'00"N	112°02'00"W	4700
Idahome	Cassia	42°24'56"N	113°23'56"W	4422
Idavada	Twin Falls	41°59'50"N	114°38'29"W	5346
Idmon	Clark	44°21'39"N	111°54'39"W	6279
Ilo (historical)	Lewis	46°14'18"N	116°29'48"W	3800
Independence (historical)	Madison	43°46'35"N	111°54'58"W	4819
Indian Grove	Cassia	42°06'46"N	113°43'39"W	7540
Indian Head Rock	Camas	43°33'19"N	114°48'02"W	6587
Indian Jim Place	Twin Falls	42°10'04"N	114°54'46"W	4575
Indian Valley	Adams	44°33'26"N	116°25'59"W	3002
Ingard	Payette	44°02'01"N	116°55'20"W	2190
Inkom	Bannock	42°47'47"N	112°15'12"W	4547
Iona	Bonneville	43°31'35"N	111°55'56"W	4782
Ireland Springs	Oneida	42°09'54"N	112°29'41"W	5300
Irwin	Bonneville	43°24'31"N	111°17'55"W	5325
Isabella Landing	Clearwater	46°50'59"N	115°37'41"W	1680
Island Park	Fremont	44°25'28"N	111°22'13"W	6290
Iversons	Cassia	42°18'17"N	114°00'36"W	4660
Jackson	Cassia	42°38'02"N	113°34'11"W	4160
Jacques	Nez Perce	46°22'18"N	116°43'35"W	1380
James Place	Owyhee	42°16'37"N	115°46'45"W	4825
Jamestown (historical)	Latah	46°59'49"N	117°01'57"W	2616
Jaype	Clearwater	46°31'49"N	115°49'44"W	3220
Jenkins Will	Fremont	44°04'36"N	111°57'35"W	5405
Jenness	Gem	43°49'56"N	116°37'27"W	2650
Jensen	Madison	43°44'34"N	111°44'47"W	5206
Jerome (county seat)	Jerome	42°43'27"N	114°31'04"W	3765
Jim Moore Place	Idaho	45°29'11"N	115°20'18"W	2640
Joel	Latah	46°42'38"N	116°52'37"W	2600
Johnson	Clearwater	46°34'22"N	116°05'07"W	3220
Johnson	Clearwater	46°25'37"N	115°54'56"W	3160
Johnsons Mill	Clearwater	46°25'38"N	115°55'02"W	3160
Jolley	Madison	43°50'32"N	111°46'38"W	4872
Jonathan	Washington	44°16'03"N	117°03'09"W	2115
Jones Crossing	Clark	44°05'13"N	112°13'07"W	4919
Joseph	Idaho	45°47'50"N	116°28'37"W	4420
Joseph	Nez Perce	46°26'52"N	116°49'33"W	807
Josephson	Canyon	43°44'56"N	116°38'15"W	2455
Judge Town	Clearwater	46°28'32"N	115°47'46"W	3090
Judkins	Teton	43°55'19"N	111°08'56"W	6025
Juliaetta	Latah	46°34'44"N	116°42'18"W	1155
Junction	Lemhi	44°42'03"N	113°22'11"W	5920
Juniper	Oneida	42°09'12"N	112°58'09"W	5055
Kameron	Clearwater	46°46'21"N	116°15'11"W	2660
Kamiah	Lewis	46°13'38"N	116°01'42"W	1263

Cities, Towns, Historic Places

City, Town, or Historic Place	County	Latitude	Longitude	Elev.
Karcher Junction	Canyon	43°35'10"N	116°36'07"W	2455
Katka	Boundary	48°41'23"N	116°08'04"W	6208
Kaufman (historical)	Clark	44°09'37"N	112°54'38"W	6085
Keeler	Shoshone	46°57'26"N	116°18'25"W	2961
Keenan City	Bonneville	43°08'27"N	111°20'22"W	6530
Kellogg	Shoshone	47°32'18"N	116°07'06"W	2310
Kelso	Bonner	48°00'43"N	116°42'20"W	2140
Kelso	Cassia	42°04'07"N	113°18'47"W	4875
Kendrick	Latah	46°36'51"N	116°38'44"W	1230
Kenyon	Cassia	42°24'56"N	113°52'10"W	4306
Keogh	Cassia	42°13'24"N	113°19'58"W	4590
Ketchum	Blaine	43°40'51"N	114°21'46"W	2840
Keuterville	Idaho	46°02'03"N	116°26'26"W	4170
Kilgore	Clark	44°24'08"N	111°53'35"W	6325
Kimama	Lincoln	42°50'17"N	113°47'41"W	4269
Kimball	Bingham	43°16'25"N	112°13'07"W	4565
Kimberly	Twin Falls	42°32'02"N	114°21'50"W	3921
King Hill	Elmore	43°00'15"N	115°12'11"W	2533
Kings Corner	Canyon	43°31'05"N	116°32'01"W	2500
Kingston	Shoshone	47°32'57"N	116°16'11"W	2180
Kinney Point Summer Home Area	Valley	44°38'19"N	115°40'25"W	5404
Kinport	Caribou	42°45'16"N	111°56'27"W	5375
Kippen	Lewis	46°18'40"N	116°32'55"W	3555
Knowlton Heights	Canyon	43°32'43"N	116°46'19"W	2367
Knull	Twin Falls	42°31'35"N	114°33'16"W	3974
Konkolville	Clearwater	46°29'44"N	116°11'49"W	1448
Kooskia	Idaho	46°08'42"N	115°58'37"W	1257
Kootenai	Bonner	48°18'37"N	116°30'45"W	2117
Kuna	Ada	43°29'31"N	116°25'09"W	2695
Kyle (historical)	Shoshone	47°19'41"N	115°45'24"W	3260
Labelle	Jefferson	43°42'12"N	111°51'36"W	4880
Laclede	Bonner	48°10'13"N	116°45'18"W	2100
Lacon	Kootenai	47°26'21"N	116°44'39"W	2782
Lago	Caribou	42°27'05"N	111°41'45"W	5100
Lake	Fremont	44°40'01"N	111°23'34"W	6500
Lake Fork	Valley	44°49'58"N	116°05'02"W	4975
Lakeshore Summer Home Area	Valley	44°38'46"N	115°39'39"W	5321
Lakeview	Bonner	47°58'13"N	116°26'44"W	2266
Lamb Creek	Bonner	48°30'59"N	116°55'35"W	2555
Lamont	Fremont	43°58'11"N	111°12'55"W	6030
Lanark	Bear Lake	42°16'54"N	111°25'40"W	5953
Landmark	Valley	44°39'23"N	115°32'39"W	6620
Landore	Adams	45°07'24"N	116°37'39"W	5340
Lane	Kootenai	47°30'25"N	116°32'07"W	2160
Lapwai	Nez Perce	46°24'18"N	116°48'14"W	940
Larson	Shoshone	47°28'15"N	115°44'44"W	3500
Last Chance	Fremont	44°22'00"N	111°24'05"W	6155
Lava Hot Springs	Bannock	42°37'10"N	112°00'37"W	5040

City, Town, or Historic Place	County	Latitude	Longitude	Elev.
Lawman Ford	Valley	45°06'36"N	115°36'32"W	3320
Leadore	Lemhi	44°40'49"N	113°21'26"W	5980
Leadville	Lemhi	44°42'21"N	113°18'26"W	6400
Leduc (historical)	Blaine	43°19'45"N	114°03'54"W	4960
Leesburg	Lemhi	45°13'26"N	114°06'47"W	6680
Leland	Nez Perce	46°34'41"N	116°36'21"W	2240
Lemhi	Lemhi	44°51'06"N	113°37'08"W	5180
Len Landing	Kootenai	47°31'45"N	116°50'18"W	2200
Lenore	Nez Perce	46°30'31"N	116°33'00"W	1000
Lenville	Latah	46°38'18"N	116°50'22"W	2370
Leon	Nez Perce	46°31'55"N	117°02'21"W	2577
Leone	Ada	43°24'28"N	116°05'34"W	3230
Leonia	Boundary	48°37'00"N	116°02'55"W	2000
Leprouse	Idaho	45°44'37"N	115°23'22"W	4363
Leslie	Custer	43°51'57"N	113°28'00"W	5700
Letha	Gem	43°53'40"N	116°38'48"W	2285
Level (historical)	Jefferson	43°53'45"N	112°29'28"W	4783
Lewiston (county seat)	Nez Perce	46°25'00"N	117°01'00"W	755
Lewiston Orchards	Nez Perce	46°22'50"N	116°58'28"W	1480
Lewisville	Jefferson	43°41'45"N	112°00'35"W	4795
Liberty	Bear Lake	42°19'02"N	111°27'05"W	5980
Liberty	Bingham	43°10'08"N	112°33'16"W	4459
Lidy Hot Springs	Clark	44°07'35"N	112°32'56"W	4978
Lidyville (historical)	Latah	46°35'44"N	116°48'01"W	2390
Lifton	Bear Lake	42°07'24"N	111°18'45"W	5927
Lightfoot	Idaho	45°43'07"N	115°21'36"W	4397
Lignite	Bonner	48°13'44"N	116°32'24"W	2060
Lincoln	Bonneville	43°30'47"N	111°57'49"W	4754
Linden	Latah	46°40'15"N	116°29'50"W	2720
Lindsayville	Bear Lake	42°25'48"N	111°25'17"W	5945
Linfor	Shoshone	47°36'30"N	116°14'05"W	2199
Little Rock	Gem	43°55'01"N	116°41'34"W	2257
Little Sugar Loaf	Owyhee	43°05'46"N	116°42'13"W	6450
Lodi (historical)	Fremont	44°05'23"N	111°27'38"W	5205
Lone Pine	Clark	44°11'10"N	112°56'03"W	6210
Lone Rock	Power	42°46'03"N	112°49'13"W	4809
Long Creek Summit Home	Boise	44°07'58"N	115°34'38"W	4682
Lorenzo	Jefferson	43°43'42"N	111°52'14"W	4862
Lost River	Butte	43°41'21"N	113°22'14"W	5422
Lotus	Benewah	47°13'48"N	116°36'59"W	2204
Lowell	Idaho	46°08'39"N	115°35'41"W	1480
Lower Stanley	Custer	44°13'37"N	114°55'40"W	6210
Lowman	Boise	44°05'01"N	115°37'11"W	3820
Lucile	Idaho	45°32'07"N	116°18'34"W	1640
Lund	Caribou	42°38'34"N	111°53'16"W	5485
Lyman	Madison	43°44'21"N	111°49'01"W	4880
Lyondale (historical)	Kootenai	47°36'28"N	116°51'00"W	2160
Mace	Shoshone	47°31'06"N	115°49'16"W	3770

City, Town, or Historic Place	County	Latitude	Longitude	Elev.
MacGregor (historical)	Valley	44°41'31"N	116°04'44"W	4855
Mackay	Custer	43°54'53"N	113°36'45"W	5900
Mackay Bar	Idaho	45°22'34"N	115°30'19"W	2180
Macks Inn PO	Fremont	44°29'57"N	111°20'13"W	6405
Macon	Camas	43°19'36"N	114°32'39"W	4995
Maddens	Canyon	43°39'43"N	116°35'49"W	2439
Magic	Camas	43°15'19"N	114°23'09"W	4822
Magic City	Blaine	43°17'26"N	114°21'53"W	4820
Magic Resort	Blaine	43°16'57"N	114°23'20"W	4800
Malad City (county seat)	Oneida	42°11'30"N	112°15'00"W	4600
Malta	Cassia	42°18'23"N	113°22'06"W	4525
Manson	Bear Lake	42°34'11"N	111°29'53"W	5831
Mapleton	Franklin	42°05'02"N	111°44'53"W	4691
Marble Creek	Shoshone	47°15'10"N	116°01'51"W	2301
Marion	Cassia	42°17'12"N	113°54'37"W	4477
Mark	Madison	43°46'31"N	111°49'52"W	4854
Marley	Lincoln	43°00'56"N	114°13'49"W	4188
Marsh (historical)	Gem	43°55'07"N	116°20'20"W	2520
Marsh Valley	Bannock	42°31'04"N	112°09'18"W	4896
Marsing	Owyhee	43°32'44"N	116°48'44"W	2280
Martin	Butte	43°31'08"N	113°33'59"W	5640
Marysville	Fremont	44°04'17"N	111°25'10"W	5285
Mashburn	Benewah	47°10'45"N	116°29'42"W	2580
Masonia (historical)	Shoshone	47°27'06"N	116°10'47"W	2780
May	Lemhi	44°36'16"N	113°54'40"W	5068
May Place	Twin Falls	42°12'04"N	114°53'23"W	4230
Mayfield	Elmore	43°25'05"N	115°54'02"W	3580
Mc Arthur	Boundary	48°31'25"N	116°25'57"W	2170
McCall	Valley	44°54'40"N	116°05'52"W	5031
McCammon	Bannock	42°39'02"N	112°11'32"W	4768
McCarthy	Shoshone	47°29'35"N	115°54'32"W	2940
McCrea Place	Idaho	45°18'14"N	116°28'35"W	4015
McDonalds	Shoshone	47°02'04"N	115°09'05"W	5000
McDonaldville	Bingham	43°15'36"N	112°22'27"W	4500
McGuire	Kootenai	47°42'45"N	116°59'25"W	2130
McHenry	Jerome	42°35'44"N	114°03'05"W	4080
McMillan	Twin Falls	42°32'36"N	114°25'14"W	3810
Meadow Creek	Boundary	48°49'12"N	116°09'21"W	2406
Meadowhurst	Benewah	47°19'20"N	116°33'37"W	2125
Meadows	Adams	44°57'40"N	116°14'34"W	3981
Meadowville	Caribou	42°45'49"N	111°36'11"W	6080
Medimont	Kootenai	47°28'34"N	116°36'12"W	2140
Melba	Canyon	43°22'32"N	116°31'41"W	2680
Melrose	Nez Perce	46°24'30"N	116°27'13"W	3184
Menan	Jefferson	43°43'15"N	111°59'21"W	4800
Meridian	Ada	43°36'44"N	116°23'26"W	2600
Mesa	Adams	44°37'44"N	116°26'59"W	3243
Meteor	Owyhee	42°06'47"N	114°40'48"W	5370
Mexican Place	Idaho	45°48'46"N	116°30'20"W	4100
Mica	Kootenai	47°37'14"N	116°52'13"W	2575
Michaud	Power	42°53'25"N	112°36'49"W	4440
Midasville	Idaho	45°33'37"N	115°28'09"W	5760
Middleton	Canyon	43°42'25"N	116°37'09"W	2398
Midnight	Valley	44°55'23"N	115°19'47"W	6320
Midvale	Washington	44°28'17"N	116°44'01"W	2545
Midway	Canyon	43°37'32"N	116°37'53"W	2422
Midway	Jefferson	43°42'25"N	112°00'15"W	4797
Miller Creek Settlement	Owyhee	42°04'54"N	116°07'38"W	5320
Milltown	Benewah	47°20'10"N	116°33'10"W	2140
Milltown	Benewah	47°19'27"N	116°32'32"W	2180
Milner	Twin Falls	42°31'07"N	114°01'19"W	4155
Milo	Bonneville	43°36'39"N	111°52'59"W	4872
Mineral	Washington	44°33'55"N	117°04'36"W	3080
Minidoka	Minidoka	42°45'14"N	113°29'22"W	4286
Mink Creek	Franklin	42°13'42"N	111°42'54"W	5140
Mitchell	Bingham	43°23'50"N	112°06'36"W	4637
Mohler	Lewis	46°17'23"N	116°20'43"W	3230
Monteview	Jefferson	43°58'19"N	112°32'08"W	4789
Montour	Gem	43°55'30"N	116°19'40"W	2510
Montpelier	Bear Lake	42°19'20"N	111°17'49"W	5964
Moody	Madison	43°50'13"N	111°38'03"W	5145
Moore	Butte	43°44'09"N	113°21'56"W	5473
Moose City	Clearwater	46°47'13"N	115°06'27"W	4120
Mora	Ada	43°27'35"N	116°21'08"W	2762
Moravia	Boundary	48°38'54"N	116°22'41"W	1842
Moreland	Bingham	43°13'22"N	112°26'30"W	4460
Morgan	Bingham	43°03'38"N	111°55'36"W	5744
Morgan Place	Owyhee	43°06'29"N	116°16'26"W	2331
Morrow	Lewis	46°07'37"N	116°31'21"W	4105
Morton	Bonner	48°12'05"N	116°41'29"W	2080
Moscow (county seat)	Latah	46°43'57"N	116°59'57"W	2583
Moss	Canyon	43°36'57"N	116°37'09"W	2430
Mound Valley	Franklin	42°21'26"N	111°42'42"W	4940
Mount Idaho	Idaho	45°54'14"N	116°04'52"W	3670
Mountain Home (county seat)	Elmore	43°07'59"N	115°41'25"W	3143
Mountain Home (historical)	Latah	47°01'09"N	116°55'38"W	2740
Mowry	Benewah	47°18'34"N	116°59'42"W	2736
Moyie Springs	Boundary	48°43'35"N	116°11'15"W	2204
Mozart	Kootenai	47°24'40"N	116°56'09"W	2600
Mud Lake	Jefferson	43°50'29"N	112°28'31"W	4787
Mud Springs	Idaho	45°51'49"N	115°35'07"W	4820
Mullan	Shoshone	47°28'13"N	115°48'03"W	3277
Murphy (county seat)	Owyhee	43°13'06"N	116°33'05"W	2823
Murphy Hot Springs	Owyhee	42°01'49"N	115°22'05"W	5150
Murray	Shoshone	47°37'38"N	115°51'27"W	2755
Murtaugh	Twin Falls	42°29'33"N	114°09'41"W	4082

Cities, Towns, Historic Places

City, Town, or Historic Place	County	Latitude	Longitude	Elev.
Musselshell	Clearwater	46°21'26"N	115°44'34"W	3187
Myers	Minidoka	42°36'32"N	113°45'14"W	4150
Myrtle	Nez Perce	46°29'49"N	116°43'30"W	860
Naf	Cassia	42°00'37"N	113°17'05"W	5290
Nampa	Canyon	43°32'27"N	116°33'45"W	2490
Naples	Boundary	48°34'16"N	116°23'29"W	2038
Nashville	Franklin	42°02'58"N	111°48'28"W	4597
Neal (historical)	Ada	43°28'59"N	115°59'50"W	3680
Neeley	Power	42°44'01"N	112°54'51"W	4317
Nelson (historical)	Oneida	42°04'51"N	112°39'20"W	4672
Nelson (historical)	Shoshone	47°39'15"N	116°05'20"W	2300
Neva	Clearwater	46°47'55"N	116°17'48"W	3200
New Centerville	Boise	43°52'52"N	115°54'33"W	4122
New Meadows	Adams	44°58'17"N	116°16'59"W	3868
New Plymouth	Payette	43°58'12"N	116°49'05"W	2250
Newdale	Fremont	43°53'00"N	111°36'20"W	5080
Newman	Bonner	48°12'28"N	116°26'03"W	2296
Newsome	Idaho	45°54'27"N	115°37'45"W	4020
Nezperce (county seat)	Lewis	46°14'06"N	116°14'23"W	3202
Nicholia	Lemhi	44°21'36"N	113°00'39"W	6920
Niter	Caribou	42°30'13"N	111°43'48"W	5405
Nora	Latah	46°45'50"N	116°42'03"W	2560
Nordman	Bonner	48°38'02"N	116°56'41"W	269
Norland	Minidoka	42°47'41"N	113°40'21"W	4295
North Fork	Lemhi	45°24'22"N	113°59'35"W	3620
North Kenyon	Cassia	42°25'56"N	113°51'37"W	4290
North Lapwai	Nez Perce	46°26'38"N	116°50'06"W	800
North Lewiston	Nez Perce	46°25'29"N	117°00'11"W	735
North Pole	Kootenai	47°54'16"N	116°44'28"W	2389
North South Ski Bowl	Benewah	47°07'00"N	116°39'32"W	3740
Norwood	Valley	44°48'31"N	116°06'08"W	4915
Notus	Canyon	43°43'32"N	116°48'01"W	2310
Nounan	Bear Lake	42°28'40"N	111°27'01"W	5970
Novene	Bear Lake	42°29'50"N	111°24'45"W	5885
Nuckols	Shoshone	47°31'08"N	115°56'13"W	5304
Oakley	Cassia	42°14'36"N	113°52'52"W	4584
Obsidian	Custer	44°04'44"N	114°50'47"W	6620
Oden	Bonner	48°19'20"N	116°24'42"W	2121
Ola	Gem	44°10'42"N	116°17'30"W	3010
Old Beaver	Clark	44°24'38"N	112°11'47"W	6020
Old Fort	Valley	44°48'42"N	116°09'24"W	4850
Old Golden	Idaho	45°47'21"N	115°39'10"W	4420
Old Williamsburg (historical)	Caribou	42°56'34"N	111°20'23"W	6700
Oldtown	Bonner	48°10'49"N	117°02'34"W	2180
Olsen	Bingham	43°17'07"N	112°37'16"W	4615
Olsons	Shoshone	47°00'10"N	115°07'45"W	5200
Omega	Benewah	47°20'02"N	116°28'39"W	2140
Omill	Clearwater	46°29'38"N	115°55'33"W	2600
Onaway	Latah	46°55'42"N	116°53'24"W	2640
Ora	Fremont	44°05'10"N	111°31'57"W	5175
Orchard	Ada	43°18'53"N	116°01'31"W	3150
Orchard Homes (historical)	Latah	46°35'48"N	116°43'44"W	1400
Oreana	Owyhee	43°03'13"N	116°23'39"W	2800
Oro Fino (historical)	Clearwater	46°28'56"N	115°47'48"W	3120
Orofino (county seat)	Clearwater	46°28'46"N	116°15'15"W	1020
Orogrande	Idaho	45°42'20"N	115°32'33"W	4580
Orvin	Bonneville	43°32'00"N	112°00'31"W	4750
Osburn	Shoshone	47°30'22"N	115°59'54"W	2529
Osgood	Bonneville	43°34'13"N	112°06'08"W	4788
Outlet Bay	Bonner	48°29'40"N	116°53'32"W	2480
Ovid	Bear Lake	42°17'20"N	111°23'51"W	5935
Owinza	Lincoln	42°53'58"N	114°03'21"W	4200
Owyhee	Ada	43°25'06"N	116°11'59"W	2963
Owyhee Heights	Owyhee	43°37'29"N	116°59'23"W	2387
Oxford	Franklin	42°15'32"N	112°01'14"W	4798
Pagari	Lincoln	43°07'05"N	114°04'19"W	4446
Page	Shoshone	47°31'58"N	116°12'10"W	2470
Palisades Corner	Payette	43°58'31"N	116°54'42"W	2279
Palisades Townsite	Bonneville	43°21'07"N	111°13'01"W	5380
Paradise Hot Springs	Elmore	43°33'14"N	115°16'22"W	4420
Paradise Valley Summer Home Area	Valley	44°40'11"N	115°39'29"W	5358
Pardee	Idaho	46°17'55"N	116°07'28"W	1140
Pardee Corner	Idaho	46°18'53"N	116°05'45"W	2942
Paris (county seat)	Bear Lake	42°13'38"N	111°24'01"W	5968
Parker	Fremont	43°57'34"N	111°45'25"W	4924
Parkinson	Madison	43°48'58"N	111°39'30"W	5190
Parma	Canyon	43°47'07"N	116°56'32"W	2240
Patterson	Lemhi	44°31'25"N	113°42'41"W	6000
Paul	Minidoka	42°36'29"N	113°46'57"W	4147
Pauline	Power	42°34'15"N	112°33'34"W	5020
Payette (county seat)	Payette	44°04'42"N	116°55'58"W	2150
Payette Heights	Payette	44°04'19"N	116°54'50"W	2250
Payne	Bonneville	43°34'22"N	112°03'09"W	4743
Pearl	Gem	43°51'21"N	116°18'59"W	4100
Pearson	Shoshone	47°21'16"N	115°43'44"W	3193
Peavey	Twin Falls	42°35'01"N	114°39'07"W	3740
Pebble	Caribou	42°44'39"N	111°59'59"W	5280
Peck	Clearwater	46°29'56"N	116°25'57"W	960
Peck	Nez Perce	46°28'31"N	116°25'24"W	1140
Pedee	Benewah	47°21'04"N	116°46'26"W	2320
Pegram	Bear Lake	42°08'34"N	111°07'41"W	6030
Pella	Cassia	42°28'31"N	113°50'16"W	4230
Perkins	Ada	43°36'18"N	116°16'22"W	2700
Perrine	Jerome	42°39'14"N	114°18'58"W	3870
Peterson (historical)	Bingham	43°14'24"N	112°22'31"W	4497
Petersons Crossing	Owyhee	43°07'21"N	116°52'32"W	6550

City, Town, or Historic Place	County	Latitude	Longitude	Elev.
Picabo	Blaine	43°18'21"N	114°04'01"W	4834
Pierce	Clearwater	46°29'28"N	115°47'53"W	3087
Pilgrim Stage Station Historic Site	Elmore	42°52'44"N	115°07'51"W	2963
Pine	Elmore	43°29'03"N	115°18'40"W	4218
Pine Creek	Shoshone	47°32'55"N	116°13'17"W	2190
Pine Ridge	Adams	44°56'20"N	116°22'54"W	4120
Pinehurst	Adams	45°15'10"N	116°20'00"W	2700
Pinehurst	Shoshone	47°32'20"N	116°14'11"W	2220
Pineview	Fremont	44°17'11"N	111°18'36"W	6130
Pingree	Blaine	43°07'00"N	112°35'52"W	4450
Pioneerville	Boise	43°58'08"N	115°50'45"W	4420
Placerville	Boise	43°56'36"N	115°56'46"W	4300
Plano	Madison	43°53'28"N	111°53'27"W	4845
Player Place	Twin Falls	42°01'12"N	114°49'00"W	5630
Plaza	Gem	43°54'41"N	116°25'55"W	2480
Pleasant Valley	Ada	43°26'41"N	116°14'46"W	2930
Pleasant Valley Place	Owyhee	42°32'26"N	116°50'11"W	5470
Pleasant View	Kootenai	47°41'13"N	117°00'41"W	2200
Pleasantview	Oneida	42°09'22"N	112°20'13"W	4485
Plummer	Benewah	47°20'07"N	116°53'15"W	2722
Plummer Junction	Benewah	47°20'24"N	116°52'07"W	2170
Pocatello (county seat)	Bannock	42°52'17"N	112°26'41"W	4464
Pocono	Shoshone	47°14'48"N	116°02'52"W	2280
Pollock	Idaho	45°18'45"N	116°21'29"W	2352
Ponderay	Bonner	48°18'20"N	116°31'58"W	2120
Poplar	Bonneville	43°37'13"N	111°41'06"W	5025
Porthill	Boundary	48°59'54"N	116°29'49"W	1840
Portneuf	Bannock	42°47'42"N	112°21'46"W	4494
Post Falls	Kootenai	47°43'05"N	116°57'02"W	2200
Potlatch	Latah	46°55'18"N	116°53'50"W	2600
Potlatch Junction	Latah	46°55'46"N	116°55'57"W	2484
Potter Place	Idaho	45°18'28"N	116°26'23"W	3810
Pottsville (historical)	Shoshone	47°28'00"N	115°42'19"W	3800
Powell Junction	Idaho	46°34'45"N	114°43'04"W	5872
Prairie	Elmore	43°30'18"N	115°34'23"W	4780
Preston (county seat)	Franklin	42°05'47"N	111°52'33"W	4716
Prichard	Shoshone	47°39'23"N	115°58'31"W	2405
Prichard Y	Shoshone	47°38'24"N	115°58'42"W	2409
Priest River	Bonner	48°10'51"N	117°02'09"W	2077
Princeton	Latah	46°54'51"N	116°49'59"W	2500
Punkin Corner	Camas	43°18'48"N	114°54'33"W	5062
Pyke	Fremont	43°57'34"N	111°43'24"W	4947
Quartzburg	Boise	43°57'40"N	115°59'15"W	4380
Quigley	Power	42°44'18"N	113°04'09"W	4449
Raft River	Cassia	42°35'50"N	113°14'07"W	4215
Ramey	Payette	44°00'10"N	116°56'13"W	2145
Ramsdell	Benewah	47°21'10"N	116°40'32"W	2160
Ramsey	Kootenai	47°51'52"N	116°48'26"W	2322

City, Town, or Historic Place	County	Latitude	Longitude	Elev.
Rands	Camas	43°20'40"N	114°39'06"W	5750
Rankin Mill	Adams	45°15'59"N	116°32'41"W	6000
Rathdrum	Kootenai	47°48'45"N	116°53'44"W	2220
Raymond	Bear Lake	42°16'29"N	111°03'29"W	6140
Rea	Fremont	44°28'11"N	111°26'04"W	6370
Rebecca	Washington	44°14'27"N	116°51'25"W	2160
Reclamation Village	Elmore	43°20'08"N	115°29'16"W	3830
Red River Hot Springs	Idaho	45°47'16"N	115°11'57"W	4960
Red Rock Junction	Bannock	42°21'16"N	112°02'57"W	4785
Regina	Ada	43°23'25"N	115°59'28"W	3340
Renfrew	Benewah	47°08'39"N	116°26'15"W	2705
Reno	Lemhi	44°16'16"N	112°58'49"W	6535
Reubens	Lewis	46°19'20"N	116°32'30"W	3514
Reverse	Elmore	43°01'33"N	115°36'21"W	3099
Rexburg (county seat)	Madison	43°49'34"N	111°47'20"W	4865
Reynolds	Owyhee	43°12'05"N	116°44'36"W	3920
Richfield	Lincoln	43°02'55"N	114°09'17"W	4324
Rickard Crossing	Owyhee	42°15'32"N	116°40'54"W	4450
Riddle	Owyhee	42°11'13"N	116°06'34"W	5367
Ridgedale	Oneida	42°01'23"N	112°28'45"W	5010
Rigby (county seat)	Jefferson	43°40'21"N	111°54'51"W	4850
Riggins	Idaho	45°25'20"N	116°18'52"W	1800
Riggins Hot Springs	Idaho	45°25'01"N	116°10'15"W	1840
Ririe	Jefferson	43°37'55"N	111°46'22"W	4962
Rising River	Bingham	43°19'26"N	112°19'51"W	4565
Ritz	Boundary	48°46'44"N	116°22'07"W	1793
Riverdale	Benewah	47°19'47"N	116°33'17"W	2115
Riverdale	Franklin	42°09'41"N	111°50'13"W	4550
Riverside	Bingham	43°11'49"N	112°26'04"W	4455
Riverside	Clearwater	46°29'44"N	116°17'47"W	996
Riverside	Owyhee	43°30'07"N	116°45'57"W	2309
Riverside (historical)	Bonner	48°10'44"N	116°45'20"W	2180
Roberts	Jefferson	43°43'13"N	112°07'32"W	4775
Robin	Bannock	42°34'38"N	112°14'45"W	4872
Robinson Bar	Custer	44°14'49"N	114°40'34"W	5923
Rock Creek	Twin Falls	42°25'56"N	114°18'17"W	4110
Rock House Place	Twin Falls	42°10'22"N	114°46'50"W	5230
Rockaway Beach	Kootenai	47°47'19"N	116°41'56"W	2250
Rockford	Bingham	43°11'23"N	112°32'02"W	4463
Rockford Bay	Kootenai	47°30'19"N	116°52'39"W	2200
Rockland	Power	42°34'24"N	112°52'35"W	4560
Rockville	Owyhee	43°20'34"N	117°00'01"W	3971
Rocky Bar	Elmore	43°41'21"N	115°17'21"W	5270
Rocky Point	Benewah	47°21'21"N	116°44'43"W	2260
Rogerson	Twin Falls	42°13'05"N	114°35'36"W	4897
Roland	Shoshone	47°22'48"N	115°40'05"W	4147
Rookstool Corner	Canyon	43°46'42"N	116°51'09"W	2393
Rose	Bingham	43°15'16"N	112°19'31"W	4518

Cities, Towns, Historic Places

City, Town, or Historic Place	County	Latitude	Longitude	Elev.
Rose	Caribou	42°36'13"N	111°32'13"W	5799
Rose Lake	Kootenai	47°32'20"N	116°28'15"W	2145
Roseberry	Valley	44°43'51"N	116°02'56"W	4871
Roseworth	Twin Falls	42°22'01"N	114°55'17"W	4625
Roswell	Canyon	43°44'57"N	116°57'39"W	2260
Rothas	Custer	43°53'15"N	113°32'08"W	5787
Rouse	Bingham	43°13'36"N	112°28'42"W	4468
Rover	Benewah	47°12'12"N	116°34'31"W	2460
Roy	Power	42°21'50"N	112°49'49"W	5104
Roy Summit	Oneida	42°19'03"N	112°48'05"W	5481
Rubicon	Adams	44°58'56"N	116°22'04"W	4180
Ruby	Cassia	42°30'14"N	113°49'19"W	4207
Rudo	Clearwater	46°29'21"N	116°00'41"W	2040
Rudy	Jefferson	43°39'24"N	111°49'57"W	4915
Rupert (county seat)	Minidoka	42°37'09"N	113°40'35"W	4158
Sage Junction	Jefferson	43°49'51"N	112°11'40"W	4912
Sagle	Bonner	48°12'09"N	116°32'48"W	2160
Saint Anthony (county seat)	Fremont	43°57'59"N	111°40'53"W	4972
Saint Charles	Bear Lake	42°06'50"N	111°23'17"W	5944
Saint Joe	Benewah	47°18'40"N	116°21'07"W	2160
Saint Johns	Oneida	42°12'50"N	112°17'28"W	4702
Saint Maries (county seat)	Benewah	47°18'52"N	116°33'42"W	2216
Salem	Madison	43°52'36"N	111°46'20"W	4885
Salmon (county seat)	Lemhi	45°10'33"N	113°53'42"W	3950
Salubria	Washington	44°33'49"N	116°39'55"W	2640
Sam	Teton	43°43'09"N	111°18'11"W	6500
Samaria	Oneida	42°07'04"N	112°20'10"W	4480
Sampson	Idaho	45°31'04"N	116°00'27"W	6015
Samuels	Bonner	48°25'46"N	116°29'34"W	2145
Sand Hollow	Payette	43°48'28"N	116°44'53"W	2420
Sanders	Benewah	47°06'23"N	116°47'43"W	2682
Sandpoint (county seat)	Bonner	48°16'36"N	116°33'08"W	2085
Sands (historical)	Owyhee	43°29'20"N	116°54'50"W	2529
Santa	Benewah	47°09'01"N	116°26'53"W	2681
Sarilda	Fremont	44°06'37"N	111°34'23"W	5285
Sawtooth City	Blaine	43°53'48"N	114°50'22"W	7335
Sawyer	Bonner	48°08'40"N	116°45'17"W	2157
Schiller	Power	42°52'47"N	112°39'27"W	4425
Schnoors	Boundary	48°41'23"N	116°20'19"W	1790
Schodde	Jerome	42°36'26"N	113°59'25"W	4165
Schow	Minidoka	42°35'13"N	113°42'38"W	4150
Scoville	Butte	43°28'50"N	112°59'43"W	5075
Seaburg	Idaho	45°28'34"N	116°01'48"W	6527
Sebree	Elmore	43°11'09"N	115°47'23"W	3165
Selby	Camas	43°20'38"N	114°42'35"W	5035
Selle	Bonner	48°21'25"N	116°29'19"W	2176
Seneacquoteen	Bonner	48°09'06"N	116°45'16"W	2101
Setters	Kootenai	47°28'05"N	117°00'26"W	2580

City, Town, or Historic Place	County	Latitude	Longitude	Elev.
Sharon	Bear Lake	42°21'02"N	111°28'42"W	6270
Shelley	Bingham	43°22'53"N	112°07'21"W	4629
Shells Lick	Idaho	45°42'24"N	116°07'37"W	5700
Sherwin	Latah	46°56'52"N	116°20'49"W	3220
Sherwood Beach	Bonner	48°29'48"N	116°50'22"W	2460
Shiloh	Boundary	48°32'42"N	116°23'40"W	2133
Shoreline Drive Summer Home Area	Valley	44°38'51"N	115°40'36"W	5480
Shoshone (county seat)	Lincoln	42°56'10"N	114°24'18"W	3965
Shoup	Lemhi	45°22'37"N	114°16'34"W	3384
Silver Beach	Kootenai	47°56'54"N	116°53'28"W	2460
Silver City	Owyhee	43°01'01"N	116°43'56"W	6180
Silver Sands Beach	Kootenai	47°53'32"N	116°53'01"W	2320
Silver Star	Franklin	42°04'55"N	111°57'15"W	4703
Silvertip Landing	Benewah	47°21'42"N	116°41'30"W	2100
Silverton	Shoshone	47°29'35"N	115°57'16"W	2700
Simplot	Canyon	43°40'11"N	116°43'57"W	2337
Sinclair	Boundary	48°55'45"N	116°10'33"W	2547
Sinker (historical)	Owyhee	43°07'15"N	116°30'35"W	3080
Slabtown	Latah	46°52'22"N	116°23'34"W	2900
Slacks Corner	Owyhee	43°04'55"N	116°44'28"W	6400
Slate Creek	Idaho	45°38'18"N	116°16'46"W	1540
Slickpoo	Lewis	46°19'01"N	116°42'36"W	1780
Slocum	Clearwater	46°33'20"N	116°00'55"W	3191
Smelter Heights	Shoshone	47°32'21"N	116°09'31"W	2330
Smelterville	Shoshone	47°32'34"N	116°10'50"W	2230
Smith Corrals	Butte	43°19'36"N	113°14'43"W	5137
Smith Springs	Blaine	42°38'15"N	113°18'38"W	4215
Smiths Ferry	Valley	44°18'05"N	116°05'19"W	4540
Snow (historical)	Nez Perce	46°18'00"N	116°47'27"W	2200
Soda Springs (county seat)	Caribou	42°39'16"N	111°36'14"W	5760
Soldier	Camas	43°22'13"N	114°47'27"W	5118
Sonna	Ada	43°36'35"N	116°27'21"W	2540
South Boise (historical)	Ada	43°34'35"N	116°09'56"W	2750
South Mountain	Owyhee	42°45'08"N	116°55'21"W	6560
Southwick	Nez Perce	46°36'13"N	116°28'16"W	2967
Spalding	Nez Perce	46°26'49"N	116°48'59"W	800
Spanish Town (historical)	Elmore	43°42'35"N	115°15'15"W	5540
Spencer	Clark	44°21'38"N	112°11'10"W	5860
Spink	Valley	44°47'25"N	116°02'50"W	4900
Spirit Lake	Kootenai	47°57'59"N	116°52'03"W	2567
Spiry Place	Owyhee	42°54'37"N	116°24'23"W	3340
Springdale	Cassia	42°30'56"N	113°41'24"W	4190
Springfield	Bingham	43°04'54"N	112°40'52"W	4425
Springston	Kootenai	47°28'42"N	116°43'55"W	2135
Squirrel	Fremont	44°01'38"N	111°17'54"W	5612
Staley Springs	Fremont	44°39'37"N	111°26'04"W	6480
Stanford	Latah	46°49'58"N	116°38'37"W	2746
Stanley	Custer	44°12'41"N	114°56'42"W	6260

City, Town, or Historic Place	County	Latitude	Longitude	Elev.
Stanton Crossing	Blaine	43°19'40"N	114°18'42"W	4840
Star	Ada	43°41'32"N	116°29'33"W	2467
Star Ranch Summer Home Area	Boise	43°54'46"N	115°58'03"W	4364
Starkey	Adams	44°51'02"N	116°26'51"W	3173
State Line Village	Kootenai	47°42'16"N	117°02'11"W	2100
Steamboat (historical)	Shoshone	47°40'32"N	116°08'38"W	2280
Steamboat Rock	Elmore	43°26'54"N	115°37'35"W	4090
Steamboat Rocks	Shoshone	47°54'53"N	116°07'32"W	3820
Sterling	Bingham	43°02'17"N	112°43'51"W	4371
Stetson	Shoshone	47°17'06"N	115°46'24"W	2690
Stevens	Adams	44°50'55"N	116°22'40"W	3555
Stevens	Valley	44°51'00"N	116°22'06"W	3625
Stibnite	Valley	44°53'54"N	115°20'18"W	6560
Stites	Idaho	46°05'30"N	115°58'31"W	1319
Stone	Oneida	42°00'59"N	112°41'40"W	4564
Stout Crossing	Elmore	43°09'16"N	115°18'33"W	4930
Strevell	Cassia	42°00'22"N	113°12'10"W	5260
Striker	Twin Falls	42°28'30"N	114°18'15"W	4072
Stull	Shoshone	47°28'20"N	115°51'49"W	2947
Sturgeon	Kootenai	47°50'59"N	116°51'35"W	2313
Sublett	Cassia	42°18'44"N	113°08'10"W	5002
Sugar City	Madison	43°52'23"N	111°44'51"W	4890
Sugar Loaf	Jerome	42°40'53"N	114°22'37"W	3845
Sun Valley	Blaine	43°41'50"N	114°21'03"W	5940
Sunbeam	Custer	44°16'16"N	114°44'00"W	6000
Sunnydell	Madison	43°41'04"N	111°45'15"W	4960
Sunnyside	Bonner	48°16'49"N	116°23'46"W	2200
Sunnyslope	Canyon	43°35'19"N	116°47'32"W	2402
Sunset (historical)	Shoshone	47°33'45"N	115°50'04"W	6424
Swan Falls	Ada	43°14'30"N	116°22'24"W	2340
Swan Lake	Bannock	42°18'28"N	111°59'53"W	4780
Swan Valley	Bonneville	43°27'22"N	111°20'27"W	5276
Swanlake	Bannock	42°18'47"N	112°00'11"W	4780
Swartz Corner	Canyon	43°30'05"N	116°33'15"W	2571
Sweeney	Shoshone	47°32'25"N	116°10'07"W	2340
Sweet	Gem	43°58'23"N	116°19'25"W	2560
Sweet Sage (historical)	Butte	43°49'40"N	112°55'37"W	4805
Sweetwater	Nez Perce	46°22'21"N	116°47'31"W	1100
Syringa	Idaho	46°09'03"N	115°43'34"W	1440
Taber	Bingham	43°19'05"N	112°41'17"W	4727
Tahoe	Idaho	46°05'47"N	115°48'22"W	3320
Talache	Bonner	48°08'27"N	116°29'03"W	2380
Talache Landing	Bonner	48°07'43"N	116°28'40"W	2140
Talmage	Caribou	42°40'44"N	111°47'12"W	5584
Tamarack	Adams	44°57'19"N	116°23'05"W	4104
Tammany	Shoshone	47°26'00"N	115°39'03"W	4000
Taney (historical)	Nez Perce	46°35'58"N	116°32'34"W	2630
Taplin	Nez Perce	46°09'08"N	116°55'49"W	900

City, Town, or Historic Place	County	Latitude	Longitude	Elev.
Taylor	Idaho	45°41'08"N	116°22'57"W	3270
Taylorville	Bonneville	43°22'58"N	112°02'09"W	4655
Teakean	Clearwater	46°33'04"N	116°22'47"W	2993
Telegraph Hill	Owyhee	43°00'56"N	116°44'48"W	6600
Telluride	Caribou	42°32'12"N	111°46'45"W	5200
Tendoy	Lemhi	44°57'34"N	113°38'38"W	4836
Tenmile Summer Home Area	Boise	43°53'50"N	115°42'47"W	4980
Tensed	Benewah	47°09'36"N	116°55'15"W	2557
Terreton	Jefferson	43°50'30"N	112°26'08"W	4787
Teton	Fremont	43°53'12"N	111°40'37"W	4949
Tetonia	Teton	43°49'09"N	111°09'43"W	6060
Thama	Bonner	48°09'12"N	116°50'48"W	2077
Thatcher	Franklin	42°24'32"N	111°43'34"W	4900
The Cedars	Clearwater	46°52'22"N	115°04'34"W	3680
The String	Teton	43°35'05"N	111°05'16"W	6330
Thiard	Shoshone	47°36'56"N	115°54'15"W	3000
Thomas	Bingham	43°10'58"N	112°30'14"W	4446
Thorenson	Franklin	42°05'00"N	111°59'30"W	4759
Thornton	Madison	43°45'30"N	111°50'40"W	4860
Three Creek	Owyhee	42°04'17"N	115°09'31"W	5480
Three Forks	Owyhee	42°06'06"N	115°50'16"W	5170
Threemile Corner	Boundary	48°43'53"N	116°17'55"W	2320
Threemile Crossing	Twin Falls	42°14'52"N	114°51'50"W	5050
Tiegs Corner	Canyon	43°30'50"N	116°30'47"W	2508
Tikura	Blaine	43°12'53"N	114°00'43"W	4679
Tindall (historical)	Owyhee	42°15'58"N	115°52'52"W	5140
Tipperary Corner	Jerome	42°34'44"N	114°17'44"W	3925
Tipton (historical)	Cassia	42°16'50"N	113°09'35"W	4930
Topaz (historical)	Bannock	42°37'38"N	112°05'05"W	4960
Torreys	Custer	44°15'13"N	114°36'05"W	5742
Trailton (historical)	Bingham	43°07'48"N	111°54'40"W	5557
Tramway	Idaho	46°17'05"N	116°06'25"W	1143
Transfer	Nez Perce	46°25'45"N	117°01'46"W	741
Travers	Minidoka	42°36'33"N	113°44'10"W	4145
Treasureton	Franklin	42°15'55"N	111°50'44"W	5076
Trestle Creek	Bonner	48°17'00"N	116°20'59"W	2180
Triangle	Owyhee	42°47'02"N	116°37'28"W	5110
Triumph	Blaine	43°38'42"N	114°15'12"W	6120
Trout	Cassia	42°19'21"N	113°54'01"W	4415
Troy	Latah	46°44'13"N	116°46'07"W	2400
Trude	Fremont	44°27'40"N	111°20'43"W	6325
Tuanna Crossing	Twin Falls	42°25'39"N	114°58'48"W	4230
Tunupa	Lincoln	42°57'18"N	114°34'42"W	3710
Turner	Caribou	42°34'36"N	111°49'05"W	5471
Turner	Idaho	45°27'23"N	115°56'13"W	2120
Turnpike	Bear Lake	42°06'36"N	111°15'46"W	5947
Tuttle	Gooding	42°51'30"N	114°50'21"W	3309
Twin Beaches	Kootenai	47°37'12"N	116°47'50"W	2160

Cities, Towns, Historic Places

City, Town, or Historic Place	County	Latitude	Longitude	Elev.
Twin Falls (county seat)	Twin Falls	42°33'47"N	114°27'36"W	3729
Twin Forks	Teton	43°43'55"N	111°11'35"W	5980
Twin Groves	Fremont	43°58'28"N	111°37'37"W	5010
Twin Springs	Boise	43°40'10"N	115°42'01"W	3340
Twinlow	Kootenai	47°52'38"N	116°51'19"W	2400
Two Forks	Teton	43°42'47"N	111°10'35"W	5980
Tyhee	Bannock	42°57'06"N	112°27'56"W	4457
Tyson Creek Station	Benewah	47°08'02"N	116°24'48"W	2745
Ucon	Bonneville	43°35'47"N	111°57'47"W	4808
Ulysses (historical)	Lemhi	45°27'28"N	114°09'02"W	4080
Underkoflers Corner	Canyon	43°37'09"N	116°40'18"W	2432
Unity	Cassia	42°31'07"N	113°44'33"W	4192
Upper Crossing	Idaho	45°38'40"N	114°42'22"W	4460
Ustick	Ada	43°38'01"N	116°19'15"W	2650
Utida	Franklin	41°59'57"N	111°58'28"W	4571
Valley View Heights	Nez Perce	46°25'24"N	116°58'11"W	800
Vans Corner	Bonner	48°31'31"N	116°56'16"W	2580
Vassar	Latah	46°49'45"N	116°36'58"W	2715
Vay	Bonner	48°06'56"N	116°46'37"W	2184
Vernon	Ada	43°35'19"N	116°09'36"W	2830
Victor	Teton	43°36'10"N	111°06'38"W	6207
View	Cassia	42°27'03"N	113°41'47"W	4230
Viola	Latah	46°50'19"N	117°01'25"W	2630
Virginia	Bannock	42°29'39"N	112°09'53"W	4805
Vollmer	Lewis	46°14'34"N	116°27'55"W	3727
Vosburg Place	Twin Falls	42°11'28"N	114°49'00"W	5349
Waha	Nez Perce	46°12'48"N	116°51'04"W	3420
Walker	Madison	43°46'20"N	111°42'28"W	5267
Wallace (county seat)	Shoshone	47°28'27"N	115°55'37"W	2744
Walters Ferry	Canyon	43°20'26"N	116°35'55"W	2260
Wapello	Bingham	43°14'49"N	112°15'22"W	4542
Wapi	Blaine	42°42'56"N	113°10'56"W	4390
Wardboro	Bear Lake	42°15'21"N	111°16'35"W	5948
Wardner	Shoshone	47°31'22"N	116°07'59"W	2620
Warm Lake	Valley	44°39'15"N	115°40'00"W	5300
Warm River	Fremont	44°06'54"N	111°19'12"W	5300
Warren	Idaho	45°15'51"N	115°40'32"W	5907
Warrens	Canyon	43°19'43"N	116°32'21"W	2400
Washington Mill	Boise	43°53'32"N	115°45'27"W	5000
Washoe	Payette	44°03'16"N	116°56'50"W	2145
Wayan	Caribou	42°58'42"N	111°22'34"W	6437
Wayland	Benewah	47°10'09"N	116°28'09"W	2609
Webb	Nez Perce	46°20'38"N	116°49'56"W	1336
Webb	Shoshone	47°29'18"N	115°53'26"W	3000
Weippe	Clearwater	46°22'34"N	115°56'14"W	3020
Weiser (county seat)	Washington	44°15'04"N	116°58'06"W	2130
Weitz	Canyon	43°40'10"N	116°45'31"W	2325
Wendell	Gooding	42°46'33"N	114°42'12"W	3435
Westlake	Idaho	46°07'19"N	116°30'22"W	4180
Westma	Canyon	43°24'57"N	116°32'28"W	2650
Westmond	Bonner	48°08'24"N	116°36'09"W	2210
Weston	Franklin	42°02'14"N	111°58'43"W	4743
Whipsaw Saddle	Idaho	45°47'34"N	116°20'54"W	3460
White Bird	Idaho	45°45'42"N	116°17'59"W	1560
White Knob (historical)	Custer	43°53'33"N	113°40'10"W	7480
Whitney	Franklin	42°03'57"N	111°50'13"W	4600
Wickahoney	Owyhee	42°27'36"N	115°58'57"W	5190
Wilder	Canyon	43°40'36"N	116°54'39"W	2424
Wildhorse	Adams	44°51'20"N	116°49'51"W	2360
Wilford	Fremont	43°54'46"N	111°40'37"W	4940
Willard	Benewah	47°14'26"N	117°02'17"W	2510
Williams Creek Summit Home Area	Valley	44°09'49"N	116°07'12"W	3940
Williams Lake Resort	Lemhi	45°00'43"N	113°57'54"W	5400
Williamsburg (historical)	Caribou	42°56'52"N	111°15'36"W	6760
Willola	Nez Perce	46°30'18"N	116°34'09"W	1000
Wilson	Owyhee	43°21'41"N	116°39'07"W	2330
Winchester	Lewis	46°14'24"N	116°37'24"W	3968
Winder	Franklin	42°10'35"N	111°54'54"W	4756
Winona	Idaho	46°07'56"N	116°06'25"W	3020
Winsper	Clark	44°07'50"N	112°30'52"W	4890
Wolf Lodge	Kootenai	47°38'29"N	116°36'57"W	2160
Wolverine	Bingham	43°16'53"N	111°56'07"W	5600
Wood	Payette	44°07'32"N	116°54'22"W	2130
Woodland	Adams	44°55'49"N	116°22'55"W	4140
Woodland	Idaho	46°18'40"N	116°04'03"W	3010
Woodlawn Park	Shoshone	47°29'10"N	115°54'04"W	2920
Woodruff	Oneida	42°02'10"N	112°12'50"W	4489
Woods Crossing	Madison	43°46'27"N	111°37'47"W	5250
Woodside	Nez Perce	46°12'23"N	116°37'06"W	4250
Woodside (historical)	Lewis	46°16'24"N	116°35'57"W	3800
Woodville	Bingham	43°25'01"N	112°08'31"W	4635
Wooleys	Bear Lake	42°27'07"N	111°23'10"W	5920
Worley	Kootenai	47°24'03"N	116°54'58"W	2650
Wrencoe	Bonner	48°13'44"N	116°42'07"W	2086
Wright (historical)	Kootenai	47°58'52"N	116°43'51"W	2340
Yale	Latah	46°51'44"N	116°41'44"W	2900
Yellow Dog (historical)	Latah	46°58'45"N	117°01'58"W	2550
Yellow Pine	Valley	44°57'54"N	115°29'34"W	4762
Yellowjacket	Lemhi	44°58'47"N	114°31'51"W	5900
York	Idaho	45°54'33"N	115°28'43"W	4808
Zaza	Nez Perce	46°03'44"N	116°50'43"W	5076
Zenda	Bannock	42°22'40"N	112°03'28"W	4800

Cliff	County	Latitude	Longitude	Elev.
Aparejo Point	Lemhi	44°55'28"N	114°43'45"W	4569
Arctic Point	Idaho	45°28'28"N	115°02'21"W	7518
Ballinger Point	Idaho	46°06'28"N	115°08'50"W	
Bartlett Point	Custer	44°01'22"N	113°56'57"W	
Bear Creek Point	Owyhee	42°23'47"N	116°51'14"W	
Bear Creek Point	Valley	45°10'40"N	115°39'18"W	8084
Bell Point	Idaho	46°13'50"N	114°44'59"W	6768
Bernard Point	Kootenai	47°57'17"N	116°31'47"W	
Bernier Point	Shoshone	47°18'23"N	115°33'01"W	5620
Big Bluff	Camas	43°17'09"N	114°52'18"W	5254
Big Bluff	Elmore	43°18'01"N	115°38'12"W	
Black Point	Adams	45°10'44"N	116°41'20"W	
Black Rock	Owyhee	42°03'25"N	115°13'29"W	5500
Black Rocks	Owyhee	42°44'30"N	115°55'10"W	3000
Bluffs, The	Payette	43°59'27"N	116°45'33"W	
Bonneville Point	Ada	43°29'32"N	116°02'24"W	
Buckshot Point	Idaho	45°37'37"N	116°03'00"W	
Bug Slope	Idaho	45°54'20"N	116°24'05"W	
Bull Trout Point	Custer	44°19'48"N	115°17'12"W	8876
California Point	Idaho	46°20'15"N	114°58'42"W	6967
Castle Rock	Owyhee	42°02'28"N	115°13'54"W	5565
Castle Rock	Shoshone	47°39'24"N	116°07'18"W	
Cathedral Rocks	Shoshone	47°55'59"N	116°06'30"W	
China Point	Clark	44°20'16"N	112°10'49"W	
Chinese Wall	Custer	44°07'43"N	114°37'06"W	
City of Rocks	Owyhee	42°47'25"N	115°50'26"W	
Clay Point	Payette	44°03'44"N	116°54'26"W	
Clugs Jumpoff	Clearwater	46°54'47"N	115°46'20"W	3690
Coal Bank	Owyhee	43°14'26"N	116°50'31"W	
Cox Point	Valley	44°52'47"N	115°10'33"W	
Crown Point	Blaine	43°49'39"N	114°30'43"W	
Crown Point	Gem	43°51'59"N	116°18'02"W	5163
Cub Point	Idaho	45°59'59"N	114°39'07"W	6831
Dead Elk Point	Idaho	46°16'00"N	114°32'16"W	7569
Devils Ladder	Idaho	45°19'42"N	116°15'35"W	
Devils Slide	Washington	44°47'36"N	116°42'47"W	
Dike, The	Adams	44°49'20"N	116°13'07"W	
Dizzy Head	Idaho	46°04'39"N	115°58'26"W	1602
Doe Point	Idaho	46°15'08"N	115°02'18"W	5982
Dog Point	Idaho	46°05'42"N	114°48'45"W	5744
Drop-off, The	Idaho	45°47'36"N	114°54'44"W	7666
Drumlummen Point	Idaho	45°32'00"N	115°38'52"W	
Dry Point	Idaho	46°14'10"N	115°32'55"W	
Eagle Bluff	Twin Falls	42°37'38"N	114°47'55"W	
Eagle Cliff	Shoshone	47°07'47"N	115°11'08"W	7543
Eagle Rock	Custer	44°39'28"N	114°04'37"W	4800
Echo Rock	Bonner	48°06'01"N	116°25'43"W	
Elk Point	Camas	43°42'54"N	114°51'20"W	9214

Cliff	County	Latitude	Longitude	Elev.
Falls Point	Idaho	46°02'07"N	115°20'00"W	5069
Feltham Creek Point	Custer	44°30'34"N	114°58'11"W	9000
Fir Bluff	Nez Perce	46°30'43"N	116°39'10"W	
Fire Creek Point	Idaho	46°10'27"N	115°22'03"W	6266
Flea Ridge Point	Idaho	46°12'27"N	115°18'16"W	6850
Footstool Point	Idaho	46°16'23"N	114°35'25"W	6269
Grandview Point	Teton	43°48'41"N	111°20'44"W	7653
Gravelly Point	Canyon	43°40'16"N	116°57'58"W	2435
Hoover Point	Lewis	46°03'55"N	116°39'37"W	4583
Hot Point	Idaho	46°06'06"N	115°35'57"W	3800
Huddleson Bluff	Idaho	45°48'59"N	115°49'49"W	
Hungry Rock	Idaho	46°19'50"N	114°46'16"W	7722
Idaho Point	Idaho	46°09'50"N	115°30'10"W	6027
Indian Creek Point	Valley	44°51'37"N	115°19'02"W	8782
Iron Creek Point	Custer	44°22'42"N	113°29'04"W	10736
Jungle Point	Idaho	46°12'02"N	115°41'06"W	
Kinney Point	Adams	45°09'59"N	116°39'47"W	7126
Lake Creek Point	Idaho	45°20'15"N	116°15'09"W	7639
Lightning Point	Valley	44°59'07"N	116°04'19"W	
Limestone Point	Nez Perce	46°04'55"N	116°57'05"W	
Lockwood Point	Adams	45°14'27"N	116°23'27"W	6967
Lone Pine Point	Idaho	46°04'57"N	115°01'01"W	5338
Lotus Point	Benewah	47°15'34"N	116°37'39"W	
Louse Point	Idaho	46°10'41"N	115°18'31"W	7020
Maple Cliff	Shoshone	47°39'21"N	116°05'25"W	
Marble Point	Idaho	45°43'00"N	115°56'38"W	5983
McFadden Point	Valley	45°10'29"N	115°17'08"W	8551
McNabbs Point	Custer	44°35'17"N	114°11'26"W	
Meeker Point	Idaho	46°07'43"N	115°00'45"W	4577
Meyers Cove Point	Lemhi	44°49'41"N	114°26'31"W	7918
Mocus Point	Idaho	46°25'31"N	115°01'18"W	5579
Moughmer Point	Idaho	45°56'27"N	116°24'41"W	4163
Murphy Rim	Owyhee	43°14'06"N	116°31'56"W	
Old Man Point	Idaho	46°16'06"N	115°21'53"W	
Orofino Creek Point	Clearwater	46°28'23"N	115°55'23"W	3713
Orphan Point	Shoshone	47°02'31"N	115°56'25"W	6254
Packsack Point	Shoshone	47°52'16"N	116°09'15"W	4494
Painter Point	Idaho	45°25'32"N	115°27'08"W	
Penny Cliffs	Idaho	46°09'09"N	115°54'40"W	
Peterson Point	Idaho	46°03'58"N	115°32'44"W	3848
Plummer Point	Idaho	45°34'01"N	115°50'05"W	7918
Puzzle Point	Idaho	46°05'51"N	114°58'27"W	6055
Rattlesnake Point	Bonneville	43°36'25"N	111°30'28"W	
Rattlesnake Point	Valley	44°18'06"N	115°47'53"W	7250
Red Point	Adams	44°50'51"N	116°17'26"W	6285
Red Point	Lemhi	44°42'59"N	113°53'05"W	
Reno Point	Clark	44°01'01"N	112°45'05"W	
Ring Creek Point	Idaho	45°37'14"N	115°00'43"W	7818

Cliffs

Cliff	County	Latitude	Longitude	Elev.
Rock Creek Point	Idaho	45°31'17"N	115°55'06"W	7308
Rocky Point	Washington	44°18'33"N	117°02'35"W	
Ross Point	Kootenai	47°42'40"N	116°53'33"W	
Salmon River Breaks	Idaho	45°34'08"N	115°11'23"W	
Scotts Bluff	Idaho	45°35'57"N	115°41'25"W	
Shasta Point	Idaho	46°14'28"N	114°55'53"W	5422
Sheep Creek Point	Idaho	45°26'25"N	116°33'17"W	5200
Shissler Peak	Idaho	46°09'57"N	114°56'20"W	5375
Slick Rock	Valley	44°57'00"N	115°57'02"W	
Snowside Point	Bear Lake	42°23'25"N	111°09'38"W	
Soda Creek Point	Idaho	45°43'43"N	115°13'29"W	
Soda Point	Caribou	42°38'28"N	111°42'06"W	
Split Creek Point	Idaho	46°13'59"N	115°21'35"W	5660
Stillman Point	Idaho	46°03'29"N	115°27'50"W	5070
Stoney Point	Shoshone	47°48'56"N	116°11'46"W	5284
Tent Creek Point	Owyhee	42°02'48"N	116°56'10"W	5141
Three Links Point	Idaho	46°10'01"N	115°07'17"W	5530
Tony Point	Idaho	46°05'36"N	114°56'03"W	6136
Tool Point	Twin Falls	42°09'12"N	114°18'00"W	
Twin Crags	Kootenai	47°26'10"N	116°21'28"W	
Uleda Point	Bonner	48°22'43"N	116°41'34"W	4953
West Fork Point	Idaho	46°00'21"N	115°33'00"W	6056
White Licks	Adams	44°40'53"N	116°13'56"W	
Wilson Bluff	Owyhee	43°19'32"N	116°45'50"W	
Windmill Bluff	Camas	43°17'00"N	114°51'04"W	5224
Windy Point	Owyhee	42°30'12"N	115°38'51"W	4290
Windy Point	Owyhee	42°11'17"N	116°29'37"W	
Wolf Point	Idaho	46°04'15"N	115°05'54"W	4925

Cliff	County	Latitude	Longitude	Elev.

Dam	County	Latitude	Longitude	Elev.
Albeni Falls Dam	Bonner	48°10'48"N	117°00'00"W	
Albeni Falls Dam	Bonner	48°10'44"N	116°59'54"W	
Albrethsen Dam	Blaine	43°26'30"N	113°56'18"W	
Albrethsen Dam	Blaine	43°18'54"N	114°12'30"W	
Alder Dam	Owyhee	42°09'32"N	116°00'18"W	6035
Alexander Dam	Caribou	42°38'42"N	111°41'44"W	5720
American Falls Dam	Power	42°46'51"N	112°52'29"W	
Anderson Ranch Dam	Elmore	43°21'27"N	115°26'52"W	
Andreason Number Three Dam	Franklin	42°19'30"N	111°45'30"W	
Ansley Dam	Boundary	48°48'42"N	116°19'12"W	
Arcadia-Lower Dam	Fremont	44°08'06"N	111°35'18"W	5400
Arcadia-Upper Dam	Fremont	44°08'32"N	111°35'42"W	5429
Armstrong Dam	Lemhi	44°49'36"N	113°43'24"W	
Arrowrock Dam	Boise	43°35'44"N	115°55'17"W	
Ashton Dam	Fremont	44°04'44"N	111°29'44"W	5200
Bar D Dam	Custer	44°26'24"N	114°08'00"W	
Barber Dam	Ada	43°33'39"N	116°07'13"W	2767
Barton Dam	Washington	44°19'12"N	116°55'18"W	
Beal Number Three Dam	Valley	44°13'42"N	116°08'06"W	
Beker Tailings Number Three Dam	Caribou	42°44'30"N	111°33'06"W	
Bettis Dam	Payette	44°05'00"N	116°33'48"W	
Bettis Number Three Dam	Payette	44°03'36"N	116°33'18"W	
Bettis Number Two Dam	Payette	44°03'42"N	116°32'06"W	
Big Blue Creek Dam	Owyhee	42°18'30"N	116°10'54"W	
Billings Dam	Owyhee	42°03'06"N	115°46'24"W	
Billy Creek Dam	Lemhi	45°08'48"N	113°56'36"W	
Birch Creek Dam	Bonneville	43°20'12"N	111°47'30"W	
Black Canyon Dam	Gem	43°55'50"N	116°26'08"W	
Blackbird Tailings Dam	Lemhi	45°06'18"N	114°18'24"W	
Blackfoot Dam	Caribou	43°00'17"N	111°42'59"W	
Blackfoot-Equalizing Dam	Bingham	43°10'00"N	112°17'48"W	
Blacks Creek Dam	Ada	43°27'45"N	116°08'40"W	
Blanchard Creek Dam	Bonner	48°00'18"N	117°00'06"W	
Bliss Dam	Gooding	42°54'49"N	115°04'09"W	
Bliss Dam	Gooding	42°54'49"N	115°04'11"W	2655
Blue Creek - Upper Dam	Fremont	44°13'26"N	111°36'39"W	
Blue Creek Number Four Dam	Fremont	44°12'08"N	111°36'52"W	5511
Blue Creek Number One Dam	Fremont	44°13'00"N	111°36'48"W	
Blue Creek Number Two Dam	Fremont	44°12'42"N	111°36'36"W	
Boggeman Dam	Lemhi	44°51'42"N	114°27'06"W	
Bohannon Dam	Lemhi	45°13'42"N	113°40'06"W	
Bohn Dam	Bonneville	43°20'54"N	111°48'36"W	
Boise Diversion Dam	Ada	43°32'18"N	116°05'36"W	
Boulder Lake Dam	Valley	44°52'04"N	115°56'44"W	
Boulder Meadow Dam	Valley	44°52'09"N	115°58'11"W	
Box Lake Dam	Valley	45°01'30"N	115°58'54"W	
Brace Dam	Owyhee	42°21'12"N	116°42'24"W	
Bray Lake Dam	Gooding	43°02'00"N	114°52'36"W	

Dam	County	Latitude	Longitude	Elev.
Brown Dam	Owyhee	43°29'30"N	117°00'00"W	
Brown Dam	Owyhee	43°17'42"N	116°58'42"W	
Brown Dam	Valley	44°43'12"N	115°59'30"W	
Brownlee Dam	Washington	44°50'10"N	116°54'00"W	
Browns Pond Dam	Valley	44°55'00"N	115°57'48"W	
Bruce Dam	Washington	44°25'18"N	116°32'12"W	
Brundage Dam	Adams	45°02'30"N	116°07'49"W	
Buckaroo Dam	Owyhee	42°46'48"N	115°42'49"W	
Bunker Hill Tailings Number One Dam	Shoshone	47°32'18"N	116°09'18"W	
Bunker Hill Tailings Number Three Dam	Shoshone	47°33'06"N	116°08'42"W	
Bunker Hill Tailings Number Two Dam	Shoshone	47°32'12"N	116°08'48"W	
Burton Dam	Owyhee	42°35'30"N	116°37'12"W	
Buster Lake Dam	Custer	44°26'24"N	114°24'54"W	
Bybee Dam	Owyhee	42°15'38"N	116°15'44"W	
C Ben Ross Dam	Adams	44°31'22"N	116°26'48"W	
C J Strike Dam	Elmore	42°56'50"N	115°58'28"W	2455
C J Strike Dam	Owyhee	42°56'48"N	115°58'30"W	
Cabinet Gorge Dam	Bonner	48°05'12"N	116°03'48"W	
Callender Dam	Valley	44°33'16"N	115°59'39"W	4860
Campbell Dam	Blaine	43°28'30"N	114°00'06"W	
Cascade Dam	Valley	44°31'27"N	116°02'59"W	
Casperson Dam	Franklin	42°14'36"N	111°53'00"W	
Cedar Creek Dam	Twin Falls	42°13'27"N	114°52'44"W	
Cedar Creek Holding Dam	Twin Falls	42°20'30"N	114°54'42"W	
Challis Creek Dam	Custer	44°33'12"N	114°30'48"W	
Cherry Dam	Fremont	44°07'36"N	111°21'30"W	
Chilco Dam	Kootenai	47°51'48"N	116°43'54"W	
China Cap Dam	Caribou	42°50'36"N	111°36'12"W	
Clarendon Hot Springs Dam	Blaine	43°33'30"N	114°24'42"W	
Clayton Mine Number One-Upper Dam	Custer	44°17'06"N	114°24'30"W	
Clayton Mine Number Two-Lower Dam	Custer	44°17'00"N	114°24'24"W	
Clear Lake Dam	Gooding	42°40'06"N	114°46'36"W	
Cocolalla Creek Fish Weir	Bonner	48°09'47"N	116°39'00"W	2110
Conda Site Tailings Dam	Caribou	42°43'48"N	111°32'00"W	
Condie Dam	Franklin	42°12'30"N	111°52'18"W	
Copper Cliffs Tailings Dam	Adams	45°05'48"N	116°40'42"W	
Corral Creek Dam	Valley	44°27'47"N	115°55'40"W	
Courtright Dam	Washington	44°28'48"N	116°32'12"W	
Cove Dam	Caribou	42°31'54"N	111°47'42"W	
Cow Creek Dam	Elmore	43°20'33"N	115°05'30"W	
Craig Dam	Washington	44°19'12"N	116°32'30"W	
Crane Creek Dam	Washington	44°21'25"N	116°36'57"W	
Cross Cut Diversion Dam	Fremont	44°01'06"N	111°35'00"W	
Crowther Dam	Oneida	42°12'18"N	112°15'24"W	
Cummings Dam	Lemhi	45°28'42"N	114°02'24"W	
Curlew Valley Dam	Oneida	42°04'30"N	112°32'30"W	
Current Creek Dam	Bannock	42°20'00"N	111°50'36"W	
Daniels Dam	Oneida	42°20'43"N	112°26'35"W	5180

Dams

Dam	County	Latitude	Longitude	Elev.
Davis Dam	Valley	44°31'17"N	115°59'28"W	4820
Day Mine Tailings Dam	Shoshone	47°31'30"N	115°53'30"W	
Deadwood Dam	Valley	44°17'36"N	115°38'42"W	
Deep Creek Dam	Oneida	42°12'40"N	112°10'18"W	
Deep Creek Number One Dam	Twin Falls	42°18'00"N	114°37'42"W	
Deep Creek Number Two Dam	Twin Falls	42°16'42"N	114°36'55"W	4690
Deer Flat-Lower Dam	Canyon	43°35'00"N	116°44'12"W	2535
Deer Flat-Middle Dam	Canyon	43°35'36"N	116°42'48"W	2535
Deer Flat-Upper Dam	Canyon	43°33'36"N	116°39'30"W	2535
Delamar Tailings Dam	Owyhee	43°01'12"N	116°49'48"W	
Devil Creek Dam	Oneida	42°17'24"N	112°12'24"W	
Dewey Dam	Cassia	42°30'06"N	113°31'00"W	
Diamond A Dam	Owyhee	42°05'42"N	115°34'30"W	
Dog Creek Dam	Gooding	43°01'30"N	114°44'36"W	
Dry Wash Dam	Camas	43°15'00"N	114°29'36"W	
Dworshak Dam	Clearwater	46°30'54"N	116°17'45"W	1620
Easterday Dam	Twin Falls	42°35'42"N	114°54'36"W	
Eld Dam	Valley	44°42'18"N	116°00'00"W	
Elk River Dam	Clearwater	46°46'09"N	116°09'59"W	
Elkhorn Dam	Oneida	42°18'26"N	112°24'49"W	5259
Ellsworth-Lower Dam	Lemhi	44°50'18"N	113°33'48"W	
Ellsworth-Middle Dam	Lemhi	44°51'30"N	116°33'00"W	
Ellsworth-Upper Dam	Lemhi	44°53'18"N	113°32'00"W	
Empire Mine Number One - Upper Dam	Custer	43°53'30"N	113°40'12"W	
Empire Mine Number Two Dam	Custer	43°53'36"N	113°39'30"W	
Enders Dam	Caribou	42°55'48"N	111°26'30"W	
Erickson Dam	Boundary	48°38'42"N	116°15'30"W	
Fairchild Dam	Washington	44°27'48"N	116°54'24"W	
Felt Dam	Teton	43°54'28"N	111°17'02"W	
Fife Dam	Bear Lake	42°24'36"N	111°29'12"W	
Fish Creek Dam	Blaine	43°25'23"N	113°49'54"W	5340
Five Mile Erosion Control Dam	Franklin	42°07'06"N	111°57'42"W	
Foland Dam	Owyhee	42°37'18"N	116°59'24"W	
Foreman Dam	Owyhee	43°01'18"N	116°19'51"W	2615
Foster Dam	Franklin	42°07'29"N	111°50'33"W	
Four Corners Dam	Owyhee	42°24'30"N	116°06'17"W	
Francis Dam	Elmore	43°20'24"N	115°07'48"W	
Fraser Dam (historical)	Elmore	43°08'25"N	115°53'34"W	3112
Galena Mill Number One-Upper Dam	Shoshone	47°29'00"N	115°57'42"W	
Galena Mill Number Three-Lower Dam	Shoshone	47°29'00"N	115°57'36"W	
Galena Mill Number Two-Middle Dam	Shoshone	47°29'00"N	115°57'42"W	
Galloway Dam	Washington	44°15'03"N	116°46'26"W	
Gardiner-Lower Dam	Bear Lake	42°18'18"N	111°06'48"W	
Gardiner-Upper Dam	Bear Lake	42°18'24"N	111°07'12"W	
Gatfield Number Two Dam	Gem	43°55'12"N	116°17'18"W	
Geertson Dam	Lemhi	45°14'18"N	113°40'00"W	
Glendale Dam	Franklin	42°07'42"N	111°48'36"W	
Glenn Gallant Dam	Adams	44°37'48"N	116°32'42"W	
Goose Lake Dam	Adams	45°04'12"N	116°10'06"W	
Gorley Creek Dam	Lemhi	45°07'42"N	113°55'30"W	
Grace Dam	Caribou	42°35'18"N	111°43'42"W	
Granite Creek Dam	Bonneville	43°28'48"N	111°29'36"W	
Granite Lake Dam	Valley	45°06'06"N	116°04'48"W	
Grasmere Dam	Owyhee	42°21'48"N	115°54'30"W	
Grays Lake Dam	Bonneville	43°07'30"N	111°29'24"W	
Grays Lake Dam	Caribou	42°59'36"N	111°29'24"W	
Greene Tree Dam	Elmore	43°49'00"N	115°06'30"W	
Groefsema Dam	Elmore	43°06'06"N	115°38'06"W	
Hagberry Dam	Owyhee	42°36'05"N	116°41'23"W	5260
Hait-Upper Dam	Valley	44°50'30"N	116°10'24"W	
Hall Dam	Elmore	43°06'42"N	115°29'59"W	
Hammond Dam	Bear Lake	42°15'33"N	111°28'00"W	6555
Hancock Dam	Adams	44°54'18"N	116°12'55"W	4655
Hardesty-Cherry Creek Dam	Owyhee	42°39'25"N	117°00'09"W	5040
Hardisty And Wilson Dam	Owyhee	42°38'51"N	116°57'11"W	5240
Harris Dam	Owyhee	42°46'35"N	115°43'02"W	
Haw Creek Dam	Gem	43°57'30"N	116°28'48"W	
Hawkins Dam	Bannock	42°30'36"N	112°19'53"W	5145
Hayden Lake Dam	Kootenai	47°45'01"N	116°45'26"W	
Heil Dam	Twin Falls	42°12'30"N	115°00'00"W	
Hells Canyon Dam	Adams	45°14'00"N	116°42'00"W	
Henrys Lake Dam	Fremont	44°35'50"N	111°21'09"W	
Herrick Dam	Valley	44°22'36"N	115°58'45"W	
Hidden Lake Dam	Valley	44°14'48"N	116°11'30"W	
Hinkley Dam	Franklin	42°01'18"N	111°52'18"W	
Hobo Creek Splash Dam	Shoshone	47°05'46"N	116°06'02"W	3860
Hoffman And Duffy Dam	Elmore	43°11'12"N	115°35'18"W	3600
Holbrook Dam	Valley	44°14'42"N	116°08'12"W	
Hollenbeak Dam	Valley	44°50'30"N	116°02'36"W	
Hoodoo Mine Number One Dam	Custer	44°10'06"N	114°37'24"W	
Hornet Creek-Lower Dam	Washington	44°47'54"N	116°43'30"W	
Hornet Creek-Upper Dam	Washington	44°47'52"N	116°43'55"W	
Horsethief Basin Dam	Valley	44°30'20"N	115°55'15"W	
Hot Springs Number Two Dam	Elmore	43°04'57"N	115°29'53"W	
Howard Dam	Blaine	43°26'18"N	113°58'42"W	
Hubbard Dam	Ada	43°31'00"N	116°21'30"W	
Hughes Dam	Valley	44°46'43"N	116°01'12"W	4970
Hulet Dam	Owyhee	43°05'34"N	116°32'36"W	3320
Icehouse Creek Dam	Fremont	44°27'18"N	111°36'12"W	
Idaho Diversion Dam	Jefferson	43°38'00"N	112°03'30"W	
Island Park Dam	Fremont	44°25'10"N	111°23'44"W	
Jemima K Dam	Valley	44°34'30"N	116°00'24"W	
Jenkins Creek Dam	Washington	44°21'22"N	116°59'26"W	
Johns Creek Dam	Idaho	45°52'48"N	116°08'30"W	
Johnson Dam	Franklin	42°06'30"N	111°48'18"W	4880
Josephine Creek Dam	Owyhee	42°45'24"N	116°40'48"W	

Dam	County	Latitude	Longitude	Elev.
Jug Creek Dam	Valley	44°50'30"N	116°00'00"W	
Juniper Creek Dam	Owyhee	42°37'24"N	116°57'00"W	
Jussila-Bow Dam	Valley	44°49'24"N	116°02'12"W	
Kelly Dam	Camas	43°18'24"N	114°37'07"W	5040
King Dam	Owyhee	43°16'54"N	116°57'30"W	
Knox Meadow Dam	Valley	44°53'56"N	116°01'54"W	
Koskella Dam	Valley	44°40'06"N	116°01'24"W	
Lake Cleveland Dam	Cassia	42°19'23"N	113°38'48"W	
Lamb-Weston Dam	Power	42°44'30"N	112°56'48"W	
Lamont Dam	Franklin	42°06'18"N	111°48'56"W	4880
Land Dam	Washington	44°19'12"N	116°32'12"W	
Last Chance Dam	Caribou	42°36'26"N	111°42'20"W	
Le Fevre Dam	Franklin	42°01'42"N	111°49'06"W	
Leatham Dam	Madison	43°42'30"N	111°39'36"W	
Lemhi River Weir	Lemhi	44°43'45"N	113°26'00"W	5730
Linderman Dam	Teton	43°55'41"N	111°23'05"W	
Little Blue Creek Dam	Owyhee	42°17'33"N	116°07'16"W	5425
Little Camas Dam	Elmore	43°21'10"N	115°23'05"W	4930
Little Canyon Dam	Elmore	43°00'14"N	115°19'56"W	2815
Little Crane Creek Dam	Washington	44°22'37"N	116°39'45"W	3500
Little Dam	Gem	44°01'06"N	116°30'00"W	
Little Dam	Washington	44°13'30"N	116°24'48"W	
Little Payette Lake Dam	Valley	44°54'13"N	116°02'26"W	5121
Little Valley Dam	Bonneville	43°05'35"N	111°31'15"W	6741
Little Wood Dam	Blaine	43°25'30"N	114°01'35"W	5240
Lone Tree Dam	Owyhee	42°53'54"N	117°00'18"W	
Long Tom Dam	Elmore	43°17'10"N	115°34'40"W	4420
Lost Valley Dam	Adams	44°57'24"N	116°27'57"W	4777
Louie Lake Dam	Valley	44°51'17"N	115°57'55"W	7020
Louisa Creek Dam	Owyhee	42°45'12"N	116°36'54"W	
Lower Idaho Falls Dam	Bonneville	43°28'18"N	112°03'36"W	
Lower Salmon Falls Dam	Gooding	42°50'30"N	114°54'24"W	
Lucky Friday Number One Dam	Shoshone	47°27'54"N	115°48'18"W	
Lucky Friday Number Three Dam	Shoshone	47°28'00"N	115°45'30"W	
Lucky Peak Dam	Ada	43°31'42"N	116°03'08"W	3080
MacKay Dam	Custer	43°57'12"N	113°40'24"W	
Mackay Dam	Custer	43°57'11"N	113°40'24"W	6061
Magic Dam	Blaine	43°15'16"N	114°21'40"W	4808
Mann Creek Dam	Washington	44°23'32"N	116°53'41"W	2883
Marble Creek Splash Dam	Shoshone	47°12'12"N	116°03'27"W	2920
McArthur Dam	Boundary	48°31'12"N	116°26'30"W	2090
McHan Dam	Camas	43°13'59"N	114°39'31"W	5445
McMullen Creek Dam	Twin Falls	42°24'42"N	114°25'30"W	
Middle Idaho Falls Dam	Bonneville	43°29'48"N	112°02'30"W	
Mikesell Number One - Lower Dam	Fremont	44°07'30"N	111°34'32"W	5340
Mikesell Number Two - Upper Dam	Fremont	44°07'34"N	111°34'55"W	
Mill Creek Lake Dam	Lemhi	44°39'22"N	113°39'24"W	8760
Milner Dam	Twin Falls	42°31'28"N	114°00'32"W	4134
Minidoka Dam	Cassia	42°40'06"N	113°29'03"W	4204
Molony Dam	Valley	44°52'36"N	115°54'30"W	
Monroe Dam	Washington	44°24'12"N	116°59'06"W	
Montpelier Creek Dam	Bear Lake	42°20'57"N	111°10'25"W	6520
Moore Diversion	Butte	43°47'11"N	113°21'30"W	
Moose Creek Dam	Latah	46°52'25"N	116°25'07"W	
Mormon Dam	Camas	43°16'51"N	114°47'57"W	5050
Morrison Dam	Franklin	42°00'00"N	111°48'12"W	
Mosquito Flat Dam	Custer	44°31'19"N	114°26'00"W	6940
Mountain Home Dam	Elmore	43°09'16"N	115°39'36"W	3283
Mountain View Dam	Owyhee	42°03'20"N	116°10'06"W	5325
Moyie Dam	Boundary	48°44'00"N	116°10'24"W	
Mud Lake Dam	Jefferson	43°52'24"N	112°24'24"W	
Munsee Dam	Twin Falls	42°34'48"N	114°55'30"W	
Murtaugh Lake Dam	Twin Falls	42°28'10"N	114°10'02"W	4130
Nash Dam	Franklin	42°05'30"N	111°59'00"W	
Nettleton Dam	Owyhee	42°50'42"N	116°36'00"W	
Nicholson Dam	Ada	43°24'36"N	116°19'00"W	
Nielson Dam	Franklin	42°04'30"N	111°58'12"W	
Nixon Number One Dam	Boundary	48°46'28"N	116°15'39"W	2985
Nixon Number Two Dam	Boundary	48°46'18"N	116°15'48"W	
Nougue Dam	Owyhee	42°04'30"N	116°58'00"W	
Oakley Dam	Cassia	42°11'48"N	113°54'58"W	4780
Olson Dam	Elmore	43°15'42"N	115°08'00"W	
Oneida Dam	Franklin	42°16'34"N	111°44'56"W	4895
Orchard Dam	Ada	43°23'18"N	116°01'00"W	3340
Osborn Tailings Dam	Shoshone	47°29'36"N	115°59'00"W	
Outlet Dam	Bonner	48°29'26"N	116°54'10"W	
Oxbow Dam	Adams	44°58'06"N	116°50'30"W	
Oxford Dam	Franklin	42°16'57"N	112°00'33"W	4895
Pacific Land Company Dam	Owyhee	43°11'12"N	116°25'00"W	
Paddock Valley Dam	Washington	44°11'53"N	116°35'52"W	3213
Paine Creek Dam	Owyhee	42°15'17"N	116°05'14"W	
Palisades Dam	Bonneville	43°19'58"N	111°12'06"W	5639
Park Dam	Clark	44°09'06"N	112°30'30"W	
Patterson Dam	Lincoln	43°07'24"N	114°31'54"W	
Paul Dam	Clark	44°27'47"N	112°20'10"W	7619
Payette Lake Dam	Valley	44°54'42"N	116°07'30"W	
Pershall Dam	Owyhee	42°47'48"N	116°54'12"W	
Pettit Lake Fish Barrier Dam	Blaine	43°59'00"N	114°51'44"W	6700
Pioneer Dam	Gooding	42°59'42"N	115°03'48"W	3100
Pitkin Dam	Valley	44°23'44"N	115°58'26"W	4825
Pops Dam	Fremont	44°06'30"N	111°22'18"W	
Poro Dam	Valley	44°51'08"N	116°00'01"W	
Portneuf Dam	Caribou	42°52'42"N	111°56'35"W	
Post Falls Dam	Kootenai	47°42'32"N	116°57'10"W	
Pot Hole Creek Dam	Owyhee	42°42'28"N	115°28'34"W	3611
Priest Lake Dam	Bonner	48°29'26"N	116°54'09"W	2445

Dams

Dam	County	Latitude	Longitude	Elev.
Rainbow Dam	Bear Lake	42°15'01"N	111°17'16"W	5941
Rattlesnake Creek Dam	Owyhee	42°17'04"N	115°59'44"W	5750
Red Pine Canyon Dam	Bear Lake	42°29'48"N	111°22'18"W	
Reservoir A Dam	Nez Perce	46°22'18"N	116°51'24"W	
Rhead Dam	Elmore	43°23'36"N	115°24'48"W	
Richards Dam	Kootenai	47°41'48"N	117°00'24"W	
Rio Vista Dam	Valley	44°55'00"N	116°05'36"W	
Ririe Dam	Bonneville	43°34'54"N	111°44'30"W	5120
Rock Creek Dam	Owyhee	42°46'50"N	116°36'47"W	5144
Sage Hen Dam	Gem	44°19'34"N	116°11'40"W	4928
Saint John Dam	Oneida	42°13'39"N	112°17'19"W	4733
Salmon Falls Creek Dam	Twin Falls	42°35'12"N	114°54'00"W	
Salmon Falls Dam	Twin Falls	42°12'44"N	114°44'00"W	5030
Samaria Number One Lower Dam	Oneida	42°11'42"N	112°21'34"W	4690
Samaria Number Two Upper Dam	Oneida	42°12'24"N	112°22'00"W	
Scott Creek Dam	Washington	44°17'48"N	117°01'36"W	
Scrappy Creek Dam	Owyhee	42°31'52"N	116°21'55"W	
Shaw Twins-Upper Dam	Valley	44°53'18"N	115°57'24"W	
Sheep Creek Dam	Bear Lake	42°14'33"N	111°08'35"W	6120
Sheridan Dam	Clark	44°27'40"N	111°41'10"W	6452
Shoofly Dam	Owyhee	42°15'42"N	116°18'30"W	
Shoshone Falls Dam	Jerome	42°35'42"N	114°24'00"W	
Silver King Mines Number One Dam	Adams	45°06'00"N	116°40'30"W	
Simplot Effluent Irrigation Dam	Bannock	42°56'42"N	112°25'48"W	
Simplot Number Six Tailings Dam	Caribou	42°44'00"N	111°28'36"W	
Skein Lake Dam	Valley	44°28'40"N	116°06'37"W	6896
Slack Dam	Owyhee	42°03'23"N	116°27'32"W	5050
Smalley Dam	Valley	44°35'35"N	116°00'42"W	4971
Snow Creek Dam	Owyhee	42°07'10"N	115°58'24"W	
Soda Creek Dam	Caribou	42°41'12"N	111°36'30"W	
Soda Dam	Caribou	42°38'39"N	111°41'46"W	
Soldiers Meadow Dam	Nez Perce	46°10'03"N	116°44'09"W	4525
Sommerville Dam	Owyhee	42°27'54"N	116°45'30"W	
Sonner Dam	Blaine	43°14'06"N	114°11'42"W	
Spencer Dam	Owyhee	42°49'05"N	116°38'34"W	5103
Spring Creek Dam	Camas	43°15'54"N	114°39'00"W	5097
Spring Valley Dam	Latah	46°47'00"N	116°45'06"W	
Squaw Creek Dam	Owyhee	42°11'41"N	116°05'30"W	5460
Stanley Lake Fish Barrier Dam	Custer	44°14'58"N	115°02'28"W	6500
Star Tailings Number Five Dam	Shoshone	47°29'42"N	115°53'12"W	
Star Tailings Number Four Dam	Shoshone	47°29'30"N	115°53'00"W	
Star Tailings Number One Dam	Shoshone	47°29'36"N	115°53'18"W	
Star Tailings Number Six Dam	Shoshone	47°28'48"N	115°53'24"W	
Star Tailings Number Three Dam	Shoshone	47°29'24"N	115°53'06"W	
Star Tailings Number Two Dam	Shoshone	47°29'30"N	115°53'12"W	
Stewart Dam	Bear Lake	42°15'09"N	111°17'16"W	5943
Stewart Ranch Dam	Caribou	42°33'18"N	111°25'42"W	
Stock Valley Dam	Franklin	42°19'48"N	111°50'12"W	

Dam	County	Latitude	Longitude	Elev.
Stoneridge Dam	Bonner	48°00'42"N	117°00'18"W	
Strickland Dam	Owyhee	42°26'28"N	115°47'59"W	
Strongarm Reservoir Number One Dam	Franklin	42°14'02"N	111°51'40"W	4960
Strongarm Reservoir Number Two Dam	Franklin	42°13'06"N	111°53'54"W	
Sublett Dam	Cassia	42°19'29"N	113°02'45"W	5341
Summit Dam	Custer	44°18'23"N	113°30'27"W	6593
Sunbeam Dam (historical)	Custer	44°16'12"N	114°43'57"W	5920
Sunset Palisades Dam	Nez Perce	46°23'30"N	117°02'12"W	
Sunshine Tailings Number One Dam	Shoshone	47°31'36"N	116°03'06"W	
Sunshine Tailings Number Two Dam	Shoshone	47°31'24"N	116°03'24"W	
Surmeier Dam	Payette	43°57'42"N	116°55'30"W	
Swain Number One Dam	Washington	44°16'30"N	116°40'00"W	
Swain Number Two Dam	Washington	44°15'18"N	116°40'00"W	
Swan Falls Dam	Ada	43°14'36"N	116°22'18"W	
Taylor Dam	Elmore	42°56'54"N	115°05'42"W	
Teapot-Lower Dam	Elmore	43°01'20"N	115°29'13"W	3134
Teapot-Upper Dam	Elmore	43°01'52"N	115°29'14"W	3145
Teton Dam	Madison	43°54'35"N	111°32'18"W	5030
Texas Basin Dam	Owyhee	43°10'42"N	116°58'24"W	
Thompson Dam	Nez Perce	46°28'24"N	116°23'06"W	
Thompson Lake Dam	Kootenai	47°29'24"N	116°44'36"W	
Thorn Creek Dam	Gooding	43°11'24"N	114°35'48"W	5505
Tindall Dam	Owyhee	42°05'10"N	115°57'59"W	6360
Tingey Dam	Franklin	42°02'48"N	111°57'30"W	
Toliver Dam	Bonneville	43°27'00"N	111°26'12"W	
Tom J Dam	Valley	44°32'36"N	116°00'00"W	
Trail Creek Dam	Blaine	43°41'30"N	114°21'00"W	
Trail Dam	Elmore	43°03'18"N	115°19'30"W	
Trail Diversion Dam	Elmore	43°03'00"N	115°17'06"W	
Tripod Dam	Valley	44°17'24"N	116°05'56"W	4980
Troy Dam	Latah	46°48'15"N	116°48'37"W	3246
Turner Butte Dam	Owyhee	42°23'51"N	116°09'12"W	
Twenty-Four Mile Dam	Caribou	42°54'12"N	111°53'10"W	
Twin Buttes Number One Dam	Bingham	43°30'00"N	112°35'00"W	
Twin Falls Dam	Jerome	42°35'18"N	114°21'24"W	
Twin Granite Dam	Adams	45°06'12"N	116°10'48"W	
Twin Lakes Dam	Kootenai	47°51'26"N	116°51'55"W	2306
Twin Lakes Northwest Dam	Franklin	42°12'00"N	111°58'54"W	4779
Twin Lakes Southwest Dam	Franklin	42°11'12"N	111°58'24"W	
Tyson Dam	Owyhee	43°14'54"N	116°44'18"W	
Unit Three Haulroad Dam	Caribou	42°50'30"N	111°25'06"W	
Upper Deep Creek Dam	Oneida	42°11'34"N	112°08'45"W	5340
Upper Duncan Creek Dam	Owyhee	42°24'23"N	116°03'59"W	
Upper Idaho Falls Number One Dam	Bonneville	43°33'24"N	112°02'54"W	
Upper Idaho Falls Number Two Dam	Bonneville	43°33'18"N	112°03'00"W	
Upper Payette Lake Dam	Valley	45°06'40"N	116°01'32"W	5558
Upper Salmon Falls Dam	Gooding	42°45'54"N	114°53'42"W	
Van Orden Dam	Franklin	42°00'24"N	111°51'06"W	

Dam	County	Latitude	Longitude	Elev.
Walker Dam	Elmore	43°01'18"N	115°11'24"W	2710
Warm Springs Creek Dam	Lemhi	45°07'48"N	113°52'06"W	
Warren Diversion Dam	Valley	44°29'12"N	115°57'48"W	
Watson Dam	Elmore	42°59'18"N	115°43'42"W	
Welling Number Two Dam	Bear Lake	42°24'30"N	111°21'24"W	
Weston Dam	Oneida	42°07'10"N	112°06'50"W	5520
Williams Dam	Twin Falls	42°16'20"N	114°25'18"W	5748
Wilson Lake Dam	Jerome	42°37'44"N	114°10'17"W	4012
Winchester Dam	Lewis	46°14'15"N	116°37'06"W	3910
Winder Dam	Franklin	42°11'00"N	111°53'08"W	4874
Wiregrass Dam	Bannock	42°28'57"N	112°12'14"W	4732
Woodall Dam	Valley	44°12'48"N	116°09'12"W	
Yellow Belly Lake Fish Barrier Dam	Custer	44°00'01"N	114°51'20"W	7060

Farm or Ranch	County	Latitude	Longitude	Elev.
45 Ranch	Owyhee	42°10'22"N	116°52'21"W	
Al Sadie Ranch	Owyhee	42°45'06"N	116°00'13"W	
Allemans Ranch	Caribou	42°33'15"N	111°10'30"W	
Allen Ranch	Caribou	42°47'04"N	111°25'20"W	
Allison Ranch	Idaho	45°33'59"N	115°13'21"W	
Antelope Creek Ranch	Bonneville	43°32'11"N	111°31'51"W	
Antelope Spring Ranch	Twin Falls	42°14'55"N	114°46'51"W	
Arbaugh Ranch	Boise	43°57'38"N	115°50'39"W	
Asay Ranch	Custer	43°55'03"N	113°28'55"W	
Austin Ranch	Caribou	43°00'32"N	111°54'13"W	5961
Bacon Reservoir Ranch (historical)	Jerome	42°47'31"N	114°24'10"W	
Badley Ranch	Idaho	45°21'28"N	115°31'13"W	
Barkdull Ranch	Power	42°36'15"N	113°07'05"W	
Battle Creek Ranch	Owyhee	42°31'44"N	116°20'42"W	
Beartrap Ranch	Idaho	45°52'17"N	115°29'13"W	4180
Beaver Lake Ranch	Bonner	48°00'33"N	116°43'12"W	
Bemis Ranch	Idaho	45°25'58"N	115°40'07"W	2320
Bennett Ranch	Caribou	42°44'51"N	111°27'43"W	
Bennett Ranch	Elmore	43°21'47"N	115°47'30"W	
Berry Ranch	Elmore	43°05'01"N	115°15'08"W	
Big Creek Ranch	Twin Falls	42°02'56"N	114°22'58"W	
Big Springs Ranch	Owyhee	42°34'44"N	116°26'29"W	
Biggs Ranch	Custer	44°43'04"N	114°47'59"W	
Bitton Ranch	Caribou	42°41'30"N	111°21'45"W	
Blackstone Ranch	Owyhee	42°26'34"N	115°47'55"W	4766
Bonus Cove Ranch	Owyhee	43°04'19"N	116°12'25"W	
Bowman Ranch	Canyon	43°42'59"N	116°52'02"W	
Boyle Ranch	Custer	44°32'57"N	114°51'02"W	
Brace Ranch	Owyhee	42°21'35"N	116°41'36"W	
Brassey Ranch	Boise	43°54'56"N	115°59'32"W	
Bray Lake Ranch	Gooding	43°00'41"N	114°54'53"W	
Brockie Ranch	Butte	43°32'36"N	113°38'31"W	
Broken Circle Ranch	Cassia	42°18'21"N	113°07'11"W	
Buckeye Ranch	Gooding	42°46'42"N	114°55'36"W	
Buckhorn Ranch	Cassia	42°21'45"N	114°04'09"W	
Bush Ranch	Owyhee	43°17'00"N	116°57'42"W	
Butte Ranch	Gem	44°08'11"N	116°25'05"W	
C Ranch	Owyhee	42°23'38"N	117°01'27"W	5250
Calvin Ranch	Idaho	45°33'00"N	116°24'00"W	4840
Cameron Ranch	Valley	44°45'17"N	114°52'52"W	5940
Campbell Ranch	Power	42°36'59"N	113°00'04"W	
Carlson Ranch (historical)	Custer	44°17'25"N	113°42'30"W	6800
Castro Ranch	Owyhee	42°24'49"N	116°40'21"W	
China Creek Ranch	Twin Falls	42°03'41"N	114°48'05"W	
Christensen Ranch	Bonneville	43°30'34"N	111°33'14"W	
Circle C Ranch	Adams	45°02'19"N	116°15'33"W	
Circle C Ranch	Idaho	45°38'13"N	116°28'05"W	
Circle C Ranch	Idaho	45°36'08"N	116°16'38"W	1700
Circle C Ranch	Idaho	45°26'41"N	116°20'46"W	2250
Clark Ranch	Bonneville	43°35'19"N	111°36'35"W	
Clark Ranch	Twin Falls	42°21'22"N	114°55'41"W	
Clements Ranch	Caribou	42°48'21"N	111°32'59"W	
Clover Flat Ranch	Owyhee	42°27'16"N	115°22'27"W	4210
Conover Ranch	Twin Falls	42°21'51"N	114°57'51"W	
Cook Ranch	Idaho	45°38'17"N	115°17'32"W	
Cougar Creek Ranch	Custer	44°44'19"N	114°55'11"W	
Cove Ranch	Idaho	45°25'12"N	115°31'22"W	5660
Crofoot Ranch	Idaho	45°33'46"N	115°08'47"W	3801
CX Ranch	Bonner	48°06'58"N	116°10'35"W	2080
Davidson Ranch	Boise	43°44'23"N	116°00'42"W	
Davis Ranch	Idaho	45°34'34"N	116°26'22"W	4020
Deep Creek Meadows Ranch	Twin Falls	42°13'39"N	114°35'26"W	
Deer Creek Ranch	Blaine	43°33'53"N	114°21'58"W	
Degan Ranch	Lemhi	45°02'43"N	114°21'02"W	
Devil Creek Ranch	Twin Falls	42°06'47"N	115°01'59"W	5780
Devils Farm	Idaho	45°17'47"N	116°33'46"W	
Dixie Ranch	Elmore	42°59'18"N	116°04'35"W	
Dobson Ranch	Boise	43°50'34"N	116°10'15"W	
Doublespring Ranch	Custer	44°20'04"N	113°45'09"W	
Dougal Ranch	Owyhee	42°39'18"N	117°00'15"W	5015
Doulls Ranch	Caribou	42°48'29"N	111°32'31"W	

Farms & Ranches

Farm or Ranch	County	Latitude	Longitude	Elev.
Duck Springs Ranch	Twin Falls	42°14'02"N	114°33'17"W	
Edie Ranch	Clark	44°22'44"N	112°36'15"W	
Elk Creek Ranch	Fremont	44°26'44"N	111°21'20"W	
Elkhead Ranch	Power	42°43'17"N	112°31'55"W	
Elkhorn Ranch	Custer	44°03'43"N	113°49'47"W	
Emery Ranch	Cassia	42°00'59"N	114°01'41"W	
Erickson Ranch	Shoshone	47°50'16"N	116°16'40"W	
Falconberry Ranch	Lemhi	44°41'54"N	114°45'49"W	4808
Farmers Union	Kootenai	47°39'02"N	116°51'37"W	
Faulkner Ranch	Caribou	42°55'53"N	112°00'09"W	
Fenstermaker Ranch	Cassia	42°37'28"N	113°13'40"W	
Finlayson Ranch	Caribou	42°43'29"N	111°26'53"W	
Fir Grove Ranch	Camas	43°12'27"N	114°45'05"W	
Flying H Ranch	Owyhee	42°12'33"N	116°10'53"W	
Flynn Creek Ranch	Idaho	45°54'51"N	116°42'24"W	
Fox Ranch	Caribou	42°47'44"N	111°28'03"W	
Frish Ranch	Cassia	42°02'29"N	114°05'17"W	
G H Hall Ranch	Bear Lake	42°19'05"N	111°03'29"W	6131
Gabica Ranch	Owyhee	43°09'20"N	116°25'51"W	
Garden Ranch	Power	42°44'11"N	112°48'05"W	
Gattin Ranch	Lemhi	45°29'13"N	114°32'29"W	
Ghost Ranch	Owyhee	43°00'33"N	116°39'15"W	
Gilmore Ranch	Idaho	45°46'33"N	115°52'53"W	
Gilmore Ranch	Owyhee	43°04'25"N	116°34'12"W	3420
Gordy Ranch	Owyhee	42°50'45"N	116°26'46"W	
Hancock Ranch	Bingham	43°12'20"N	111°44'26"W	
Hansen Ranch	Oneida	42°15'07"N	112°49'10"W	
Hart Ranch	Owyhee	42°20'38"N	115°52'54"W	
Henry Baker Ranch	Idaho	45°33'08"N	116°22'54"W	4600
Hibbs Ranch	Idaho	45°20'16"N	116°38'48"W	2007
Hood Ranch Campground	Valley	44°43'50"N	114°59'37"W	4353
Hot Springs Ranch	Blaine	43°20'00"N	113°55'00"W	4764
Hotzel Ranch	Idaho	45°22'28"N	115°11'12"W	
House Creek Ranch	Twin Falls	42°08'42"N	115°00'25"W	5677
Howard Ranch	Idaho	45°24'08"N	116°05'45"W	
Howell Ranch	Fremont	44°05'16"N	111°11'59"W	
Idaho County Poor Farm	Idaho	45°54'19"N	116°04'30"W	3460
Idaho Ranch	Caribou	42°40'11"N	111°42'44"W	5783
Idaho Rocky Mountain Ranch	Custer	44°06'23"N	114°50'51"W	
Idaho Youth Ranch	Minidoka	42°48'21"N	113°35'28"W	
Indian Creek Guest Ranch	Lemhi	45°27'32"N	114°09'02"W	4300
Indian Creek Ranch	Idaho	45°25'22"N	115°35'32"W	
Jacoby Ranch	Clark	44°11'13"N	112°01'07"W	
James Ranch	Idaho	45°24'07"N	115°37'11"W	
Jarvis Ranch (historical)	Custer	44°13'06"N	113°56'23"W	6960
Jensen Ranch	Cassia	42°31'23"N	113°09'23"W	
Jerome Game Farm	Jerome	42°42'30"N	114°30'17"W	3815
Johnson Ranch	Fremont	44°37'48"N	111°26'59"W	

Farm or Ranch	County	Latitude	Longitude	Elev.
Jones Ranch	Owyhee	42°00'16"N	115°31'22"W	
Jougelard Ranch	Caribou	42°44'23"N	111°28'50"W	
Juniper Basin Ranch	Owyhee	42°04'44"N	116°28'20"W	
Juniper Ranch	Owyhee	42°22'49"N	115°19'58"W	4430
Kempner Ranch	Boise	43°58'23"N	115°34'35"W	5300
Kingsbury Ranch	Lemhi	44°58'20"N	114°20'25"W	
Knudsen Ranch	Caribou	42°41'55"N	111°21'12"W	
KV Farm Spur	Boundary	48°58'51"N	116°29'28"W	
Lake Creek Ranch	Lemhi	45°00'27"N	113°56'07"W	
Lakeview Farm	Minidoka	42°32'33"N	113°55'32"W	
Lamb Ranch	Idaho	45°47'54"N	115°57'47"W	
Lazy F C Ranch	Cassia	42°30'40"N	113°16'08"W	
Leacock Ranch	Lemhi	45°08'16"N	114°12'48"W	
Linderman Ranch	Teton	43°55'17"N	111°22'34"W	
Little Ponderosa Ranch	Idaho	45°44'57"N	115°23'18"W	4421
Lone Pine Ranch (historical)	Custer	44°25'33"N	113°44'32"W	
Lone Tree Ranch	Owyhee	42°48'14"N	117°01'31"W	
Long Tom Ranch	Elmore	43°20'25"N	115°35'03"W	
Lowrey Ranch	Owyhee	42°32'30"N	116°58'05"W	
Lynes Ranch	Adams	45°07'52"N	116°44'23"W	
Mallard Creek Ranch	Idaho	45°36'54"N	115°19'28"W	
Marble Creek Ranch	Valley	44°46'21"N	114°59'12"W	
McComas Ranch (historical)	Idaho	45°54'25"N	115°55'05"W	3220
McCrea Ranch	Fremont	44°27'12"N	111°24'30"W	
McGinness Ranch	Elmore	42°58'43"N	115°24'40"W	
McLead Ranch	Idaho	45°28'17"N	116°33'10"W	1600
McMeekin Ranch	Idaho	45°28'49"N	115°55'50"W	3080
Merritt Ranch	Lemhi	44°59'11"N	114°20'51"W	
Mile High Ranch	Valley	45°08'45"N	115°00'31"W	5680
Miller Ranch	Caribou	42°45'12"N	111°03'50"W	6293
Miller Ranch	Lemhi	45°04'41"N	114°15'36"W	
Milo Dry Farm	Bonneville	43°34'36"N	111°47'22"W	
Monasterio Ranch	Owyhee	42°24'13"N	116°14'32"W	
Montini Ranch	Owyhee	43°09'54"N	116°23'26"W	
Mormon Ranch	Lemhi	44°56'58"N	114°43'47"W	
Morrow Ranch	Owyhee	42°28'02"N	116°09'38"W	
Mumford Ranch	Bear Lake	42°16'14"N	111°04'35"W	
Nate Ranch	Caribou	42°34'30"N	111°09'14"W	
Neil Ranch	Bingham	43°17'30"N	112°14'31"W	
Nez Perce County Farm	Nez Perce	46°22'29"N	116°53'36"W	
North Star Ranch	Idaho	45°56'13"N	114°49'39"W	
Orchard Ranch (historical)	Ada	43°20'14"N	116°03'22"W	
Oyler Ranch	Custer	44°46'09"N	114°15'08"W	
Parker Ranch	Owyhee	42°46'24"N	116°01'48"W	
Perkins Ranch	Caribou	42°51'09"N	111°43'58"W	
Petersen Ranch	Bonneville	43°31'01"N	111°30'22"W	
Peterson Ranch	Lemhi	45°03'54"N	114°02'35"W	
Petersons Ranch	Caribou	42°46'22"N	111°03'59"W	6200

Farm or Ranch	County	Latitude	Longitude	Elev.
Petterson Ranch	Caribou	42°39'52"N	111°20'31"W	
Pine Creek Ranch (historical)	Lemhi	45°20'33"N	114°15'38"W	
Pistol Creek Ranch	Valley	44°43'55"N	115°08'51"W	4796
Point Ranch	Twin Falls	42°09'29"N	114°39'32"W	
Polly Bemis Ranch	Idaho	45°25'58"N	115°40'06"W	2160
Portlock Ranch	Owyhee	42°15'32"N	116°08'41"W	
Priest Ranch	Owyhee	43°16'38"N	116°24'24"W	
R Allen Ranch	Caribou	42°47'00"N	111°23'12"W	
Railroad Ranch	Fremont	44°20'20"N	111°28'06"W	
Reed Ranch	Bonneville	43°34'30"N	111°35'43"W	
Rehn Ranch	Cassia	42°33'11"N	113°11'46"W	
Reno Ranch	Clark	44°01'42"N	112°42'53"W	
Rock Creek Ranch	Blaine	43°21'37"N	114°23'19"W	
Romine Ranch	Idaho	45°22'55"N	115°37'13"W	
Rood Ranch (historical)	Lemhi	45°16'21"N	114°20'02"W	
Root Ranch	Idaho	45°18'56"N	115°01'30"W	
Rosenkranze Ranch	Custer	43°53'51"N	113°58'50"W	
Ross Point Bird Farm	Kootenai	47°42'36"N	116°52'34"W	3340
Rowley Ranch	Blaine	43°20'54"N	114°28'54"W	
Roy Johnson Ranch (historical)	Owyhee	42°56'35"N	115°57'24"W	
Running Creek Ranch	Idaho	45°54'58"N	114°49'42"W	
Rutherford Ranch	Owyhee	42°00'48"N	115°31'25"W	
Seven L Ranch	Elmore	43°24'08"N	115°56'20"W	
Shepp Ranch	Idaho	45°26'03"N	115°39'57"W	5508
Shoofly Ranch	Owyhee	42°57'55"N	116°03'24"W	
Simplot Ranch	Lemhi	44°47'57"N	114°48'12"W	4040
Six-Mile Ranch	Cassia	42°07'15"N	113°11'33"W	5740
Soelberg Ranch	Butte	43°35'44"N	113°22'34"W	
Spaulding Ranch	Bonneville	43°36'12"N	111°34'12"W	
Spencer Ranch	Idaho	45°51'07"N	116°39'39"W	
Splettstosser Ranch	Lemhi	44°49'09"N	114°25'49"W	
Spring Camp Ranch	Idaho	45°51'50"N	116°34'32"W	
Spring Ranch	Owyhee	43°03'56"N	116°32'48"W	3700
Stanley Ranch	Shoshone	47°15'15"N	115°55'17"W	
Star Lake Ranch	Lincoln	42°50'40"N	114°12'18"W	
Star Ranch	Boise	43°54'34"N	115°58'22"W	
Star Ranch	Owyhee	42°27'09"N	116°45'08"W	
Star Valley Ranch	Owyhee	42°02'26"N	116°56'20"W	
Stephens Fish Creek Ranch	Fremont	44°07'03"N	111°17'02"W	
Stewart Ranch	Blaine	43°25'34"N	114°23'42"W	
Stewart Ranch	Caribou	42°44'20"N	111°19'37"W	
Stocking Ranch	Caribou	42°51'45"N	111°23'22"W	
Stocking Ranch	Caribou	42°49'31"N	111°20'59"W	
Stonebraker Ranch	Idaho	45°23'22"N	115°11'48"W	
Strong Ranch	Caribou	42°45'19"N	111°26'53"W	
Sulphur Creek Ranch	Valley	44°32'18"N	115°22'15"W	
Summit Ranch	Custer	44°18'34"N	113°31'36"W	
Swauger Ranch (historical)	Custer	44°21'21"N	113°36'12"W	
Sweet Ranch	Caribou	42°42'29"N	111°22'02"W	
Tappen Ranch	Lemhi	44°52'24"N	114°45'58"W	
Thomson Ranch	Elmore	43°23'35"N	115°50'07"W	
Thorn Creek Ranch	Latah	46°36'30"N	117°00'20"W	
Tindall Ranch	Owyhee	42°16'00"N	115°52'11"W	
Tindall Ranch	Owyhee	42°15'57"N	115°52'49"W	
Tokembamy Ranch	Owyhee	42°03'49"N	115°47'26"W	
Tollgate Ranch	Idaho	45°49'10"N	116°03'48"W	
Twin Peaks Ranch	Lemhi	44°56'57"N	113°59'15"W	
Twin Springs Ranch	Owyhee	42°17'40"N	116°41'14"W	4790
Van Duesan Ranch	Gem	43°59'48"N	116°30'19"W	
Vent Ranch	Caribou	42°43'37"N	111°22'40"W	
Walters Ranch	Idaho	45°41'24"N	116°27'35"W	
Warm Creek Ranch	Cassia	42°20'44"N	114°01'27"W	
Webster Ranch	Fremont	44°37'53"N	111°25'48"W	
West Steadman Ranch	Caribou	42°49'32"N	111°29'44"W	
Western Home Ranch	Power	42°47'28"N	112°57'31"W	
Wheeler Ranch	Bonneville	43°32'20"N	111°35'29"W	
Whitewater Ranch	Idaho	45°31'55"N	115°18'19"W	
Whitson Ranch	Owyhee	42°55'51"N	116°05'43"W	
Wild Rose Ranch	Fremont	44°40'09"N	111°23'58"W	
Wiley Ranch	Owyhee	42°14'03"N	116°32'39"W	
Willow Ranch	Elmore	43°17'10"N	115°50'39"W	
Wilson Ranch	Idaho	45°34'00"N	116°29'49"W	1400
Wood Ranch	Custer	44°28'30"N	113°48'50"W	5420
Woodall Ranch	Caribou	42°45'46"N	111°32'22"W	
Yellowjacket Ranch	Lemhi	44°58'53"N	114°30'18"W	
Yorks Ranch (historical)	Valley	44°55'18"N	116°04'45"W	5030
Young Ranch	Caribou	42°43'30"N	111°18'49"W	

Highway Intersections

Intersection	County	Latitude	Longitude
I 15 I 15B	Bannock	42°53'51"N	112°26'07"W
I 15 I 15B	Bannock	42°50'23"N	112°24'38"W
I 15 I 15B	Bannock	42°47'56"N	112°21'48"W
I 15 I 15B	Bannock	42°47'55"N	112°15'46"W
I 15 I 15B	Bannock	42°47'37"N	112°14'37"W
I 15 I 15B	Bingham	43°11'56"N	112°21'50"W
I 15 I 15B	Bonneville	43°29'48"N	112°03'13"W
I 15 I 15B	Bonneville	43°26'21"N	112°07'17"W
I 15 I 86	Bannock	42°54'52"N	112°26'25"W
I 15 US 20	Bonneville	43°30'16"N	112°03'21"W
I 15 US 20	Bonneville	43°29'48"N	112°03'13"W
I 15 US 26	Bingham	43°11'56"N	112°21'50"W
I 15 US 26	Bonneville	43°26'21"N	112°07'17"W
I 15 US 30	Bannock	42°39'39"N	112°12'41"W
I 15 US 91	Bannock	42°30'08"N	112°10'10"W
I 15 US 91	Bingham	43°08'60"N	112°23'19"W
I 15 SR 22	Clark	44°10'35"N	112°14'25"W
I 15 SR 33	Jefferson	43°49'51"N	112°11'42"W
I 15 SR 36	Oneida	42°14'04"N	112°13'50"W
I 15 SR 38	Oneida	42°11'00"N	112°13'58"W
I 15 SR 40	Bannock	42°25'50"N	112°10'56"W
I 15 SR 48	Jefferson	43°43'02"N	112°08'04"W
I 15B US 26	Bonneville	43°29'27"N	112°02'22"W
I 15B US 30	Bannock	42°52'37"N	112°27'02"W
I 15B US 30	Bannock	42°39'38"N	112°11'39"W
I 15B US 91	Bannock	42°53'30"N	112°27'05"W
I 15B US 91	Bannock	42°29'44"N	112°09'46"W
I 15B US 91	Bingham	43°11'19"N	112°20'44"W
I 15B US 91	Bonneville	43°26'21"N	112°04'45"W
I 84 I 84B	Canyon	43°40'48"N	116°41'21"W
I 84 I 84B	Canyon	43°35'57"N	116°30'50"W
I 84 I 84B	Elmore	43°10'25"N	115°44'48"W
I 84 I 84B	Elmore	43°05'16"N	115°36'52"W
I 84 I 86	Cassia	42°34'14"N	113°31'04"W
I 84 I 184	Ada	43°35'50"N	116°17'42"W
I 84 US 20	Ada	43°33'59"N	116°11'46"W
I 84 US 20	Canyon	43°41'24"N	116°41'35"W
I 84 US 20	Canyon	43°39'44"N	116°39'36"W
I 84 US 20	Elmore	43°08'21"N	115°39'50"W
I 84 US 26	Ada	43°33'59"N	116°11'46"W
I 84 US 26	Canyon	43°41'24"N	116°41'35"W
I 84 US 26	Canyon	43°39'44"N	116°39'36"W
I 84 US 26	Gooding	42°56'16"N	115°00'02"W
I 84 US 26	Gooding	42°55'25"N	114°56'01"W
I 84 US 30	Cassia	42°34'14"N	113°31'04"W
I 84 US 30	Gooding	42°56'16"N	115°00'02"W
I 84 US 30	Minidoka	42°34'08"N	113°44'12"W
I 84 US 30	Payette	43°54'24"N	116°48'56"W
I 84 US 93	Jerome	42°38'31"N	114°26'45"W
I 84 US 95	Payette	43°58'16"N	116°54'51"W
I 84 SR 21	Ada	43°32'32"N	116°09'41"W
I 84 SR 24	Minidoka	42°34'08"N	113°44'12"W
I 84 SR 25	Jerome	42°43'27"N	114°32'47"W
I 84 SR 25	Jerome	42°34'36"N	114°04'10"W
I 84 SR 25	Minidoka	42°34'35"N	113°55'57"W
I 84 SR 27	Minidoka	42°34'06"N	113°47'17"W
I 84 SR 44	Canyon	43°42'20"N	116°41'58"W
I 84 SR 46	Gooding	42°45'59"N	114°42'13"W
I 84 SR 50	Jerome	42°34'38"N	114°17'47"W
I 84 SR 51	Elmore	43°08'22"N	115°39'50"W
I 84 SR 55	Ada	43°35'46"N	116°21'16"W
I 84 SR 55	Ada	43°35'36"N	116°23'37"W
I 84 SR 55	Canyon	43°36'01"N	116°34'23"W
I 84 SR 69	Ada	43°35'36"N	116°23'37"W
I 84 SR 77	Cassia	42°34'08"N	113°37'26"W
I 84 SR 78	Elmore	42°56'47"N	115°25'42"W
I 84 SR 79	Jerome	42°41'20"N	114°31'06"W
I 84B SR 19	Canyon	43°40'10"N	116°42'12"W
I 84B SR 45	Canyon	43°34'40"N	116°33'46"W
I 84B SR 51	Elmore	43°07'56"N	115°41'44"W
I 84B SR 51	Elmore	43°07'35"N	115°41'38"W
I 84B SR 55	Canyon	43°36'17"N	116°36'08"W
I 84B SR 55	Canyon	43°35'08"N	116°34'23"W
I 86 US 30	Power	42°54'45"N	112°31'45"W
I 86 US 91	Bannock	42°54'46"N	112°27'58"W
I 86 SR 37	Power	42°44'57"N	112°52'47"W
I 86 SR 39	Power	42°47'34"N	112°49'43"W
I 90 I 90B	Kootenai	47°41'59"N	116°48'27"W
I 90 I 90B	Kootenai	47°40'05"N	116°45'05"W
I 90 US 95	Kootenai	47°41'55"N	116°47'31"W
I 90 SR 3	Kootenai	47°34'25"N	116°26'32"W
I 90 SR 4	Shoshone	47°28'18"N	115°55'06"W
I 90 SR 41	Kootenai	47°42'39"N	116°53'43"W
I 90 SR 97	Kootenai	47°37'26"N	116°38'55"W
I 90B US 95	Kootenai	47°41'14"N	116°47'49"W
I 184 US 20	Ada	43°37'08"N	116°14'02"W
I 184 US 26	Ada	43°37'08"N	116°14'02"W
US 2 US 95	Bonner	48°16'34"N	116°33'11"W
US 2 US 95	Bonner	48°16'23"N	116°32'52"W
US 2 US 95	Boundary	48°43'53"N	116°17'60"W
US 2 SR 2	Bonner	48°11'02"N	117°02'23"W
US 2 SR 57	Bonner	48°10'50"N	116°54'56"W
US 2 SR 200	Bonner	48°17'26"N	116°32'56"W
US 12 US 95	Nez Perce	46°26'52"N	116°51'10"W
US 12 US 95	Nez Perce	46°25'51"N	116°59'24"W
US 12 SR 3	Nez Perce	46°28'18"N	116°46'13"W
US 12 SR 3	Nez Perce	46°27'11"N	116°48'60"W
US 12 SR 7	Clearwater	46°28'40"N	116°15'32"W
US 12 SR 7	Clearwater	46°28'23"N	116°15'15"W
US 12 SR 11	Lewis	46°23'31"N	116°10'45"W
US 12 SR 13	Idaho	46°08'53"N	115°58'59"W
US 12 SR 62	Lewis	46°13'45"N	116°01'46"W
US 12 SR 128	Nez Perce	46°25'33"N	117°00'02"W
US 20 US 20B	Fremont	43°58'18"N	111°38'06"W
US 20 US 20B	Fremont	43°57'44"N	111°40'55"W
US 20 US 20B	Jefferson	43°43'37"N	111°52'22"W
US 20 US 20B	Jefferson	43°40'57"N	111°54'19"W
US 20 US 20B	Jefferson	43°39'38"N	111°55'13"W
US 20 US 26	Blaine	43°18'07"N	113°56'46"W
US 20 US 26	Butte	43°30'23"N	112°53'21"W
US 20 US 26	Elmore	43°08'21"N	115°39'50"W
US 20 US 30	Elmore	43°08'21"N	115°39'50"W
US 20 US 93	Blaine	43°18'07"N	113°56'46"W
US 20 US 93	Butte	43°38'12"N	113°18'00"W
US 20 US 95	Canyon	43°52'23"N	116°57'17"W
US 20 US 95	Canyon	43°46'00"N	116°54'43"W
US 20 SR 18	Canyon	43°47'14"N	116°56'50"W
US 20 SR 22	Butte	43°36'26"N	113°09'57"W
US 20 SR 32	Fremont	44°04'18"N	111°27'22"W
US 20 SR 33	Madison	43°49'34"N	111°48'41"W
US 20 SR 43	Bonneville	43°35'50"N	111°57'56"W
US 20 SR 44	Ada	43°38'57"N	116°16'48"W
US 20 SR 46	Camas	43°20'32"N	114°42'45"W
US 20 SR 47	Fremont	44°20'23"N	111°25'55"W
US 20 SR 48	Jefferson	43°40'13"N	111°54'43"W
US 20 SR 51	Elmore	43°08'21"N	115°39'50"W
US 20 SR 55	Ada	43°39'47"N	116°21'16"W
US 20 SR 75	Blaine	43°19'56"N	114°16'44"W
US 20 SR 84	Fremont	44°29'55"N	111°20'17"W
US 20 SR 87	Fremont	44°38'04"N	111°19'38"W
US 20B SR 48	Jefferson	43°40'13"N	111°54'54"W
US 26 US 30	Gooding	42°55'25"N	114°56'22"W

Intersection	County	Latitude	Longitude
US 26 US 91	Bonneville	43°26'21"N	112°04'45"W
US 26 US 93	Lincoln	42°56'07"N	114°24'25"W
US 26 US 95	Canyon	43°52'23"N	116°57'17"W
US 26 US 95	Canyon	43°46'00"N	116°54'43"W
US 26 SR 18	Canyon	43°47'14"N	116°56'50"W
US 26 SR 22	Butte	43°36'26"N	113°09'57"W
US 26 SR 29	Bonneville	43°36'33"N	111°48'16"W
US 26 SR 31	Bonneville	43°27'05"N	111°20'23"W
US 26 SR 39	Bingham	43°12'03"N	112°22'05"W
US 26 SR 43	Bonneville	43°32'24"N	111°57'56"W
US 26 SR 44	Ada	43°38'57"N	116°16'48"W
US 26 SR 46	Gooding	42°55'26"N	114°42'46"W
US 26 SR 55	Ada	43°39'47"N	116°21'16"W
US 26 SR 75	Lincoln	42°56'10"N	114°24'22"W
US 26 SR 75	Lincoln	42°56'10"N	114°24'22"W
US 30 US 89	Bear Lake	42°19'20"N	111°17'52"W
US 30 US 89	Bear Lake	42°19'03"N	111°17'52"W
US 30 US 93	Twin Falls	42°33'50"N	114°34'29"W
US 30 US 93	Twin Falls	42°33'47"N	114°28'44"W
US 30 US 95	Payette	44°01'31"N	116°55'24"W
US 30 US 95	Payette	43°58'28"N	116°54'51"W
US 30 US 95	Washington	44°14'46"N	116°58'13"W
US 30 SR 24	Minidoka	42°34'08"N	113°44'12"W
US 30 SR 27	Cassia	42°32'17"N	113°47'36"W
US 30 SR 34	Caribou	42°39'17"N	111°35'41"W
US 30 SR 34	Caribou	42°38'57"N	111°43'47"W
US 30 SR 50	Twin Falls	42°32'53"N	114°21'53"W
US 30 SR 72	Payette	43°57'09"N	116°48'45"W
US 30 SR 74	Twin Falls	42°33'21"N	114°28'12"W
US 30 SR 81	Cassia	42°32'14"N	113°46'10"W
US 89 SR 36	Bear Lake	42°17'18"N	111°23'54"W
US 89 SR 61	Bear Lake	42°21'33"N	111°02'60"W
US 91 SR 36	Franklin	42°06'33"N	111°52'37"W
US 91 SR 36	Franklin	42°05'46"N	111°52'36"W
US 91 SR 40	Bannock	42°25'50"N	112°07'49"W
US 93 SR 24	Lincoln	42°55'47"N	114°24'37"W
US 93 SR 25	Jerome	42°43'26"N	114°26'35"W
US 93 SR 28	Lemhi	45°10'22"N	113°53'14"W
US 93 SR 43	Lemhi	45°41'37"N	113°56'53"W
US 93 SR 74	Twin Falls	42°30'46"N	114°34'29"W
US 93 SR 75	Custer	44°28'31"N	114°12'31"W
US 93 SR 75	Lincoln	42°56'10"N	114°24'22"W
US 93 SR 79	Jerome	42°49'34"N	114°26'02"W
US 93 SR 79	Jerome	42°36'32"N	114°27'08"W

Intersection	County	Latitude	Longitude
US 95 US 95B	Idaho	46°03'38"N	116°20'34"W
US 95 US 95B	Idaho	46°02'44"N	116°20'40"W
US 95 US 95B	Lewis	46°15'18"N	116°36'15"W
US 95 US 95B	Lewis	46°14'20"N	116°34'41"W
US 95 US 95B	Payette	44°04'55"N	116°55'24"W
US 95 US 95B	Payette	44°03'21"N	116°55'32"W
US 95 US 195	Nez Perce	46°28'27"N	117°02'12"W
US 95 US 195N	Nez Perce	46°28'51"N	117°02'21"W
US 95 SR 1	Boundary	48°53'33"N	116°21'28"W
US 95 SR 5	Benewah	47°20'06"N	116°53'16"W
US 95 SR 6	Latah	46°55'43"N	116°56'02"W
US 95 SR 6	Latah	46°55'12"N	116°57'04"W
US 95 SR 7	Idaho	45°56'22"N	116°09'37"W
US 95 SR 8	Latah	46°43'57"N	117°00'06"W
US 95 SR 8	Latah	46°43'36"N	117°00'07"W
US 95 SR 13	Idaho	45°55'37"N	116°07'51"W
US 95 SR 18	Canyon	43°47'14"N	116°56'50"W
US 95 SR 19	Canyon	43°40'07"N	116°54'43"W
US 95 SR 19	Owyhee	43°37'03"N	116°56'03"W
US 95 SR 52	Payette	44°04'21"N	116°55'24"W
US 95 SR 53	Kootenai	47°49'04"N	116°47'17"W
US 95 SR 54	Kootenai	47°56'52"N	116°41'56"W
US 95 SR 55	Adams	44°58'17"N	116°16'53"W
US 95 SR 55	Owyhee	43°32'42"N	116°51'07"W
US 95 SR 58	Kootenai	47°25'40"N	116°59'09"W
US 95 SR 60	Benewah	47°13'43"N	116°55'52"W
US 95 SR 62	Lewis	46°14'27"N	116°28'03"W
US 95 SR 66	Latah	46°51'20"N	117°01'17"W
US 95 SR 71	Washington	44°34'18"N	116°40'37"W
US 95 SR 200	Bonner	48°17'25"N	116°32'56"W
US 95B SR 52	Payette	44°04'21"N	116°56'07"W

Intersection	County	Latitude	Longitude

Hospitals

Hospital	County	Latitude	Longitude	Elev.
Bannock Memorial Hospital	Bannock	42°52'19"N	112°25'50"W	
Bingham Memorial Hospital	Bingham	43°11'05"N	112°20'35"W	
Bonner General Hospital	Bonner	48°16'40"N	116°32'55"W	
Boundary Cnty Comm. Hospital & Nursing Home	Boundary	48°42'10"N	116°20'15"W	
Breast Cancer Detection Centers	Ada	43°37'05"N	116°23'36"W	
Canyon View Hospital	Twin Falls	42°33'57"N	114°29'00"W	3710
Cassia Memorial Hospital	Cassia	42°31'28"N	113°48'06"W	
Eastern Idaho Regional Behavioral Health Cntr	Bonneville	43°28'30"N	111°59'50"W	
Elks Hospital	Ada	43°37'00"N	116°11'26"W	2700
Elmore Memorial Hospital	Elmore	43°08'24"N	115°41'32"W	
Family Emergency Center	Bonneville	43°28'58"N	111°59'10"W	
Family Emergency Center West	Bonneville	43°29'45"N	112°03'45"W	
Franklin County Medical Center	Franklin	42°05'50"N	111°52'21"W	4720
Gritman Medical Center	Latah	46°43'44"N	116°59'57"W	
Harms Memorial Hospital and Nursing Home	Power	42°46'56"N	112°50'56"W	4420
Kootenai Medical Center	Kootenai	47°41'52"N	116°47'48"W	2220
Latter Day Saint Hospital	Bonneville	43°29'55"N	112°02'25"W	
Lost River Hospital	Butte	43°38'22"N	113°17'40"W	
Magic Valley Memorial Hospital	Twin Falls	42°33'56"N	114°29'39"W	
Magic Valley Regional Medical Center	Twin Falls	42°33'56"N	114°29'37"W	3690
Magic Valley Speech and Hearing Clinic	Twin Falls	42°34'14"N	114°27'15"W	3708
Mercy Medical Center	Canyon	43°34'08"N	116°33'15"W	2500
Minidoka Memorial Hospital	Minidoka	42°37'15"N	113°41'09"W	4156
New Hope Center Hospital	Ada	43°37'28"N	116°16'30"W	2700
North Idaho Cancer Treatment Center	Kootenai	47°41'40"N	116°47'25"W	
North Idaho Day Surgery and Laser Center	Kootenai	47°41'50"N	116°47'51"W	
North Idaho Immediate Care Center	Kootenai	47°41'48"N	116°47'57"W	
Northview Hospital	Ada	43°37'36"N	116°16'29"W	2680
Pine Crest Hospital	Kootenai	47°41'53"N	116°47'57"W	
Rigby Hospital	Madison	43°49'25"N	111°46'55"W	
River Crest Hospital	Nez Perce	46°24'10"N	117°01'13"W	
Sacred Heart Hospital	Bonneville	43°28'30"N	112°01'54"W	
Saint Alphonsus Hospital	Ada	43°36'48"N	116°15'19"W	2715
Saint Anthony Hospital (historical)	Bannock	42°52'35"N	112°26'42"W	
Saint Benedicts Family Medical Center	Jerome	42°43'48"N	114°31'08"W	
Saint Lukes Hospital	Ada	43°36'48"N	116°11'30"W	2700
Shoshone Medical Center	Shoshone	47°32'38"N	116°07'40"W	
State Hospital North	Clearwater	46°29'30"N	116°15'34"W	
State Hospital South	Bingham	43°11'15"N	112°20'02"W	
Veterans Hospital	Ada	43°37'17"N	116°11'25"W	2700
Western Rehabilitation Institute	Bonneville	43°28'15"N	111°59'40"W	
Wood River Medical Center	Blaine	43°41'54"N	114°21'08"W	5980

Island	County	Latitude	Longitude	Elev.
Argy Island	Owyhee	43°22'06"N	116°37'19"W	
Banks Island	Payette	44°04'57"N	116°58'15"W	2137
Bartoo Island	Bonner	48°32'45"N	116°52'34"W	
Bay Island	Caribou	42°56'04"N	111°37'57"W	6125
Bayha Island	Owyhee	43°22'10"N	116°37'28"W	
Bear Island	Bonneville	43°03'35"N	111°24'52"W	6660
Beaver Island	Power	42°40'55"N	112°59'19"W	
Becky Island	Owyhee	43°22'06"N	116°37'19"W	
Big Cottonwood Island	Canyon	43°49'40"N	117°01'36"W	2180
Big Foot Island	Idaho	45°49'05"N	116°18'22"W	
Big Island	Madison	43°51'40"N	111°52'42"W	
Big Island (historical)	Clearwater	46°39'32"N	116°00'15"W	1340
Big Rocky Island	Canyon	43°19'57"N	116°35'47"W	
Billy Goat Island	Canyon	43°45'41"N	117°01'28"W	
Bird Island	Blaine	42°40'03"N	113°22'05"W	
Birding Island	Payette	44°00'50"N	116°49'03"W	
Bishop Island	Bonneville	43°02'55"N	111°24'02"W	6427
Blackburn Island	Canyon	43°26'39"N	116°43'52"W	
Blacks Island (historical)	Owyhee	42°56'26"N	115°44'20"W	
Blackwell Island	Kootenai	47°40'54"N	116°48'09"W	
Blind Island	Canyon	43°20'57"N	116°36'13"W	
Bridge Island	Canyon	43°51'23"N	117°00'25"W	
Brooks Island	Canyon	43°21'19"N	116°36'32"W	
Buck Island	Caribou	42°55'18"N	111°33'29"W	6210
Butler Island	Jefferson	43°39'29"N	111°44'27"W	
Chain Island	Caribou	42°54'35"N	111°35'57"W	6130
Cinder Island	Caribou	42°52'50"N	111°36'01"W	6190
Clarks Island	Canyon	43°35'45"N	116°52'23"W	
Cormorant Island	Caribou	42°58'30"N	111°39'27"W	6110
Cottage Island	Bonner	48°13'28"N	116°19'20"W	
Cottontail Island	Owyhee	43°27'30"N	116°45'01"W	2225
Cougar Island	Valley	44°58'13"N	116°04'38"W	
Crane Island	Caribou	42°55'19"N	111°37'49"W	6135
Crow Island	Minidoka	42°32'04"N	113°45'19"W	
Crow Island	Payette	43°56'07"N	116°57'49"W	
Current Island	Canyon	43°23'03"N	116°39'27"W	
Custer Island	Minidoka	42°32'53"N	113°48'42"W	
Derr Island	Bonner	48°08'46"N	116°13'29"W	
Dilley Island	Canyon	43°28'20"N	116°46'19"W	
Dolman Island	Twin Falls	42°46'06"N	114°54'23"W	
Dredge Island	Canyon	43°26'17"N	116°43'36"W	
Eagle Island	Ada	43°40'59"N	116°22'40"W	
Eagle Island	Adams	44°51'51"N	116°52'32"W	1801
Edgington Island (historical)	Owyhee	42°56'29"N	115°46'49"W	
Eightmile Island	Bonner	48°34'49"N	116°50'31"W	
Fir Island	Nez Perce	46°31'02"N	116°39'55"W	
Fischer Island	Canyon	43°22'58"N	116°39'03"W	
Fisherman Island	Bonner	48°17'02"N	116°25'02"W	2114

Island	County	Latitude	Longitude	Elev.
Foglers Island	Canyon	43°27'08"N	116°44'21"W	
Fourmile Island	Bonner	48°31'42"N	116°51'25"W	
Fruit Island	Canyon	43°28'31"N	116°46'36"W	
Fuller Island	Gem	43°53'38"N	116°28'32"W	
Gamble Island	Payette	43°54'45"N	116°58'08"W	
Goat Island	Canyon	43°38'59"N	116°57'41"W	
Goat Island	Cassia	42°32'19"N	113°45'44"W	
Goat Island	Power	42°40'40"N	112°59'27"W	
Gold Island	Canyon	43°50'10"N	117°01'04"W	2188
Gold Isle	Owyhee	42°59'05"N	116°04'53"W	
Goldeneye Island	Canyon	43°25'43"N	116°42'46"W	
Goose Egg Island	Canyon	43°45'57"N	117°01'31"W	
Gridley Island	Gooding	42°45'36"N	114°52'36"W	
Guffey Island	Canyon	43°18'00"N	116°32'10"W	2610
Gull Island	Canyon	43°30'54"N	116°36'09"W	2535
Gull Island	Caribou	42°53'52"N	111°36'46"W	6250
Harbor Island	Kootenai	47°41'48"N	116°52'12"W	
Heltons Island	Owyhee	43°37'31"N	116°56'02"W	
Hermit Island	Owyhee	43°24'07"N	116°40'41"W	
Heron Island	Canyon	43°49'46"N	117°01'10"W	2185
Hog Island	Cassia	42°33'12"N	113°46'31"W	
Hog Island	Nez Perce	46°26'51"N	116°52'23"W	
Holbrook Island	Nez Perce	46°25'18"N	117°00'58"W	
I P Bills Island	Fremont	44°25'40"N	111°26'18"W	
Kalispell Island	Bonner	48°34'06"N	116°53'56"W	
Kellers Island	Jefferson	43°46'04"N	112°01'53"W	
Kelly Island	Bonner	48°11'36"N	117°02'16"W	
Kid Island	Kootenai	47°38'49"N	116°47'42"W	
Little Banks Island	Payette	44°05'10"N	116°58'21"W	
Little Island	Madison	43°51'01"N	111°53'20"W	
Little Rocky Island	Canyon	43°19'47"N	116°35'45"W	
Long Island	Caribou	42°54'02"N	111°36'16"W	6130
Long Island	Washington	44°12'52"N	116°58'06"W	
Loveridge Island (historical)	Owyhee	42°56'14"N	115°45'59"W	
McConnel Island	Canyon	43°45'23"N	116°55'34"W	2220
McRea Island	Washington	44°14'50"N	117°01'03"W	
McTucker Island	Blaine	43°01'52"N	112°38'25"W	
Memaloose Island	Bonner	48°12'13"N	116°17'46"W	2094
Millet Island	Gooding	42°46'14"N	114°55'51"W	
Nettle Island	Caribou	43°01'01"N	111°42'17"W	6105
Noble Island	Canyon	43°18'34"N	116°34'36"W	
O'Callaghans Island	Boundary	48°41'50"N	116°18'30"W	
Papoose Island	Bonner	48°33'19"N	116°53'27"W	
Parees Island	Cassia	42°31'32"N	113°40'27"W	
Pearl Island	Bonner	48°13'04"N	116°19'48"W	
Pleasant Valley Island	Idaho	45°38'31"N	116°29'00"W	
Poison Creek Island	Bingham	43°01'22"N	111°42'24"W	6110
Pool Island	Payette	44°06'42"N	116°55'18"W	
Popcorn Island	Kootenai	47°31'32"N	116°34'03"W	2220
Popike Island	Owyhee	43°21'50"N	116°37'07"W	
Rabbit Island	Canyon	43°36'19"N	116°54'32"W	
Raccoon Island	Canyon	43°24'42"N	116°41'33"W	
Rail Island	Canyon	43°17'48"N	116°31'02"W	
Rat Island	Bonner	48°11'56"N	117°02'20"W	
Richards Island	Canyon	43°22'39"N	116°38'41"W	
Rippee Island	Owyhee	43°23'42"N	116°40'27"W	
Rock Island	Idaho	46°04'53"N	115°25'21"W	
Rock Island	Washington	44°20'19"N	117°11'15"W	2120
Ross Island	Gem	43°52'53"N	116°29'37"W	
Schoffs Island	Elmore	42°56'10"N	115°25'56"W	
Settlers Island	Canyon	43°49'47"N	117°01'48"W	180
Sheep Island	Caribou	42°54'29"N	111°36'56"W	6220
Shellworth Island	Valley	44°57'35"N	116°03'37"W	
Ship Island	Lemhi	45°10'01"N	114°37'31"W	
Sign Island	Canyon	43°17'37"N	116°30'46"W	
Smith Island	Washington	44°11'49"N	116°58'02"W	
Smiths Island	Canyon	43°35'34"N	116°52'05"W	
Snells Island	Clearwater	46°29'52"N	116°23'03"W	
Sparlin Island	Elmore	42°55'48"N	115°29'37"W	
Split Island	Caribou	42°58'35"N	111°39'55"W	6130
Spring Island	Caribou	42°56'28"N	111°37'10"W	6125
Strong Island	Bonner	48°10'35"N	117°05'35"W	
Sugarloaf	Valley	44°36'47"N	116°05'02"W	
Tule Island	Power	42°37'34"N	113°08'47"W	
Twin Islands	Bonner	48°39'53"N	116°51'55"W	
Tyrrel Island (historical)	Elmore	42°57'40"N	115°47'17"W	
Walters Island	Owyhee	43°20'17"N	116°36'07"W	
Ware Island	Canyon	43°22'46"N	116°38'50"W	
Warren Island	Bonner	48°14'12"N	116°19'33"W	
Wildcat Island	Washington	44°14'26"N	116°58'18"W	
Wilkins Island	Owyhee	42°56'13"N	115°43'16"W	
Willow Island	Caribou	42°53'29"N	111°36'32"W	6130
Wright Island	Owyhee	43°22'18"N	116°37'46"W	

Lakes, Ponds, Reservoirs

Lake, Pond, or Reservoir	County	Latitude	Longitude	Elev.
26 Mile Lake	Jerome	42°46'17"N	114°10'41"W	
Aikers Reservoir	Owyhee	42°17'17"N	115°08'47"W	5175
Airplane Lake	Lemhi	45°09'22"N	114°35'51"W	8402
Alexander Reservoir	Caribou	42°38'39"N	111°41'46"W	5719
Alice Lake	Blaine	43°56'21"N	114°56'35"W	8596
Alidade Lake	Elmore	43°57'42"N	115°09'52"W	7936
Allan Lake	Lemhi	45°35'19"N	114°02'59"W	7810
Alligator Lake	Teton	43°42'37"N	111°15'45"W	
Alpine Lake	Custer	44°10'53"N	115°03'11"W	7823
Alpine Lake	Custer	44°03'51"N	115°00'38"W	8331
Alpine Lake	Kootenai	47°48'51"N	116°45'51"W	2430
Alpine Lake	Lemhi	45°04'39"N	114°36'57"W	8416
Alturas Lake	Blaine	43°54'55"N	114°51'30"W	7016
Amber Lakes	Blaine	43°50'22"N	114°28'25"W	
American Falls Reservoir	Gooding	42°45'54"N	114°53'42"W	
American Falls Reservoir	Oneida	42°46'51"N	112°52'29"W	4354
American Hill Lake	Idaho	45°48'47"N	115°25'56"W	
Anderson Lake	Kootenai	47°28'02"N	116°44'53"W	
Anderson Lake	Valley	44°53'10"N	115°55'49"W	7306
Anderson Ranch Reservoir	Elmore	43°21'27"N	115°26'52"W	
Anderson Reservoir	Owyhee	42°38'30"N	116°57'08"W	
Andys Lakes	Idaho	46°09'35"N	115°29'06"W	
Anns Lake	Valley	45°11'31"N	115°53'41"W	
Antelope Flat Reservoir	Owyhee	42°36'08"N	116°22'53"W	
Antelope Lake	Blaine	43°14'36"N	113°08'14"W	
Antelope Lake	Bonner	48°08'07"N	116°09'18"W	
Antelope Lakes	Clark	44°16'53"N	112°38'21"W	
Antelope Reservoir	Lincoln	42°55'40"N	113°58'06"W	
Antelope Reservoir	Minidoka	42°54'52"N	113°25'51"W	
Antelope Reservoir	Owyhee	42°30'57"N	116°43'46"W	
Appendix Lake	Idaho	45°19'48"N	116°33'15"W	
Ardeth Lake	Boise	43°57'58"N	115°00'44"W	8228
Arrow Point Reservoir	Owyhee	42°28'51"N	116°05'04"W	
Arrowhead Lake	Custer	43°44'41"N	114°05'36"W	9895
Arrowhead Lake	Elmore	43°58'11"N	115°07'00"W	8770
Arrowrock Reservoir	Boise	43°35'44"N	115°55'17"W	
Artillery Lake	Valley	44°40'05"N	115°13'42"W	
Ashton Reservoir	Fremont	44°04'44"N	111°29'44"W	5154
Aspitarte Lake	Minidoka	43°09'07"N	113°35'49"W	
Atwater Lake	Latah	46°41'19"N	116°38'52"W	
Avery Reservoir	Owyhee	42°30'44"N	116°39'45"W	5241
Avondale Lake	Kootenai	47°46'03"N	116°45'25"W	
Azure Lake	Elmore	43°57'58"N	115°07'54"W	8265
Bacon Lake	Lincoln	43°09'59"N	113°46'29"W	
Bacon Lake	Shoshone	46°58'04"N	115°16'09"W	
Badger Reservoir	Owyhee	42°12'18"N	115°55'29"W	
Bailing Reservoir	Lincoln	43°10'29"N	114°26'05"W	
Baird Reservoir	Minidoka	43°10'41"N	113°41'41"W	

Lake, Pond, or Reservoir	County	Latitude	Longitude	Elev.
Baker Lake	Blaine	43°41'27"N	114°40'30"W	8796
Baker Lake	Custer	44°03'23"N	114°33'50"W	8472
Bald Mountain Lake	Idaho	46°27'34"N	115°13'29"W	5732
Bald Mountain Reservoir	Owyhee	42°18'50"N	116°55'25"W	
Baldwin Lake	Blaine	43°13'46"N	113°42'38"W	
Baldy Lake	Idaho	45°18'50"N	116°34'16"W	7190
Ball Lakes	Boundary	48°47'36"N	116°37'13"W	6708
Banner Reservoir	Lincoln	42°54'56"N	114°00'14"W	
Baptie Lake	Custer	43°46'36"N	114°00'50"W	
Barinaga Lake	Owyhee	42°22'44"N	115°41'20"W	4790
Barking Fox Lake	Lemhi	45°07'49"N	114°36'37"W	8335
Baron Lake	Boise	44°04'53"N	114°59'31"W	8312
Baron Lakes	Boise	44°05'00"N	115°02'00"W	
Bartlett Lake	Bear Lake	42°24'21"N	111°29'02"W	7093
Barton Reservoir	Washington	44°19'12"N	116°55'18"W	
Basco Lake	Minidoka	43°03'58"N	113°40'45"W	
Basin Lake	Idaho	45°20'44"N	116°33'20"W	
Basin Lake	Idaho	45°19'07"N	114°47'22"W	7688
Basin Lake	Lemhi	44°50'30"N	113°51'08"W	8890
Bass Lake	Boundary	48°58'00"N	116°28'43"W	1746
Bass Lake	Twin Falls	42°36'22"N	114°27'54"W	
Bass Lakes	Camas	43°44'35"N	115°00'14"W	
Battle Creek Lakes	Owyhee	42°16'19"N	116°25'15"W	
Battle Lake	Idaho	46°13'05"N	114°35'42"W	6625
Bayhorse Lake	Custer	44°24'43"N	114°24'09"W	8584
Bead Lakes	Boise	44°04'15"N	115°01'19"W	
Bear Creek Lake	Custer	43°59'37"N	113°30'41"W	
Bear Creek Lake	Valley	44°36'19"N	115°35'59"W	
Bear Den Lake	Minidoka	43°10'05"N	113°29'58"W	
Bear Lake	Bear Lake	42°07'20"N	111°18'50"W	5923
Bear Lake	Custer	44°08'16"N	114°41'37"W	8899
Bear Lake	Fremont	44°07'13"N	111°11'18"W	
Bear Lake	Idaho	45°35'00"N	115°39'57"W	7395
Bear Lake	Idaho	45°26'11"N	115°50'38"W	
Bear Lake	Valley	45°10'03"N	115°42'45"W	7611
Bear Lake	Valley	45°02'35"N	115°16'56"W	
Bear Pete Lake	Idaho	45°18'27"N	115°57'18"W	
Bear Valley Lakes	Lemhi	44°48'15"N	113°52'00"W	9135
Bearsden Waterhole	Butte	43°23'42"N	113°29'55"W	
Beaver Lake	Bonner	48°12'20"N	116°24'51"W	
Beaver Lake	Bonner	48°01'59"N	116°42'18"W	
Beaver Lake	Boundary	48°31'12"N	116°22'25"W	
Beaver Lake	Fremont	44°08'00"N	111°06'20"W	
Beaver Lake	Idaho	46°33'41"N	114°26'07"W	
Beaver Pond	Fremont	44°12'40"N	111°16'41"W	
Beaver Ponds	Benewah	47°09'41"N	116°41'05"W	
Beaver Ponds	Fremont	44°38'02"N	111°14'11"W	
Beck Reservoir	Owyhee	42°07'49"N	116°22'39"W	

Lake, Pond, or Reservoir	County	Latitude	Longitude	Elev.
Beech Lake	Gem	44°15'29"N	116°17'02"W	
Beehive lake	Boundary	48°39'10"N	116°38'45"W	5975
Beehive Lakes	Boundary	48°39'21"N	116°38'57"W	
Bell Lake	Idaho	46°02'53"N	114°29'50"W	
Bellas Lakes	Custer	43°46'53"N	113°58'48"W	
Bells Lake	Benewah	47°20'19"N	116°32'07"W	
Ben Ross Reservoir	Adams	44°31'22"N	116°26'48"W	3146
Bench Lake	Custer	44°06'40"N	114°57'49"W	
Bench Lake	Custer	43°46'23"N	114°00'08"W	
Bend Reservoir	Jerome	42°48'09"N	114°05'02"W	
Benedict Lake	Boise	43°57'51"N	115°03'30"W	
Benewah Lake	Benewah	47°20'45"N	116°41'19"W	
Bennett Reservoir	Owyhee	42°35'25"N	116°36'48"W	
Berel Lake	Shoshone	46°56'58"N	115°15'28"W	5960
Berger Reservoir	Twin Falls	42°25'03"N	114°41'33"W	4515
Bernard Lake	Valley	44°26'57"N	115°30'52"W	
Berry Lake	Lemhi	45°13'00"N	113°37'32"W	
Bert Monday Reservoir	Owyhee	42°04'35"N	116°22'09"W	
Bettis Reservoir	Payette	44°05'00"N	116°33'48"W	
Betty Lake	Custer	43°47'10"N	114°00'40"W	10379
Big Bend Reservoir	Owyhee	42°15'52"N	116°49'03"W	
Big Blue Creek Reservoir	Owyhee	42°18'30"N	116°10'54"W	
Big Boulder Lakes	Custer	44°06'24"N	114°36'52"W	
Big Butte Reservoir	Butte	43°26'30"N	113°06'06"W	
Big Butte Reservoir Number Two	Butte	43°24'48"N	113°04'51"W	
Big Clear Lake	Lemhi	45°09'58"N	114°34'16"W	8562
Big Creek Lake	Franklin	42°04'46"N	111°33'02"W	
Big Draw Reservoir	Lincoln	43°07'11"N	114°29'35"W	4575
Big Fall Creek Lake	Custer	43°52'26"N	114°14'47"W	
Big Fill Reservoir	Bingham	43°05'22"N	112°45'54"W	
Big Fisher Lake	Boundary	48°50'25"N	116°34'00"W	6732
Big Fog Lake	Idaho	46°08'51"N	115°10'55"W	
Big Hazard Lake	Idaho	45°12'54"N	116°08'09"W	6968
Big Lake	Butte	43°23'46"N	113°07'05"W	
Big Lake	Custer	43°43'52"N	113°51'26"W	
Big Lake	Owyhee	42°31'58"N	115°40'00"W	
Big Lake	Owyhee	42°22'02"N	115°41'55"W	
Big Lakes	Owyhee	42°20'27"N	115°14'38"W	4790
Big Lookout Lake	Elmore	43°36'13"N	115°25'16"W	
Big Pass Reservoir	Owyhee	43°26'02"N	116°52'37"W	
Big Pool	Shoshone	47°46'57"N	116°04'28"W	2650
Big Rainbow Lake	Elmore	43°35'35"N	115°24'49"W	8126
Big Roaring River Lake	Elmore	43°37'10"N	115°26'40"W	
Big Sagehen Reservoir	Owyhee	42°39'22"N	116°18'00"W	
Big Sand Lake	Idaho	46°18'53"N	114°31'06"W	5285
Big Sink Waterhole	Butte	43°26'09"N	113°33'03"W	
Big Sinks Reservoir	Minidoka	42°55'19"N	113°27'48"W	
Big Talk Lake	Shoshone	46°59'21"N	115°48'36"W	

Lake, Pond, or Reservoir	County	Latitude	Longitude	Elev.
Big Trinity Lake	Elmore	43°37'27"N	115°25'38"W	
Bigfoot Lake	Idaho	46°00'31"N	114°35'01"W	6253
Bilk Lake	Idaho	45°56'13"N	115°03'02"W	
Bill Reservoir	Lincoln	42°47'20"N	114°00'51"W	
Billings Reservoir	Owyhee	42°03'06"N	115°46'24"W	
Bills Lake	Idaho	45°58'14"N	114°35'28"W	6761
Bills Lake	Valley	44°47'31"N	115°48'44"W	
Billy Snipe Reservoir	Oneida	42°08'55"N	112°17'28"W	
Binder Lake	Blaine	43°14'17"N	113°06'00"W	
Birdbill Lake	Lemhi	45°09'02"N	114°35'11"W	
Bishop Lake	Fremont	44°23'06"N	111°35'08"W	
Bitch Lake	Idaho	46°02'35"N	114°57'08"W	
Bitterbrush Reservoir	Owyhee	42°14'40"N	115°55'03"W	
Black Canyon Reservoir	Gem	43°55'50"N	116°26'08"W	2437
Black Canyon Reservoir	Owyhee	42°20'14"N	116°33'32"W	
Black Lake	Adams	45°11'16"N	116°33'31"W	
Black Lake	Clearwater	46°52'37"N	115°32'50"W	
Black Lake	Idaho	45°19'51"N	114°49'33"W	6438
Black Lake	Idaho	45°14'44"N	116°11'54"W	
Black Lake	Kootenai	47°27'43"N	116°39'34"W	
Black Lake	Valley	44°59'36"N	116°00'15"W	
Black Lake	Valley	44°48'24"N	115°23'35"W	
Black Leg Reservoir	Owyhee	42°03'22"N	115°51'00"W	
Blackbird Reservoir	Lincoln	43°05'17"N	113°43'35"W	
Blackfoot Reservoir	Bingham	43°00'17"N	111°42'59"W	
Blackfoot Reservoir	Caribou	42°50'36"N	111°36'12"W	
Blackmare Lake	Valley	44°46'12"N	115°48'13"W	
Blacks Creek Reservoir	Ada	43°27'45"N	116°08'40"W	
Blackstone Reservoir	Owyhee	42°26'28"N	115°47'59"W	4790
Blacktail Lake	Bonner	48°20'34"N	116°08'15"W	
Blackwell Lake	Valley	44°58'16"N	115°59'58"W	6589
Blair Trail Reservoir	Elmore	43°03'21"N	115°19'10"W	3472
Blanchard Lake	Bonner	48°00'31"N	116°58'11"W	2252
Bleak Creek Lake	Idaho	45°39'05"N	115°01'22"W	7205
Bliss Reservoir	Elmore	42°54'49"N	115°04'09"W	
Bloom Lake	Boundary	48°30'08"N	116°24'58"W	
Bloomington Lake	Bear Lake	42°08'42"N	111°34'29"W	
Blowout Reservoir	Blaine	43°17'42"N	113°33'08"W	
Blue Creek Reservoir	Fremont	44°13'26"N	111°36'39"W	5574
Blue Creek Reservoir	Owyhee	42°18'36"N	116°10'52"W	5416
Blue Jay Lake	Elmore	43°55'32"N	115°05'37"W	8500
Blue Jay Lake	Elmore	43°53'50"N	115°19'03"W	7675
Blue Lake	Bonner	48°16'38"N	116°49'38"W	2238
Blue Lake	Boundary	48°34'56"N	116°23'14"W	
Blue Lake	Custer	44°37'21"N	114°54'18"W	
Blue Lake	Franklin	42°02'42"N	111°34'25"W	8515
Blue Lake	Kootenai	47°29'27"N	116°41'13"W	
Blue Lake	Nez Perce	46°12'52"N	116°50'32"W	

Lakes, Ponds, Reservoirs

Lake, Pond, or Reservoir	County	Latitude	Longitude	Elev.
Blue Lake	Shoshone	46°57'22"N	115°37'08"W	
Blue Lake	Valley	45°07'54"N	115°42'53"W	7647
Blue Lake	Valley	44°24'34"N	116°08'01"W	
Blue Lakes	Jerome	42°36'48"N	114°28'15"W	
Blue Rock Lake	Boise	44°03'52"N	115°01'48"W	8205
Bluebird Reservoir	Minidoka	43°05'36"N	113°42'03"W	
Bluegill Lake	Twin Falls	42°36'58"N	114°53'37"W	
Bluegrass Lake	Minidoka	43°08'23"N	113°35'38"W	
Boise Cascade Lake	Ada	43°37'58"N	116°14'11"W	2660
Bonanza Lake	Power	42°42'41"N	113°04'23"W	
Bond Lake	Boundary	48°32'05"N	116°26'25"W	2114
Bonner Lake	Boundary	48°43'33"N	116°06'27"W	2489
Borden Lake	Elmore	42°57'10"N	115°58'39"W	2475
Border Reservoir	Lincoln	43°11'20"N	114°25'44"W	
Born Lakes	Custer	44°03'33"N	114°36'58"W	
Boston Mountain Lake	Idaho	45°38'30"N	115°10'52"W	
Bottle Lake	Bonner	48°42'53"N	116°52'30"W	
Bottleneck Lake	Boundary	48°39'30"N	116°36'04"W	
Bottleneck Lake	Minidoka	43°05'14"N	113°36'39"W	
Boulder Chain Lakes	Custer	44°04'20"N	114°34'35"W	
Boulder Lake	Custer	44°05'54"N	114°36'49"W	
Boulder Lake	Custer	43°47'52"N	114°07'47"W	
Boulder Lake	Valley	44°52'04"N	115°56'44"W	6973
Boulder Lakes	Blaine	43°50'38"N	114°30'44"W	
Boulder Meadows Reservoir	Valley	44°52'09"N	115°58'11"W	
Bower Reservoir	Owyhee	42°29'40"N	116°38'37"W	
Bowknot Lake	Custer	43°57'59"N	114°57'30"W	
Bowl Reservoir	Blaine	43°14'35"N	113°36'07"W	
Box Lake	Valley	45°01'30"N	115°58'54"W	
Boyd Lake	Idaho	46°16'13"N	114°32'06"W	6698
Boyle Creek Reservoir	Owyhee	42°01'48"N	116°08'59"W	5317
Brace Reservoir	Owyhee	42°21'12"N	116°42'24"W	
Bradley Lake	Boundary	48°32'55"N	116°26'14"W	2104
Brandon Lakes	Idaho	45°31'31"N	115°43'06"W	
Braxon Lake	Boise	44°05'16"N	115°00'43"W	8272
Bray Lake	Gooding	43°02'00"N	114°52'36"W	
Breakdown Reservoir	Owyhee	42°00'29"N	116°51'13"W	
Brigham Lake Reservoir	Minidoka	43°10'40"N	113°38'14"W	
Brigham Point Lake	Minidoka	42°55'23"N	113°30'00"W	
Brockie Lake	Custer	43°41'21"N	113°51'01"W	
Brockie Lake	Minidoka	42°56'24"N	113°25'30"W	
Broken Wagon Flat Reservoir	Owyhee	42°39'09"N	115°47'14"W	
Brokie Lake	Minidoka	43°05'57"N	113°41'20"W	
Brooks Lake	Boundary	48°41'30"N	116°36'47"W	
Brown Lake	Idaho	45°35'06"N	115°01'49"W	
Brownlee Reservoir	Washington	44°50'10"N	116°54'00"W	
Browns Lake	Elmore	43°55'58"N	115°08'35"W	8278
Browns Pond	Valley	44°55'00"N	115°57'48"W	5235
Brundage Reservoir	Adams	45°02'30"N	116°07'49"W	6218
Brush Corral Reservoir Number Two	Owyhee	42°08'26"N	115°53'18"W	
Brush Lake	Boundary	48°53'32"N	116°19'41"W	
Brush Lake	Valley	45°03'05"N	115°59'13"W	
Brushy Fork Lake	Idaho	45°58'50"N	114°34'16"W	
Buck Brush Reservoir	Camas	43°13'40"N	114°26'45"W	
Buck Lake	Idaho	45°59'08"N	115°06'59"W	7011
Buck Lake	Idaho	45°13'57"N	116°14'22"W	6883
Buck Lake	Lemhi	45°05'52"N	114°35'39"W	
Buck Lake	Valley	44°47'17"N	115°16'02"W	
Buck Lakes	Lemhi	44°46'54"N	113°50'09"W	8474
Buckbrush Reservoir	Camas	43°13'39"N	114°26'43"W	5350
Buckhorn Lake	Valley	44°54'00"N	115°51'28"W	6963
Buckhorn Mountain Lake	Valley	44°52'57"N	115°52'51"W	
Buckhorn Reservoir	Owyhee	42°07'10"N	115°58'24"W	6115
Buffalo Lake	Fremont	44°19'49"N	111°04'22"W	
Buffulo Skull Lake	Lemhi	44°39'45"N	113°41'00"W	
Bugler Reservoir	Lincoln	43°03'34"N	113°43'26"W	
Bull Camp Reservoir	Owyhee	42°00'52"N	116°46'07"W	
Bull Creek Reservoir	Owyhee	42°05'10"N	115°57'59"W	6361
Bull Frame Reservoirs	Owyhee	43°02'39"N	116°43'53"W	
Bull Trout Lake	Boise	44°17'56"N	115°15'08"W	
Bullhead Reservoir	Owyhee	42°34'44"N	116°37'01"W	
Bullshot Reservoir	Blaine	43°17'00"N	113°32'38"W	
Buncel Reservoir	Owyhee	42°22'53"N	116°06'26"W	
Bunting Reservoir	Minidoka	42°57'09"N	113°35'26"W	
Burchett Lake	Caribou	42°49'37"N	111°35'09"W	6112
Burkhart Reservoir	Owyhee	42°37'09"N	116°39'01"W	
Burnside Lake	Valley	45°02'57"N	115°57'17"W	
Burnt Creek Lake	Custer	44°08'34"N	113°37'39"W	
Burnt Knob Lakes	Idaho	45°42'16"N	114°59'12"W	
Burnt Lake	Owyhee	42°02'08"N	115°58'16"W	
Burts Reservoir	Owyhee	42°13'55"N	116°32'57"W	
Bussard Lake	Boundary	48°54'25"N	116°10'52"W	
Buster Lake	Custer	44°26'24"N	114°24'54"W	
Butch Reservoir	Owyhee	42°13'27"N	117°00'09"W	5330
Butte Slough	Jefferson	43°46'33"N	112°00'13"W	
Buttermilk Slough	Washington	44°11'22"N	116°54'12"W	
Bybee Reservoir	Owyhee	42°15'38"N	116°15'44"W	5432
C J Strike Reservoir	Elmore	42°56'48"N	115°58'30"W	
Cabinet Gorge Reservoir	Bonner	48°05'12"N	116°03'48"W	
Cache Creek Lakes	Lemhi	44°46'53"N	114°41'03"W	
Calendar Reservoir	Valley	44°33'16"N	115°59'39"W	4860
California Lake	Idaho	46°20'32"N	114°59'58"W	
California Lake	Idaho	45°19'48"N	115°49'44"W	
Cameron Reservoir	Blaine	43°26'34"N	113°56'19"W	5844
Camp Four Reservoir	Bingham	43°13'24"N	112°49'10"W	
Camp Lake	Elmore	43°56'20"N	115°00'10"W	

Lake, Pond, or Reservoir	County	Latitude	Longitude	Elev.
Camp Two Lake	Jerome	42°45'57"N	114°14'00"W	
Campbell Ponds	Clearwater	46°32'43"N	115°52'16"W	
Campbell Reservoir	Blaine	43°28'46"N	114°00'20"W	5649
Canter Spring Reservoir	Owyhee	43°21'26"N	116°57'23"W	
Canyon Lake	Boundary	48°54'09"N	116°37'26"W	5872
Canyon Lake	Idaho	45°51'48"N	114°25'02"W	6925
Cape Horn Lakes	Custer	44°24'26"N	115°07'56"W	
Carey Lake	Blaine	43°18'44"N	113°55'06"W	
Caribou Lake	Bonner	48°26'00"N	116°39'57"W	5192
Caribou Lakes	Boundary	48°50'32"N	116°45'37"W	
Carlotta Reservoir	Lincoln	43°04'32"N	113°44'36"W	
Carlson Lake	Custer	44°16'57"N	113°45'04"W	
Carpenter Reservoir	Blaine	43°12'49"N	113°43'25"W	
Carraway Reservoir	Blaine	43°14'04"N	113°41'56"W	
Cascade Reservoir	Valley	44°31'27"N	116°02'59"W	4828
Casino Lakes	Custer	44°10'34"N	114°48'59"W	
Castle Lake	Custer	44°02'49"N	114°34'34"W	9419
Castle Lake	Teton	43°39'11"N	111°17'57"W	
Cat Lakes	Boise	44°15'50"N	115°23'17"W	
Cathedral Lake	Lemhi	45°07'50"N	114°32'09"W	8282
Catherine Lake	Valley	45°01'58"N	115°15'59"W	
Catholic Lake	Owyhee	42°24'25"N	115°40'24"W	
Caton Lake	Valley	44°53'12"N	115°25'39"W	6106
Cave Lake	Kootenai	47°27'56"N	116°36'27"W	
Cayuse Lake	Clearwater	46°36'11"N	114°51'18"W	
CCC Rock Pond	Owyhee	42°06'57"N	115°17'28"W	5450
Cedar Creek Reservoir	Twin Falls	42°13'27"N	114°52'44"W	5225
Cedar Draw Lake	Twin Falls	42°32'10"N	114°36'24"W	
Cedar Lake	Shoshone	47°19'56"N	115°49'54"W	
Cedar Mesa Reservoir	Twin Falls	42°20'34"N	114°54'41"W	4696
Cedar Tree Trail Lakes	Owyhee	42°15'47"N	115°41'03"W	
Cedar Well	Power	42°37'14"N	112°32'45"W	5514
Center Creek Lakes	Idaho	45°34'22"N	115°02'29"W	
Center Lake	Idaho	45°14'06"N	116°03'22"W	
Center Lambert Reservoir	Owyhee	42°17'34"N	116°42'57"W	
Chain Lakes	Fremont	44°02'03"N	111°05'25"W	
Challis Creek Lakes	Custer	44°33'12"N	114°30'48"W	
Champion Lakes	Custer	44°00'19"N	114°41'24"W	
Chase Lake	Bonner	48°27'32"N	116°49'22"W	2495
Chatcolet Lake	Benewah	47°21'50"N	116°45'15"W	
Cheatgrass Reservoir	Jerome	42°45'04"N	114°03'36"W	
Chesterfield Reservoir	Caribou	42°52'42"N	111°56'35"W	
Chickadee Lake	Elmore	43°55'52"N	115°03'34"W	
Chilco Lake	Kootenai	47°51'48"N	116°43'54"W	
Chilcoot Lake	Valley	44°46'41"N	115°25'10"W	
Chimney Lake	Idaho	46°11'49"N	115°17'47"W	6115
Chimney Pot Reservoir	Owyhee	43°01'55"N	116°57'21"W	
Choke Cherry Reservoir	Camas	43°13'12"N	114°27'53"W	

Lake, Pond, or Reservoir	County	Latitude	Longitude	Elev.
Chokecherry Reservoir	Camas	43°13'12"N	114°27'52"W	5550
Cinnabar Reservoir	Owyhee	42°14'05"N	115°53'55"W	
Cirque Lake	Custer	44°06'25"N	114°37'14"W	
Cirque Reservoir	Camas	43°12'48"N	114°28'24"W	
Clark Lake	Fremont	44°44'06"N	111°22'27"W	
Clay Bottom Reservoir	Owyhee	42°11'05"N	116°37'04"W	
Clear Lake	Custer	43°46'20"N	113°58'35"W	
Clear Lake	Gooding	42°40'06"N	114°46'36"W	
Clear Lake Reservoir	Blaine	43°10'20"N	113°02'06"W	
Clear Lakes	Gooding	42°40'06"N	114°46'36"W	
Cleveland Reservoir	Owyhee	42°02'08"N	116°19'04"W	
Cliff Lake	Clearwater	46°52'49"N	115°29'39"W	
Cliff Lake	Elmore	43°56'29"N	115°07'25"W	
Cliff Lake	Lincoln	43°11'51"N	113°47'21"W	
Cly Lakes	Valley	45°01'28"N	115°54'24"W	
Cocolalla Lake	Bonner	48°07'35"N	116°37'03"W	2203
Coeur d'Alene Lake	Kootenai	47°42'32"N	116°57'10"W	
Coffee Cup Lake	Idaho	45°10'14"N	116°13'00"W	
Coffee Point Reservoir	Bingham	43°09'27"N	112°58'06"W	
Colburn Lake	Bonner	48°22'57"N	116°37'27"W	
Cold Lake	Clearwater	46°47'28"N	115°18'06"W	
Collie Lake	Custer	44°24'31"N	115°13'32"W	
Colt Lake	Idaho	46°24'05"N	114°43'38"W	6978
Combe Reservoir	Butte	43°43'22"N	113°10'26"W	
Community Lake	Minidoka	42°57'54"N	113°42'43"W	
Condie Reservoir	Franklin	42°12'30"N	111°52'18"W	4886
Confusion Lake	Elmore	43°56'21"N	115°03'10"W	
Continental Lake	Boundary	48°55'19"N	116°54'28"W	
Cony Lake	Boise	44°03'44"N	115°03'38"W	8769
Cooks Lake	Boundary	48°41'58"N	116°35'18"W	
Cooks Lake	Idaho	45°20'32"N	115°44'40"W	
Coolwater Lake	Idaho	46°09'38"N	115°27'13"W	6017
Copper Lake	Boundary	48°59'39"N	116°06'23"W	
Copper Lake	Shoshone	46°56'30"N	115°16'04"W	5757
Coquina Lake	Idaho	46°05'13"N	114°28'41"W	6900
Corbus Lake	Elmore	43°44'58"N	115°13'16"W	
Corky Lake	Idaho	46°19'54"N	114°31'09"W	6625
Corn Lake	Lemhi	45°23'43"N	114°36'55"W	7829
Corner Lake	Boundary	48°41'00"N	116°36'49"W	
Corner Reservoir	Blaine	43°12'50"N	113°14'42"W	
Corner Reservoir	Lincoln	42°58'55"N	114°14'09"W	4141
Cornice Lake	Custer	44°03'11"N	114°36'23"W	
Corral Creek Reservoir	Valley	44°27'47"N	115°55'40"W	
Corral Lake	Idaho	45°07'15"N	116°10'59"W	
Corral Reservoir	Lincoln	42°55'30"N	113°59'24"W	
Correction Reservoir	Owyhee	42°13'52"N	115°55'25"W	
Cottle Lake	Owyhee	43°14'30"N	116°49'36"W	
Cottontail Lake	Lincoln	42°58'38"N	113°45'20"W	

Lakes, Ponds, Reservoirs

Lake, Pond, or Reservoir	County	Latitude	Longitude	Elev.
Cottontail Waterhole	Bingham	43°13'02"N	112°56'27"W	
Cottontail Waterhole	Butte	43°28'18"N	112°51'10"W	
Cottonwood Lake	Idaho	45°17'01"N	114°47'14"W	8342
Cottonwood Reservoir	Lincoln	43°09'13"N	114°21'57"W	4586
Cottonwood Trail Pond	Owyhee	42°57'10"N	116°56'11"W	
Cougar Lake	Lemhi	44°50'32"N	114°19'02"W	
Cougar Lakes	Valley	44°50'14"N	115°48'37"W	
Cougar Reservoir	Camas	43°12'36"N	114°30'06"W	
Cove Arm Lake	Owyhee	42°58'15"N	115°51'44"W	
Cove Arm Lake	Owyhee	42°58'15"N	115°51'45"W	2460
Cove Lake	Custer	44°06'07"N	114°36'31"W	9842
Cove Lakes	Idaho	46°09'13"N	115°13'56"W	
Cow Creek Reservoir	Elmore	43°20'33"N	115°05'30"W	5239
Cow Lake	Valley	45°01'45"N	115°49'43"W	
Cow Pasture Reservoir Number 1	Bingham	43°16'52"N	112°55'06"W	
Cow Pasture Reservoir Number 2	Bingham	43°17'02"N	112°54'51"W	
Cowan Reservoir	Owyhee	42°05'42"N	115°34'30"W	
Coyote Flat Reservoir	Owyhee	42°09'26"N	116°47'17"W	
Coyote Hole Reservoir	Owyhee	42°08'36"N	116°47'49"W	
Coyote Water Hole	Butte	43°20'44"N	113°12'34"W	
Crag Lake	Shoshone	46°56'55"N	115°35'30"W	
Crane Creek Reservoir	Washington	44°21'25"N	116°36'57"W	3191
Crane Falls Lake	Owyhee	42°57'57"N	115°50'30"W	2460
Crane Falls Lake	Owyhee	42°57'54"N	115°50'27"W	
Cranes Reservoir	Bear Lake	42°34'27"N	111°25'36"W	
Crater Lake	Custer	44°08'33"N	114°36'29"W	8919
Crater Lake	Idaho	45°27'05"N	116°27'51"W	
Crater Lake	Lemhi	45°09'47"N	114°34'41"W	8707
Crater Lake	Lincoln	43°05'29"N	114°00'21"W	
Crater Lake	Minidoka	42°58'22"N	113°26'57"W	
Crater Lake	Owyhee	42°04'04"N	115°53'58"W	
Crater Lake	Shoshone	47°02'09"N	115°58'54"W	
Crater Lake	Valley	45°02'41"N	115°25'41"W	8103
Crater Reservoir	Lincoln	43°10'59"N	114°29'41"W	
Crater Reservoir	Lincoln	42°56'53"N	114°12'43"W	
Crawfish Spring	Owyhee	42°07'08"N	115°15'24"W	5295
Cream Can Lake	Minidoka	43°11'03"N	113°36'33"W	
Crescent Lake	Idaho	45°35'29"N	115°39'28"W	7609
Crimson Lake	Custer	44°25'14"N	114°53'29"W	8330
Cross Road Reservoir	Lincoln	43°09'34"N	114°29'15"W	
Crossing Lake	Minidoka	43°08'40"N	113°39'39"W	
Crossroad Lake	Blaine	42°54'42"N	113°23'33"W	
Crossroad Reservoir	Lincoln	43°09'34"N	114°29'16"W	5090
Crossroads Reservoir	Lincoln	42°50'14"N	114°07'47"W	
Crow Lake	Shoshone	47°06'45"N	115°55'30"W	
Crows Nest Lake	Owyhee	42°33'53"N	115°16'44"W	4234
Crows Nest Reservoir	Lincoln	43°11'54"N	114°34'02"W	
Crowthers Reservoir	Oneida	42°12'18"N	112°15'24"W	4716
Cruickshank Reservoir	Valley	44°53'56"N	116°01'54"W	
Cruzen Pond	Valley	44°46'32"N	116°06'41"W	
Crystal Lake	Adams	45°13'24"N	116°33'07"W	8090
Crystal Lake	Benewah	47°22'36"N	116°22'56"W	
Crystal Lake	Idaho	46°14'34"N	115°11'21"W	
Crystal Lake	Idaho	45°37'42"N	115°40'55"W	7315
Crystal Lake	Valley	44°57'09"N	115°57'49"W	
Crystle Lake	Teton	43°38'14"N	111°22'53"W	
Cub Lake	Fremont	44°07'25"N	111°10'57"W	
Cub Lake	Idaho	46°00'29"N	114°29'52"W	6390
Culvert Reservoir	Blaine	43°14'07"N	114°20'44"W	4745
Cummings Lake	Lemhi	45°28'42"N	114°02'24"W	4616
Curlew Valley Reservoir	Oneida	42°04'30"N	112°32'30"W	
Curtis Lake	Valley	44°35'14"N	115°48'13"W	7624
Curtis Reservoir	Twin Falls	42°19'04"N	114°24'21"W	
Cut Off Reservoir	Lincoln	43°09'51"N	113°46'54"W	
Cutoff Lake	Boundary	48°51'12"N	116°40'53"W	
Cutthroat Lake	Custer	44°32'48"N	115°13'14"W	
Cutthroat Lake	Idaho	45°22'33"N	115°20'02"W	7388
D Bar Reservoir	Owyhee	42°33'55"N	116°36'44"W	
Dan Lake	Idaho	46°28'37"N	114°27'29"W	6722
Dandy Lake	Elmore	43°55'02"N	115°05'10"W	8494
Daniels Reservoir	Oneida	42°20'43"N	112°26'35"W	5162
Danskin Lake	Elmore	43°25'49"N	115°39'44"W	
Darkhorse Lake	Lemhi	45°09'46"N	113°34'56"W	
Darling, Lake	Bonner	48°24'16"N	116°09'04"W	
Darrah Reservoir	Camas	43°13'04"N	114°29'23"W	
Daves Reservoir	Lincoln	43°10'32"N	114°27'53"W	
Daves Reservoir	Owyhee	42°31'32"N	116°42'36"W	
Davis Lake	Fremont	43°58'46"N	111°47'48"W	
Davis Lake Number One	Fremont	44°20'10"N	111°41'47"W	6703
Davis Lake Number Three	Fremont	44°21'22"N	111°42'18"W	
Davis Lake Number Two	Fremont	44°20'56"N	111°42'25"W	
Davis Reservoir	Valley	44°31'17"N	115°59'28"W	4810
Dawson Lake	Boundary	48°46'43"N	116°14'18"W	2959
Dead Tree Reservoir	Owyhee	42°15'34"N	116°35'07"W	
Deadeye Reservoir	Cassia	42°21'13"N	114°14'10"W	
Deadeye Reservoir	Lincoln	43°09'03"N	114°25'48"W	
Deadhorse Reservoir	Blaine	43°14'38"N	113°15'58"W	
Deadwood Reservoir	Valley	44°17'36"N	115°38'42"W	5311
Deep Creek Reservoir	Oneida	42°12'40"N	112°10'18"W	5155
Deep Creek Reservoir	Twin Falls	42°16'42"N	114°36'55"W	4685
Deep Lake	Minidoka	43°07'40"N	113°38'07"W	
Deep Lake	Valley	45°09'50"N	115°55'51"W	
Deer Lake	Idaho	45°32'53"N	115°41'15"W	7304
Deer Lake	Lemhi	45°08'59"N	114°30'19"W	8330
Deer Lakes	Custer	43°56'28"N	114°37'47"W	
Deer Reservoir	Lincoln	43°08'22"N	114°27'01"W	

Lake, Pond, or Reservoir	County	Latitude	Longitude	Elev.
Dennick Lake	Bonner	48°29'50"N	116°23'39"W	
Dennis Lakes	Idaho	45°32'16"N	114°52'40"W	
Denton Slough	Bonner	48°12'07"N	116°15'17"W	
Devils Creek Reservoir	Oneida	42°17'42"N	112°29'52"W	5140
Devils Lake	Lemhi	44°36'08"N	113°32'20"W	8920
Devils Lake	Shoshone	46°57'25"N	115°38'27"W	
Diamond Lake	Elmore	43°55'37"N	115°09'16"W	8055
Diamond Lake	Idaho	46°05'38"N	114°32'19"W	6300
Dickshooter Reservoir	Owyhee	42°24'13"N	116°31'35"W	
Dierkes Lake	Twin Falls	42°35'47"N	114°23'07"W	3445
Dierkes Lake	Twin Falls	42°35'43"N	114°23'12"W	
Dike Lake	Custer	44°07'35"N	114°36'44"W	
Disappointment Lake	Idaho	45°11'00"N	116°12'25"W	
Dishpan Lake	Idaho	46°11'50"N	115°13'00"W	
Dismal Lake	Shoshone	47°07'03"N	115°38'02"W	
Dismal Lake	Valley	44°49'07"N	115°49'18"W	
District Boundary Lake	Blaine	42°58'28"N	113°23'42"W	
Divide Creek Lake	Clark	44°24'11"N	112°49'11"W	
Divide Lake	Boundary	48°31'12"N	116°10'21"W	5820
Divide Reservoir	Camas	43°12'53"N	114°34'23"W	
Dodge Lake	Idaho	46°28'06"N	114°26'56"W	6948
Dodge Lake	Idaho	46°21'14"N	114°51'34"W	6176
Doe Lake	Idaho	46°00'09"N	115°05'17"W	6705
Doe Lake	Lemhi	45°05'54"N	114°35'54"W	
Dog Creek Reservoir	Gooding	43°01'30"N	114°44'36"W	
Dog Lake	Idaho	45°17'43"N	116°33'21"W	
Dollar Lake	Blaine	43°41'18"N	114°25'03"W	
Dome Lake	Lemhi	45°15'38"N	114°31'08"W	6112
Don Reservoir	Jerome	42°48'09"N	114°03'37"W	
Dougal Reservoir	Owyhee	42°39'25"N	117°00'09"W	5035
Doves Waterholes	Butte	43°27'53"N	113°31'54"W	
Dry Canyon Reservoir	Owyhee	42°30'04"N	116°07'09"W	
Dry Cottonwood Reservoir	Twin Falls	42°18'21"N	114°26'54"W	
Dry Creek Reservoir	Custer	44°08'47"N	113°32'45"W	7460
Dry Creek Reservoir	Owyhee	42°34'12"N	116°22'38"W	5930
Dry Creek Reservoir	Owyhee	42°31'52"N	116°21'55"W	5815
Dry Creek Reservoir Number Eight	Owyhee	42°35'24"N	116°22'05"W	
Dry Creek Reservoir Number Eleven	Owyhee	42°33'19"N	116°24'05"W	
Dry Creek Reservoir Number Five	Owyhee	42°29'16"N	116°22'32"W	
Dry Creek Reservoir Number Four	Owyhee	42°29'42"N	116°21'44"W	
Dry Creek Reservoir Number Nine	Owyhee	42°31'34"N	116°22'29"W	
Dry Creek Reservoir Number Seven	Owyhee	42°29'23"N	116°20'43"W	
Dry Creek Reservoir Number Six	Owyhee	42°27'52"N	116°20'19"W	
Dry Creek Reservoir Number Ten	Owyhee	42°32'20"N	116°23'05"W	
Dry Creek Reservoir Number Three	Owyhee	42°30'43"N	116°22'22"W	
Dry Creek Reservoir Number Two	Owyhee	42°33'19"N	116°22'09"W	
Dry Fork Reservoir	Twin Falls	42°19'30"N	114°26'52"W	
Dry Gulch Reservoir	Lincoln	43°04'57"N	114°28'08"W	4375

Lake, Pond, or Reservoir	County	Latitude	Longitude	Elev.
Dry Lake	Idaho	45°25'30"N	115°54'40"W	
Dry Lake	Owyhee	42°08'20"N	116°02'08"W	
Dry Pond	Twin Falls	42°06'18"N	114°55'37"W	
Dry Rock Lake	Lincoln	43°08'22"N	113°52'13"W	
Dry Wash Reservoir	Bingham	43°10'29"N	112°58'43"W	
Dryden Reservoir	Owyhee	43°10'31"N	116°42'50"W	
Duck Lake	Idaho	46°18'17"N	114°36'20"W	6613
Duck Lake	Idaho	45°06'52"N	116°09'21"W	
Duck Lake	Lincoln	42°51'50"N	114°06'56"W	
Duck Lake	Valley	45°03'36"N	115°55'51"W	
Duck Lake Reservoir	Bingham	43°10'26"N	112°46'50"W	
Dug Reservoir	Minidoka	42°53'38"N	113°25'57"W	
Duncan Creek Reservoir	Owyhee	42°25'26"N	116°04'20"W	
Dunes Lake	Owyhee	42°53'43"N	115°41'34"W	2470
Dutch Lake	Custer	44°20'42"N	115°09'58"W	6980
Dworshak Reservoir	Clearwater	46°30'54"N	116°17'45"W	1600
Eagle Mountain Lake	Idaho	46°20'30"N	115°06'43"W	6643
Eagle Reservoir	Bingham	43°12'05"N	112°57'53"W	
East Artifact Reservoir	Jerome	42°48'26"N	114°08'16"W	
East Basin Lake	Custer	44°20'01"N	114°47'34"W	
East Goose Creek Pond	Owyhee	42°55'36"N	116°54'47"W	
East Gospel Lake	Idaho	45°38'05"N	115°56'21"W	7425
East Horse Basin Reservoir	Owyhee	42°00'59"N	116°31'38"W	
East Lake	Valley	45°07'25"N	115°55'07"W	
East Monument Reservoir	Lincoln	42°54'56"N	114°04'58"W	
East Mountain Reservoir	Valley	44°23'44"N	115°58'26"W	4820
East Peak Lake	Idaho	46°12'28"N	115°10'14"W	
Eaton Lake	Bonner	48°11'51"N	116°24'31"W	
Echo Lake	Idaho	45°19'43"N	116°34'07"W	
Echo Lake	Lemhi	45°06'32"N	114°36'28"W	
Echo Lake	Twin Falls	42°34'55"N	114°21'21"W	
Eddy Lake	Custer	44°37'51"N	114°25'54"W	
Eden Lake	Idaho	45°14'39"N	116°12'18"W	
Edith Lake	Custer	43°58'38"N	114°57'31"W	
Edna Lake	Boise	43°58'19"N	114°59'28"W	8404
Eds Defeat Reservoir	Owyhee	42°21'10"N	116°16'26"W	
Edwards Lake	Fremont	44°44'22"N	111°22'55"W	
Egin Lakes	Fremont	43°57'37"N	111°51'29"W	
Elden Reservoir	Lincoln	42°50'15"N	114°03'39"W	
Elizabeth Lake	Clearwater	46°47'21"N	115°16'16"W	
Elizabeth Lake	Custer	44°15'57"N	115°09'02"W	8090
Elizabeth Lake	Idaho	46°11'55"N	115°12'33"W	5868
Elk Creek Reservoir	Clearwater	46°46'09"N	116°09'59"W	2816
Elk Lake	Boise	44°01'31"N	115°04'14"W	6650
Elk Lake	Custer	44°13'45"N	114°44'49"W	
Elk Lake	Idaho	45°50'35"N	115°04'46"W	6636
Elk Lake	Idaho	45°17'27"N	116°14'48"W	6998
Elk Reservoir	Owyhee	42°40'11"N	116°15'15"W	

Lakes, Ponds, Reservoirs

Lake, Pond, or Reservoir	County	Latitude	Longitude	Elev.
Elk Track Lakes	Idaho	45°37'43"N	114°48'49"W	
Ellie Lake	Clearwater	46°47'34"N	115°16'37"W	
Ellis Lake	Valley	45°07'43"N	116°05'44"W	
Elmore, Lake	Ada	43°39'49"N	116°16'20"W	
Elsie Lake	Shoshone	47°25'37"N	116°01'21"W	
Emerald Lake	Adams	45°12'44"N	116°34'08"W	
Emerald Lake	Custer	44°03'06"N	114°36'34"W	9910
Emerald Lake	Idaho	45°57'53"N	114°36'16"W	
Emerald Lake	Idaho	45°32'32"N	115°55'08"W	7110
Emerald Lake	Minidoka	42°34'15"N	113°44'25"W	4141
Emerald Lake	Minidoka	42°34'15"N	113°44'25"W	4141
Enos Lake	Valley	45°06'04"N	115°51'28"W	
Equalizing Reservoir	Bingham	43°10'42"N	112°16'59"W	4523
Equalizing Reservoir	Bingham	43°10'00"N	112°17'48"W	
Erwin Reservoir	Blaine	43°15'29"N	113°31'15"W	
Estelle, Lake	Bonner	48°22'39"N	116°07'03"W	
Everly Lake	Boise	43°57'23"N	115°04'56"W	8628
Everson Lake	Lemhi	44°37'33"N	113°36'47"W	8920
F Eighty-Two Lake	Custer	44°23'52"N	115°04'22"W	6827
Fairchild Reservoir	Washington	44°27'48"N	116°54'24"W	
Falconberry Lake	Lemhi	44°45'07"N	114°45'30"W	
Farley Lake	Custer	43°58'48"N	114°55'53"W	7745
Fault Lake	Boundary	48°33'54"N	116°41'43"W	5980
Fawn Lake	Idaho	45°59'42"N	115°06'32"W	6410
Fawn Lake	Idaho	45°32'08"N	115°38'06"W	
Fawn Lake	Lemhi	45°05'59"N	114°35'56"W	
Fawn Lake	Shoshone	46°57'29"N	115°31'03"W	5912
Fawn Reservoir	Camas	43°13'48"N	114°28'13"W	
Feather Lakes	Boise	44°04'01"N	115°01'13"W	
Felix Reservoir	Owyhee	42°24'49"N	115°11'03"W	4695
Fernan Lake	Kootenai	47°40'30"N	116°43'29"W	
Fiddle Lake	Elmore	43°36'21"N	115°25'26"W	
Fife Reservoir	Bear Lake	42°24'36"N	111°29'12"W	
Finger Lakes	Custer	44°29'42"N	115°09'31"W	
Fingers Butte Reservoir	Butte	43°25'39"N	113°16'17"W	
Fire Lake	Idaho	46°09'11"N	115°26'58"W	
Fish Creek Reservoir	Camas	43°25'23"N	113°49'54"W	5300
Fish Lake	Adams	44°54'18"N	116°12'55"W	
Fish Lake	Clearwater	46°49'03"N	114°54'42"W	5945
Fish Lake	Custer	44°26'33"N	114°55'38"W	
Fish Lake	Idaho	46°19'59"N	115°03'01"W	5631
Fish Lake	Idaho	45°36'36"N	115°37'14"W	5684
Fish Lake	Idaho	45°23'18"N	115°19'06"W	7128
Fish Lake	Shoshone	47°06'04"N	115°57'43"W	5457
Fish Lake	Valley	45°01'21"N	115°25'45"W	
Fish Pond	Fremont	44°18'45"N	111°25'21"W	6104
Fishpole Lake	Custer	43°38'36"N	113°50'59"W	
Fivemile Lake	Owyhee	42°30'32"N	115°46'20"W	4488

Lake, Pond, or Reservoir	County	Latitude	Longitude	Elev.
Flat Draw Reservoir	Owyhee	42°14'39"N	115°25'52"W	5035
Flat Tire Reservoir	Lincoln	43°11'59"N	114°24'54"W	
Flat Top Reservoir	Lincoln	42°56'07"N	113°43'29"W	
Flat Top Reservoir	Owyhee	42°17'19"N	116°45'34"W	
Flea Lake	Idaho	46°12'18"N	115°17'45"W	6074
Florence Lake	Idaho	46°10'40"N	115°12'55"W	6288
Flossie Lake	Idaho	45°23'39"N	115°16'26"W	
Floyds Reservoir	Camas	43°15'17"N	114°28'38"W	
Fly Lake	Clearwater	46°52'57"N	115°13'21"W	
Fogg Lake	Valley	44°50'32"N	115°54'27"W	7349
Forage Lake	Shoshone	46°58'44"N	115°16'19"W	
Foremans Reservoir	Owyhee	43°01'18"N	116°19'51"W	
Forster Reservoir	Owyhee	42°37'25"N	117°00'04"W	4949
Foster Reservoir	Franklin	42°07'29"N	111°50'33"W	4857
Four Corners Reservoir	Owyhee	42°06'36"N	116°36'54"W	
Fourth Of July Lake	Custer	44°02'37"N	114°37'49"W	9365
Fraser Reservoir (historical)	Elmore	43°08'25"N	115°53'34"W	
Freeland Watering	Owyhee	42°31'20"N	115°07'06"W	4000
Freeman Lake	Bonner	48°13'27"N	117°01'45"W	
French Creek Lakes	Idaho	45°10'50"N	116°04'13"W	7595
Frog Lake	Custer	44°04'47"N	114°32'43"W	8855
Frog Lake	Idaho	46°20'12"N	114°27'38"W	
Frog Lake	Idaho	45°24'47"N	115°17'00"W	
Frog Lake	Idaho	45°09'36"N	116°11'54"W	
Frog Lake	Shoshone	47°05'28"N	115°09'38"W	
Frog Lake	Valley	45°09'47"N	115°56'52"W	
Frog Lakes	Custer	44°04'44"N	114°32'31"W	
Frog Pond	Idaho	45°24'06"N	115°16'45"W	
Fuller Reservoir	Cassia	42°21'17"N	114°13'13"W	
Gabes Bathtub	Gem	44°26'00"N	116°10'05"W	
Gabettas Reservoir	Minidoka	42°55'42"N	113°25'50"W	
Gamlin Lake	Bonner	48°13'23"N	116°23'12"W	2083
Gardner Lake	Clark	44°20'56"N	112°02'13"W	
Garland Lakes	Custer	44°09'56"N	114°47'31"W	
Garnet Lake	Idaho	46°25'01"N	114°23'34"W	6700
Gay Lake	Idaho	45°18'31"N	116°14'24"W	6608
Gem Lake	Bonner	48°22'41"N	116°08'09"W	
Gem Lake	Clearwater	46°35'44"N	115°12'49"W	
Gem Lake	Idaho	45°20'10"N	116°33'14"W	
Gentian Lake	Custer	44°05'51"N	114°36'42"W	
Gentian Lake	Lemhi	45°09'17"N	114°35'08"W	8630
Gentle Horse Lake	Blaine	43°11'19"N	113°02'41"W	
Gerity Reservoir	Blaine	43°14'45"N	113°30'00"W	
German Lake	Minidoka	42°49'30"N	113°31'40"W	
Geyselman Lake	Boundary	48°57'40"N	116°10'35"W	
Gibson Lakes	Franklin	42°02'15"N	111°37'48"W	
Glacier Lake	Custer	44°02'54"N	114°36'39"W	
Glacier Lake	Elmore	43°56'36"N	115°07'49"W	8595

Lake, Pond, or Reservoir	County	Latitude	Longitude	Elev.
Glacier Lake	Lemhi	45°10'08"N	114°35'06"W	8945
Glendale Reservoir	Franklin	42°07'42"N	111°48'36"W	4951
Gnat Lake	Shoshone	46°57'49"N	115°35'39"W	
Goal Lake	Custer	43°46'44"N	114°01'08"W	10438
Goat Lake	Camas	43°43'50"N	115°00'23"W	8739
Goat Lake	Clearwater	46°39'38"N	114°49'15"W	6492
Goat Lake	Custer	44°10'18"N	115°01'03"W	8220
Goat Lake	Custer	44°05'56"N	114°34'52"W	
Goat Lake	Idaho	45°57'54"N	115°00'14"W	7158
Goat Lake	Idaho	45°35'44"N	115°05'38"W	
Goat Lake	Idaho	45°18'54"N	116°13'18"W	6642
Goat Lake	Lemhi	45°13'22"N	114°31'54"W	7988
Goat Lake	Lemhi	44°44'05"N	113°52'53"W	
Goat Lakes	Idaho	46°15'57"N	114°37'20"W	6396
Gold Lake	Shoshone	46°56'35"N	115°16'30"W	6120
Gold Pan Lake	Idaho	45°38'45"N	114°48'20"W	
Golden Lake	Custer	43°43'53"N	113°51'45"W	
Golden Lake	Fremont	44°21'20"N	111°28'46"W	6133
Golden Lake	Valley	44°58'00"N	115°56'01"W	7442
Golden Trout Lake	Lemhi	45°06'43"N	114°31'14"W	8149
Gooding Reservoir	Lincoln	43°04'02"N	114°29'58"W	
Gooding Reservoir	Lincoln	43°03'59"N	114°29'58"W	4110
Goose Heaven Lake	Benewah	47°21'25"N	116°39'55"W	2116
Goose Lake	Adams	45°04'12"N	116°10'06"W	6362
Goose Lake	Clearwater	46°54'58"N	114°56'03"W	5765
Goose Lake	Jerome	42°37'25"N	114°15'19"W	3877
Gooseneck Lake	Lemhi	45°09'54"N	114°34'53"W	
Gospel Lakes	Idaho	45°37'50"N	115°57'03"W	
Govenors Punchbowl	Blaine	43°53'53"N	114°41'58"W	
Granite Butte Reservoir	Camas	43°14'42"N	114°27'17"W	5430
Granite Lake	Bonner	48°00'14"N	116°41'12"W	
Granite Lake	Idaho	46°39'18"N	114°24'48"W	
Granite Lake	Idaho	46°08'11"N	114°33'32"W	6698
Granite Lake	Valley	45°06'06"N	116°04'48"W	6734
Grante Butte Reservoir	Camas	43°14'41"N	114°27'18"W	
Grasmere Reservoir	Owyhee	42°21'48"N	115°54'30"W	
Grass Mountain Lakes	Idaho	45°10'02"N	116°11'21"W	
Grassy Hills Reservoir	Owyhee	42°12'49"N	115°07'38"W	5650
Grassy Hills Reservoir Number Three	Owyhee	42°07'48"N	115°04'14"W	5750
Grassy Hills Reservoir Number Two	Owyhee	42°09'04"N	115°04'06"W	5670
Grays Lake	Bonneville	43°07'30"N	111°29'24"W	
Grays Lake	Bonneville	43°04'02"N	111°26'10"W	6385
Grays Lake	Caribou	42°59'36"N	111°29'24"W	
Green Island Lake	Elmore	43°36'40"N	115°24'55"W	
Green Lake	Custer	43°42'22"N	113°51'44"W	
Green Lake	Minidoka	42°52'46"N	113°33'16"W	
Green Spring Reservoir	Lincoln	43°06'30"N	114°28'34"W	4575
Greybull Reservoir	Lincoln	43°06'07"N	114°27'46"W	4525
Greystone Lake	Idaho	46°23'55"N	115°04'19"W	
Grimes Lake	Idaho	46°34'04"N	114°22'16"W	
Groner Reservoir	Washington	44°24'21"N	116°29'49"W	
Gronewell Lake	Caribou	42°49'59"N	111°35'07"W	6112
Grouse Creek Lake	Custer	44°24'03"N	113°57'46"W	
Grouse Lake	Idaho	46°24'22"N	114°41'34"W	6422
Grouse Lake	Lemhi	44°49'55"N	114°41'32"W	8215
Grouse Lakes	Elmore	43°42'56"N	115°04'21"W	8185
Grouse Reservoir	Camas	43°15'55"N	114°28'41"W	5330
Grover Lake Reservoir	Bingham	43°17'26"N	112°48'09"W	
Gunsight Lake	Custer	44°07'38"N	114°36'26"W	
Guts Lake	Minidoka	43°10'07"N	113°33'58"W	
Haas Lake	Idaho	45°16'44"N	116°33'36"W	
Hackberry Reservoir	Owyhee	42°36'05"N	116°41'23"W	5260
Hager Lake	Bonner	48°35'50"N	116°58'14"W	
Hait Reservoir	Valley	44°50'30"N	116°10'24"W	4955
Half Acre Lake	Idaho	45°34'04"N	115°43'01"W	
Halfway Lake	Lincoln	43°04'40"N	113°47'14"W	4570
Halfway Lake	Minidoka	42°52'43"N	113°38'36"W	
Halo Lake	Shoshone	46°58'38"N	115°15'33"W	
Halogeton Reservoir	Minidoka	42°55'53"N	113°33'48"W	
Halverson Lake	Ada	43°17'32"N	116°29'46"W	2270
Hamer Lake	Jefferson	43°56'08"N	112°12'22"W	
Hamilton Lakes	Idaho	45°32'44"N	114°49'37"W	7464
Hancock Lake	Clark	44°31'54"N	111°51'14"W	
Hansen Lake	Lincoln	42°58'28"N	113°44'51"W	
Hanson Lakes	Custer	44°12'35"N	115°07'07"W	8102
Hanson Lakes	Idaho	45°18'52"N	116°32'11"W	
Harbor Lake	Lemhi	45°08'33"N	114°35'30"W	8922
Hard Butte Lake	Idaho	45°15'55"N	116°12'34"W	
Hard Creek Lake	Idaho	45°10'20"N	116°08'37"W	
Harold Reservoir	Butte	43°17'49"N	113°14'23"W	
Harris Creek Reservoir	Owyhee	42°19'38"N	116°14'58"W	
Harris Lake	Owyhee	42°11'38"N	115°54'04"W	
Harris Lake	Owyhee	42°07'28"N	115°59'53"W	
Harris Reservoir	Blaine	43°16'44"N	113°59'13"W	4840
Harrison Lake	Boundary	48°40'45"N	116°39'06"W	6182
Harrison Slough	Kootenai	47°28'08"N	116°46'20"W	
Harveys Canyon Reservoir	Owyhee	42°31'42"N	116°08'44"W	
Harveys V Reservoir	Owyhee	42°30'29"N	116°06'23"W	
Hat Creek Lakes	Lemhi	44°52'43"N	114°12'22"W	8758
Hatchet Lake	Custer	44°04'10"N	114°33'46"W	8884
Hauser Lake	Kootenai	47°46'42"N	117°01'04"W	2187
Hawk Reservoir	Blaine	43°15'31"N	113°29'59"W	
Hawkins Reservoir	Bannock	42°30'36"N	112°19'53"W	5142
Hayden Lake	Kootenai	47°45'01"N	116°45'26"W	
Hazard Lake	Idaho	45°12'13"N	116°08'23"W	7060
He Devil Lake	Idaho	45°19'33"N	116°33'45"W	

Lakes, Ponds, Reservoirs

Lake, Pond, or Reservoir	County	Latitude	Longitude	Elev.
Headwall Lake	Custer	44°04'27"N	114°35'51"W	9755
Heart Lake	Camas	43°32'42"N	115°02'22"W	
Heart Lake	Custer	44°01'47"N	114°40'37"W	
Heart Lake	Elmore	43°56'25"N	114°59'50"W	8562
Heart Lake	Elmore	43°35'57"N	115°25'07"W	
Heart Lake	Lemhi	45°08'08"N	114°35'38"W	8623
Heart Lake	Shoshone	46°56'28"N	115°35'05"W	
Heart Lake	Valley	45°02'42"N	115°59'52"W	
Heath Lake	Bonner	48°10'49"N	116°34'56"W	2173
Heather Lake	Shoshone	46°57'06"N	115°15'53"W	
Heifer Reservoir	Butte	43°21'29"N	113°15'55"W	
Hell Roaring Lake	Custer	44°01'29"N	114°56'07"W	7407
Helldiver Lake	Custer	44°32'05"N	115°10'20"W	
Hells Canyon Reservoir	Adams	45°14'00"N	116°42'00"W	
Henry Lake	Owyhee	42°13'47"N	116°18'34"W	
Henrys Lake	Fremont	44°35'50"N	111°21'09"W	6472
Herd Lake	Custer	44°05'21"N	114°10'19"W	7176
Herman Lake	Boundary	48°41'38"N	116°03'52"W	2481
Hero Lake	Shoshone	46°58'07"N	115°35'41"W	
Herrick Reservoir	Valley	44°22'36"N	115°58'45"W	
Heslip Lake	Idaho	46°20'55"N	115°02'06"W	
Hewitt Watering	Owyhee	42°34'20"N	115°05'26"W	3860
Hidden Lake	Benewah	47°22'58"N	116°45'26"W	
Hidden Lake	Boise	44°00'03"N	114°59'50"W	
Hidden Lake	Boise	43°59'53"N	114°59'51"W	8563
Hidden Lake	Boundary	48°52'58"N	116°45'25"W	5443
Hidden Lake	Custer	44°17'45"N	115°06'59"W	7024
Hidden Lake	Custer	44°04'46"N	114°35'44"W	9517
Hidden Lake	Franklin	42°01'07"N	111°34'13"W	8814
Hidden Lake	Idaho	46°21'00"N	114°31'25"W	5805
Hidden Lake	Idaho	45°35'12"N	115°48'53"W	
Hidden Lake	Idaho	45°28'38"N	114°40'34"W	
Hidden Lake	Idaho	45°08'55"N	116°09'04"W	
Hidden Lake	Minidoka	43°09'04"N	113°40'14"W	
Hidden Lake	Nez Perce	46°25'22"N	116°57'23"W	
Hidden Lake	Valley	44°27'17"N	116°06'59"W	
Hidden Lake	Valley	44°14'48"N	116°11'30"W	
Hidden Lake Reservoir	Fremont	44°15'25"N	111°35'46"W	6550
Hidden Reservoir	Jerome	42°47'10"N	114°08'51"W	
Hideway Lake	Elmore	43°35'27"N	115°24'00"W	
Highline Lakes	Idaho	46°01'29"N	115°07'37"W	
Hillside Lake	Blaine	43°13'30"N	113°38'50"W	
Hillside Reservoir	Lincoln	43°05'11"N	114°27'27"W	4460
Hindman Lake	Custer	44°23'10"N	114°55'15"W	8058
Hinton Reservoir	Owyhee	42°54'43"N	116°02'34"W	
Hiway Reservoir	Blaine	43°14'22"N	114°19'08"W	4815
Hjort Lake	Idaho	46°10'58"N	115°12'34"W	
Hodges Slough	Bonneville	43°31'41"N	111°56'58"W	

Lake, Pond, or Reservoir	County	Latitude	Longitude	Elev.
Hole in Rock Lake	Owyhee	42°39'55"N	115°58'05"W	
Hole in the Rock	Caribou	42°50'01"N	111°36'39"W	
Home Lake	Boise	43°52'08"N	116°14'54"W	
Homer Wells Reservoir	Owyhee	42°01'00"N	116°37'37"W	4829
Honeymoon Lake	Valley	44°35'08"N	115°23'33"W	
Hoodoo Lake	Bonner	48°03'17"N	116°49'12"W	
Hoodoo Lake	Custer	44°10'02"N	114°38'28"W	8677
Hoodoo Lake	Idaho	46°19'14"N	114°39'01"W	5787
Hook Lake	Custer	44°06'18"N	114°36'26"W	
Hope Reservoir	Jerome	42°47'49"N	114°08'41"W	
Hornet Reservoir	Washington	44°47'52"N	116°43'55"W	
Horse Butte Storage Tank	Owyhee	42°25'24"N	115°14'22"W	5005
Horse Heaven Lake	Idaho	45°17'15"N	116°33'28"W	
Horse Lake	Bannock	42°47'55"N	112°30'34"W	
Horse Lake	Caribou	42°50'16"N	112°04'45"W	
Horse Lake	Custer	44°20'46"N	113°22'27"W	
Horse Lake	Franklin	42°01'41"N	111°33'23"W	8282
Horse Lake	Owyhee	42°04'21"N	115°18'27"W	5650
Horseshoe Lake	Custer	44°26'59"N	114°56'16"W	7994
Horseshoe Lake	Fremont	44°09'43"N	111°06'11"W	
Horseshoe Lake	Idaho	46°32'59"N	115°04'04"W	
Horseshoe Lake	Idaho	45°31'56"N	115°43'07"W	
Horseshoe Reservoir	Lincoln	42°49'26"N	114°08'25"W	
Horseshoe Reservoir	Owyhee	42°23'56"N	116°03'29"W	
Horsethief Reservoir	Valley	44°30'20"N	115°55'15"W	
Horton Lake	Valley	45°07'44"N	116°05'12"W	
Hot Springs Creek Reservoir	Elmore	43°06'42"N	115°29'59"W	
Houghland Reservoir Number 1	Blaine	43°11'39"N	113°06'35"W	
Houghland Reservoir Number 2	Blaine	43°11'50"N	113°06'16"W	
Hourglass Lake	Custer	44°04'41"N	114°35'18"W	
Howard Reservoir	Blaine	43°26'18"N	113°58'41"W	5750
Hubbard Reservoir	Ada	43°31'00"N	116°21'30"W	2720
Huff Lake	Blaine	43°19'48"N	113°50'40"W	4815
Hulet-Sinker Creek Reservoir	Owyhee	43°05'34"N	116°32'36"W	3310
Hull Creek Reservoir	Lemhi	45°28'44"N	114°02'33"W	4920
Hum Lake	Valley	45°04'12"N	115°54'20"W	
Hummock Lake	Custer	44°04'44"N	114°35'32"W	9514
Hump Lake	Idaho	45°36'55"N	115°41'17"W	7850
Hungry Lake	Idaho	46°19'36"N	114°45'53"W	6682
Hunt Lake	Boundary	48°34'44"N	116°42'12"W	5813
Hunt Waterhole	Lincoln	42°56'42"N	114°10'29"W	
Hurst Lake	Idaho	45°30'53"N	115°44'11"W	
Ice Lake	Clearwater	46°46'50"N	115°17'56"W	
Idaho Lake	Owyhee	42°00'45"N	115°27'59"W	5995
Idler Lakes	Valley	44°55'11"N	115°53'05"W	
Iest Reservoir	Lincoln	43°04'34"N	114°20'26"W	
Imogene Lake	Custer	43°59'44"N	114°57'02"W	8436
Independence Lakes	Cassia	42°11'53"N	113°39'55"W	9168

Lake, Pond, or Reservoir	County	Latitude	Longitude	Elev.
Indian Battleground Reservoir	Owyhee	42°35'55"N	116°46'41"W	
Indian Creek Reservoir	Ada	43°23'22"N	116°00'55"W	3334
Indian Creek Reservoir	Owyhee	42°10'33"N	116°06'29"W	
Indian Hole Pond	Elmore	42°56'19"N	115°19'00"W	2498
Indian Lake	Blaine	43°06'11"N	113°16'33"W	
Indian Lake	Fremont	44°04'10"N	111°02'28"W	
Indian Lake	Idaho	46°08'43"N	114°42'21"W	6130
Indian Lake	Minidoka	43°08'35"N	113°33'29"W	
Indian Lake	Owyhee	42°28'23"N	116°38'57"W	
Indian Postoffice Lake	Idaho	46°32'17"N	114°59'04"W	
Indigo Lake	Idaho	45°33'10"N	115°55'56"W	7316
Ingeborg, Lake	Elmore	43°56'57"N	115°02'29"W	8890
Inside Lakes	Owyhee	42°17'54"N	115°34'15"W	
Inside Reservoir	Owyhee	42°11'38"N	117°00'53"W	5255
Iris Lake	Custer	44°31'00"N	115°12'06"W	
Iron Bog Lake	Custer	43°39'06"N	113°50'42"W	
Iron Lake	Lemhi	44°54'25"N	114°11'38"W	8806
Isaac Lake	Idaho	46°16'07"N	114°48'21"W	6236
Island Lake	Custer	44°28'24"N	115°08'47"W	
Island Lake	Custer	44°05'43"N	114°35'37"W	
Island Lake	Elmore	43°56'52"N	115°07'25"W	
Island Park Reservoir	Fremont	44°25'10"N	111°23'44"W	
Jack Brockie Lake	Minidoka	42°58'18"N	113°27'05"W	
Jack Lake	Clearwater	46°44'17"N	115°23'54"W	5900
Jack Reservoir	Lincoln	42°49'32"N	114°01'36"W	
Jacks Creek Reservoir	Owyhee	42°26'36"N	116°08'34"W	
Jackson Reservoir	Owyhee	42°05'03"N	116°20'03"W	
James Lake	Owyhee	42°17'16"N	115°45'25"W	
Jarvis Pasture Reservoir Number One	Owyhee	42°10'58"N	116°28'30"W	
Jarvis Pasture Reservoir Number Three	Owyhee	42°10'18"N	116°25'28"W	
Jarvis Pasture Reservoir Number Two	Owyhee	42°09'30"N	116°26'31"W	
Jean Reservoir	Minidoka	43°04'38"N	113°42'42"W	
Jeanette Lake	Idaho	46°18'38"N	114°34'55"W	5898
Jefferson Reservoir	Jefferson	43°57'16"N	112°27'35"W	4792
Jemima K Reservoir	Valley	44°34'30"N	116°00'24"W	
Jenkins Creek Reservoir	Washington	44°21'22"N	116°59'26"W	3060
Jennie Lake	Boise	44°01'04"N	115°24'13"W	7850
Jensen Lake	Canyon	43°19'27"N	116°33'21"W	
Jensens Lake	Bingham	43°12'17"N	112°21'16"W	4477
Jerome Slu	Boundary	48°55'49"N	116°26'47"W	
Jewel Lake	Bonner	48°08'28"N	116°43'05"W	
Jim Bob Reservoir	Owyhee	42°02'57"N	115°19'34"W	5745
Jimmy Smith Lake	Custer	44°10'15"N	114°24'14"W	6326
Jims Reservoir	Owyhee	42°33'17"N	116°44'41"W	
Joe Lake	Boundary	48°53'19"N	116°46'28"W	5588
Joes Folly Reservoir	Owyhee	42°31'26"N	116°47'30"W	
John G Reservoir	Owyhee	42°07'13"N	116°50'02"W	
John Hoffman Reservoir	Elmore	43°11'12"N	115°35'18"W	3586
John Lake	Idaho	45°18'13"N	116°13'52"W	6698
Johns Pond	Twin Falls	42°05'35"N	114°56'53"W	
Johns Reservoir	Camas	43°13'48"N	114°30'28"W	
Johnson Lake	Elmore	43°56'44"N	115°08'32"W	7998
Johnson Reservoir	Franklin	42°06'30"N	111°48'18"W	4878
Johnson Reservoir	Lincoln	43°11'49"N	114°11'33"W	
Johnson Reservoir	Owyhee	42°37'23"N	116°34'19"W	
Johnston Lake	Jefferson	43°56'52"N	112°25'51"W	
Johnston Lakes	Owyhee	43°06'03"N	116°50'01"W	
Jones Reservoir	Lincoln	43°08'29"N	114°24'23"W	
Josephine Lake	Idaho	45°13'26"N	115°58'21"W	
Josephus Lake	Custer	44°32'52"N	115°08'40"W	
Junction Reservoir	Blaine	43°17'24"N	113°37'31"W	
Jungle Lake	Valley	45°06'23"N	115°49'35"W	
Juniper Basin Reservoir	Owyhee	42°15'27"N	116°55'59"W	
Juniper Basin Reservoir	Owyhee	42°03'23"N	116°27'32"W	5042
Juniper Lake	Owyhee	42°15'14"N	115°17'58"W	5135
Juniper Lake	Owyhee	42°12'32"N	116°18'04"W	
Jus Reservoir	Owyhee	42°19'39"N	116°32'30"W	
Kane Lake	Custer	43°47'08"N	114°09'34"W	
Karney Lakes	Boise	43°40'34"N	116°03'01"W	
Kathryn, Lake	Custer	44°02'15"N	115°02'15"W	8996
Kecks Reservoir	Owyhee	42°31'28"N	116°44'23"W	
Keith Reservoir	Cassia	42°20'11"N	114°14'42"W	
Keith Reservoir	Elmore	43°11'27"N	115°33'19"W	
Kelly Lake	Clearwater	46°46'00"N	114°50'24"W	6180
Kelly Lake	Custer	44°16'53"N	115°09'28"W	7842
Kelly Lake	Idaho	45°38'34"N	115°40'20"W	
Kelly Reservoir	Camas	43°18'24"N	114°37'07"W	5035
Kelso Lake	Bonner	48°00'37"N	116°42'23"W	
Kennally Lakes	Valley	44°50'07"N	115°53'07"W	
Kenneth Lake	Idaho	45°13'04"N	116°07'01"W	
Kent Lake	Boundary	48°43'36"N	116°40'07"W	
Keokee Lake	Bonner	48°25'19"N	116°39'18"W	5558
Kerosene Lake	Elmore	43°24'02"N	115°16'37"W	
Kerr Lake	Bonner	48°39'09"N	116°56'02"W	
Kerr Lake	Boundary	48°53'49"N	116°26'00"W	1747
Kettle Lake	Idaho	46°11'35"N	115°13'55"W	
Kid Lake	Clearwater	46°46'43"N	114°48'52"W	
Kidney Lake	Custer	44°31'22"N	114°58'17"W	
Kidney Lake	Idaho	46°20'58"N	114°43'29"W	6764
Killarney Lake	Kootenai	47°31'15"N	116°33'48"W	
Kilroy Lakes	Bonner	48°07'14"N	116°22'17"W	
Kimama Marsh Reservoir	Lincoln	42°55'18"N	113°43'14"W	
Kimball Basin Reservoir	Owyhee	42°11'43"N	116°32'13"W	
Kincaid Reservoir	Owyhee	42°18'52"N	116°34'12"W	
King Lakes	Blaine	43°14'15"N	113°04'33"W	
Kinzie Butte Reservoir	Lincoln	43°05'59"N	114°20'08"W	

Lakes, Ponds, Reservoirs

Lake, Pond, or Reservoir	County	Latitude	Longitude	Elev.
Knapp Lakes	Custer	44°25'39"N	114°55'59"W	
Knob Lakes	Idaho	45°37'35"N	115°54'49"W	
Knox Lake	Valley	44°41'45"N	115°37'17"W	
Konkols Pond	Clearwater	46°32'30"N	115°56'49"W	3140
Koskella Reservoir	Valley	44°40'09"N	116°01'26"W	4960
Lahtinen Spring Reservoir	Owyhee	42°36'35"N	116°17'59"W	
Laidlaw Lake	Minidoka	43°05'43"N	113°38'27"W	
Lake Cleveland	Cassia	42°19'23"N	113°38'48"W	8263
Lake Creek Lakes	Blaine	43°46'24"N	114°19'32"W	
Lake Creek Lakes	Idaho	45°36'34"N	115°03'54"W	
Lake Lowell	Canyon	43°35'00"N	116°44'12"W	2531
Lake Pit Reservoir	Owyhee	42°11'02"N	115°25'13"W	5375
Lake Reservoir	Lincoln	43°07'04"N	114°24'35"W	4685
Lake Rock Lake	Idaho	45°12'28"N	115°55'05"W	
Lake San Souci	Bonner	48°00'42"N	117°00'18"W	
Lake Walcott	Blaine	42°40'06"N	113°29'03"W	4135
Lakey Reservoir	Caribou	42°36'27"N	111°27'45"W	
Lamberton Reservoir	Elmore	43°11'38"N	115°34'51"W	
Lambertson Lake	Bonner	48°03'07"N	116°43'04"W	
Lamont Reservoir	Franklin	42°06'18"N	111°48'56"W	4873
Landmark Reservoir	Lincoln	43°09'36"N	114°25'58"W	
Langer Lake	Custer	44°28'50"N	115°08'02"W	
Lapwai Lake	Lewis	46°14'15"N	116°37'06"W	3902
Larkins Lake	Shoshone	46°57'02"N	115°36'39"W	
Last Chance Reservoir	Gooding	42°58'52"N	114°53'38"W	
Last Chance Reservoir	Owyhee	42°04'19"N	116°50'29"W	
Lau Lakes	Bingham	43°11'15"N	112°57'32"W	
Lava Butte Lakes	Idaho	45°16'50"N	116°07'25"W	
Lava Lake	Blaine	43°23'04"N	113°43'02"W	5158
Lava Lake Reservoir	Bingham	43°16'31"N	112°48'53"W	4804
Leavitz Pond	Jefferson	43°55'33"N	112°14'52"W	
Lee Lake	Bonner	48°28'36"N	116°49'14"W	2475
Leeks Pond	Adams	44°43'21"N	116°25'25"W	2945
Legend Lake	Idaho	46°09'17"N	115°11'16"W	
Leggit Lake	Elmore	43°46'21"N	115°02'29"W	8526
Lemhi Lake	Idaho	45°23'09"N	115°22'08"W	
Lemon Lake	Fremont	44°03'51"N	111°33'36"W	5122
Leo Lake	Clearwater	46°44'13"N	114°46'24"W	6415
Leo Waterhole	Owyhee	42°16'34"N	115°24'12"W	4970
Liberty Lakes	Lemhi	44°45'25"N	114°38'54"W	
Lick Lake	Valley	45°04'35"N	115°18'18"W	
Lightning Lake	Custer	44°27'55"N	114°47'10"W	
Lily Lake	Custer	44°05'50"N	114°57'25"W	
Lily Lake	Idaho	46°37'54"N	114°23'00"W	6376
Lily Pad Lake	Idaho	45°21'12"N	116°33'35"W	
Lily Pond	Fremont	44°17'05"N	111°17'38"W	
Limber Lake	Boise	44°02'44"N	115°03'11"W	
Limekiln Lake	Bear Lake	42°07'57"N	111°32'33"W	
Lincoln Reservoir	Cassia	42°20'33"N	114°15'07"W	
Linda Reservoir	Blaine	43°14'17"N	113°37'16"W	
Line Lake	Idaho	45°34'21"N	114°34'30"W	7615
Little Baron Lake	Boise	44°05'17"N	114°59'23"W	8141
Little Bayhorse Lake	Custer	44°24'45"N	114°23'19"W	8338
Little Bear Lake	Camas	43°44'42"N	114°58'33"W	8811
Little Blue Creek Reservoir	Owyhee	42°17'33"N	116°07'16"W	5418
Little Blue Table Reservoir	Owyhee	42°17'17"N	116°04'23"W	
Little Camas Reservoir	Elmore	43°21'10"N	115°23'05"W	4924
Little Cottonwood Reservoir	Lincoln	43°06'52"N	114°22'23"W	
Little Crane Creek Reservoir	Washington	44°22'37"N	116°39'45"W	3494
Little Deer Lakes	Minidoka	43°07'39"N	113°38'17"W	
Little Draw Reservoir	Lincoln	43°07'48"N	114°29'26"W	
Little Frog Lake	Custer	44°04'47"N	114°32'29"W	
Little Grassy Reservoir	Owyhee	42°09'19"N	115°06'48"W	5630
Little Horse Basin Reservoir	Owyhee	42°05'33"N	116°38'58"W	
Little Jarvis Lake	Owyhee	42°05'44"N	116°13'35"W	
Little Juniper Basin Reservoir	Owyhee	42°05'06"N	116°31'00"W	
Little Lake	Bonner	48°14'40"N	116°40'20"W	
Little Lake	Clark	44°29'46"N	112°09'12"W	
Little Lake	Minidoka	42°52'38"N	113°38'20"W	
Little Lookout Lake	Elmore	43°36'26"N	115°25'10"W	
Little Lost Lake	Blaine	43°44'12"N	114°39'34"W	9078
Little Lost Lake	Shoshone	47°03'58"N	115°56'25"W	5765
Little Magic Reservoir	Camas	43°13'51"N	114°23'13"W	
Little Muskrat Lake	Bonner	48°15'02"N	116°40'24"W	
Little Payette Lake	Valley	44°54'13"N	116°02'26"W	5119
Little Prairie Waterhole	Butte	43°23'34"N	113°29'44"W	
Little Rainbow Lake	Elmore	43°35'50"N	115°24'22"W	
Little Redfish Lake	Custer	44°09'19"N	114°54'47"W	6489
Little Redfish Lake	Custer	44°06'14"N	114°32'09"W	
Little Roaring River Lake	Elmore	43°37'47"N	115°26'44"W	7810
Little Sheepeater Lake	Idaho	45°22'50"N	115°23'46"W	1636
Little Spangle Lake	Elmore	43°56'37"N	115°01'49"W	
Little Summit Lake	Valley	45°11'20"N	115°58'29"W	
Little Trinity Lake	Elmore	43°37'50"N	115°25'53"W	7780
Little Valley Reservoir	Bear Lake	42°15'33"N	111°28'00"W	6552
Little Valley Reservoir	Bonneville	43°05'35"N	111°31'15"W	
Little Wood River Reservoir	Blaine	43°25'30"N	114°01'42"W	
Little Wood River Reservoir	Camas	43°25'30"N	114°01'35"W	5235
Livermore Lake	Bonner	48°12'42"N	116°23'42"W	
Lizard Lake	Teton	43°40'44"N	111°14'28"W	
Lizard Lakes	Idaho	46°13'02"N	115°09'15"W	6387
Lloyd Lake	Idaho	46°11'21"N	115°13'01"W	6206
Lloyds Lake	Idaho	45°11'34"N	116°09'47"W	
Loa Lake	Caribou	42°25'19"N	111°37'16"W	
Lodgepole Lake	Custer	44°04'16"N	114°34'31"W	9008
Lodgepole Lake	Elmore	43°53'34"N	115°18'04"W	

Lake, Pond, or Reservoir	County	Latitude	Longitude	Elev.
Logan Lake	Valley	45°05'11"N	115°22'10"W	
Lola Lakes	Custer	44°23'29"N	115°13'51"W	
Lone Lake	Shoshone	47°26'00"N	115°46'23"W	
Lone Lakes	Idaho	46°13'24"N	115°10'30"W	
Lone Pine Lake	Boise	44°04'52"N	116°07'10"W	
Lone Rock Reservoir	Camas	43°13'30"N	114°24'18"W	
Lonesome Lake	Custer	44°04'28"N	114°36'22"W	
Lonesome Lake	Idaho	45°59'25"N	114°57'35"W	
Long Lake	Custer	43°43'32"N	113°50'09"W	
Long Lake	Idaho	46°20'44"N	115°08'54"W	5858
Long Lake	Owyhee	42°17'08"N	115°43'27"W	
Long Lake	Valley	45°07'15"N	115°55'58"W	
Long Lake	Valley	44°29'51"N	115°39'15"W	
Long Mountain Lake	Boundary	48°50'00"N	116°36'34"W	
Long Pool	Shoshone	47°47'20"N	116°04'15"W	
Long Pull Reservoir	Owyhee	42°16'27"N	116°58'17"W	
Long Tom Reservoir	Elmore	43°17'10"N	115°34'40"W	4418
Lookout Lake	Boundary	48°46'46"N	116°46'43"W	
Lookout Lake	Clearwater	46°36'33"N	115°13'16"W	
Loon Lake	Valley	45°09'49"N	115°50'22"W	
Lost Lake	Bonner	48°20'48"N	116°24'59"W	
Lost Lake	Bonner	48°11'15"N	116°23'24"W	
Lost Lake	Custer	44°31'43"N	115°09'32"W	
Lost Lake	Idaho	45°42'55"N	115°44'20"W	6462
Lost Lake	Kootenai	47°46'12"N	116°59'58"W	
Lost Lake	Lemhi	45°06'56"N	114°37'19"W	8356
Lost Lake	Shoshone	47°27'28"N	115°59'49"W	
Lost Lake	Shoshone	47°04'25"N	115°57'25"W	5536
Lost Lake	Valley	44°27'27"N	116°07'19"W	
Lost Lake	Valley	44°17'17"N	115°25'47"W	7346
Lost Lakes	Valley	44°16'40"N	115°24'54"W	
Lost Packer Lake	Lemhi	45°28'18"N	114°46'35"W	7370
Lost Valley Reservoir	Adams	44°57'24"N	116°27'57"W	4766
Lottie Lake	Idaho	46°16'02"N	115°15'02"W	
Louie Lake	Valley	44°51'17"N	115°57'55"W	7004
Louse Lake	Idaho	46°10'55"N	115°17'46"W	6603
Low Pass Lake	Elmore	43°56'48"N	115°02'58"W	
Lower Antelope Reservoir	Owyhee	42°02'17"N	116°22'01"W	
Lower Arcadia Reservoir	Fremont	44°08'06"N	111°35'18"W	5398
Lower Battle Creek Crossing Reservoir No. 1	Owyhee	42°23'07"N	116°23'11"W	5590
Lower Bear Lake	Idaho	46°06'22"N	114°28'58"W	
Lower Box Canyon Lake	Blaine	43°43'10"N	114°03'05"W	8827
Lower Cannon Lake	Idaho	45°19'25"N	116°30'41"W	
Lower Cramer Lake	Custer	44°02'00"N	114°59'37"W	
Lower Deep Creek Reservoir	Twin Falls	42°18'00"N	114°37'42"W	
Lower Glidden Lake	Shoshone	47°31'00"N	115°43'47"W	
Lower Goose Creek Reservoir	Cassia	42°11'48"N	113°54'58"W	4704
Lower Goose Lake	Fremont	44°05'38"N	111°07'32"W	
Lower Gospel Lake	Idaho	45°38'14"N	115°56'58"W	6725
Lower Granite Lake	Whitman	46°39'38"N	117°25'37"W	
Lower Hornet Reservoir	Washington	44°47'54"N	116°43'30"W	6890
Lower Jug Creek Reservoir	Valley	44°51'08"N	116°00'01"W	5550
Lower Knob Lake	Idaho	45°37'48"N	115°55'04"W	6616
Lower Nichol Flat Reservoir	Owyhee	42°19'48"N	116°14'15"W	
Lower Palisades Lake	Bonneville	43°25'50"N	111°09'33"W	6131
Lower Pleasantview Reservoir	Oneida	42°11'42"N	112°21'34"W	4685
Lower Standard Lake	Boundary	48°40'58"N	116°43'04"W	5102
Lower Stevens Lake	Shoshone	47°26'01"N	115°45'30"W	5553
Lower Teapot Reservoir	Elmore	43°01'20"N	115°29'13"W	3138
Lower Thumb Reservoir	Minidoka	43°05'51"N	113°33'12"W	
Lower Trilby Lake	Idaho	45°39'29"N	114°59'45"W	
Lower Twin Lake	Idaho	45°16'10"N	116°12'05"W	7625
Lucky Peak Lake	Ada	43°31'42"N	116°03'08"W	3060
Lulu Reservoir	Blaine	42°55'24"N	113°21'53"W	
Lye Lake	Gooding	43°02'50"N	114°55'45"W	3301
Mabel Reservoir	Minidoka	42°57'37"N	113°33'04"W	
Mable Lakes	Custer	44°27'21"N	115°09'07"W	
Mac Han Lake	Idaho	45°11'10"N	116°05'14"W	7090
Mac Rae Lake	Lincoln	43°05'39"N	113°43'21"W	
Macarthur Lake	Idaho	45°43'14"N	114°58'33"W	
MacKay Reservoir	Custer	43°57'12"N	113°40'24"W	
Mackay Reservoir	Custer	43°57'11"N	113°40'24"W	
Mackerts Pond	Fremont	43°59'36"N	111°42'15"W	
Macon Lake	Camas	43°18'44"N	114°31'35"W	
Magic Reservoir	Blaine	43°15'17"N	114°21'25"W	4797
Magpie Reservoir	Camas	43°16'23"N	114°29'37"W	
Maki Lake	Valley	44°58'29"N	115°54'53"W	7283
Mallard Lake	Lincoln	42°51'25"N	114°05'39"W	
Mallard Lake	Shoshone	46°56'21"N	115°30'45"W	
Mallard Slough	Jefferson	43°55'22"N	112°14'50"W	
Malony Lake	Valley	44°52'28"N	115°54'01"W	7210
Mann Creek Reservoir	Washington	44°23'32"N	116°53'41"W	2883
Mann Lake	Nez Perce	46°22'18"N	116°51'24"W	1820
Maple Lake	Idaho	46°17'16"N	114°45'54"W	7128
Marge Lake	Valley	45°04'35"N	115°57'30"W	
Marie, Lake	Fremont	44°34'17"N	111°29'25"W	
Marion Lake	Valley	44°53'45"N	115°31'33"W	
Market Lake	Jefferson	43°46'17"N	112°07'51"W	4762
Market Lake Slough	Jefferson	43°47'21"N	112°09'28"W	
Marsh Lake	Boundary	48°56'32"N	116°47'15"W	5758
Marsh Lake	Idaho	45°39'45"N	114°55'50"W	
Marshall Lake	Custer	44°09'31"N	114°59'09"W	7715
Marshall Lake	Idaho	45°22'22"N	115°53'44"W	
Marten Lake	Custer	44°17'20"N	115°09'48"W	7556
Martin Lake	Blaine	43°13'54"N	113°08'34"W	
Martin Lake	Boise	44°18'14"N	115°15'45"W	

Lakes, Ponds, Reservoirs

Lake, Pond, or Reservoir	County	Latitude	Longitude	Elev.
Martin Reservoir	Lincoln	43°11'18"N	114°24'10"W	
Mary Alice Lake	Twin Falls	42°35'25"N	114°26'15"W	3640
Mary Alice Lake	Twin Falls	42°35'23"N	114°26'14"W	
Mary Lake	Idaho	45°18'27"N	116°14'09"W	6603
Maud Lake	Idaho	46°28'14"N	114°24'19"W	
Maude Lake	Idaho	46°15'33"N	115°15'17"W	6078
May Lake	Idaho	46°15'30"N	114°51'56"W	6655
McArthur Lake	Boundary	48°31'12"N	116°26'30"W	2085
McCormick Lake	Boundary	48°33'25"N	116°41'28"W	6073
McDonald Lake	Custer	43°59'58"N	114°53'22"W	7097
McGown Lakes	Custer	44°10'54"N	115°04'40"W	
McGregor Reservoir	Owyhee	42°15'34"N	115°12'12"W	5110
McHan Reservoir	Camas	43°13'59"N	114°39'31"W	5438
McRenolds Reservoir	Teton	43°57'55"N	111°03'42"W	6731
McWillards Lake	Boise	44°03'26"N	115°02'34"W	
Meadow Creek Lake	Valley	44°51'48"N	115°22'25"W	
Meadow Lake	Boise	44°03'23"N	115°03'21"W	
Meadow Lake	Lemhi	44°26'02"N	113°18'48"W	
Meadowlark Reservoir	Lincoln	42°47'52"N	114°02'14"W	
Medicine Lake	Kootenai	47°28'00"N	116°35'05"W	
Melton Reservoir	Valley	44°46'43"N	116°01'12"W	4965
Merkley Lake	Bear Lake	42°10'35"N	111°16'23"W	5925
Merles Pond	Twin Falls	42°07'13"N	114°57'06"W	
Merriam Lake	Custer	44°06'58"N	113°45'09"W	
Meteor Hole	Bingham	43°02'07"N	111°41'06"W	6260
Middle Bench Reservoir	Owyhee	42°30'10"N	116°24'54"W	
Middle Cramer Lake	Custer	44°01'52"N	114°59'23"W	
Middle Fork Lake	Valley	44°53'39"N	115°52'45"W	
Middle Knob Lake	Idaho	45°37'34"N	115°54'49"W	6939
Middle Lake	Valley	45°02'06"N	115°25'53"W	
Middle Rainbow Lake	Elmore	43°35'39"N	115°24'34"W	
Middle Trilby Lake	Idaho	45°39'29"N	114°59'45"W	
Midnight Lake	Bear Lake	42°16'12"N	111°33'21"W	8477
Mike Reservoir	Minidoka	42°58'25"N	113°33'59"W	
Mikesell Reservoir Number 2	Fremont	44°07'34"N	111°34'55"W	5373
Mikesell Reservoir Number One	Fremont	44°07'30"N	111°34'32"W	5337
Milk Lake	Custer	44°22'33"N	114°33'50"W	
Milk Lake	Valley	45°02'45"N	115°05'07"W	7045
Mill Creek Lake	Lemhi	44°23'28"N	113°20'34"W	
Mill Lake	Blaine	43°46'47"N	114°38'20"W	8222
Mill Lake	Lemhi	44°39'22"N	113°39'24"W	8760
Millin Reservoir	Owyhee	42°08'31"N	116°21'47"W	
Millward Slough	Madison	43°51'43"N	111°53'29"W	
Milner Lake	Twin Falls	42°31'28"N	114°00'32"W	4134
Miner Lake	Blaine	43°45'35"N	114°39'51"W	8770
Mirror Lake	Bonner	48°09'51"N	116°29'58"W	2370
Mirror Lake	Idaho	45°37'12"N	115°42'17"W	
Mirror Lake	Idaho	45°20'14"N	116°31'29"W	

Lake, Pond, or Reservoir	County	Latitude	Longitude	Elev.
Mirror Lake	Kootenai	47°26'09"N	116°21'36"W	
Mirrow Lake	Lemhi	45°09'16"N	114°34'35"W	
Moe Lake	Idaho	46°05'43"N	114°33'06"W	6594
Mokins Slough	Kootenai	47°47'07"N	116°40'07"W	
Moline Reservoir	Blaine	43°14'35"N	113°30'34"W	
Mollies Lake	Boundary	48°51'04"N	116°49'40"W	
Molony Lake	Valley	44°52'36"N	115°54'30"W	
Monasterio-Jacks Creek Reservoir	Owyhee	42°24'43"N	116°12'11"W	
Monroe Lake	Clearwater	46°32'53"N	115°12'21"W	
Montpelier Reservoir	Bear Lake	42°20'57"N	111°10'25"W	6517
Montrose Reservoir	Lincoln	42°49'03"N	113°56'31"W	
Monument Lake	Minidoka	42°53'10"N	113°31'04"W	
Monument Lake	Owyhee	42°25'14"N	115°42'38"W	
Monument Reservoir	Blaine	43°13'06"N	113°36'36"W	
Monument Reservoir	Lincoln	42°54'45"N	114°06'47"W	
Moores Lake	Idaho	45°37'24"N	115°53'49"W	6325
Moose Creek Reservoir	Latah	46°52'25"N	116°25'07"W	2883
Moose Lake	Bonner	48°21'12"N	116°06'29"W	
Moose Lake	Custer	43°48'58"N	114°03'41"W	9345
Moose Lake	Idaho	46°36'12"N	114°21'27"W	
Moose Lake	Idaho	46°13'02"N	114°29'36"W	
Moose Reservoir	Bingham	43°15'32"N	112°52'58"W	
Moran Lake	Blaine	43°14'46"N	113°40'54"W	
Morehead Lake	Valley	44°34'48"N	115°20'10"W	7476
Morgan Lake	Idaho	45°10'26"N	116°14'36"W	
Morgans Waterhole	Power	42°38'59"N	113°06'24"W	
Mormon Reservoir	Camas	43°16'51"N	114°47'57"W	
Morrow Reservoir	Elmore	43°00'14"N	115°19'56"W	2813
Mosquito Flat Reservoir	Custer	44°31'19"N	114°26'00"W	6931
Mosquito Lake	Idaho	45°17'15"N	115°20'51"W	
Mosquito Lake Reservoir	Owyhee	42°13'11"N	115°17'19"W	5250
Mountain Home Reservoir	Elmore	43°09'16"N	115°39'36"W	3280
Mountain View Lake	Owyhee	42°03'20"N	116°10'06"W	5315
Mud Flat Lake	Lincoln	43°06'17"N	113°45'42"W	4595
Mud Lake	Bear Lake	42°12'31"N	111°20'22"W	5923
Mud Lake	Bear Lake	42°07'58"N	111°17'47"W	5923
Mud Lake	Blaine	43°12'52"N	114°10'33"W	4625
Mud Lake	Boise	44°08'27"N	116°08'28"W	
Mud Lake	Bonner	48°11'35"N	116°23'55"W	
Mud Lake	Butte	43°59'02"N	113°25'13"W	7434
Mud Lake	Idaho	46°15'37"N	115°14'07"W	
Mud Lake	Idaho	45°56'07"N	114°59'09"W	
Mud Lake	Jefferson	43°52'24"N	112°24'24"W	
Mud Lake	Nez Perce	46°13'03"N	116°50'42"W	
Mud Lake	Shoshone	46°57'27"N	115°36'22"W	
Mud Lake	Teton	43°36'59"N	111°04'27"W	
Mud Lake	Valley	44°39'07"N	115°29'47"W	
Mud Reservoir	Owyhee	42°04'27"N	116°17'55"W	

Lake, Pond, or Reservoir	County	Latitude	Longitude	Elev.
Mud Springs Pond	Owyhee	42°56'55"N	116°56'22"W	
Mud Springs Reservoir	Lewis	46°11'58"N	116°35'01"W	
Muddy Reservoir	Butte	43°22'01"N	113°12'55"W	
Mule Shoe Lake	Lincoln	42°50'11"N	114°07'03"W	
Murtaugh Lake	Twin Falls	42°28'10"N	114°10'02"W	4128
Muskrat Lake	Bonner	48°14'48"N	116°40'19"W	2065
Muskrat Pond	Jefferson	43°56'16"N	112°14'59"W	
Mustard Lake	Lincoln	42°58'49"N	113°46'18"W	
Mustard Reservoir	Lincoln	42°49'40"N	113°55'08"W	
Myrtle Lake	Boundary	48°45'15"N	116°37'48"W	
Mystery Lake	Custer	44°29'20"N	114°47'36"W	9070
Narrows, The	Blaine	43°35'40"N	114°10'35"W	
Narrows, The	Caribou	42°58'04"N	111°39'05"W	6100
Neck Lake	Custer	44°06'49"N	114°36'58"W	
Neilsen Reservoir	Camas	43°15'58"N	114°41'55"W	
Nethker Lake	Idaho	45°14'34"N	115°56'52"W	
New Road Lake	Lincoln	42°54'45"N	113°45'09"W	
New Years Lake	Lincoln	42°49'57"N	114°06'09"W	
Newman Lake	Custer	44°21'19"N	115°12'16"W	
Nez Perce Lake	Lemhi	44°30'36"N	113°23'23"W	8860
Nichols Reservoir	Butte	43°34'45"N	113°28'09"W	
Nickel Creek Reservoir	Owyhee	42°33'37"N	116°48'42"W	
Nipple Reservoir	Lincoln	43°06'42"N	114°27'31"W	4625
Noisy Lake	Custer	44°03'29"N	114°34'57"W	8997
Nolan Lake	Custer	44°01'20"N	113°32'13"W	
North Big Springs Reservoir	Owyhee	42°30'22"N	116°27'02"W	
North Blowout Reservoir	Blaine	43°18'58"N	113°36'06"W	
North Farms Lakes	Minidoka	42°53'49"N	113°40'37"W	
North Fork Lake	Custer	43°54'46"N	114°23'46"W	9354
North Lake	Jefferson	43°55'00"N	112°23'45"W	
North Lake	Valley	45°08'50"N	115°54'50"W	
North Lake	Valley	45°07'32"N	115°55'31"W	
North Lambert Reservoir	Owyhee	42°18'33"N	116°43'21"W	
North Lone Lake	Idaho	46°13'33"N	115°10'42"W	
North Lookout Reservoir	Owyhee	42°08'16"N	116°32'26"W	
North Pole Lake	Blaine	43°19'19"N	113°35'02"W	
North Star Lake	Elmore	43°36'39"N	115°30'48"W	
North Three Links Lakes	Idaho	46°11'02"N	115°11'15"W	
Northbound Lake	Shoshone	46°56'24"N	115°34'13"W	5436
Northside Reservoir	Blaine	43°20'31"N	113°34'56"W	
Northwest Pasture Reservoir	Lincoln	43°11'33"N	114°15'42"W	4715
Norton Lake	Valley	44°53'23"N	114°54'08"W	8199
Norton Lakes	Blaine	43°45'06"N	114°39'19"W	
Noseeum Lake	Shoshone	47°01'20"N	115°46'40"W	
No-see-um Lake	Shoshone	47°01'14"N	115°46'32"W	
Notch Lake	Blaine	42°58'29"N	113°23'50"W	
Nub Lakes	Clearwater	46°51'40"N	115°29'29"W	
Nut Basin Lake	Idaho	45°31'52"N	116°09'56"W	

Lake, Pond, or Reservoir	County	Latitude	Longitude	Elev.
O X Lake	Owyhee	42°38'32"N	116°13'24"W	5628
Ocalkens Lake	Custer	44°07'27"N	114°38'21"W	
Old Man Lake	Idaho	46°12'27"N	115°14'16"W	
Olson Pond	Clearwater	46°28'29"N	115°57'40"W	
One Hour Reservoir	Camas	43°14'09"N	114°27'03"W	
One Shot Lake	Lincoln	42°49'27"N	114°03'04"W	
Oneida Narrows Reservoir	Franklin	42°16'34"N	111°44'56"W	4887
Opal Lake	Lemhi	44°53'57"N	114°16'50"W	
Open Crossing Reservoir	Gooding	42°56'06"N	114°49'19"W	
Oreamnos Lake	Boise	44°03'18"N	115°02'58"W	8151
Oregon Butte Lake	Idaho	45°31'37"N	115°40'01"W	7375
Otter Reservoir	Owyhee	42°09'32"N	116°00'18"W	6029
Owens Reservoir	Owyhee	42°18'03"N	116°21'12"W	
Owinza Butte Lake	Lincoln	42°50'52"N	114°04'08"W	
Oxbow Reservoir	Adams	44°58'06"N	116°50'30"W	
Oxbow Slough	Jefferson	43°45'21"N	112°03'25"W	
Oxford Reservoir	Franklin	42°16'57"N	112°00'33"W	4882
Packrat Lake	Boise	44°02'45"N	115°02'22"W	8656
Packsaddle Lake	Blaine	43°14'43"N	113°03'18"W	
Packsaddle Lake	Teton	43°46'15"N	111°20'18"W	7346
Packsaddle Reservoir	Owyhee	42°10'15"N	116°34'29"W	
Paddelford Flat Lake	Blaine	43°18'54"N	113°47'39"W	4813
Paddock Reservoir	Camas	43°13'15"N	114°29'25"W	
Paddock Valley Reservoir	Washington	44°11'53"N	116°35'52"W	3213
Paddy Lake	Fremont	44°03'22"N	111°07'25"W	
Palisades Reservoir	Bonneville	43°19'58"N	111°12'06"W	5620
Pancho Lake	Elmore	43°56'18"N	115°04'47"W	8905
Papoose Lake	Idaho	46°08'10"N	114°46'58"W	6562
Papoose Lake	Idaho	45°23'07"N	116°28'48"W	
Papoose Lake	Idaho	45°14'38"N	114°50'02"W	8198
Papoose Lakes	Valley	44°47'40"N	115°16'45"W	
Parachute Lake	Idaho	46°24'53"N	114°25'18"W	5921
Paradise Lake	Camas	43°44'01"N	114°50'44"W	
Paradise Lake	Idaho	45°18'18"N	116°12'17"W	7438
Paragon Lake	Lemhi	45°04'59"N	114°37'07"W	8554
Park Lake	Idaho	46°10'39"N	114°37'23"W	6765
Parker Lake	Boundary	48°51'47"N	116°35'55"W	
Parleys Reservoir	Bingham	43°10'52"N	111°42'59"W	6496
Parrot Lake	Lemhi	45°10'53"N	114°37'04"W	8178
Parson Reservoir	Butte	43°26'52"N	113°10'13"W	
Parsons Lake	Idaho	46°08'36"N	115°09'04"W	
Partridge Creek Lake	Idaho	45°16'10"N	116°12'04"W	7172
Pass Creek Lake	Butte	44°04'36"N	113°01'00"W	
Pass Lake	Custer	44°05'26"N	113°45'18"W	
Pats Lake	Elmore	43°58'19"N	115°07'24"W	
Paul Reservoir	Clark	44°27'47"N	112°20'10"W	7615
Paws Water	Owyhee	42°41'00"N	116°23'26"W	
Payette Lake	Valley	44°57'15"N	116°05'03"W	

Lakes, Ponds, Reservoirs

Lake, Pond, or Reservoir	County	Latitude	Longitude	Elev.
Payette Lake	Valley	44°54'42"N	116°07'30"W	
Payne Creek Reservoir	Owyhee	42°15'17"N	116°05'14"W	5462
Pearl Lake	Valley	45°04'42"N	115°58'36"W	
Pend Oreille, Lake	Bonner	48°09'39"N	116°20'25"W	
Penny Lake	Blaine	43°41'10"N	114°25'03"W	
Perkins Lake	Blaine	43°55'43"N	114°50'33"W	7014
Perkins Lake	Boundary	48°45'26"N	116°05'32"W	2632
Perkons Lake	Camas	43°45'14"N	114°58'22"W	8710
Pete Ott Lake	Clearwater	46°46'30"N	115°17'35"W	
Petes Lake	Valley	44°51'19"N	115°51'03"W	
Petition Reservoir	Owyhee	42°13'04"N	116°33'39"W	
Pettit Lake	Blaine	43°58'46"N	114°52'33"W	6996
Phyllis Lake	Custer	44°01'23"N	114°38'52"W	
Pine Creek Reservoir	Oneida	42°16'27"N	112°56'00"W	
Pintail Pond	Canyon	43°48'38"N	116°59'32"W	2190
Pintail Reservoir	Lincoln	42°50'42"N	114°08'02"W	
Pinyon Lake	Custer	44°35'00"N	114°54'33"W	
Pinyon Lake	Custer	44°25'49"N	114°43'47"W	
Pioneer Reservoir	Gooding	42°59'42"N	115°03'48"W	3082
Piper Lake	Idaho	45°18'10"N	116°14'27"W	7135
Pistol Lake	Valley	44°43'23"N	115°24'22"W	7812
Piute Basin Reservoir	Owyhee	42°12'04"N	116°31'38"W	
Piute Creek Reservoir	Owyhee	42°10'01"N	116°30'47"W	
Plateau Lake	Lemhi	45°04'33"N	114°37'35"W	7964
Platinum Lake	Clearwater	46°55'52"N	115°15'46"W	5800
Plummer Lake	Elmore	43°57'06"N	115°04'50"W	
Point Lake	Minidoka	43°08'19"N	113°34'03"W	
Poison Creek Reservoir	Owyhee	42°10'57"N	115°26'40"W	5230
Poison Creek Reservoir	Owyhee	42°02'07"N	115°18'47"W	5718
Poison Lake	Valley	44°38'48"N	116°10'10"W	
Pole Creek Reservoir Number Five	Owyhee	42°36'29"N	116°22'08"W	
Pole Creek Reservoir Number Four	Owyhee	42°35'02"N	116°24'06"W	
Pole Creek Reservoir Number One	Owyhee	42°35'44"N	116°23'01"W	
Pole Creek Reservoir Number Three	Owyhee	42°35'18"N	116°22'28"W	
Pole Creek Reservoir Number Two	Owyhee	42°35'24"N	116°23'04"W	
Pole Lake	Lemhi	44°45'56"N	114°39'24"W	8003
Poleline Reservoir	Jerome	42°46'27"N	114°08'23"W	
Pollywog Lake	Idaho	46°21'08"N	115°06'31"W	6571
Pollywog Lake	Lemhi	45°10'57"N	114°36'33"W	8283
Ponds, The	Caribou	42°36'32"N	111°20'56"W	
Pony Lake	Lemhi	45°11'05"N	114°06'40"W	
Porcupine Lake	Bonner	48°14'32"N	116°11'03"W	
Porcupine Lake	Fremont	44°04'21"N	111°10'07"W	
Porcupine Reservoir	Owyhee	42°18'35"N	116°41'35"W	
Porphyry Lake	Idaho	46°18'33"N	114°45'01"W	6929
Porrett Lake	Benewah	47°19'36"N	116°34'38"W	
Porters lake	Kootenai	47°32'40"N	116°27'46"W	2141
Portneuf River Slough	Bannock	42°47'19"N	112°17'59"W	
Portuguese Reservoir	Lincoln	43°10'15"N	114°30'39"W	
Post Office Reservoir	Owyhee	42°12'53"N	115°27'28"W	5135
Pot Hole Reservoir	Owyhee	42°42'28"N	115°28'34"W	3610
Pot Holes	Cassia	42°13'30"N	113°42'36"W	
Pot Lake	Clearwater	46°44'36"N	115°24'15"W	5980
Pot Lake	Valley	45°01'04"N	115°58'08"W	
Pothole Lake	Lemhi	45°10'01"N	114°34'34"W	
Pothole Reservoir	Owyhee	42°07'34"N	116°51'13"W	
Potter Lake	Elmore	43°36'53"N	115°29'55"W	
Potter Reservoir	Owyhee	42°21'05"N	116°48'19"W	
Potters Pond	Valley	44°25'10"N	116°07'51"W	
Poverty Flat Reservoir	Blaine	43°23'23"N	114°19'24"W	
Prairie Lakes	Blaine	43°45'14"N	114°40'33"W	8701
Pratt Lake	Butte	43°19'37"N	113°14'45"W	
Priest Lake	Bonner	48°29'26"N	116°54'09"W	2438
Profile Lake	Custer	44°00'57"N	114°58'21"W	
Profile Lake	Valley	45°04'24"N	115°28'08"W	
Pronghorn Reservoir	Minidoka	43°07'00"N	113°35'07"W	
Providence Lake	Bonner	48°13'07"N	116°26'14"W	
Purdy Reservoir	Blaine	43°13'50"N	113°31'11"W	
Purgatory Lake	Idaho	45°19'00"N	116°33'14"W	7940
Pyramid Lake	Boundary	48°48'00"N	116°36'46"W	6050
Quad Lake	Idaho	45°19'27"N	116°34'05"W	
Quaking Aspen Reservoir	Butte	43°24'22"N	113°11'23"W	
Quartzite Lake	Custer	44°07'25"N	114°36'21"W	
Quartzite Lake	Idaho	45°31'28"N	115°44'45"W	7192
Quayles Lake	Madison	43°55'47"N	111°54'08"W	4849
Queen Lake	Boundary	48°52'37"N	116°11'57"W	
Quiet Lake	Custer	44°03'19"N	114°35'34"W	9242
Quigley Pond	Blaine	43°32'21"N	114°16'09"W	8553
Rabbit Lake Reservoir	Jerome	42°47'31"N	114°10'50"W	
Raft Lake	Valley	44°29'31"N	116°07'15"W	
Raft Reservoir	Lincoln	43°04'03"N	114°21'35"W	
Rainbow Lake	Custer	43°59'11"N	114°43'17"W	8495
Rainbow Lake	Idaho	46°10'34"N	115°13'43"W	
Rainbow Lake	Idaho	45°39'56"N	115°37'43"W	6995
Rainbow Lake	Idaho	45°15'15"N	116°11'43"W	7050
Rainbow Lake	Valley	44°49'49"N	115°34'36"W	7742
Rainbow Lakes	Valley	44°54'49"N	115°50'57"W	
Ralphs Pond	Canyon	43°48'58"N	117°00'40"W	2192
Ramshorn Lake	Lemhi	45°05'06"N	114°36'44"W	8114
Ranger Dip	Bear Lake	42°00'51"N	111°27'48"W	
Ranger Lake	Idaho	46°30'53"N	114°24'57"W	
Rapid Lake	Valley	44°51'21"N	115°54'32"W	7238
Rat Farm Pond	Jefferson	43°56'23"N	112°14'28"W	
Rat Lake	Bingham	43°07'03"N	111°40'59"W	6520
Rattler Reservoir	Gooding	43°04'19"N	114°56'49"W	4028
Rattler Reservoir	Lincoln	43°10'49"N	113°47'05"W	

Lake, Pond, or Reservoir	County	Latitude	Longitude	Elev.
Rattlesnake Draw Reservoir	Owyhee	42°12'10"N	115°20'59"W	5330
Rattlesnake Lake	Idaho	45°34'57"N	115°04'30"W	
Rattlesnake Reservoir	Owyhee	42°17'04"N	115°59'44"W	5750
Rattlesnake Water Hole Number One	Blaine	43°16'06"N	113°13'46"W	
Rattlesnake Water Hole Number Two	Blaine	43°15'56"N	113°13'02"W	
Rays Lake	Jefferson	43°54'28"N	112°16'25"W	4784
Rays Reservoir	Owyhee	42°31'52"N	116°49'14"W	
Red Lake	Custer	44°19'57"N	114°10'35"W	
Red Lake	Idaho	45°58'08"N	115°05'03"W	6656
Red Mountain Lakes	Boise	44°15'17"N	115°23'49"W	
Redfish Lake	Custer	44°06'35"N	114°55'58"W	6547
Redhead Pond	Canyon	43°49'07"N	116°59'56"W	2195
Reed Lake	Lincoln	42°51'12"N	114°04'47"W	
Reeder Lake	Bonner	48°39'20"N	116°58'13"W	2920
Reflection Lake	Lemhi	45°06'20"N	114°36'12"W	8109
Regan Lake	Boise	44°09'58"N	115°04'31"W	
Reserve Reservoir	Camas	43°14'03"N	114°25'09"W	
Reservoir Number Ninety-Three	Blaine	43°13'44"N	114°19'01"W	4757
Reservoir, The	Caribou	42°43'05"N	111°37'22"W	
Revett Lake	Shoshone	47°33'37"N	115°45'02"W	
Reynolds Lake	Lemhi	45°33'27"N	114°32'41"W	7523
Rice Lake	Valley	44°30'32"N	115°38'31"W	
Rim Lake	Minidoka	42°51'58"N	113°32'20"W	
Rim Rock Reservoir	Owyhee	42°24'10"N	116°22'51"W	
Ring Lake	Clearwater	46°46'02"N	115°17'12"W	
Riordan Lake	Valley	44°51'01"N	115°26'14"W	6533
Ripples Reservoir	Bingham	43°12'11"N	112°55'53"W	
Ririe Reservoir	Bonneville	43°34'54"N	111°44'30"W	5120
Road Lake Waterhole	Blaine	43°14'37"N	113°10'23"W	
Road Reservoir	Camas	43°12'25"N	114°27'36"W	
Roaring Creek Lakes	Lemhi	45°12'22"N	114°33'02"W	8135
Roaring Lakes	Valley	44°44'33"N	115°38'00"W	
Roberts Slough	Jefferson	43°42'57"N	112°06'34"W	
Robinson Lake	Boundary	48°58'11"N	116°12'30"W	
Robinson Lake	Fremont	44°10'05"N	111°04'15"W	
Robinson Lake	Latah	46°45'18"N	116°54'35"W	2686
Robison Reservoir	Bingham	43°13'48"N	111°42'09"W	6290
Rock Cut	Clearwater	46°44'59"N	116°12'07"W	
Rock Island Lake	Elmore	43°57'28"N	115°07'48"W	8545
Rock Island Lake	Idaho	45°19'55"N	116°33'14"W	
Rock Lake	Blaine	43°06'43"N	113°05'19"W	
Rock Lake	Custer	44°03'00"N	114°36'30"W	
Rock Lake	Idaho	46°16'17"N	115°14'07"W	5897
Rock Lake	Lincoln	42°57'45"N	114°03'56"W	
Rock Lake	Owyhee	42°36'23"N	115°43'36"W	
Rock Lake	Valley	45°03'59"N	115°32'34"W	
Rock Lake	Valley	45°01'58"N	115°32'26"W	
Rock Lake	Valley	44°30'03"N	116°09'21"W	

Lake, Pond, or Reservoir	County	Latitude	Longitude	Elev.
Rock Lakes	Jefferson	44°01'53"N	112°20'02"W	
Rock Lakes	Lemhi	44°45'17"N	114°40'09"W	
Rock Point Reservoir	Owyhee	42°14'50"N	116°33'59"W	
Rock Slide Lake	Boise	43°57'13"N	115°03'07"W	8668
Rocks Reservoir	Lincoln	43°11'04"N	114°33'10"W	
Rocky Lake	Butte	43°22'21"N	113°02'42"W	
Rocky Lake	Custer	44°29'11"N	115°08'03"W	
Rocky Lake	Owyhee	42°12'30"N	115°36'54"W	
Rocky Point Water Hole	Butte	43°22'02"N	113°08'09"W	
Rocky Reservoir	Gooding	42°58'57"N	114°50'42"W	
Rocky Ridge Lake	Idaho	46°26'30"N	115°29'31"W	
Rocky Ridge Lake	Jerome	42°46'50"N	114°06'04"W	
Rodeo Lake Reservoir	Owyhee	42°39'38"N	116°14'55"W	
Roger Smith Reservoir	Owyhee	41°59'53"N	116°20'05"W	
Roman Nose Lakes	Boundary	48°38'05"N	116°35'18"W	
Roosevelt Lake	Valley	44°58'05"N	115°10'18"W	
Rose Lake	Kootenai	47°33'09"N	116°28'05"W	
Ross Fork Lakes	Camas	43°45'12"N	115°01'52"W	8758
Ross Lake	Owyhee	42°09'04"N	116°18'35"W	
Ross Lake	Owyhee	41°59'45"N	116°18'08"W	
Rough Lake	Custer	44°11'04"N	114°48'01"W	
Rough Lake	Custer	43°43'22"N	113°50'44"W	
Rough Lake	Owyhee	42°09'24"N	116°15'11"W	
Round Lake	Benewah	47°21'48"N	116°43'21"W	
Round Lake	Bonner	48°09'44"N	116°38'14"W	
Round Lake	Bonner	48°00'37"N	116°41'49"W	
Round Lake	Custer	43°46'32"N	113°59'45"W	
Round Lake	Custer	43°43'35"N	113°49'45"W	
Round Lake	Idaho	46°20'09"N	115°08'09"W	6185
Round Lake	Idaho	45°32'33"N	115°40'26"W	7370
Ruby Lake	Idaho	45°34'15"N	115°39'48"W	6985
Rudd-Moore Lakes	Idaho	46°35'35"N	114°27'02"W	
Ruffneck Lake	Custer	44°28'34"N	115°08'33"W	
Ruiz Lake	Lincoln	42°59'28"N	114°14'30"W	
Running Lake	Idaho	45°54'54"N	115°02'50"W	6541
Rush Lake	Washington	44°46'53"N	116°43'48"W	
Russian Lake	Jerome	42°45'19"N	114°00'53"W	
Ruth Lake	Adams	45°14'32"N	116°33'21"W	
Rye Grass Reservoir	Bingham	43°14'10"N	112°54'35"W	
Saddle Lake	Boundary	48°57'07"N	116°45'52"W	
Saddle Lake	Idaho	45°37'32"N	115°01'37"W	7344
Saddle Reservoir	Camas	43°13'50"N	114°28'57"W	
Saddleback Lakes	Custer	44°03'50"N	114°58'18"W	8393
Sage Brush Reservoir	Butte	43°29'17"N	113°16'23"W	
Sage Hen Reservoir	Gem	44°19'34"N	116°11'40"W	4925
Sage Lake	Blaine	43°13'13"N	113°38'29"W	
Sagebrush Spring Reservoir	Fremont	44°18'11"N	111°35'54"W	
Sailor Creek Lake	Owyhee	42°26'37"N	115°09'02"W	4530

Lakes, Ponds, Reservoirs

Lake, Pond, or Reservoir	County	Latitude	Longitude	Elev.
Saint Joe Lake	Shoshone	47°01'04"N	115°04'48"W	
Saint Johns Reservoir	Oneida	42°13'39"N	112°17'19"W	4709
Salmon Falls Creek Reservoir	Twin Falls	42°12'44"N	114°44'00"W	5007
Salvador Lake	Owyhee	42°35'26"N	115°40'57"W	
Samaria Lake	Oneida	42°06'46"N	112°19'12"W	4396
Sams Pond	Owyhee	42°56'23"N	116°57'28"W	
Sand Creek Reservoir	Fremont	44°12'08"N	111°36'52"W	5511
Sand Lake	Boundary	48°30'03"N	116°23'12"W	
Sand Lake	Lincoln	42°52'06"N	114°05'17"W	
Sand Lake Reservoir	Blaine	43°07'10"N	113°01'38"W	
Sandhole Lake	Jefferson	43°55'27"N	112°14'38"W	4786
Sandy Reservoir	Owyhee	42°30'32"N	116°47'53"W	
Sapphire Lake	Custer	44°06'14"N	114°36'54"W	9888
Satan Lake	Adams	45°12'04"N	116°33'11"W	
Sauce Pan Reservoir	Bingham	43°13'43"N	112°57'11"W	
Sawtooth Lake	Custer	44°10'21"N	115°03'46"W	8430
Scarborough Waterhole	Owyhee	42°06'54"N	115°56'53"W	
Scenic Lake	Elmore	43°53'56"N	115°08'31"W	8390
Schlicht Lake	Bonner	48°06'29"N	116°05'05"W	
School Section Lake	Minidoka	42°56'58"N	113°32'40"W	
Schrike Reservoir	Lincoln	42°49'25"N	113°51'43"W	
Scoop Lake	Custer	44°04'22"N	114°35'40"W	9643
Scree Lake	Custer	44°03'42"N	114°35'42"W	
Scribner Lake	Idaho	45°15'01"N	116°06'45"W	7154
Scurvy Lake	Clearwater	46°41'16"N	115°07'28"W	
Seafoam Lake	Custer	44°30'37"N	115°07'08"W	
Search Lake	Boundary	48°55'12"N	116°48'40"W	
Section Corner Reservoir	Minidoka	42°54'08"N	113°29'32"W	
Section Eight Reservoir	Blaine	42°55'00"N	113°23'39"W	
Section Line Reservoir	Minidoka	42°55'18"N	113°36'36"W	
Section Twenty-Eight Reservoir	Lincoln	42°57'30"N	113°43'19"W	
Section Twenty-Five Lake	Minidoka	42°57'59"N	113°40'57"W	
Seed lake	Shoshone	46°56'56"N	115°15'58"W	6280
Seeding Reservoir	Lincoln	42°51'10"N	114°09'19"W	
Seep Reservoir	Lincoln	43°05'33"N	114°26'47"W	4535
Serene, Lake	Idaho	45°11'37"N	116°11'36"W	
Seven Devils Lake	Idaho	45°20'42"N	116°30'59"W	
Seven Hundred Reservoir	Lincoln	43°06'30"N	114°29'57"W	4555
Seven Lakes	Idaho	46°15'21"N	115°13'37"W	6484
Seventy-One Draw Reservoir	Owyhee	42°09'54"N	115°08'41"W	5995
Sewell Reservoir	Owyhee	42°20'34"N	116°06'24"W	
Shadow Lakes	Custer	44°02'26"N	113°33'06"W	
Shale Butte Lake	Lincoln	42°59'44"N	113°47'54"W	4412
Shallow Lake	Custer	44°03'49"N	114°35'52"W	9635
Shasta Lake	Idaho	46°15'03"N	115°12'04"W	6430
Shaw Twin Lakes	Valley	44°53'21"N	115°57'25"W	7311
Shaw Twin Lakes Upper	Valley	44°53'18"N	115°57'24"W	
Shea Reservoir	Owyhee	42°11'02"N	116°50'43"W	

Lake, Pond, or Reservoir	County	Latitude	Longitude	Elev.
Shearing Corral Reservoir	Owyhee	42°42'20"N	116°11'52"W	
Sheep Creek Pond	Owyhee	42°40'11"N	116°29'43"W	
Sheep Creek Reservoir	Bear Lake	42°14'33"N	111°08'35"W	6117
Sheep Lake	Camas	43°14'06"N	114°31'39"W	
Sheep Lake	Custer	44°06'50"N	114°36'39"W	9875
Sheep Lake	Idaho	45°19'55"N	116°32'20"W	7875
Sheep Lake Reservoir	Bingham	43°10'07"N	112°47'57"W	
Sheep Spring Lake	Idaho	45°35'04"N	115°05'33"W	
Sheepeater Lake	Idaho	45°23'05"N	115°20'10"W	7678
Sheepeater Lake	Lemhi	45°08'59"N	114°36'10"W	8653
Sheepherder Lake	Valley	44°29'59"N	115°34'56"W	
Shelf Lake	Custer	44°04'15"N	114°34'05"W	8939
Shelf Lake	Idaho	45°20'35"N	116°32'59"W	
Shell Rock Lake	Valley	44°48'39"N	115°35'34"W	7938
Shepherd Lake	Bonner	48°11'08"N	116°31'35"W	2280
Sheridan Reservoir	Clark	44°27'40"N	111°41'10"W	6452
Shewag Lake	Lemhi	45°26'58"N	113°51'40"W	
Shining Lake	Idaho	45°33'27"N	115°43'04"W	7410
Ship Island Lake	Lemhi	45°09'57"N	114°37'23"W	
Shirk Reservoir	Owyhee	42°14'42"N	115°08'38"W	5405
Shirts Lake	Valley	44°28'10"N	116°07'35"W	
Shoban Lake	Lemhi	45°09'11"N	114°36'06"W	
Shoofly Reservoir	Owyhee	42°15'42"N	116°18'30"W	
Shoshone Reservoir	Lincoln	43°09'07"N	114°29'02"W	
Siah Lake	Idaho	46°31'23"N	114°26'37"W	
Sid Lake	Idaho	46°07'49"N	114°41'53"W	
Sid Lake Number One	Lincoln	42°54'43"N	113°58'49"W	
Sid Lake Number Two	Lincoln	42°54'39"N	113°58'59"W	
Sid Reservoir	Lincoln	42°50'51"N	113°54'54"W	
Silver Lake	Blaine	43°50'19"N	114°32'52"W	
Silver Lake	Fremont	44°19'37"N	111°27'58"W	6119
Silver Lake	Shoshone	46°56'17"N	115°16'16"W	
Sinclair Lake	Boundary	48°55'35"N	116°10'37"W	
Sinks Lake	Minidoka	42°54'57"N	113°29'47"W	
Siphon Lakes	Boise	43°51'52"N	116°15'19"W	
Sisters Lakes	Valley	45°01'51"N	115°59'21"W	
Six Lakes	Custer	44°01'30"N	114°40'24"W	
Sixmile Reservoir	Cassia	42°07'12"N	113°10'10"W	6000
Skein Lake	Valley	44°28'40"N	116°06'37"W	6896
Skookum Lake	Idaho	46°38'11"N	114°21'45"W	6635
Skyhigh Lake	Lemhi	45°07'10"N	114°36'28"W	8632
Skyland Lake	Shoshone	46°56'58"N	115°32'04"W	4799
Skytop Lake	Lemhi	45°12'00"N	113°36'34"W	
Slab Barn Lake	Custer	44°31'00"N	114°21'42"W	
Slate Lake	Idaho	45°35'51"N	115°56'30"W	6735
Slate Lakes	Idaho	45°35'53"N	115°56'25"W	
Slate Reservoir	Owyhee	42°14'31"N	115°53'58"W	
Slick Reservoir	Lincoln	42°49'36"N	114°04'08"W	

Lake, Pond, or Reservoir	County	Latitude	Longitude	Elev.
Slide Lake	Clearwater	46°52'40"N	115°29'46"W	
Slide Lake	Custer	44°06'45"N	114°37'10"W	
Slide Lake	Elmore	43°56'23"N	115°07'31"W	
Slide Rock Lake	Idaho	45°17'26"N	116°34'02"W	
Sliderock Lake	Custer	44°04'16"N	114°34'16"W	8978
Sliman Reservoir	Gooding	42°58'34"N	114°36'48"W	
Smalley Reservoir	Valley	44°35'35"N	116°00'42"W	4971
Smith Creek Lake	Elmore	43°36'56"N	115°29'25"W	
Smith Lake	Bonner	48°25'10"N	116°06'12"W	
Smith Lake	Boundary	48°50'32"N	116°40'28"W	
Smith Lake	Boundary	48°46'28"N	116°15'39"W	2981
Smokey Lake	Clearwater	46°44'38"N	114°46'21"W	
Smoky Dome Lakes	Camas	43°30'15"N	114°56'54"W	
Smoky Lake	Blaine	43°44'22"N	114°39'31"W	
Snake Reservoir	Lincoln	43°11'55"N	114°31'45"W	
Snow Lake	Boundary	48°38'36"N	116°35'42"W	
Snow Lake	Custer	44°05'46"N	114°36'48"W	
Snowbank Lake	Elmore	43°56'50"N	115°07'17"W	
Snowdrift Reservoir	Blaine	43°19'06"N	113°32'41"W	
Snowslide Lake	Valley	44°59'00"N	115°56'00"W	7175
Snowslide Lakes	Camas	43°44'20"N	114°49'37"W	
Soldier Lakes	Custer	44°31'44"N	115°11'51"W	
Soldiers Meadow Reservoir	Nez Perce	46°10'03"N	116°44'09"W	4522
Solomon Lake	Boundary	48°47'55"N	116°06'17"W	
Sonners Reservoir	Blaine	43°14'28"N	114°13'48"W	4770
Sorrel Reservoir	Owyhee	42°06'46"N	116°58'55"W	
Soulen Reservoir	Washington	44°14'51"N	116°34'14"W	3475
South Bench Reservoir	Owyhee	42°29'20"N	116°25'51"W	
South Big Springs Reservoir	Owyhee	42°28'38"N	116°25'03"W	
South End Lake	Minidoka	43°05'26"N	113°32'48"W	
South Fork Lake	Lemhi	45°04'03"N	114°38'26"W	
South Lake	Blaine	43°10'48"N	113°04'08"W	
South Lake	Valley	45°07'16"N	115°55'33"W	
South Lambert Reservoir	Owyhee	42°17'00"N	116°44'47"W	
South Lone Lake	Idaho	46°13'10"N	115°10'28"W	
South Three Links Lakes	Idaho	46°10'02"N	115°11'53"W	
South Wildhorse Lake	Lincoln	43°00'43"N	113°46'41"W	4470
Spangle Lake	Elmore	43°56'50"N	115°01'58"W	8585
Spangle Lakes	Boise	43°56'30"N	115°02'00"W	
Sparrow Reservoir	Minidoka	43°07'39"N	113°34'59"W	
Spencer Lake	Blaine	43°16'01"N	113°30'39"W	
Spencer Reservoir	Owyhee	42°49'05"N	116°38'34"W	5103
Spirit Lake	Kootenai	47°56'28"N	116°53'38"W	
Split Butte Lake	Blaine	42°54'02"N	113°24'29"W	
Split Road Lake	Minidoka	42°53'38"N	113°25'56"W	
Sponge Lake	Idaho	46°21'31"N	115°06'07"W	6025
Spray Reservoir Number Two	Lincoln	43°10'52"N	114°28'59"W	
Spread Point Lake	Idaho	45°38'39"N	115°00'32"W	7182

Lake, Pond, or Reservoir	County	Latitude	Longitude	Elev.
Spring Creek Reservoir	Camas	43°15'54"N	114°39'00"W	5097
Spring Hill Reservoir	Latah	46°47'02"N	116°45'07"W	2750
Spring Lake	Clearwater	46°47'30"N	115°17'54"W	
Spring Valley Reservoir	Latah	46°47'00"N	116°45'06"W	
Spruce Creek Lakes	Idaho	46°34'09"N	114°23'29"W	
Spruce Gulch Lake	Custer	44°36'14"N	114°27'01"W	
Spruce Lake	Boundary	48°55'06"N	116°05'51"W	
Spruce Lake	Idaho	46°05'05"N	114°30'10"W	6645
Square Lake	Blaine	43°15'39"N	114°18'40"W	
Square Lake	Idaho	45°35'55"N	115°38'09"W	7364
Square Mountain Lake	Idaho	45°36'22"N	115°51'52"W	6960
Square Rock Lake	Idaho	45°59'09"N	115°00'50"W	6130
Square Top Lake	Valley	44°45'19"N	115°47'20"W	
Squaw Creek Reservoir	Owyhee	42°11'41"N	116°05'30"W	5460
Squaw Lake	Valley	44°59'33"N	115°59'57"W	
Squaw Meadows Reservoir Number One	Owyhee	42°22'45"N	116°29'06"W	
Squaw Meadows Reservoir Number Three	Owyhee	42°22'09"N	116°26'31"W	
Squaw Meadows Reservoir Number Two	Owyhee	42°20'36"N	116°28'48"W	
Stampede Lake	Boundary	48°33'19"N	116°24'14"W	
Standard Lakes	Boundary	48°41'00"N	116°42'16"W	
Stanley Lake	Custer	44°14'42"N	115°03'18"W	6513
Star Lake	Lincoln	42°49'53"N	114°10'55"W	
Star Reservoir	Owyhee	42°27'52"N	116°45'18"W	
State Line Reservoir	Owyhee	42°00'02"N	116°21'10"W	
Steamboat Lake	Minidoka	43°02'28"N	113°42'48"W	
Steamboat Lake	Shoshone	47°01'24"N	115°47'26"W	
Steamboat Lake	Valley	45°10'50"N	115°43'21"W	7472
Steele Lake	Fremont	44°06'41"N	111°03'05"W	
Steep Lakes	Clearwater	46°52'33"N	114°56'40"W	6085
Steve Lake	Clearwater	46°52'35"N	115°29'26"W	
Stevens Lakes	Custer	44°05'50"N	114°57'51"W	
Stewart Reservoir	Lincoln	42°49'57"N	114°02'05"W	
Still Lake	Bonner	48°14'38"N	116°09'05"W	
Stillman Lake	Idaho	45°42'45"N	114°59'33"W	6857
Stingray Lake	Idaho	45°59'11"N	114°33'01"W	6712
Stoddard Lake	Idaho	45°15'38"N	114°45'59"W	8148
Stone Reservoir	Lemhi	44°34'51"N	113°28'02"W	
Stone Reservoir	Oneida	42°04'06"N	112°41'41"W	4595
Storm Lake	Idaho	46°33'22"N	114°21'44"W	
Storm Peak Lake	Valley	45°07'37"N	115°54'27"W	
Strong Arm Reservoir	Franklin	42°13'07"N	111°53'59"W	4827
Strongarm Reservoir Number One	Franklin	42°14'00"N	111°51'36"W	
Student Lake	Blaine	43°14'10"N	113°31'04"W	
Stump Lake	Valley	44°45'19"N	115°49'22"W	
Stump Reservoir	Owyhee	42°03'04"N	116°49'45"W	
Sublett Reservoir	Cassia	42°19'29"N	113°02'45"W	5331
Succor Creek Reservoir	Owyhee	43°10'51"N	116°58'16"W	
Sullivan Lake	Blaine	43°18'52"N	114°09'22"W	4850

Lakes, Ponds, Reservoirs

Lake, Pond, or Reservoir	County	Latitude	Longitude	Elev.
Sullivan Lake	Custer	44°13'25"N	114°26'38"W	6731
Summit Lake	Valley	45°10'16"N	115°54'48"W	
Summit Lake	Valley	45°02'08"N	115°56'01"W	
Summit Lake	Valley	44°51'43"N	115°54'08"W	7555
Summit Lake	Valley	44°38'49"N	115°35'16"W	
Summit Reservoir	Camas	43°14'46"N	114°29'18"W	
Summit Reservoir	Custer	44°18'23"N	113°30'27"W	6593
Sundown Reservoir	Owyhee	42°00'48"N	116°54'25"W	
Sunset Lake	Butte	43°20'15"N	113°05'12"W	
Surprise Lake	Elmore	43°56'09"N	115°03'17"W	8580
Surprise Waterhole	Butte	43°28'30"N	113°29'46"W	
SW Gate Gulch Reservoir	Owyhee	43°08'01"N	116°59'25"W	
Swamp Angel Lake	Idaho	45°19'07"N	114°45'32"W	8054
Swamp Lake	Custer	44°18'03"N	115°10'33"W	
Swamp Lake	Idaho	46°22'42"N	114°43'11"W	6990
Swan Lake	Bannock	42°17'41"N	111°59'28"W	4756
Swan Lake	Bear Lake	42°00'04"N	111°29'28"W	8315
Swan Lake	Benewah	47°19'33"N	116°33'10"W	
Swan Lake	Caribou	42°36'31"N	111°28'26"W	
Swan Lake	Fremont	44°17'26"N	111°27'36"W	6118
Swan Lake	Fremont	44°06'59"N	111°04'52"W	
Swan Lake	Kootenai	47°28'18"N	116°38'08"W	
Swan Lakes	Caribou	42°36'27"N	111°28'26"W	
Swan Pond	Canyon	43°49'15"N	117°00'15"W	2195
Swartz Pond	Idaho	45°48'05"N	116°16'17"W	
Swauger Lakes	Custer	44°04'54"N	113°34'52"W	
Sweeten Reservoir	Oneida	42°07'02"N	112°38'05"W	4750
Swet Lake	Idaho	45°31'38"N	114°49'01"W	
Swet Lake Pond	Idaho	45°31'55"N	114°49'47"W	8014
Swimm Lake	Custer	44°08'58"N	114°40'01"W	8866
Swimming Bear Lake	Shoshone	47°06'31"N	115°18'12"W	
Table Reservoir	Owyhee	42°17'41"N	116°02'04"W	
Tadpole Lake	Ada	43°11'02"N	116°10'40"W	
Tadpole Lake	Idaho	46°20'29"N	114°28'26"W	6812
Tapper Lake	Blaine	43°12'33"N	114°11'05"W	4635
Taylor Slough	Jefferson	43°43'25"N	112°09'12"W	
Tea Kettle Reservoir	Gooding	42°59'07"N	114°51'06"W	
Teal Reservoir	Lincoln	42°50'29"N	114°09'35"W	
Teardrop Lake	Valley	45°04'50"N	115°59'23"W	
Templeman Lake	Boundary	48°47'51"N	116°16'30"W	
Tent Creek Reservoir	Owyhee	42°04'39"N	116°57'38"W	
Terrace Lakes	Lemhi	45°08'10"N	114°36'36"W	8458
Texas Basin Reservoir	Owyhee	43°10'42"N	116°58'24"W	
Texas Lake	Lincoln	42°51'00"N	114°03'05"W	
Theriault Lake	Shoshone	47°09'17"N	116°01'38"W	
Thirteen Lakes	Idaho	45°31'18"N	114°45'35"W	
Thirtythree Lake	Valley	45°00'17"N	115°53'38"W	
Thomas Lakes	Owyhee	42°59'37"N	116°39'19"W	

Lake, Pond, or Reservoir	County	Latitude	Longitude	Elev.
Thompson Hole	Fremont	44°05'21"N	111°03'57"W	6180
Thompson Lake	Kootenai	47°29'24"N	116°44'36"W	
Thorn Creek Reservoir	Camas	43°11'24"N	114°35'48"W	5505
Three Forks Lake	Minidoka	43°05'49"N	113°36'38"W	
Three Island Lake	Boise	43°57'05"N	115°03'40"W	8598
Three Lake	Boise	44°03'02"N	115°03'22"W	8474
Three Point Reservoir	Lincoln	43°11'15"N	114°27'52"W	
Threemile Reservoir	Clark	44°22'24"N	112°07'33"W	
Thrush Reservoir	Lincoln	42°49'32"N	113°53'29"W	
Tille Lake	Clearwater	46°47'40"N	115°16'26"W	
Timber Draw Reservoir	Owyhee	42°14'37"N	115°56'18"W	
Timberline Lake	Lemhi	45°12'40"N	113°37'21"W	
Timpa Lake	Elmore	43°55'32"N	115°03'10"W	7911
Tin Cup Lake	Butte	43°27'46"N	113°12'07"W	
Tin Cup Lake	Custer	44°07'22"N	114°36'35"W	
Tin lake	Shoshone	46°56'28"N	115°15'56"W	5720
Tindall Reservoir	Owyhee	42°11'11"N	115°53'21"W	
Tindall Water Hole Two	Owyhee	42°06'09"N	115°53'43"W	
Tiny Lake	Custer	44°04'38"N	114°35'23"W	
Tip Top Lake	Lemhi	45°05'29"N	114°36'41"W	8361
Titus Lake	Blaine	43°51'19"N	114°42'37"W	
Tolo Lake	Idaho	45°54'56"N	116°14'06"W	3232
Tom Gooding Lake	Lincoln	43°06'19"N	114°29'05"W	4650
Tom J Reservoir	Valley	44°32'36"N	116°00'00"W	
Toxaway Lake	Custer	43°57'45"N	114°58'10"W	8323
Trail Creek Lakes	Boise	44°09'31"N	115°04'47"W	8225
Trail Lake	Valley	45°09'29"N	115°56'27"W	
Trailer Lakes	Custer	44°09'44"N	115°04'02"W	
Trap Lake	Boundary	48°54'57"N	116°53'37"W	
Treasureton Reservoir	Franklin	42°14'02"N	111°51'40"W	4960
Trestle Reservoir	Blaine	43°14'28"N	114°20'56"W	4745
Triangle Lake	Elmore	43°55'36"N	115°08'57"W	8267
Triangle Lake	Idaho	45°19'14"N	116°33'48"W	
Triangle Reservoir	Owyhee	42°46'50"N	116°36'47"W	5150
Triguero Lake	Owyhee	42°09'24"N	115°40'54"W	
Trilby Lakes	Idaho	45°39'30"N	114°59'46"W	7046
Trinity Lake	Lincoln	42°50'12"N	114°04'31"W	
Triple Lakes	Idaho	45°56'55"N	114°26'17"W	
Tripod Reservoir	Valley	44°17'24"N	116°05'56"W	4980
Trout Lake	Boundary	48°49'19"N	116°35'15"W	
Tule Lake	Fremont	44°04'13"N	111°07'52"W	
Tule Lake	Lemhi	44°59'04"N	113°46'32"W	
Tule Lake	Valley	44°37'46"N	115°40'58"W	
Tumbleweed Reservoir	Lincoln	42°46'34"N	114°01'57"W	
Turkey Lake	Gooding	43°03'27"N	114°39'49"W	
Turner Reservoir	Owyhee	42°36'43"N	116°17'40"W	
Turquoise Lake	Lemhi	45°06'42"N	114°36'43"W	8612
Turtle Lake	Benewah	47°20'26"N	116°29'26"W	

Lake, Pond, or Reservoir	County	Latitude	Longitude	Elev.
Twentyfour Mile Reservoir	Caribou	42°54'12"N	111°53'10"W	
Twentymile Lake	Idaho	45°42'11"N	115°44'29"W	6482
Twentymile Lakes	Valley	45°07'20"N	115°55'28"W	
Twin Cove Lake	Lemhi	45°06'02"N	114°36'12"W	8364
Twin Creek Lakes	Custer	44°34'47"N	114°28'40"W	
Twin Lakes	Adams	45°09'14"N	116°31'09"W	
Twin Lakes	Adams	45°06'10"N	116°11'05"W	7164
Twin Lakes	Blaine	43°56'18"N	114°57'20"W	
Twin Lakes	Blaine	43°14'44"N	113°37'53"W	
Twin Lakes	Butte	43°58'03"N	113°25'48"W	
Twin Lakes	Idaho	45°37'37"N	115°49'44"W	6660
Twin Lakes	Idaho	45°37'19"N	115°50'15"W	
Twin Lakes	Idaho	45°16'25"N	114°49'14"W	8302
Twin Lakes	Kootenai	47°51'26"N	116°51'55"W	2306
Twin Lakes	Owyhee	42°39'52"N	116°13'27"W	
Twin Lakes	Owyhee	42°33'45"N	115°35'45"W	
Twin Lakes	Owyhee	42°15'23"N	115°43'22"W	
Twin Lakes	Shoshone	47°04'43"N	115°29'20"W	
Twin Lakes	Twin Falls	42°02'52"N	114°57'51"W	
Twin Lakes Reservoir	Franklin	42°12'00"N	111°58'54"W	4763
Twin Lakes Reservoir	Owyhee	42°40'05"N	116°14'10"W	5814
Twin Oaks Reservoir	Lincoln	43°09'38"N	114°33'16"W	
Twin Reservoir	Camas	43°13'41"N	114°26'49"W	
Twin Reservoirs	Owyhee	42°31'09"N	116°48'05"W	
Twin Sisters Lakes	Elmore	43°37'33"N	115°27'42"W	
Twin Valley Reservoir	Owyhee	42°48'40"N	116°37'57"W	
Two Bottom Lake	Blaine	43°15'45"N	113°35'23"W	
Two Lakes	Idaho	46°16'39"N	115°07'52"W	6509
Two Mile Limit Reservoir	Owyhee	42°14'19"N	115°52'48"W	
Two Mouth Lakes	Boundary	48°42'24"N	116°38'35"W	
Two-way Pond	Jefferson	43°55'50"N	112°15'46"W	4788
U P Lake	Lemhi	45°14'08"N	114°00'50"W	
Unique Reservoir	Lincoln	43°11'50"N	114°14'38"W	4700
Upper Antelope Reservoir	Owyhee	42°02'30"N	116°20'34"W	
Upper Arcadia Reservoir	Fremont	44°08'32"N	111°35'42"W	5429
Upper Baron Lake	Boise	44°04'40"N	114°59'46"W	8505
Upper Bear Lake	Idaho	46°05'44"N	114°28'11"W	
Upper Box Canyon Lake	Blaine	43°43'46"N	114°04'06"W	9670
Upper Burnt Creek Reservoir	Twin Falls	42°18'36"N	114°27'36"W	
Upper Cannon Lake	Idaho	45°19'23"N	116°31'34"W	
Upper Cramer Lake	Custer	44°01'49"N	114°59'10"W	8381
Upper Deep Creek Reservoir	Oneida	42°11'34"N	112°08'45"W	5340
Upper East Fork Shoofly Reservoir	Owyhee	42°40'57"N	116°15'33"W	
Upper Glidden Lake	Shoshone	47°31'02"N	115°43'03"W	
Upper Goose Lake	Fremont	44°06'12"N	111°06'40"W	
Upper Gospel Lake	Idaho	45°37'49"N	115°57'03"W	7199
Upper Hazard Lake	Idaho	45°10'25"N	116°08'03"W	7428
Upper John G Reservoir	Owyhee	42°07'02"N	116°50'25"W	

Lake, Pond, or Reservoir	County	Latitude	Longitude	Elev.
Upper Jug Creek Reservoir	Valley	44°50'35"N	115°59'54"W	
Upper Knob Lake	Idaho	45°37'09"N	115°55'11"W	
Upper Palisades Lake	Bonneville	43°26'15"N	111°07'11"W	6633
Upper Payette Lake	Valley	45°06'40"N	116°01'32"W	5555
Upper Piute Creek Reservoir	Owyhee	42°08'32"N	116°34'31"W	
Upper Pleasantview Reservoir	Oneida	42°12'23"N	112°21'57"W	4730
Upper Priest Lake	Bonner	48°47'09"N	116°53'18"W	
Upper Rattlesnake Reservoir	Owyhee	42°18'36"N	116°00'40"W	
Upper Redfish Lakes	Custer	44°02'46"N	115°01'37"W	
Upper Seven-Hundred Reservoir	Lincoln	43°11'25"N	114°29'50"W	
Upper Slate Lake	Idaho	45°35'34"N	115°56'51"W	7125
Upper Slope Reservoir	Owyhee	42°40'29"N	116°12'20"W	
Upper Standard Lake	Boundary	48°41'03"N	116°42'11"W	5318
Upper Stevens Lake	Shoshone	47°25'41"N	115°45'41"W	
Upper Stump Canyon Reservoir	Power	42°33'00"N	112°41'05"W	
Upper Teapot Reservoir	Elmore	43°01'52"N	115°29'14"W	3142
Upper Trilby Lake	Idaho	45°39'23"N	115°00'09"W	7136
Upper Twin Lake	Idaho	45°16'04"N	116°12'27"W	7747
Upper Twin Lakes	Owyhee	42°11'21"N	115°43'30"W	
Upper Walcot Basin Reservoir	Owyhee	42°05'19"N	116°52'05"W	
Urquhart Lake	Benewah	47°19'07"N	116°18'45"W	
Valley Creek Lake	Custer	44°22'20"N	114°57'04"W	8163
Valley Waterhole	Owyhee	42°15'23"N	115°26'38"W	5000
Vanity Lakes	Custer	44°29'21"N	115°03'26"W	
Vernon Lake	Boise	43°57'49"N	114°59'27"W	8460
Vics Lake	Valley	44°50'57"N	115°54'21"W	7590
Victor Lake	Valley	45°08'55"N	115°53'59"W	
Village of Troy Reservoir	Latah	46°48'15"N	116°48'37"W	3246
Vineyard Lake	Jerome	42°35'28"N	114°20'42"W	
Virginia Lake	Boise	43°58'35"N	114°59'31"W	8254
Waha, Lake	Nez Perce	46°12'25"N	116°50'05"W	3389
Walcot Basin Reservoir	Owyhee	42°02'53"N	116°53'07"W	
Walker Lake	Custer	44°06'27"N	114°35'49"W	9239
Walker Reservoir	Elmore	43°01'18"N	115°11'24"W	2710
Walker Waterhole	Gooding	42°59'06"N	114°47'00"W	
Walking Fish Lake	Clark	44°30'37"N	111°47'31"W	
Wallace Lake	Lemhi	45°14'46"N	114°00'20"W	
Wally Reservoir	Lincoln	43°03'05"N	113°43'29"W	
Walsh Lake	Bonner	48°25'58"N	116°29'44"W	
Walton Lakes	Idaho	46°26'02"N	114°42'26"W	6502
Warbonnet Lake	Boise	44°03'50"N	115°01'34"W	8916
Warm Lake	Valley	44°38'47"N	115°40'08"W	5298
Warm Springs Reservoir	Owyhee	42°16'11"N	116°43'15"W	
Warner Pond	Valley	44°29'12"N	115°57'36"W	
Warren Pond	Valley	44°29'12"N	115°57'48"W	
Warrior Lakes	Elmore	43°52'55"N	115°17'42"W	
Washington Lake	Custer	44°01'57"N	114°37'12"W	9362
Washington Lake	Custer	43°47'52"N	114°08'40"W	

Lakes, Ponds, Reservoirs

Lake, Pond, or Reservoir	County	Latitude	Longitude	Elev.
Water Reservoir	Lincoln	43°09'04"N	114°26'52"W	
Waterdog Lake	Custer	44°04'21"N	114°33'27"W	
Weather Station Reservoir	Owyhee	42°09'03"N	116°30'07"W	
Weigle Pond	Blaine	43°25'20"N	113°56'00"W	6060
Welcome Lake	Lemhi	45°07'44"N	114°35'24"W	8332
Well Number 2 Reservoir	Bingham	43°17'54"N	112°55'57"W	
West Fork Lake	Boundary	48°50'14"N	116°44'56"W	
West Fork Lakes	Custer	44°44'04"N	114°23'31"W	
West Fork Lakes	Lemhi	44°47'55"N	114°39'31"W	
West Horse Basin Reservoir	Owyhee	42°04'21"N	116°39'16"W	
West Monument Reservoir	Lincoln	42°54'51"N	114°08'02"W	4117
West Puddin Lake	Lemhi	45°07'10"N	114°38'06"W	7717
West Rim Reservoir Number 1	Owyhee	42°32'48"N	116°27'38"W	
West Side Reservoir	Butte	43°24'20"N	113°13'32"W	
West Wildhorse Lake	Lincoln	43°02'27"N	113°50'00"W	4470
Weston Creek Reservoir	Oneida	42°07'10"N	112°06'50"W	5520
Wet Weather Reservoir	Fremont	44°16'07"N	111°34'27"W	
Wheatly Pond	Blaine	43°21'13"N	113°49'45"W	4860
Whickney Tree Reservoir	Owyhee	42°42'29"N	116°14'30"W	
Whitby Reservoir	Owyhee	42°38'51"N	116°57'11"W	5240
White Cap Lakes	Idaho	45°55'01"N	114°27'06"W	
White Goat Lake	Lemhi	44°43'41"N	114°24'57"W	
White Horse Reservoir	Owyhee	42°48'57"N	116°31'00"W	5157
White Lake	Owyhee	42°32'09"N	115°37'54"W	
White Rock Reservoir	Owyhee	42°16'52"N	116°35'47"W	
White Sand Lake	Idaho	46°26'50"N	114°25'12"W	6971
Whitehouse Pond	Idaho	46°30'36"N	114°46'40"W	
Whites Reservoir	Owyhee	42°29'01"N	116°13'29"W	
Wiggler Lake	Franklin	42°00'03"N	111°34'02"W	
Wild Horse Lake	Blaine	43°11'22"N	113°04'11"W	
Wildcat Reservoir	Lincoln	43°09'48"N	114°28'22"W	
Wildhorse Lake	Blaine	43°13'00"N	113°47'25"W	
Wildhorse Lake	Idaho	45°39'11"N	115°38'41"W	7340
Wildhorse Lakes	Custer	43°45'40"N	114°07'34"W	
Wiley Reservoir	Owyhee	42°14'59"N	116°28'52"W	
Wiley Reservoir	Owyhee	42°14'42"N	116°32'50"W	
Williams Lake	Clearwater	46°38'58"N	114°47'30"W	
Williams Lake	Lemhi	45°00'57"N	113°58'34"W	5252
Williams Lake	Minidoka	43°10'41"N	113°34'46"W	
Williams Reservoir	Twin Falls	42°16'20"N	114°25'18"W	5748
Willies Reservoir	Owyhee	42°08'07"N	116°17'39"W	
Willow Creek Reservoir	Owyhee	42°15'51"N	116°22'07"W	
Willow Creek Reservoir	Valley	44°28'10"N	116°06'02"W	6290
Willow Lake	Custer	44°04'29"N	114°33'38"W	8735
Willow Pond	Jefferson	43°56'23"N	112°14'47"W	
Wilson Lake	Lemhi	45°08'38"N	114°35'09"W	
Wilson Lake	Owyhee	42°01'02"N	115°29'25"W	5970
Wilson Lake Reservoir	Jerome	42°37'45"N	114°10'17"W	4012

Lake, Pond, or Reservoir	County	Latitude	Longitude	Elev.
Wind Butte Lakes	Ada	43°12'40"N	116°07'33"W	
Wind Lakes	Idaho	46°23'24"N	114°44'41"W	
Winder Reservoir	Franklin	42°11'00"N	111°53'08"W	4874
Window Lake	Blaine	43°51'19"N	114°30'30"W	10022
Windy Lake	Blaine	43°44'16"N	114°02'27"W	9425
Winifred, Lake	Adams	45°08'58"N	116°35'03"W	
Wiregrass Reservoir	Bannock	42°28'57"N	112°12'14"W	4732
Wiseboy Lakes	Idaho	45°37'50"N	115°41'42"W	7701
Wolf Lodge Holding Ponds	Kootenai	47°39'58"N	116°36'03"W	2220
Wood Duck Pond	Canyon	43°48'24"N	116°59'40"W	2195
Wood Lake	Idaho	46°12'27"N	115°15'10"W	6328
Wood Road Lake	Bingham	43°18'40"N	112°48'46"W	
Woodtick Lake	Lemhi	44°49'31"N	114°40'45"W	8476
Worley Reservoir	Twin Falls	42°14'38"N	115°01'18"W	5390
Worm Lake	Bear Lake	42°08'53"N	111°32'58"W	
Y P Reservoir Number 1	Owyhee	42°06'30"N	116°26'07"W	5175
Y P Reservoir Number 33	Owyhee	42°00'11"N	116°28'55"W	5250
Y Reservoir	Owyhee	42°16'50"N	116°33'32"W	
Yatahoney Reservoir Number 1	Owyhee	42°11'34"N	116°16'43"W	
Yatahoney Reservoir Number 2	Owyhee	42°12'27"N	116°17'09"W	
Yellow Belly Lake	Custer	44°00'02"N	114°52'33"W	7076
Yellowjacket Lake	Lemhi	45°04'02"N	114°33'10"W	7939
Yellowjacket Waterhole	Butte	43°24'02"N	113°30'18"W	
Zada Reservoir	Blaine	43°13'41"N	113°32'40"W	
Zero Lake	Lincoln	42°50'59"N	114°09'49"W	

Mine	County	Latitude	Longitude	Elev.
Adelmann Mine	Ada	43°35'52"N	116°02'46"W	
Afterthought Mine	Owyhee	42°59'59"N	116°41'51"W	
Ajax Mine	Idaho	45°33'47"N	115°27'59"W	
Alamance Mine	Idaho	45°50'29"N	115°25'00"W	
Alaska Mine	Adams	45°07'58"N	116°38'49"W	
Alberta Mine	Idaho	45°46'28"N	115°16'44"W	
Alberta Mine	Idaho	45°23'50"N	115°49'10"W	
Alhambra Mine	Shoshone	47°30'44"N	116°06'09"W	
Alice Mine	Shoshone	47°28'40"N	115°50'32"W	
Alma Mine	Camas	43°34'40"N	114°43'45"W	
Alta Mine	Owyhee	43°01'09"N	116°49'56"W	
Altemont Mine	Idaho	45°48'43"N	115°20'04"W	
American Eagle Mine	Idaho	45°47'20"N	115°21'19"W	
American Girl Mine	Boundary	48°55'06"N	116°19'48"W	
American Mine	Idaho	45°33'35"N	115°27'32"W	
Amicus Mine	Blaine	43°41'52"N	114°17'16"W	
Amie Number One	Owyhee	42°54'10"N	116°26'05"W	
Amie Number Two	Owyhee	42°54'16"N	116°26'08"W	
Amy-Matchless Mine	Shoshone	47°30'32"N	116°14'26"W	
Antimony Mine	Shoshone	47°32'10"N	116°14'42"W	
Argentine Mine (historical)	Shoshone	47°29'06"N	115°58'33"W	
Aspen Mine	Gem	43°51'05"N	116°18'18"W	
Atlas Mine	Idaho	45°35'29"N	115°41'10"W	
Auxor Mine	Bonner	48°16'05"N	116°14'13"W	
Axolotl Mine	Camas	43°43'59"N	114°54'20"W	
Aztec Mine	Custer	44°05'06"N	114°44'31"W	
Badger Mine	Butte	44°06'25"N	113°07'39"W	
Banner Mine	Boise	44°01'07"N	115°31'53"W	
Banner Mine	Idaho	45°28'58"N	116°02'03"W	
Banner Mine	Owyhee	43°00'32"N	116°45'25"W	
Beacon Light Mines	Shoshone	47°27'54"N	115°41'27"W	
Bear Gulch Gravel Pit	Fremont	44°09'20"N	111°17'04"W	
Beardsley Mine	Custer	44°24'12"N	114°18'25"W	6680
Bellville Mine	Bonner	47°54'16"N	116°27'49"W	
Belshazzar Mine	Boise	43°56'36"N	116°00'40"W	
Bergh Mine	Owyhee	43°09'09"N	116°49'12"W	
Bethlehem Mine	Boundary	48°53'38"N	116°18'12"W	
Big Buffalo Mine	Idaho	45°36'27"N	115°41'14"W	
Big Lead Mine	Lemhi	45°21'20"N	114°16'25"W	
Big Lead Mine	Lemhi	44°51'17"N	114°30'13"W	
Big Three Mine	Idaho	45°27'47"N	115°59'33"W	
Black Bear Mine	Idaho	45°29'33"N	116°03'47"W	
Black Diamond Mine	Idaho	45°31'52"N	115°27'06"W	
Black Hornet Mine	Ada	43°35'13"N	116°02'48"W	
Black Jack Mine	Owyhee	43°01'18"N	116°45'22"W	
Black Lady Mine	Idaho	45°47'13"N	115°30'31"W	
Black Leopard Discovery Mine	Adams	45°12'37"N	116°27'02"W	
Black Rock Mine	Custer	44°00'13"N	114°39'37"W	

Mine	County	Latitude	Longitude	Elev.
Blackbird Mine	Idaho	45°48'16"N	115°41'01"W	
Blackbird Mine	Lemhi	45°07'32"N	114°20'57"W	
Blackbird Mines	Lemhi	45°07'02"N	114°20'24"W	
Blackhawk Mine	Boise	43°57'28"N	115°59'26"W	
Blackpine Mine	Lemhi	45°02'50"N	114°12'12"W	
Blackstone Mine	Bear Lake	42°06'39"N	111°26'59"W	
Blue Jacket Mine	Idaho	45°33'24"N	116°26'28"W	
Blue Jay Mine	Lemhi	44°36'47"N	113°33'22"W	
Blue Plate Mine	Lemhi	45°29'56"N	114°25'14"W	
Blue Ribbon Mine	Idaho	45°48'16"N	115°22'06"W	
Bluebird Mine	Owyhee	42°48'19"N	116°25'04"W	
Bluebird Placer Mine	Idaho	45°50'16"N	115°39'20"W	
Bluejacket Mine	Adams	45°07'33"N	116°38'16"W	
Bluejacket Mine	Owyhee	43°00'13"N	116°41'45"W	
Bobby Anderson Mine	Shoshone	47°31'01"N	116°14'18"W	
Bodie Canyon Mine	Bonner	48°12'28"N	117°04'35"W	
Bonami Number 1 Mine	Latah	46°55'02"N	116°37'27"W	
Bonaparte Mine	Elmore	43°40'14"N	115°13'06"W	
Boulder Mine	Bear Lake	42°13'45"N	111°29'44"W	
Boulder Mine	Boise	43°54'56"N	115°47'26"W	
Boulder Mine	Boundary	48°33'46"N	116°13'09"W	
Bradbury Mine	Kootenai	47°51'52"N	116°36'08"W	
Brown Bear Mine	Teton	43°43'09"N	111°18'24"W	
Brown Bull Mine	Lemhi	44°28'16"N	113°19'35"W	
Brunzell Mine	Owyhee	43°00'58"N	116°45'48"W	
Buckhorn Mine	Boundary	48°52'24"N	116°03'19"W	
Buckhorn Mine	Idaho	45°47'17"N	115°39'01"W	
Buckhorn Mine	Lemhi	44°47'20"N	113°27'43"W	
Buckskin Mine	Custer	44°19'15"N	114°33'47"W	
Buffalo Idaho Mine	Idaho	45°50'30"N	115°40'53"W	
Bullion Mine	Idaho	45°28'02"N	115°58'50"W	
Bullion Mine	Shoshone	47°24'04"N	115°42'08"W	
Bumblebee Mine	Boise	43°39'13"N	116°03'04"W	
Bunker Hill Mine	Shoshone	47°30'42"N	116°08'39"W	
Burnt Cabin Mine	Kootenai	47°46'11"N	116°34'41"W	
Burpee Mine	Idaho	45°35'28"N	115°28'38"W	
Buster Mine	Idaho	45°50'10"N	115°25'41"W	
Buttercup Mine	Camas	43°31'04"N	114°35'37"W	
Cal-Ida Mine	Custer	44°11'08"N	114°34'07"W	
Cal-Idaho Pit	Idaho	45°48'43"N	115°27'55"W	
California Mine	Shoshone	47°30'36"N	115°54'40"W	
Camas Mine	Blaine	43°24'43"N	114°28'36"W	
Campbell Mine	Clearwater	46°35'37"N	115°39'06"W	
Cape Horn Mine	Owyhee	43°00'51"N	116°41'53"W	
Carbonate Hill Mine	Shoshone	47°27'04"N	115°46'05"W	
Caribou Mine	Kootenai	47°36'04"N	116°39'45"W	
Carpie Mine	Bonner	48°05'22"N	116°04'28"W	
Carrico Mine	Latah	46°59'45"N	116°48'27"W	

Mines

Mine	County	Latitude	Longitude	Elev.
Carrie Leonard Mine	Camas	43°36'02"N	114°41'47"W	
Center Star Mine	Idaho	45°48'37"N	115°33'29"W	
Chalk Mine	Gooding	43°05'35"N	114°52'02"W	
Champion Mine	Custer	43°51'39"N	113°40'17"W	
Chatauqua Mine (historical)	Owyhee	43°00'58"N	116°50'10"W	
Chitwood Mine	Idaho	46°12'15"N	115°37'01"W	
Circ Twins Mine	Idaho	45°41'15"N	115°31'51"W	
Clark Mine	Bear Lake	42°08'56"N	111°27'02"W	
Clayton Mine	Custer	44°16'51"N	114°24'35"W	
Clearwater Mine	Idaho	45°48'31"N	115°41'15"W	
Clearwater Mine	Shoshone	46°58'18"N	115°09'10"W	
Climax Mine	Blaine	43°29'53"N	114°20'28"W	
Clipper Bullion Mine	Lemhi	45°22'08"N	114°16'36"W	
Coal Mine	Clark	44°32'47"N	111°56'55"W	
Coeur d'Alene Mine	Idaho	45°47'29"N	115°37'17"W	
Coin Bond Mine	Boise	43°57'46"N	115°58'29"W	
Columbia Mine	Lemhi	44°58'55"N	114°32'53"W	
Comeback Mine	Boise	44°00'14"N	115°49'12"W	
Commodore Mine	Lemhi	44°47'02"N	113°25'34"W	
Commonwealth Mine	Kootenai	47°49'23"N	116°39'26"W	
Comstock Mine	Idaho	45°31'01"N	115°27'04"W	
Conda Mine	Caribou	42°43'44"N	111°31'36"W	
Conjecture Mine	Bonner	47°54'56"N	116°25'46"W	
Continental Mine	Boundary	48°55'16"N	116°54'00"W	
Copper Basin Mine	Custer	43°48'11"N	113°49'17"W	
Copper Cliff Mine	Adams	45°06'02"N	116°40'35"W	4815
Copper King Mine	Lemhi	45°18'35"N	114°16'52"W	
Copper King Mine	Shoshone	47°29'45"N	115°46'05"W	
Copper Queen Mine	Lemhi	44°58'08"N	113°28'31"W	
Corral Mine	Custer	43°56'25"N	114°15'28"W	7760
Cougar Mine	Valley	44°40'51"N	115°18'07"W	
Courier Mine	Blaine	43°39'14"N	114°15'21"W	
Crackerjack Mine	Idaho	45°37'05"N	115°39'58"W	
Crater Mine	Custer	44°08'21"N	114°36'25"W	
Crawford Mine	Clearwater	46°34'00"N	115°56'49"W	
Crescent Mine	Owyhee	42°55'46"N	116°46'05"W	
Crescent Mine	Shoshone	47°30'20"N	116°04'24"W	
Croesus Mine	Blaine	43°28'26"N	114°20'47"W	
Cuddy Mine	Washington	44°48'34"N	116°44'15"W	
Custer Mine	Shoshone	47°32'08"N	115°50'42"W	
Cyprus-Thompson C Open Pit Mine	Custer	44°18'52"N	114°33'04"W	8120
Daisy Mine	Owyhee	43°00'27"N	116°50'01"W	6100
Daley Con Mines	Elmore	43°20'27"N	115°28'30"W	
Dandy Mine	Kootenai	47°51'18"N	116°33'22"W	
Darlington Shaft	Custer	43°52'54"N	113°40'37"W	
De Lamar Mine	Owyhee	43°00'42"N	116°50'03"W	
Deadwood Mine	Valley	44°28'14"N	115°34'50"W	
Deer Creek Mine	Nez Perce	46°04'07"N	116°42'25"W	

Mine	County	Latitude	Longitude	Elev.
Del Rio Mine	Idaho	45°33'22"N	115°40'55"W	
Democrat Mine	Blaine	43°31'32"N	114°24'17"W	
Democrat Mine	Clearwater	46°34'49"N	115°56'19"W	
Dewey Mine	Idaho	45°56'03"N	116°00'36"W	
Dewey Mine	Owyhee	43°02'16"N	116°45'51"W	
Dewey Mine	Valley	44°57'34"N	115°08'42"W	
Diamond Drill Mine	Idaho	45°47'39"N	116°20'10"W	
Diamond Hitch Mine	Idaho	45°41'08"N	115°31'03"W	
Dillinger Mine	Idaho	45°31'49"N	115°26'30"W	
Dixie Queen Mine	Bonner	47°58'18"N	116°22'22"W	3540
Dixie Queen Mine (historical)	Idaho	45°32'48"N	115°28'12"W	
Dixie Royal Mine	Idaho	45°31'44"N	115°30'17"W	
Doer Mine	Latah	46°52'52"N	116°34'16"W	
Doughboy Mine	Owyhee	42°55'59"N	116°47'14"W	
Dougherty Mine	Bonner	48°24'45"N	116°08'31"W	5920
Drummond Mine	Blaine	43°40'23"N	113°55'18"W	
Durden Mine	Idaho	45°20'17"N	115°54'37"W	
Eagle Bird Mine	Blaine	43°38'35"N	113°54'57"W	
Easter Mine	Idaho	45°51'51"N	115°38'03"W	
El Oro Mine	Camas	43°43'02"N	114°57'49"W	
Elkhorn Mine	Boise	43°56'10"N	115°45'35"W	
Ella Mine	Butte	43°36'19"N	113°34'30"W	
Elmo Mine	Shoshone	47°29'43"N	115°59'24"W	
Empire Mine	Custer	43°53'33"N	113°40'12"W	
Erdman Mine	Owyhee	43°00'22"N	116°45'09"W	
Eureka Silver King Mine	Valley	44°31'36"N	115°41'53"W	
Eutopia Mine	Idaho	45°36'10"N	115°32'58"W	
Evergreen Mine	Bonneville	43°04'59"N	111°18'51"W	
Falls Creek Mine	Bonner	48°03'38"N	116°24'14"W	
Fern Mine	Valley	44°54'24"N	115°16'59"W	
Fisher Mine	Idaho	45°49'43"N	115°54'30"W	
Five Points Mine	Camas	43°31'15"N	114°49'03"W	
Forest King Mine	Boise	43°54'24"N	115°46'27"W	
Fort Hall Mine	Bannock	42°46'24"N	112°21'29"W	
Fourmile Mine	Idaho	45°46'05"N	115°36'51"W	
Fourmile Mine	Idaho	45°46'01"N	115°38'25"W	4520
Franklin Mine	Elmore	43°27'46"N	115°18'58"W	
Franklin Mine	Shoshone	47°18'10"N	115°55'48"W	
Frisco Mine	Shoshone	47°30'37"N	115°51'12"W	
G Portal	Caribou	42°43'57"N	111°29'53"W	
Galena Mine	Shoshone	47°28'40"N	115°58'09"W	
Gallagher Mine	Idaho	45°43'59"N	115°23'02"W	4478
Galore Mine	Camas	43°34'53"N	114°43'21"W	
Gambrinus Mine	Boise	43°53'15"N	115°47'32"W	
Gambrinus Surprise Mine	Boise	43°53'42"N	115°46'10"W	
Garfield Mine	Owyhee	43°01'15"N	116°50'40"W	
Gay Mine	Bingham	43°02'03"N	112°02'32"W	
Gem State Mine	Gem	43°51'25"N	116°19'16"W	

Mine	County	Latitude	Longitude	Elev.
General Mine	Shoshone	47°31'33"N	116°13'37"W	
Gilt Edge Mine	Idaho	45°46'32"N	115°38'21"W	4784
Glasgow Mine	Valley	45°03'14"N	115°24'58"W	
Glen Silver Pit Number 1	Owyhee	43°00'58"N	116°50'28"W	
Glen Silver Pit Number 2	Owyhee	43°01'10"N	116°50'57"W	
Gnome Mine	Idaho	45°44'26"N	115°31'12"W	
Golconda Mine	Owyhee	42°44'42"N	116°55'09"W	
Gold Bug Mine	Idaho	45°28'42"N	116°02'14"W	
Gold Bug Mine	Latah	46°58'01"N	116°44'54"W	
Gold Coin Mine	Bonner	48°11'00"N	116°25'36"W	
Gold Hill Mine	Boise	43°57'32"N	115°59'07"W	
Gold Hill Mine	Boise	43°50'13"N	115°49'43"W	
Gold Hill Mine	Latah	46°57'50"N	116°44'47"W	
Gold Hill Mine	Lemhi	45°22'12"N	114°17'10"W	
Gold Hunter Mine	Shoshone	47°28'29"N	115°47'38"W	
Gold Leaf Mine	Idaho	45°31'18"N	115°27'31"W	
Gold Master Mine	Idaho	45°36'15"N	115°32'44"W	
Gold Nugget Mine	Shoshone	47°03'52"N	115°11'12"W	
Gold Point Mine	Idaho	45°46'55"N	115°23'38"W	4240
Goldbug Mine	Idaho	45°36'17"N	115°32'07"W	
Golden Age Mine	Boise	44°01'15"N	115°49'38"W	
Golden Anchor Mine	Idaho	45°24'26"N	115°51'42"W	
Golden Chariot Mine	Boise	44°00'45"N	115°40'13"W	
Golden Chariot Mine	Butte	43°30'09"N	113°34'58"W	
Golden Chest Mine	Shoshone	47°37'01"N	115°50'01"W	
Golden Crown Mine	Owyhee	42°51'13"N	116°26'25"W	
Golden Cup Mine	Valley	45°08'58"N	115°20'46"W	
Golden Cycle Mine	Boise	44°00'50"N	115°39'47"W	
Golden Cycle Mine	Owyhee	43°01'03"N	116°51'03"W	
Golden Hand Mine	Idaho	45°13'21"N	115°19'15"W	
Golden Rule Placer Mine	Idaho	45°16'45"N	115°49'46"W	
Golden Star Mine	Blaine	43°24'04"N	114°27'43"W	
Goldstone Mine	Lemhi	45°08'34"N	113°34'54"W	
Golen Scepter Mine	Boundary	48°59'28"N	116°25'24"W	
Grand View Mine (historical)	Boise	44°00'20"N	115°53'59"W	
Grangeville Mine	Idaho	45°45'55"N	115°29'12"W	
Granite Mine	Shoshone	47°31'29"N	115°52'24"W	
Grave Ajax Mine	Idaho	45°35'15"N	115°41'13"W	
Gray Wolf Mine	Kootenai	47°35'52"N	116°39'24"W	2360
Green Monarch Mine	Bonner	48°07'27"N	116°20'06"W	2760
Greenback Mine	Custer	44°10'53"N	115°05'58"W	
Gregor Mine	Lemhi	45°23'12"N	114°17'14"W	
Grey Brothers Lime Quarry	Bonner	48°03'39"N	116°26'20"W	
Greyhound Mine	Custer	44°34'37"N	115°07'44"W	
Grunter Mine	Lemhi	45°22'25"N	114°17'18"W	
Haidee Mine	Lemhi	45°14'04"N	114°12'37"W	
Halley Placer Mine	Boise	43°56'20"N	115°56'52"W	
Hansey Mines	Shoshone	47°19'22"N	115°35'12"W	

Mine	County	Latitude	Longitude	Elev.
Happy Day Mine	Camas	43°26'27"N	114°31'11"W	6338
Harmony Mine	Lemhi	45°00'54"N	113°49'41"W	
Haystack Mine	Idaho	45°46'23"N	115°35'37"W	
Hazel Pine Mine	Cassia	42°04'53"N	113°01'54"W	
Hecla Mine	Shoshone	47°31'11"N	115°48'49"W	
Hematite Mine	Idaho	45°36'34"N	115°33'25"W	
Henrietta Mine	Owyhee	43°00'40"N	116°51'03"W	
Henry Mine	Caribou	42°52'31"N	111°28'08"W	
Henry Strip Mine	Caribou	42°52'31"N	111°28'05"W	6520
Hercules Mine	Idaho	45°45'48"N	115°17'51"W	
Hercules Mine	Shoshone	47°32'33"N	115°48'32"W	
Hermit Mine	Custer	44°08'50"N	114°35'12"W	
Hi Yu Mine	Idaho	45°29'04"N	116°02'21"W	
Hillman Mine	Teton	43°43'10"N	111°19'23"W	
Hilltop Mine	Lemhi	44°28'53"N	113°19'06"W	
Homebuilder Mine	Kootenai	47°41'56"N	116°33'39"W	2800
Homestake Mine	Boundary	48°37'53"N	116°04'15"W	
Homestake Mine	Idaho	45°39'55"N	115°31'26"W	
Hope And Faith Shaft	Bonner	48°09'21"N	116°30'45"W	
Hope Mine	Blaine	43°28'23"N	114°20'58"W	
Hope Mine	Bonner	48°10'02"N	116°10'20"W	
Horn Silver Mine	Camas	43°35'30"N	114°43'02"W	
Hornet Mine	Owyhee	42°56'06"N	116°47'14"W	
Horseshoe Mine	Custer	43°53'52"N	113°41'26"W	
Hub Mine	Butte	43°30'42"N	113°36'49"W	
Humboldt Mine	Idaho	45°19'58"N	115°48'36"W	
Humboldt Mine	Owyhee	43°00'55"N	116°45'22"W	
Hyatt Mine	Idaho	45°22'45"N	115°49'20"W	
Ida Belle Mine	Owyhee	43°10'29"N	116°48'17"W	
Idaho Almaden Mine	Washington	44°14'25"N	116°42'53"W	
Idaho Champion Group Prospects	Idaho	45°45'15"N	115°31'20"W	4920
Idaho Lakeview Mine	Bonner	47°54'50"N	116°28'27"W	
Idaho Mine	Owyhee	43°01'23"N	116°45'35"W	
Idaho Mine	Teton	43°43'54"N	111°18'12"W	
Idaho Muldoon Mine	Blaine	43°36'28"N	113°53'01"W	
Idamont Mine	Boundary	48°36'11"N	116°04'55"W	
Illinois Mine	Boise	43°53'13"N	115°47'00"W	
Ima Mine	Lemhi	44°32'01"N	113°41'30"W	
Independence Mine	Blaine	43°40'15"N	114°16'31"W	
Independence Mine	Valley	45°09'13"N	115°23'32"W	
Interstate Mine	Shoshone	47°32'50"N	115°50'58"W	4640
Iowa Mine	Boise	43°57'45"N	115°59'00"W	
Iron Crown Mine	Idaho	45°55'29"N	115°36'26"W	4800
Iron Dyke Mine	Lemhi	44°27'48"N	113°18'37"W	
Italian Mine	Lemhi	45°13'17"N	114°11'29"W	
IXL Mine	Adams	44°59'54"N	116°25'26"W	4481
Jack Waite Mine	Shoshone	47°39'52"N	115°44'31"W	4400
Jack White Mine	Shoshone	47°40'05"N	115°44'42"W	

Mines

Mine	County	Latitude	Longitude	Elev.
Jefferson Mine	Owyhee	42°56'20"N	116°47'23"W	
Jericho Mine	Clearwater	46°42'45"N	116°07'16"W	3700
Juanita Mine	Idaho	45°32'36"N	115°27'53"W	
Jumbo Mine	Idaho	45°33'33"N	115°40'51"W	
June Day Mine	Blaine	43°40'59"N	114°18'14"W	
Jureano Placer Diggings	Lemhi	45°10'52"N	114°14'05"W	
Keep Cool Mine	Bonner	47°54'28"N	116°26'38"W	
Kentuck Mine	Lemhi	45°22'17"N	114°17'48"W	
Keystone Mine	Custer	44°26'43"N	114°19'56"W	
Kimberly Mine	Idaho	45°24'11"N	115°52'08"W	
King of the West Mine	Camas	43°35'19"N	114°42'35"W	
Kitty Burton Mine	Lemhi	45°27'36"N	114°09'32"W	
Kline Mine	Valley	44°39'20"N	115°41'32"W	5380
Klondike Mine (historical)	Boise	44°00'35"N	115°53'49"W	
L	Idaho	45°34'24"N	115°27'45"W	
Lakeview Mines	Bonner	47°58'03"N	116°27'34"W	
Last Chance Mine	Butte	43°35'18"N	113°34'29"W	
Last Chance Mine	Camas	43°36'37"N	114°42'07"W	
Last Chance Mine	Latah	46°52'49"N	116°34'54"W	
Last Chance Mine	Shoshone	47°31'08"N	116°08'19"W	
Latest Out Mine	Lemhi	44°27'27"N	113°17'31"W	
Lawrence Mine	Bonner	48°08'52"N	116°08'59"W	
Laxey Mine	Owyhee	42°44'41"N	116°55'02"W	
Lead Bell Mine	Bannock	42°32'55"N	111°56'51"W	
Leadville Mine	Idaho	45°24'03"N	115°50'54"W	
Leary Placer Mine	Boise	43°56'44"N	115°55'52"W	
Leggett Placer Mine	Idaho	45°50'15"N	115°37'43"W	
Liberty Gem Mine	Blaine	43°27'24"N	114°26'19"W	
Liberty Mine	Gem	44°02'16"N	116°18'40"W	
Lincoln Mine	Gem	43°50'47"N	116°20'13"W	
Little Klondike Mine	Boise	43°54'44"N	115°46'31"W	
Little Leggett Placer Mine	Idaho	45°49'57"N	115°39'03"W	
Little Livingston Mine	Custer	44°08'47"N	114°35'11"W	
Livingston Mine	Custer	44°08'07"N	114°35'08"W	
Lobe Mine	Owyhee	43°00'58"N	116°45'36"W	
Lone Pine Mine	Idaho	45°48'46"N	115°41'12"W	
Lone Pine Mine	Idaho	45°28'51"N	116°03'15"W	
Long Tom Mine	Idaho	45°32'00"N	115°28'11"W	
Lost Packer Mine	Custer	44°35'54"N	114°52'49"W	
Lost Wheelbarrow Mine	Latah	46°59'49"N	116°47'02"W	
Lotspiech Mine	Valley	45°02'29"N	115°24'09"W	
Lucky Boy Mine	Boise	43°52'58"N	115°46'18"W	
Lucky Boy Mine	Custer	44°22'24"N	114°40'43"W	
Lucky Calumet Mine	Shoshone	47°29'20"N	115°44'38"W	
Lucky Friday Mine	Shoshone	47°28'17"N	115°46'50"W	
Lucky Gulch Mine	Blaine	43°38'51"N	114°14'42"W	
Lucky Lad Mine	Idaho	45°37'12"N	115°41'07"W	
Lucky Lad Mine	Valley	44°39'45"N	115°16'24"W	

Mine	County	Latitude	Longitude	Elev.
Lucky Strike Mine	Idaho	45°46'40"N	115°30'35"W	
Lucky Swede Mine (historical)	Shoshone	47°22'36"N	115°43'37"W	3860
Ludwig Mine	Valley	45°07'39"N	115°24'26"W	
Luella Mine	Latah	46°52'52"N	116°35'35"W	
Mackey Mine	Idaho	45°50'29"N	115°41'31"W	
Mammoth Mine	Boise	44°00'12"N	115°42'43"W	
Mammoth Mine	Idaho	45°27'45"N	115°29'50"W	
Marsh Creek Mine	Elmore	43°38'56"N	115°11'07"W	
Martin Mine	Butte	43°29'14"N	113°35'28"W	
Mary Blue Mine	Valley	44°21'50"N	115°36'14"W	
Mary Jane Mine	Valley	44°23'05"N	115°35'52"W	
Maryland Mine	Lemhi	44°47'15"N	113°26'35"W	
Mascot Mine	Blaine	43°41'44"N	114°05'51"W	
Mascot Mine	Boise	43°53'04"N	115°46'03"W	
Mastodon Mine	Shoshone	47°20'41"N	115°58'20"W	
Matteson Mine	Owyhee	43°08'19"N	116°37'08"W	
Mayflower Mine	Boise	43°57'16"N	115°59'39"W	
Mayflower Mine	Lemhi	45°17'00"N	114°17'03"W	
McCoy Mine	Blaine	43°27'34"N	114°21'21"W	
McCrae Mine	Valley	45°09'53"N	115°24'04"W	
McKinley Mine	Idaho	45°31'48"N	116°16'55"W	
McKinley Mine	Idaho	45°31'43"N	115°27'41"W	
McLean Mine	Boundary	48°53'21"N	116°56'16"W	
Meador Diggings	Idaho	45°49'25"N	115°39'22"W	
Meadow Creek Mine	Valley	44°54'01"N	115°19'57"W	
Mica Mine	Adams	44°39'09"N	116°15'00"W	
Middleman Mine	Gem	43°51'01"N	116°18'27"W	
Mikesell Mine	Teton	43°44'52"N	111°19'06"W	
Miller Brothers Mine	Boundary	48°56'55"N	116°17'51"W	
Miller Mine (historical)	Shoshone	47°20'38"N	115°37'19"W	
Mineral Hill Mine	Boise	43°59'09"N	115°54'31"W	
Mineral Mountain Mine	Shoshone	47°30'23"N	116°02'23"W	
Minerva Mine	Bonner	48°04'24"N	116°24'32"W	
Minerva Mine	Elmore	43°46'25"N	115°07'05"W	
Minnie Moore Mine	Blaine	43°28'09"N	114°17'35"W	
Missouri Mine	Boise	44°00'02"N	115°51'19"W	
Mizpah Mine	Latah	46°59'57"N	116°29'30"W	
Monarca Mine	Owyhee	43°08'36"N	116°49'37"W	
Monarch Mine	Elmore	43°46'51"N	115°07'03"W	
Monarch Mine	Shoshone	47°35'05"N	115°46'26"W	
Monitor Mine (historical)	Shoshone	47°21'07"N	115°34'27"W	
Monitor Mines (historical)	Shoshone	47°20'34"N	115°36'19"W	
Monroe Mine	Gem	43°51'56"N	116°18'33"W	
Monte Cristo Mine	Bonneville	43°05'49"N	111°19'53"W	
Montgomery Mine	Blaine	43°41'35"N	114°17'30"W	
Montgomery Mine	Boundary	48°57'19"N	116°23'06"W	
Moonlight Mine	Bannock	42°53'02"N	112°16'41"W	
Morning Glory Mine	Madison	43°38'55"N	111°32'02"W	

Mine	County	Latitude	Longitude	Elev.
Morning Mine	Shoshone	47°28'04"N	115°48'43"W	
Morning Mine 4	Shoshone	47°29'17"N	115°48'32"W	
Morning Mine 5	Shoshone	47°29'44"N	115°48'12"W	
Morning Star Mine	Owyhee	43°01'22"N	116°43'53"W	
Mortimer Mine	Washington	44°32'36"N	117°01'34"W	
Moscow Mine	Valley	45°06'01"N	115°24'14"W	
Mother Lode Mine	Idaho	45°35'15"N	115°40'57"W	
Mother Lode Mine	Idaho	45°28'41"N	116°03'21"W	6413
Mount Marshall Mine	Idaho	45°23'35"N	115°51'42"W	
Mountain Boy Mine	Lemhi	44°27'30"N	113°18'43"W	
Mountain Chief Mine	Boise	43°55'41"N	116°00'20"W	
Mountain Chief Mine	Bonner	48°47'44"N	116°55'23"W	
Mountain King Mine	Custer	44°36'20"N	115°01'48"W	
Muldoon Mine	Blaine	43°36'14"N	113°53'28"W	
Mule Shoe Mine	Custer	44°14'41"N	114°21'47"W	
Musgrove Mine	Lemhi	45°01'50"N	114°19'45"W	
Mutual Mine	Blaine	43°36'54"N	113°54'33"W	
Nabob Mine	Lemhi	45°22'33"N	114°23'17"W	
National Mine	Shoshone	47°29'24"N	115°45'43"W	
Navigation Mine	Bonner	48°46'25"N	116°53'09"W	
Nellie Ann Mine	Owyhee	42°56'52"N	116°47'47"W	
Nellie Mine	Shoshone	47°30'11"N	116°01'32"W	
New Rainbow Mine	Bonner	47°54'21"N	116°25'41"W	
New Rainbow Mine	Kootenai	47°26'43"N	116°31'52"W	3220
New York Mine	Idaho	45°50'19"N	115°40'20"W	
North De Lamar Pit	Owyhee	43°00'49"N	116°49'40"W	
North Hill Mine	Idaho	45°44'45"N	115°35'09"W	
North Hornet Mine	Adams	44°53'31"N	116°38'16"W	
North Star Mine	Blaine	43°39'17"N	114°15'59"W	
North Star Mine	Idaho	45°38'19"N	115°40'55"W	
North Star Mine	Idaho	45°32'08"N	115°27'00"W	
North Tunnel Mine	Valley	44°54'59"N	115°20'14"W	
Ohio Mine (historical)	Owyhee	43°00'46"N	116°49'24"W	
Old North Star Mine	Shoshone	47°21'42"N	115°41'45"W	
Old Orogrande Mine	Idaho	45°41'49"N	115°33'27"W	5040
Old Portland Mine	Idaho	45°44'56"N	115°28'17"W	
Old Rainbow Mine	Benewah	47°24'40"N	116°30'13"W	3420
Old Sam Mine	Camas	43°43'27"N	114°56'16"W	
Old Timer Mine	Idaho	45°25'12"N	116°29'49"W	6400
Old Triumph Mine	Blaine	43°39'32"N	114°16'34"W	
Olson Mine	Latah	46°52'46"N	116°34'45"W	
Ontario Mine	Idaho	45°32'41"N	115°27'58"W	
Ore Cash Mine	Lemhi	45°17'05"N	113°41'31"W	
Oregon Tipton Mnine	Idaho	45°18'50"N	116°25'56"W	4350
Oro Fino Mine	Owyhee	43°00'35"N	116°41'41"W	
Oro Mine (historical)	Boise	44°00'35"N	115°50'45"W	
Orogrande Frisco Mine	Idaho	45°42'28"N	115°32'29"W	
Oxford Mine	Clearwater	46°36'43"N	115°37'56"W	

Mine	County	Latitude	Longitude	Elev.
Oxide Lode Mine	Butte	43°35'47"N	113°34'36"W	
Pacific Mine	Custer	44°24'34"N	114°18'59"W	
Packsaddle Mine	Teton	43°44'14"N	111°20'22"W	
Page Mine	Shoshone	47°31'42"N	116°11'59"W	
Painter Mine	Idaho	45°25'24"N	115°27'38"W	
Palisade Mine	Kootenai	47°26'05"N	116°21'05"W	
Park Mine	Lemhi	45°00'46"N	114°32'33"W	
Parker Mine	Blaine	43°41'32"N	114°17'25"W	
Parker Mine	Boundary	48°53'35"N	116°51'16"W	
Parker Mine	Lemhi	44°35'56"N	114°33'29"W	
Pasadena Mine	Idaho	45°45'23"N	115°18'25"W	
Paymaster Mine	Blaine	43°28'52"N	113°39'40"W	
Pell Placer Mine	Idaho	45°57'38"N	115°40'13"W	
Penman Mine	Idaho	45°39'48"N	115°31'32"W	
Penn Dixie Mine	Idaho	45°32'21"N	115°27'35"W	
Perseverance Mine	Owyhee	42°55'16"N	116°46'42"W	
Phi Kappa Mines	Custer	43°50'47"N	114°12'12"W	
Pierce Tunnel	Blaine	43°41'37"N	114°17'15"W	
Pintar Mine	Teton	43°44'30"N	111°20'07"W	
Pioneer Mine	Clearwater	46°24'29"N	115°41'09"W	
Pistol Grip Mine	Clearwater	46°29'23"N	115°40'56"W	
Placer Mine	Owyhee	43°00'20"N	116°44'38"W	
Plowboy Mine	Bonner	48°47'47"N	116°54'51"W	
Polaris Mine	Shoshone	47°30'12"N	116°03'10"W	
Poorman Mine	Idaho	45°28'29"N	116°03'38"W	
Poorman Mine	Owyhee	43°00'16"N	116°42'36"W	
Pope Shenon Mine	Lemhi	45°04'26"N	113°51'25"W	
Portland Group Mine	Bonner	48°11'20"N	116°23'41"W	2600
Portland Mine	Lemhi	44°27'55"N	113°19'41"W	
Portland Mine	Valley	44°31'12"N	115°29'06"W	
Prichard Mine	Idaho	45°33'22"N	115°27'51"W	
Princess Blue Ribbon Mine	Camas	43°27'37"N	114°32'29"W	
Queen Mine	Ada	43°34'16"N	116°02'56"W	
Queen Mine	Boundary	48°54'23"N	116°13'21"W	
Queen of the Hills Mine	Lemhi	45°14'35"N	113°56'44"W	
Rabbit Foot Mine	Lemhi	44°53'01"N	114°20'17"W	
Ramshorn Mine	Custer	44°24'38"N	114°21'41"W	
Ray Lode Mine	Lemhi	44°35'18"N	113°33'36"W	
Reber Mine	Elmore	43°19'35"N	115°28'44"W	5100
Red Cloud Mine	Blaine	43°30'59"N	114°26'52"W	
Red Cloud Mine	Clearwater	46°28'14"N	115°42'24"W	
Red Elephant Mine	Blaine	43°29'24"N	114°25'50"W	
Red Horse Mine	Camas	43°42'06"N	115°00'11"W	
Red Horse Mine	Idaho	45°53'17"N	115°37'08"W	4160
Red Horse Mine	Kootenai	47°32'32"N	116°38'53"W	2920
Red Jacket Mine	Lemhi	44°59'39"N	114°32'01"W	
Red Ledge Mine	Adams	45°13'47"N	116°40'08"W	
Red Metals Mine	Valley	45°02'36"N	115°25'11"W	

Mines

Mine	County	Latitude	Longitude	Elev.
Red Spar Mine	Lemhi	44°51'21"N	114°32'09"W	
Redbird Mine	Custer	44°18'36"N	114°28'30"W	
Regal Mine	Boundary	48°51'02"N	116°15'35"W	
Reid Placer Mine	Boise	43°56'38"N	115°55'42"W	
Reindeer Queen Mine	Shoshone	47°27'03"N	115°45'41"W	
Reliance Mine	Butte	43°35'58"N	113°35'17"W	
Rescue Mine	Idaho	45°15'26"N	115°40'07"W	
Revelation Mine	Idaho	45°55'45"N	115°43'50"W	
Richard Allen Mine	Camas	43°28'05"N	114°48'54"W	
Richland Mine	Boise	43°56'57"N	115°59'40"W	
Rideout Mine	Boundary	48°52'47"N	116°50'20"W	5955
Ridgeway Mine	Lemhi	44°28'03"N	113°18'43"W	
Ringbone Cayuse Mine	Lemhi	45°09'38"N	114°09'21"W	
Rising Star Mine	Owyhee	42°55'10"N	116°46'35"W	
Riverview Mine	Custer	44°23'10"N	114°17'50"W	
Robinson Dike Mine	Idaho	45°31'22"N	115°28'12"W	
Ruth Mine	Cassia	42°05'27"N	113°03'08"W	
S Bridge	Shoshone	47°27'23"N	115°45'55"W	
Saint Clair Mine	Lemhi	45°27'48"N	114°26'48"W	
Saint Louis Mine	Blaine	43°41'27"N	114°17'34"W	
Saint Louis Mine	Butte	43°35'34"N	113°35'24"W	
Saint Louis Mine	Idaho	45°34'13"N	115°40'59"W	
Schafer Mine	Bonner	48°06'29"N	116°22'41"W	
Scheelite Jim Mine	Custer	44°20'02"N	114°35'06"W	6640
Seafoam Mine	Custer	44°31'46"N	115°07'43"W	
Sentinel Mine	Idaho	45°48'46"N	115°41'12"W	
Shamrock Mine	Kootenai	47°50'31"N	116°38'50"W	
Sheelite Mine	Idaho	45°48'46"N	115°39'13"W	
Sherman-Howe Mine	Idaho	45°24'35"N	115°51'58"W	
Silver Bell Mine	Butte	43°30'05"N	113°36'51"W	
Silver Bell Mine	Custer	44°19'06"N	114°21'36"W	
Silver City Mine	Owyhee	43°00'50"N	116°43'58"W	
Silver Hills Mine	Cassia	42°05'41"N	113°04'12"W	
Silver King Mine	Blaine	43°51'10"N	114°52'34"W	
Silver King Mine	Latah	46°48'28"N	116°19'53"W	
Silver Leaf Mine	Bonner	47°55'22"N	116°24'14"W	
Silver Queen Mine	Owyhee	42°55'41"N	116°46'33"W	
Silver Rock Mine	Owyhee	42°44'22"N	116°28'20"W	
Silver Rute Mine	Custer	44°10'37"N	114°34'26"W	
Silver Star Mine	Camas	43°35'48"N	114°42'33"W	
Silver Star Queen	Blaine	43°28'05"N	114°17'00"W	
Silver Summit Mine	Shoshone	47°30'25"N	116°01'30"W	
Silver Tip Mine	Kootenai	47°35'14"N	116°38'20"W	2900
Silver Vault Mine (historical)	Owyhee	43°00'20"N	116°50'26"W	
Silverbell Mine	Custer	44°32'15"N	115°08'15"W	
Silvertip Mine	Shoshone	47°33'53"N	115°50'18"W	5800
Sims Mine	Lemhi	45°18'17"N	113°59'21"W	
Singheiser Mine	Lemhi	44°51'30"N	114°23'49"W	

Mine	County	Latitude	Longitude	Elev.
Sixteen-to-One Mine	Shoshone	47°32'16"N	115°52'19"W	
Sixtyfour Mine	Idaho	45°34'46"N	115°27'35"W	
Skylark Mine	Custer	44°24'53"N	114°21'48"W	
Skylark Mine	Idaho	45°33'20"N	115°28'42"W	
Slip Easy Mine	Idaho	45°34'12"N	115°27'41"W	
Sloper Mine	Boise	44°01'32"N	115°40'57"W	
Smith Mine	Idaho	46°12'57"N	115°37'46"W	
Snoose Mine	Blaine	43°29'18"N	114°19'18"W	
Snowshoe Mine	Idaho	45°11'54"N	115°04'14"W	
Snowstorm Mine	Shoshone	47°28'52"N	115°43'43"W	4700
Snowstorm Mines	Shoshone	47°28'46"N	115°44'06"W	
Solace Mine	Blaine	43°47'41"N	114°50'48"W	
Sommercamp Mine (historical)	Owyhee	43°00'41"N	116°50'18"W	
Sommercamp Pit	Owyhee	43°00'46"N	116°49'47"W	
Sonneman Mine	Owyhee	42°44'47"N	116°55'15"W	
South Butte Mine	Custer	44°17'15"N	114°27'04"W	
South Fork Mine	Idaho	45°49'49"N	115°32'24"W	
South Peacock Mine	Adams	45°09'59"N	116°38'52"W	
Specimen Mine	Boise	44°06'16"N	115°32'54"W	
Spotted Horse Mine	Idaho	45°27'05"N	116°26'00"W	4880
Spring Creek Mines	Lemhi	45°27'19"N	114°19'16"W	
Springfield Mine	Valley	44°47'00"N	115°22'12"W	
St Joe Quartz Mine	Shoshone	47°11'49"N	115°29'39"W	3900
Standard Mammoth Mine	Shoshone	47°31'10"N	115°50'53"W	
Standard Mine	Owyhee	42°44'26"N	116°54'14"W	
Star Hope Mine	Custer	43°41'51"N	113°55'24"W	
Star Mine	Shoshone	47°29'34"N	115°49'27"W	
Steckner Mine	Idaho	45°44'46"N	115°18'58"W	
Sulphide Mine	Bonner	48°11'34"N	116°22'22"W	
Sultan Shaft	Idaho	45°50'31"N	115°25'25"W	
Summit Mine	Owyhee	43°01'06"N	116°45'58"W	
Sumner Mine	Shoshone	47°23'56"N	115°47'32"W	
Sunbeam Mine	Custer	44°26'14"N	114°43'56"W	
Sunday Mine	Valley	45°06'58"N	115°20'38"W	
Sungold Mine	Idaho	45°43'38"N	115°34'47"W	
Sunnyside Mine	Valley	44°57'22"N	115°07'49"W	
Sunrise Mine (historical)	Custer	44°34'40"N	114°53'45"W	8920
Sunset Mine	Shoshone	47°33'33"N	115°50'08"W	
Sunshine Mine	Blaine	43°41'43"N	114°06'11"W	
Sunshine Mine	Lemhi	45°07'47"N	114°21'42"W	
Sunshine Mine	Shoshone	47°30'11"N	116°04'28"W	
Superior Mine	Teton	43°42'25"N	111°18'53"W	
Swastika Mine	Idaho	45°30'26"N	115°27'54"W	
Talache Mine	Bonner	48°08'22"N	116°28'52"W	
Tallman Mine	Cassia	42°04'38"N	113°02'34"W	
Tango Mine	Custer	44°10'38"N	114°34'59"W	
Tepee Circles	Butte	43°26'36"N	113°31'35"W	5767
Texas Mine	Owyhee	42°44'31"N	116°54'39"W	

Mine	County	Latitude	Longitude	Elev.
The Post Mine	Owyhee	42°42'40"N	116°58'50"W	
Tiawaka Mine	Idaho	45°33'59"N	115°28'16"W	
Tiger-Poorman Mine	Shoshone	47°31'27"N	115°50'00"W	
Tilley Mine	Boundary	48°54'37"N	116°12'38"W	
Tin Cup Mine	Lemhi	44°58'03"N	114°31'57"W	
Tip Top Mine	Blaine	43°24'56"N	114°27'16"W	
Tip Top Mine	Camas	43°44'05"N	114°59'51"W	
Tip Top Mine	Owyhee	43°01'05"N	116°45'27"W	
Tolache Mine	Elmore	43°47'29"N	115°06'40"W	
Tormay Mine	Lemhi	45°06'19"N	113°59'07"W	
Toropah Mine	Idaho	45°33'58"N	115°29'41"W	
Trade Dollar Mine	Owyhee	43°00'30"N	116°44'16"W	
Treasure Vault Mine	Blaine	43°25'49"N	114°27'14"W	
Treasure Vault Mine	Owyhee	42°55'50"N	116°46'43"W	
Triumph Mine	Blaine	43°38'41"N	114°15'32"W	
Tungsten Hill Mine	Boundary	48°53'59"N	116°16'29"W	
Turtle Mine	Custer	44°22'12"N	114°18'19"W	
Tuttle Mine	Idaho	45°23'01"N	115°49'35"W	
Twin Apex Mine	Custer	44°18'24"N	114°30'09"W	
Twin Butte Mine	Idaho	45°40'52"N	115°32'10"W	
Twin Peaks Mine	Lemhi	44°57'14"N	114°00'09"W	
Twin Sisters Mine	Boise	43°54'39"N	115°52'58"W	
Tyrannis Mine	Camas	43°34'46"N	114°43'42"W	
Uhrig Mine	Blaine	43°18'38"N	114°21'37"W	
Ulysses Mine	Lemhi	45°27'11"N	114°08'20"W	
Umatilla Mine	Idaho	45°42'25"N	115°34'29"W	
Una Mine	Idaho	45°43'58"N	115°34'44"W	
Union Group Mines	Idaho	45°33'43"N	115°32'03"W	
Unity Mine	Idaho	45°15'29"N	115°40'55"W	
US Mine	Idaho	45°29'19"N	116°02'05"W	
Valley Arco Mine	Butte	43°35'34"N	113°34'56"W	
Valley View Mine	Butte	44°05'23"N	113°08'59"W	
Venable Mine	Valley	44°57'52"N	115°07'25"W	
Venus Mine	Owyhee	43°00'11"N	116°45'41"W	
Vesuvius Mine	Idaho	45°36'54"N	115°40'30"W	
Vienna Mine	Blaine	43°47'38"N	114°50'37"W	
Voscha Mine (historical)	Owyhee	43°00'48"N	116°49'35"W	
Vulcan Mine	Bonner	47°58'50"N	116°25'53"W	
Wagner Mine	Idaho	45°46'01"N	115°30'59"W	
Wahl Mine (historical)	Owyhee	43°00'59"N	116°49'47"W	
Walton Mine	Teton	43°37'21"N	111°14'03"W	
War Eagle Mine	Idaho	45°29'24"N	115°36'53"W	
Washington Mine	Boise	43°54'10"N	115°45'37"W	
Waverly Mine	Idaho	45°30'09"N	116°02'31"W	
Weber Mine	Bonner	47°54'11"N	116°25'55"W	
Webfoot Mine	Blaine	43°48'10"N	114°50'33"W	
Werdenhoff Mine	Valley	45°11'46"N	115°20'44"W	
Westvaco Mine	Bingham	43°01'51"N	112°03'04"W	

Mine	County	Latitude	Longitude	Elev.
White Cross Mine	Latah	46°48'12"N	116°54'51"W	
Whitedelf Mine	Bonner	48°09'50"N	116°11'09"W	
Wild Hope Mine	Idaho	45°46'37"N	115°38'22"W	
Wildhorse Mines	Custer	43°47'08"N	114°06'00"W	
Willson Mine	Valley	45°03'18"N	115°24'56"W	
Wiseboy Mine	Idaho	45°38'18"N	115°40'51"W	
Wonder Mine	Idaho	45°47'08"N	115°39'32"W	
Wonder Mine	Shoshone	47°29'08"N	115°51'07"W	4100
Wonderful Mine	Shoshone	47°24'38"N	115°43'56"W	
Woodrat Mine	Bonner	48°32'17"N	116°54'29"W	
Worthington Mine	Cassia	42°00'41"N	114°04'39"W	
X L Mine	Gem	43°52'01"N	116°17'23"W	
Yakima Mine	Idaho	45°27'28"N	116°01'41"W	
Yellow Pine Mine	Valley	44°55'44"N	115°20'12"W	
Yellowjacket Mine	Lemhi	44°59'16"N	114°31'11"W	
Zenith Mine	Idaho	45°46'32"N	115°31'01"W	

Miscellaneous Features

Type	Name	County	Latitude	Longitude	Elev.
arch	Ant and the Yellowjacket, The	Nez Perce	46°26'47"N	116°50'43"W	
arch	Arch, The	Owyhee	42°12'40"N	115°32'45"W	
arch	Bridge Of Tears	Butte	43°22'37"N	113°25'55"W	
arch	Rainbow Rock	Valley	44°59'09"N	115°35'49"W	
beach	Amos Bench	Nez Perce	46°30'08"N	116°35'25"W	
beach	Big Bench	Nez Perce	46°16'19"N	116°57'19"W	2463
beach	Bronco Beach	Power	42°56'09"N	112°41'17"W	
beach	Compressor Station Bench	Power	42°49'20"N	112°42'10"W	
beach	Egin Bench	Fremont	43°56'35"N	111°48'42"W	
beach	Honeysuckle Beach	Kootenai	47°45'13"N	116°45'17"W	
beach	Hot Springs Bench	Owyhee	42°18'56"N	115°40'53"W	
beach	Lava Point	Minidoka	43°08'34"N	113°34'47"W	
beach	North Beach	Valley	44°59'39"N	116°03'58"W	
beach	Palisades Bench	Bonneville	43°23'54"N	111°14'45"W	
beach	Preston Beach	Kootenai	47°46'16"N	116°40'34"W	
beach	Rainbow Beach	Power	42°52'53"N	112°42'53"W	
beach	Red Ledge	Adams	45°13'38"N	116°39'56"W	
beach	Sanders Beach	Kootenai	47°39'58"N	116°46'13"W	
beach	Sandy Point Beach	Ada	43°31'51"N	116°03'26"W	
beach	Silver Beach	Kootenai	47°39'35"N	116°44'28"W	
beach	Steamboat Summit	Idaho	45°14'43"N	115°44'44"W	
beach	Sylvan Beach	Valley	44°57'25"N	116°05'40"W	
beach	Wheatgrass Bench	Power	42°49'32"N	112°39'35"W	
cave	Ampitheater Cave	Butte	43°22'37"N	113°25'54"W	
cave	Bear Trap Cave	Blaine	42°59'02"N	113°21'07"W	
cave	Beauty Cave	Butte	43°26'42"N	113°31'25"W	
cave	Black Ridge Caves	Lincoln	43°01'25"N	113°57'25"W	
cave	Boy Scout Cave	Butte	43°26'42"N	113°31'32"W	
cave	Buffalo Caves	Butte	43°25'35"N	113°32'23"W	
cave	Clay Caves	Jerome	42°37'00"N	114°17'46"W	3855
cave	Creons Cave	Power	42°57'40"N	113°13'07"W	
cave	Crystal Ice Cave	Power	42°57'05"N	113°12'53"W	
cave	Dead Horse Cave	Gooding	43°00'43"N	114°50'06"W	
cave	Devils Kitchen	Twin Falls	42°32'10"N	114°50'55"W	3906
cave	Dewdrop Cave	Butte	43°26'39"N	113°31'40"W	
cave	Dynamite Cave	Lincoln	42°52'58"N	114°24'27"W	
cave	Eureka Cave	Elmore	43°07'23"N	115°34'23"W	
cave	Fissure Cave	Blaine	43°13'17"N	113°16'03"W	
cave	Formation Cave	Caribou	42°41'38"N	111°33'12"W	
cave	Great Owl Cavern	Butte	43°25'21"N	113°33'24"W	
cave	Gwinn Cave	Lincoln	43°04'30"N	114°24'33"W	4441
cave	Hidden Mouth Cave	Butte	43°57'22"N	113°26'20"W	
cave	Higby Cave	Ada	43°17'44"N	116°07'54"W	
cave	Horseshoe Cave	Butte	43°26'47"N	113°29'52"W	
cave	Indian Cave	Twin Falls	42°09'42"N	114°45'46"W	
cave	Johnson Cave	Lincoln	43°03'37"N	114°25'08"W	4288
cave	Kimama Butte Cave	Lincoln	42°46'13"N	113°52'31"W	
cave	Kuna Cave	Ada	43°24'44"N	116°26'56"W	
cave	Lariat Cave	Power	42°55'05"N	113°11'28"W	
cave	Last Chance Cave	Butte	43°26'49"N	113°29'42"W	
cave	Lava River Cave	Butte	43°26'55"N	113°29'39"W	
cave	Little Arch Cave	Lincoln	43°05'04"N	114°24'24"W	4510
cave	Lucile Caves	Idaho	45°31'28"N	116°18'03"W	
cave	Mammoth Cave	Lincoln	43°04'13"N	114°24'41"W	4422
cave	Middle Butte Cave	Bingham	43°30'24"N	112°42'48"W	
cave	Minnetonka Cave	Franklin	42°05'15"N	111°31'05"W	
cave	Moonshiners Cave	Bingham	43°30'09"N	112°38'45"W	
cave	Moss Cave	Butte	43°22'36"N	113°26'00"W	
cave	Moss Cave	Power	42°52'43"N	113°12'29"W	
cave	Needles Cave	Butte	43°26'49"N	113°30'03"W	
cave	Papoose Cavern	Idaho	45°23'17"N	116°26'10"W	
cave	Paris Ice Cave	Bear Lake	42°13'54"N	111°33'37"W	7815
cave	Roberson Cave	Owyhee	42°48'17"N	115°43'32"W	
cave	Seventeenmile Cave	Bonneville	43°32'33"N	112°21'13"W	
cave	Shoshone Ice Cave	Lincoln	43°09'53"N	114°20'45"W	
cave	Snowshoe Cave	Owyhee	42°27'34"N	116°53'27"W	
cave	South Grotto	Power	42°56'09"N	113°12'42"W	
cave	Sullivans Cave	Power	43°00'19"N	113°11'12"W	
cave	Surprise Cave	Butte	43°26'48"N	113°31'36"W	
cave	Wilson Butte Cave	Jerome	42°47'13"N	114°13'12"W	
cave	Wind Caves	Ada	43°15'16"N	116°07'53"W	
crater	Alexander Crater	Caribou	42°39'17"N	111°44'59"W	5821
crater	Big Craters	Butte	43°26'38"N	113°33'37"W	
crater	Black Butte Crater	Lincoln	43°11'03"N	114°21'01"W	4650
crater	Black Ridge Crater	Lincoln	43°00'52"N	113°57'37"W	
crater	Blow Out, The	Lincoln	43°08'44"N	113°46'16"W	
crater	Bowl Crater	Blaine	43°20'06"N	113°30'59"W	
crater	Butte Crater	Fremont	44°14'46"N	111°50'13"W	
crater	Cinder Pit	Bingham	43°21'50"N	112°42'27"W	
crater	Cottrells Blowout	Power	42°59'31"N	113°11'24"W	5056
crater	Crater Hole	Owyhee	42°09'42"N	115°31'29"W	
crater	Crater Rings	Elmore	43°11'25"N	115°51'12"W	2957
crater	Crater, The	Clark	44°17'42"N	112°09'12"W	
crater	Crater, The	Lincoln	42°47'20"N	113°48'14"W	4372
crater	Echo Crater	Butte	43°23'43"N	113°30'39"W	
crater	Horse Butte	Power	43°00'38"N	113°11'32"W	
crater	Morgan Crater	Clark	44°17'56"N	111°59'35"W	
crater	North Crater	Butte	43°27'05"N	113°33'46"W	6244
crater	Pothole, The	Camas	43°14'44"N	114°51'20"W	
crater	Snowdrift Crater	Blaine	43°20'05"N	113°33'15"W	5438
crater	Winters Blowout	Power	42°53'36"N	113°06'23"W	
falls	Albeni Falls	Bonner	48°10'40"N	117°00'01"W	
falls	Auger Falls	Twin Falls	42°37'56"N	114°31'35"W	
falls	Baron Creek Falls	Boise	44°06'01"N	114°59'14"W	
falls	Bear Creek Falls	Adams	44°58'52"N	116°42'03"W	
falls	Big Drops	Lincoln	42°57'28"N	114°23'35"W	

Type	Name	County	Latitude	Longitude	Elev.
falls	Big Falls	Boise	44°03'59"N	115°43'20"W	
falls	Bridal Veil Falls	Custer	44°12'43"N	115°06'21"W	
falls	Bull Run Creek Falls	Clearwater	46°44'14"N	116°11'48"W	
falls	Camel Falls	Owyhee	42°32'43"N	116°37'19"W	
falls	Caribou Falls	Boundary	48°49'14"N	116°46'41"W	
falls	Char Falls	Bonner	48°22'00"N	116°10'13"W	
falls	Copper Falls	Boundary	48°58'18"N	116°08'28"W	
falls	Crane Falls (historical)	Elmore	42°58'26"N	115°49'47"W	
falls	Dagger Falls	Valley	44°31'47"N	115°17'06"W	
falls	Deadman Falls	Elmore	42°53'21"N	115°21'03"W	2946
falls	Devils Washboard Falls	Gooding	42°40'09"N	114°46'40"W	
falls	Devlin Falls	Lemhi	45°13'37"N	114°05'51"W	
falls	Elk Creek Falls	Clearwater	46°44'11"N	116°10'30"W	
falls	Fall Creek Falls	Bonneville	43°26'29"N	111°22'35"W	
falls	Falls, The	Caribou	42°45'43"N	111°06'43"W	7060
falls	Falls, The	Owyhee	42°31'47"N	116°11'14"W	
falls	Fern Falls	Boise	44°02'01"N	115°05'19"W	
falls	Fern Falls	Shoshone	47°45'36"N	116°06'14"W	3280
falls	Goat Falls	Custer	44°10'35"N	115°01'01"W	
falls	Grouse Creek Falls	Bonner	48°27'30"N	116°20'16"W	
falls	Hazard Falls	Idaho	45°12'14"N	116°15'20"W	
falls	Jump Creek Falls	Owyhee	43°28'37"N	116°55'28"W	
falls	Lady Face Falls	Custer	44°13'54"N	115°05'42"W	
falls	Little Drops	Lincoln	42°58'44"N	114°25'13"W	
falls	Little Falls	Boise	44°04'21"N	115°45'42"W	
falls	Lost Creek Falls	Adams	44°55'05"N	116°22'30"W	4380
falls	Lower Mesa Falls	Fremont	44°10'32"N	111°19'08"W	
falls	Lower Salmon Falls	Gooding	42°50'35"N	114°54'12"W	
falls	Mallard Creek Falls	Idaho	45°33'58"N	115°18'01"W	
falls	McAbee Falls	Bonner	48°17'04"N	116°52'16"W	
falls	Mission Falls	Bonner	48°25'18"N	116°55'40"W	
falls	Moyie Falls	Boundary	48°43'59"N	116°10'28"W	
falls	Napias Creek Falls	Lemhi	45°08'30"N	114°12'00"W	
falls	Patsy Ann Falls	Idaho	45°57'33"N	114°27'26"W	
falls	Perrine Coulee Falls	Twin Falls	42°35'48"N	114°28'18"W	
falls	Phantom Falls	Cassia	42°05'28"N	114°07'51"W	
falls	Pillar Falls	Twin Falls	42°35'57"N	114°25'52"W	3125
falls	Pillar Falls	Twin Falls	42°35'55"N	114°25'49"W	
falls	Rambiker Falls	Shoshone	47°01'10"N	115°05'14"W	
falls	Ross Falls	Twin Falls	42°12'54"N	114°17'20"W	
falls	Rush Falls	Washington	44°43'10"N	116°41'20"W	6060
falls	Salmon Falls	Idaho	45°29'55"N	114°59'53"W	
falls	Salmon Falls	Twin Falls	42°46'00"N	114°53'47"W	2885
falls	Selway Falls	Idaho	46°03'08"N	115°18'23"W	
falls	Shadow Falls	Shoshone	47°45'33"N	116°06'14"W	3280
falls	Sheep Falls	Fremont	44°12'00"N	111°23'35"W	
falls	Sheep Falls	Fremont	44°04'52"N	111°05'28"W	
falls	Shoestring Falls	Idaho	46°16'43"N	115°23'25"W	

Type	Name	County	Latitude	Longitude	Elev.
falls	Shoshone Falls	Twin Falls	42°35'43"N	114°24'00"W	
falls	Smith Falls	Boise	44°00'01"N	115°02'23"W	
falls	Smith Falls	Boundary	48°57'36"N	116°33'25"W	
falls	Snow Creek Falls	Boundary	48°40'03"N	116°25'42"W	2280
falls	Torrelle Falls	Bonner	48°17'06"N	116°57'27"W	
falls	Twin Falls	Jerome	42°35'25"N	114°21'11"W	
falls	Upper Mesa Falls	Fremont	44°11'17"N	111°19'45"W	
falls	Upper Priest Falls	Boundary	48°59'36"N	116°56'25"W	
falls	Upper Salmon Falls	Twin Falls	42°46'00"N	114°53'47"W	
falls	Wildhorse Falls	Adams	44°50'58"N	116°49'00"W	
lava	Bear Trap Lava Tube	Blaine	42°59'25"N	113°19'55"W	
lava	Big Craters Flow	Butte	43°25'44"N	113°34'52"W	
lava	Black Flow	Butte	43°24'58"N	113°28'38"W	
lava	Blackfoot Lava Field	Caribou	42°48'47"N	111°40'58"W	
lava	Blue Dragon Flow	Butte	43°25'14"N	113°33'26"W	
lava	Buffalo Caves Flow	Butte	43°25'32"N	113°30'51"W	
lava	Crystal Fissure Flow	Butte	43°25'38"N	113°33'51"W	
lava	Derelict Flow	Butte	43°23'58"N	113°32'43"W	
lava	Devils Orchard	Butte	43°27'06"N	113°32'09"W	
lava	Fissure Butte Flows	Blaine	43°21'26"N	113°28'02"W	
lava	Lava Cascades	Butte	43°26'06"N	113°32'32"W	
lava	Little Prairie Aa Flow	Butte	43°23'06"N	113°28'48"W	
lava	North Crater Aa Flow	Butte	43°27'46"N	113°34'14"W	
lava	North Crater Flow	Butte	43°27'46"N	113°32'02"W	
lava	Sawtooth Flow	Blaine	43°18'16"N	113°29'29"W	
lava	Sawtooth Flow	Butte	43°23'01"N	113°31'12"W	
lava	Sentinel South Flow	Butte	43°22'14"N	113°27'54"W	
lava	Sentinel West Flow	Blaine	43°22'06"N	113°30'09"W	
lava	Serrate Flow	Butte	43°28'38"N	113°29'15"W	
lava	Sheep Trail Butte East Flow	Blaine	43°21'02"N	113°26'56"W	
lava	Sheep Trail Butte Southeast Flow	Blaine	43°18'44"N	113°26'30"W	
lava	Vermilion Chasm Flow	Blaine	43°20'38"N	113°25'13"W	
lava	Wapi Flow	Power	42°49'03"N	113°11'51"W	
lava	Willow Creek Lava Field	Bingham	43°06'51"N	111°42'40"W	
swamp	Deer Lake Swamp	Boundary	48°46'20"N	116°15'56"W	2966
swamp	Dingle Swamp	Bear Lake	42°10'59"N	111°18'39"W	
swamp	Dismal Swamp	Elmore	43°43'46"N	115°21'55"W	
swamp	Donabahba Yogee	Owyhee	42°05'23"N	116°08'17"W	
swamp	Frog Pond	Adams	44°46'22"N	116°19'55"W	5310
swamp	Goose Lake	Caribou	42°56'26"N	111°30'46"W	
swamp	Iron Bog Swamp	Custer	43°38'24"N	113°45'27"W	
swamp	Lily Marsh	Valley	44°56'52"N	116°04'31"W	5010
swamp	Moody Swamp	Madison	43°40'05"N	111°27'00"W	
swamp	Oxford Slough	Franklin	42°14'09"N	111°59'33"W	
swamp	Pinchot Marsh	Shoshone	47°02'29"N	115°55'21"W	
swamp	Quicksand Bog	Lemhi	45°29'34"N	114°32'30"W	
swamp	Swamps, The	Custer	43°47'51"N	113°47'48"W	
swamp	Thousand Springs	Custer	44°08'14"N	113°55'21"W	

Miscellaneous Features

Type	Name	County	Latitude	Longitude	Elev.
swamp	Tottens Pond	Kootenai	47°44'24"N	116°45'07"W	2260
woods	Cedar Grove	Shoshone	47°05'09"N	116°06'45"W	4480
woods	Coconut Grove	Idaho	45°19'06"N	116°27'58"W	5460
woods	Fir Grove	Camas	43°12'18"N	114°46'40"W	
woods	Hobo Cedar Grove	Shoshone	47°05'28"N	116°07'34"W	
woods	Lewis and Clark Grove	Idaho	46°17'49"N	115°42'39"W	3280
woods	Pine Grove	Caribou	42°54'28"N	111°51'27"W	6220
woods	Thorns, The	Gooding	43°05'28"N	114°37'43"W	

Mountain, Peak, or Summit	County	Latitude	Longitude	Elev.
45 Hill	Owyhee	42°07'43"N	116°56'47"W	5487
Abandon Mountain	Boundary	48°48'40"N	116°43'24"W	7022
Abes Knob	Latah	46°57'09"N	116°26'40"W	4220
Abrams Height	Idaho	45°25'37"N	115°28'43"W	4262
Ace Butte	Idaho	45°32'03"N	115°52'02"W	7776
Acorn Butte	Valley	45°10'46"N	115°04'27"W	7760
Adams Peak	Shoshone	47°14'58"N	115°30'45"W	5820
Admiral Peak	Clearwater	46°48'41"N	114°51'31"W	7323
Aggipah Mountain	Lemhi	45°09'14"N	114°38'13"W	9920
Ajax Peak	Lemhi	45°19'50"N	113°44'16"W	10028
Aldape Summit	Ada	43°38'25"N	116°03'43"W	4797
Alder Creek Summit	Boise	43°58'58"N	115°56'46"W	4856
Allan Mountain	Lemhi	45°35'42"N	114°03'16"W	9154
Allen Point	Shoshone	47°12'44"N	115°40'26"W	5340
Ally Point	Shoshone	47°16'45"N	115°58'10"W	4271
Alpine Peak	Custer	44°10'18"N	115°02'55"W	9861
Amethyst Peak	Shoshone	47°54'29"N	116°10'12"W	4214
Anderson Butte	Idaho	45°53'16"N	115°19'49"W	6847
Anderson Mountain	Lemhi	45°36'53"N	113°54'34"W	8034
Anderson Peak	Custer	44°11'52"N	114°01'25"W	9339
Andrew Nyman Mountain	Franklin	42°02'26"N	111°38'46"W	9460
Andys Hump	Idaho	46°09'16"N	115°29'23"W	6530
Angel Butte	Clearwater	46°34'54"N	116°02'14"W	3820
Angle Point	Shoshone	47°05'07"N	115°25'14"W	5784
Ant Butte	Adams	45°08'15"N	116°26'58"W	7048
Ant Butte	Blaine	43°16'13"N	113°37'24"W	5190
Ant Butte	Washington	44°19'20"N	116°36'11"W	3986
Ant Hill	Idaho	46°20'26"N	115°29'26"W	4439
Ant Hill	Owyhee	42°40'57"N	116°28'10"W	6345
Antelope Butte	Blaine	43°15'32"N	113°07'04"W	5380
Antelope Butte	Jefferson	43°52'21"N	112°36'22"W	4957
Antelope Hill	Cassia	42°24'45"N	113°42'55"W	4577
Antelope Mountain	Bonner	48°08'26"N	116°07'56"W	4392
Anthony Peak	Shoshone	46°58'53"N	116°11'21"W	4686
Appendicitis Hill	Butte	43°44'07"N	113°26'20"W	8523
Arange Peak	Fremont	44°31'38"N	111°27'49"W	8984
Arch Table	Owyhee	42°09'37"N	115°32'00"W	4810
Archer Mountain	Idaho	45°55'11"N	114°55'22"W	7492
Archer Point	Idaho	45°55'47"N	114°51'51"W	5061
Archie Mountain	Boise	44°02'41"N	115°28'23"W	7665
Arco Peak	Butte	43°40'00"N	113°17'56"W	7547
Arid Peak	Shoshone	47°21'32"N	115°45'46"W	5306
Arrowhead, The	Custer	44°01'30"N	114°58'20"W	1500
Artillery Dome	Valley	44°40'28"N	115°12'53"W	9318
Asbestos Peak	Idaho	45°43'45"N	116°02'29"W	5820
Ashpile Peak	Idaho	46°29'32"N	115°04'10"W	6400
Aspen Butte	Fremont	44°18'44"N	111°43'26"W	6670
Atkinson Peak	Bonneville	43°28'02"N	111°10'39"W	9366

Mountain, Peak, or Summit	County	Latitude	Longitude	Elev.
Atlasta Mountain	Bonner	48°25'39"N	116°42'46"W	6361
Attention Peak	Shoshone	47°06'56"N	115°26'48"W	6075
Aura, Mount	Idaho	45°51'26"N	114°51'05"W	6811
Austin Butte	Owyhee	42°34'27"N	115°39'08"W	4082
Avery Hill	Shoshone	47°16'06"N	115°49'32"W	4421
Avery Table	Owyhee	42°33'53"N	116°36'27"W	5749
Babel, Tower of	Idaho	45°19'50"N	116°31'40"W	9268
Baby Grand Mountain	Latah	47°00'14"N	116°30'41"W	4744
Bachelor Mountain	Custer	44°22'19"N	114°38'28"W	9498
Bacon Peak	Shoshone	46°58'55"N	115°16'44"W	6361
Bad Bear Peak	Shoshone	47°03'06"N	115°28'52"W	5685
Bad Luck Mountain	Idaho	45°53'19"N	114°43'17"W	5344
Bad Tom Mountain	Shoshone	47°24'59"N	115°56'33"W	5587
Badger Knoll	Caribou	42°58'57"N	111°38'56"W	6290
Badger Mountain	Cassia	42°07'40"N	114°09'32"W	7585
Badger Mountain	Kootenai	47°49'10"N	116°33'14"W	4762
Badger Mountain	Shoshone	47°01'41"N	115°28'44"W	6167
Badger Peak	Cassia	42°33'04"N	113°03'42"W	6435
Bailey Mountain	Idaho	46°15'21"N	114°52'20"W	7386
Baird, Mount	Bonneville	43°21'49"N	111°05'48"W	10025
Baird, Mount	Bonneville	43°21'47"N	111°05'39"W	10025
Baker Peak	Camas	43°40'12"N	114°40'44"W	10174
Baking Powder Mountain	Idaho	45°41'31"N	115°45'50"W	7593
Bald Eagle Mountain	Boundary	48°30'43"N	116°14'59"W	6098
Bald Hill	Twin Falls	42°11'37"N	114°19'38"W	6911
Bald Hill	Valley	44°58'36"N	115°26'33"W	7720
Bald Knob	Clearwater	46°49'51"N	115°25'47"W	5332
Bald Mountain	Bear Lake	42°24'38"N	111°15'03"W	9248
Bald Mountain	Blaine	43°39'18"N	114°24'30"W	9151
Bald Mountain	Boise	44°04'53"N	115°53'16"W	5148
Bald Mountain	Boise	43°44'43"N	115°45'03"W	7314
Bald Mountain	Bonner	48°19'55"N	116°41'35"W	6193
Bald Mountain	Bonneville	43°06'13"N	111°12'35"W	8488
Bald Mountain	Clark	44°05'29"N	112°58'29"W	9006
Bald Mountain	Clearwater	46°34'32"N	115°53'55"W	5036
Bald Mountain	Custer	44°24'03"N	114°42'14"W	8969
Bald Mountain	Custer	44°22'02"N	114°20'35"W	10313
Bald Mountain	Elmore	43°44'44"N	115°13'39"W	9368
Bald Mountain	Idaho	46°27'16"N	115°14'30"W	6526
Bald Mountain	Idaho	45°24'02"N	116°28'46"W	8066
Bald Mountain	Latah	47°01'54"N	116°34'15"W	5334
Bald Mountain	Lemhi	45°28'22"N	114°11'52"W	7693
Bald Mountain	Owyhee	43°02'51"N	116°41'02"W	6301
Bald Mountain	Owyhee	42°42'33"N	116°20'53"W	6719
Bald Mountain	Owyhee	42°21'06"N	116°57'48"W	6140
Bald Mountain	Shoshone	47°22'56"N	115°37'15"W	6033
Bald Peak	Fremont	44°43'28"N	111°20'50"W	10180
Baldy	Lemhi	45°08'40"N	114°00'28"W	9149

Mountain, Peak, or Summit	County	Latitude	Longitude	Elev.
Baldy Knoll	Bingham	43°20'08"N	112°26'35"W	4801
Baldy Knoll	Teton	43°50'41"N	111°20'10"W	6831
Baldy Mountain	Bannock	42°32'45"N	111°56'42"W	8330
Baldy Mountain	Bonneville	43°26'02"N	111°12'43"W	9835
Baldy Mountain	Clark	44°34'05"N	111°52'12"W	9889
Baldy Mountain	Idaho	45°57'20"N	115°43'54"W	6613
Baldy Mountain	Lemhi	44°35'59"N	113°05'18"W	10773
Baldy, Mount	Bonneville	43°26'21"N	111°34'25"W	7380
Ball Butte	Latah	46°52'40"N	116°59'12"W	3330
Bally Mountain	Idaho	45°06'48"N	116°15'30"W	6818
Bannock Peak	Power	42°36'13"N	112°42'23"W	8263
Bare Hill	Owyhee	43°04'05"N	116°52'44"W	6296
Baron Peak	Boise	44°07'07"N	114°59'45"W	10297
Barren Hill	Idaho	46°15'22"N	114°58'59"W	6220
Barren Hill	Shoshone	47°03'09"N	115°47'46"W	6020
Bartlett Point	Clearwater	46°37'44"N	116°01'24"W	3217
Barton Hump	Bonner	48°02'17"N	116°25'17"W	3724
Basalt Hill	Latah	46°50'26"N	116°46'51"W	3295
Basin Butte	Custer	44°20'18"N	114°58'37"W	8854
Bat Point	Idaho	45°34'18"N	115°15'39"W	5984
Bathtub Mountain	Shoshone	47°04'47"N	115°34'07"W	6219
Battle Ground	Idaho	45°47'12"N	116°16'53"W	2264
Battleaxe Mountain	Valley	44°50'30"N	115°08'58"W	9210
Beacon Butte	Jefferson	43°57'46"N	112°10'01"W	5055
Beacon Hill	Clark	44°22'55"N	112°11'52"W	6740
Beals Butte	Latah	46°56'38"N	116°30'16"W	4932
Bean Hill	Latah	46°41'38"N	116°36'43"W	2780
Bear Butte	Clearwater	46°41'54"N	115°53'41"W	4070
Bear Creek Point	Valley	44°53'35"N	114°48'16"W	8629
Bear Creek Summit	Custer	43°43'14"N	113°43'50"W	7660
Bear Creek Summit	Washington	44°15'26"N	116°42'09"W	3607
Bear Den Butte	Minidoka	43°10'43"N	113°30'23"W	5104
Bear Mountain	Bonner	48°18'31"N	116°04'02"W	6061
Bear Mountain	Custer	44°21'27"N	113°27'06"W	10744
Bear Mountain	Idaho	46°25'53"N	114°55'28"W	7184
Bear Peak	Blaine	43°40'32"N	114°35'19"W	9525
Bear Pete Mountain	Idaho	45°16'34"N	115°58'45"W	8751
Bear Skull	Shoshone	47°06'40"N	115°41'22"W	6346
Bear Valley Mountain	Valley	44°27'25"N	115°22'53"W	829
Bear Wallow	Idaho	46°07'37"N	115°03'06"W	5290
Bear Wallow Point	Washington	44°14'41"N	115°57'41"W	5777
Beaver Butte	Clearwater	46°49'18"N	115°41'11"W	5000
Beaver Creek Summit	Boise	44°01'35"N	115°37'21"W	6041
Beaver Jack Mountain	Idaho	45°44'35"N	114°41'25"W	7068
Beaver Mountain	Benewah	47°13'34"N	116°22'21"W	4760
Beaver Peak	Shoshone	47°59'14"N	116°10'10"W	5384
Beaver Peak	Shoshone	47°03'21"N	115°25'29"W	5655
Beaverdam Peak	Valley	45°01'40"N	115°57'53"W	8653

Mountains, Peaks, Summits

Mountain, Peak, or Summit	County	Latitude	Longitude	Elev.
Bechtel Butte	Shoshone	46°59'30"N	116°19'19"W	4660
Bed Springs Butte	Clark	44°16'16"N	112°07'52"W	5730
Bee Top Mountain	Bonner	48°14'07"N	116°08'49"W	6220
Beehive	Lemhi	45°13'16"N	114°33'00"W	9610
Beetle Hump	Shoshone	47°14'08"N	115°24'11"W	6183
Belial, Mount	Idaho	45°18'58"N	116°32'52"W	8880
Bell Mountain	Blaine	43°25'52"N	114°06'37"W	7988
Bell Mountain	Butte	44°14'12"N	113°11'40"W	11612
Bemis Point	Idaho	45°17'12"N	115°38'42"W	7630
Benchmark Hill	Shoshone	47°16'43"N	115°47'52"W	4873
Bennett Mountain	Elmore	43°14'58"N	115°26'02"W	7438
Bennett Peak	Shoshone	47°49'27"N	116°01'58"W	6209
Bennett Point	Shoshone	47°12'16"N	115°33'54"W	4605
Benning Mountain	Bonner	48°20'17"N	116°04'58"W	6596
Benson Hill	Latah	46°48'56"N	116°32'12"W	2921
Benton Butte	Clearwater	46°50'19"N	115°48'09"W	4945
Benton Peak	Washington	44°37'21"N	116°57'49"W	7226
Berg Mountain	Idaho	45°25'32"N	116°13'02"W	6020
Berge Peak	Shoshone	47°11'59"N	115°19'18"W	6186
Berger Butte	Twin Falls	42°25'36"N	114°37'45"W	4701
Bernard Mountain	Valley	44°26'35"N	115°31'07"W	8203
Bernard Peak	Kootenai	47°56'08"N	116°31'59"W	5156
Bertha Hill	Clearwater	46°45'51"N	115°47'29"W	5520
Best Hill	Kootenai	47°41'46"N	116°45'03"W	2882
Bethlehem Mountain	Boundary	48°54'01"N	116°17'13"W	4857
Between The Creeks	Owyhee	42°43'34"N	116°16'33"W	5510
Bible Back Mountain	Custer	43°59'52"N	114°38'13"W	9928
Big Baldy	Valley	44°46'58"N	115°13'05"W	9705
Big Black Dome	Custer	43°49'22"N	113°58'44"W	11353
Big Blowout Butte	Blaine	43°18'06"N	113°36'09"W	5401
Big Buck Mountain	Elmore	43°59'07"N	115°12'48"W	8777
Big Burn Point	Idaho	45°54'01"N	115°49'45"W	5700
Big Butte	Butte	43°23'46"N	113°01'21"W	7560
Big Butte	Idaho	46°08'58"N	116°08'21"W	3882
Big Cinder Butte	Butte	43°25'01"N	113°32'17"W	6515
Big Creek Peak	Lemhi	44°28'18"N	113°32'35"W	11250
Big Creek Point	Idaho	46°10'42"N	114°34'33"W	7707
Big Creek Point	Idaho	45°34'34"N	115°30'37"W	7020
Big Creek Point	Valley	45°03'46"N	115°27'19"W	8893
Big Dick Point	Shoshone	47°18'51"N	115°43'32"W	5419
Big Elk Mountain	Bonneville	43°14'10"N	111°16'27"W	9476
Big Flat Top	Custer	43°46'55"N	113°38'29"W	9350
Big Fog Mountain	Idaho	46°07'44"N	115°10'46"W	7122
Big Foot Butte	Ada	43°12'24"N	116°13'40"W	3535
Big Grassy Butte	Fremont	43°59'05"N	112°03'31"W	5147
Big Hill	Franklin	42°02'57"N	111°45'17"W	6072
Big Hill	Idaho	46°09'29"N	115°37'53"W	4067
Big Hill	Owyhee	42°37'19"N	115°50'46"W	4406
Big Hill	Valley	45°08'29"N	114°46'09"W	8094
Big Hill	Washington	44°34'29"N	117°05'21"W	4815
Big Hill, The	Owyhee	42°40'07"N	115°14'55"W	3993
Big Horse Butte	Bonneville	43°23'27"N	111°42'34"W	6519
Big Peak	Camas	43°40'09"N	114°43'39"W	10047
Big Rock Mountain	Idaho	46°12'28"N	115°02'12"W	7103
Big Soldier Mountain	Custer	44°34'10"N	115°15'33"W	8984
Big Southern Butte	Butte	43°24'05"N	113°01'23"W	7560
Big Springs Butte	Owyhee	42°28'28"N	116°27'33"W	6132
Big Table Mountain	Clark	44°32'15"N	112°01'45"W	9083
Big Windy Peak	Lemhi	44°20'32"N	113°16'26"W	10380
Bighorn Point	Clearwater	46°37'22"N	115°27'03"W	4659
Bilk Mountain	Idaho	45°56'31"N	115°02'44"W	7610
Billy Goat Hill	Shoshone	47°17'20"N	116°11'23"W	3100
Binarch Mountain	Bonner	48°30'18"N	116°57'14"W	4180
Binocular Peak	Shoshone	47°06'09"N	115°09'34"W	7266
Birch Creek Mountain	Bingham	43°16'31"N	111°51'40"W	7486
Birch Hill	Clearwater	46°52'08"N	115°06'50"W	5587
Birch Mountain	Clearwater	46°52'49"N	115°09'09"W	6054
Birdseye Butte	Fremont	44°20'19"N	111°41'46"W	6770
Bishop Mountain	Fremont	44°20'00"N	111°33'09"W	7810
Bismark Mountain	Bonner	48°35'43"N	116°59'46"W	3900
Bismark Mountain	Idaho	45°12'54"N	115°08'28"W	8128
Bitterroot Mountain	Lemhi	45°29'31"N	113°47'37"W	8581
Bitters Butte	Madison	43°50'11"N	111°35'16"W	5494
Bitters Peak	Bonneville	43°09'30"N	111°06'10"W	7195
Black Butte	Idaho	45°29'23"N	115°52'27"W	6731
Black Butte	Lincoln	43°08'07"N	114°28'37"W	5110
Black Butte	Owyhee	43°04'47"N	116°10'35"W	3145
Black Butte	Owyhee	42°47'13"N	116°28'11"W	6025
Black Butte	Owyhee	42°41'14"N	115°22'59"W	4207
Black Butte	Owyhee	42°16'58"N	115°05'39"W	5390
Black Butte	Valley	45°10'53"N	114°50'59"W	8711
Black Cap	Butte	43°31'04"N	113°37'23"W	7553
Black Cap Peak	Butte	43°39'39"N	113°36'40"W	6578
Black Hawk Mountain	Idaho	45°48'50"N	115°16'26"W	6091
Black Imp	Adams	45°14'06"N	116°33'41"W	8380
Black Jack Peak	Shoshone	47°12'45"N	115°25'24"W	6015
Black Knoll	Fremont	44°04'24"N	111°43'38"W	5391
Black Mesa	Elmore	42°54'29"N	115°12'12"W	3080
Black Mountain	Bonneville	48°36'38"N	116°15'24"W	6096
Black Mountain	Bonneville	43°07'06"N	111°07'44"W	8911
Black Mountain	Clark	44°17'21"N	112°40'12"W	8860
Black Mountain	Clearwater	46°52'44"N	115°33'04"W	7077
Black Mountain	Fremont	44°42'39"N	111°24'51"W	10237
Black Mountain	Fremont	44°21'03"N	111°11'11"W	7642
Black Mountain	Lemhi	44°50'27"N	114°20'36"W	9510
Black Mountain	Owyhee	43°07'46"N	116°42'43"W	6616

Mountain, Peak, or Summit	County	Latitude	Longitude	Elev.
Black Mountain	Twin Falls	42°13'44"N	114°19'38"W	6792
Black Nose	Blaine	43°28'50"N	113°53'20"W	6141
Black Peak	Shoshone	47°42'43"N	115°44'30"W	6546
Black Peak	Shoshone	47°10'57"N	115°16'16"W	6425
Black Pine Cone	Cassia	42°04'56"N	113°03'33"W	8008
Black Pine Mountain	Bonner	48°08'48"N	116°39'58"W	3820
Black Pine Peak	Cassia	42°07'13"N	113°07'12"W	9385
Black Rock	Owyhee	42°06'17"N	115°37'56"W	6005
Black Stone	Owyhee	42°06'22"N	115°43'45"W	5738
Black Table	Owyhee	42°28'27"N	116°40'49"W	5323
Black Tip	Idaho	45°09'18"N	116°04'32"W	8292
Black Warrior	Owyhee	42°55'14"N	116°46'04"W	6307
Blackbird Mountain	Lemhi	45°07'05"N	114°24'53"W	9096
Blackdome Peak	Shoshone	46°58'59"N	115°48'54"W	6412
Blacklead Mountain	Clearwater	46°38'22"N	114°51'01"W	7318
Blackman Peak	Custer	44°03'31"N	114°39'05"W	10300
Blackmare	Valley	44°49'14"N	115°47'19"W	8724
Blacknose Mountain	Elmore	43°58'30"N	115°06'46"W	9802
Blacks Knoll	Fremont	44°17'32"N	111°49'09"W	6518
Blacktail Butte	Idaho	46°12'31"N	115°05'23"W	7003
Blacktail Butte	Idaho	45°53'24"N	115°59'46"W	4620
Blacktail Mountain	Bonner	48°43'21"N	116°56'14"W	5495
Blacktail Mountain	Bonner	48°07'08"N	116°31'04"W	4960
Blacktop Mountain	Bonner	48°11'54"N	116°03'36"W	6505
Blackwell Hill	Kootenai	47°41'18"N	116°49'22"W	2510
Blizzard Mountain	Blaine	43°30'03"N	113°40'49"W	9276
Blizzard Mountain	Elmore	43°52'20"N	115°04'43"W	0
Blodgett Mountain	Idaho	46°16'48"N	114°28'11"W	8647
Bloom Peak	Shoshone	47°45'18"N	115°49'46"W	5863
Bloomington Peak	Bear Lake	42°10'13"N	111°35'20"W	9317
Blossom Mountain	Kootenai	47°39'33"N	116°57'46"W	4386
Blowout Mountain	Idaho	45°30'19"N	115°28'21"W	6629
Blue Bunch Mountain	Valley	44°28'42"N	115°16'27"W	8743
Blue Hill	Cassia	42°03'11"N	113°56'02"W	5109
Blue Joe Mountain	Boundary	48°59'55"N	116°47'20"W	5022
Blue Joint	Idaho	45°33'45"N	114°34'19"W	8681
Blue Mountain	Bonner	48°24'04"N	116°38'46"W	6662
Blue Mountain	Custer	44°27'41"N	114°16'20"W	8328
Blue Mountain	Idaho	45°23'59"N	116°26'49"W	6303
Blue Nose	Lemhi	45°28'22"N	114°21'28"W	8677
Blue Ribbon Mountain	Idaho	45°48'28"N	115°22'18"W	5112
Bluebird Mountain	Clark	44°13'25"N	112°47'48"W	9048
Bluebird Point	Shoshone	47°17'57"N	115°37'44"W	5525
Bluff Point	Shoshone	47°10'42"N	115°31'09"W	5256
BM Hill	Clearwater	46°38'58"N	114°41'30"W	6380
Bobcat Butte	Twin Falls	42°09'13"N	114°52'24"W	6103
Bobtail Peak	Shoshone	47°42'45"N	115°52'41"W	4128
Boehls Butte	Clearwater	46°52'05"N	115°52'25"W	3610

Mountain, Peak, or Summit	County	Latitude	Longitude	Elev.
Bogus Point	Shoshone	47°13'28"N	116°18'40"W	4130
Boise Peak	Boise	43°42'09"N	116°05'23"W	6525
Boise Peak	Kootenai	47°27'09"N	116°24'17"W	5441
Bonanza Peak	Custer	44°21'25"N	114°41'23"W	9311
Bonehead Hill	Shoshone	47°10'03"N	115°47'44"W	5420
Boni Table	Owyhee	42°31'30"N	116°44'29"W	5773
Bonneville Peak	Bannock	42°45'47"N	112°08'23"W	9271
Boone Peak	Owyhee	42°54'56"N	116°40'26"W	7004
Borah Peak	Custer	44°08'14"N	113°46'46"W	12662
Border Mountain	Boundary	48°59'33"N	116°12'47"W	4769
Border Summit	Bear Lake	42°12'43"N	111°06'33"W	6335
Boston Mountain	Idaho	45°38'44"N	115°11'18"W	7648
Bottle Point	Shoshone	47°52'34"N	115°56'04"W	6305
Bottleneck Peak	Boundary	48°39'04"N	116°36'10"W	6923
Boulder Mountain	Bonner	48°45'05"N	116°59'38"W	5654
Boulder Mountain	Boundary	48°31'35"N	116°10'48"W	6298
Boulder Mountain	Valley	44°52'53"N	115°56'15"W	8377
Boulder Peak	Blaine	43°49'58"N	114°31'02"W	10981
Boulder Summit	Valley	44°52'53"N	115°55'28"W	7975
Boundary Peak	Idaho	46°22'09"N	115°31'22"W	5243
Boundary Peak	Shoshone	47°48'53"N	116°19'21"W	4466
Boundary Point	Latah	46°54'15"N	116°38'25"W	3900
Bowery Peak	Custer	44°03'02"N	114°21'37"W	10861
Bowl Butte	Idaho	46°24'53"N	115°26'09"W	6365
Box Car Mountain	Idaho	45°55'35"N	114°58'02"W	7589
Boyle Mountain	Blaine	43°40'42"N	114°31'15"W	8962
Bradley Mountain	Power	42°33'25"N	112°27'10"W	7258
Braxon Peak	Boise	44°05'37"N	114°59'45"W	10353
Bread Loaf Rock	Boise	43°51'27"N	116°13'20"W	3540
Breezy Point	Shoshone	47°06'48"N	115°56'49"W	6482
Breitenbach, Mount	Custer	44°04'00"N	113°40'20"W	12140
Bridge Point	Shoshone	47°03'35"N	115°23'01"W	5250
Brigham Point	Blaine	42°42'43"N	113°15'53"W	4300
Brigham Point	Minidoka	42°57'05"N	113°28'44"W	4469
Broken Top	Butte	43°25'45"N	113°32'30"W	6058
Broken Top Butte	Lincoln	43°08'58"N	113°50'50"W	4957
Bromaghin Peak	Blaine	43°49'50"N	114°43'06"W	10225
Brown Butte	Lincoln	42°59'55"N	114°13'09"W	4301
Browns Peak	Elmore	43°56'04"N	115°07'59"W	9705
Bruin Hill	Clearwater	46°45'38"N	114°54'45"W	6504
Bruin Mountain	Idaho	45°10'35"N	116°07'09"W	8767
Brundage Mountain	Adams	45°01'17"N	116°07'38"W	7803
Brush Creek	Valley	44°59'30"N	114°54'17"W	8925
Brush Hill	Idaho	46°16'31"N	115°34'39"W	4620
Brush Hill	Shoshone	47°00'33"N	115°43'58"W	5003
Brush Mountain	Adams	45°07'08"N	116°24'24"W	6246
Bryan Mountain	Idaho	45°16'02"N	116°29'02"W	8358
Buck Butte	Clearwater	46°35'41"N	115°58'28"W	4334

Mountains, Peaks, Summits

Mountain, Peak, or Summit	County	Latitude	Longitude	Elev.
Buck Mountain	Boundary	48°33'51"N	116°09'34"W	5683
Buck Mountain	Caribou	42°39'52"N	111°04'22"W	7598
Buck Mountain	Gem	44°27'13"N	116°14'40"W	6051
Buck Mountain	Owyhee	43°22'37"N	116°52'51"W	4604
Buck Mountain	Valley	44°41'16"N	115°33'55"W	7780
Buck Peak	Franklin	42°11'38"N	112°03'28"W	7579
Buck Peak	Power	42°20'28"N	112°47'56"W	6410
Buck Point	Shoshone	47°03'02"N	115°33'56"W	5500
Buck Point	Valley	45°05'17"N	115°03'30"W	8232
Buckhorn Mountain	Boundary	48°52'50"N	116°03'06"W	6177
Buckhorn Mountain	Valley	44°52'04"N	115°54'03"W	8457
Buckhorn Summit	Valley	44°51'35"N	115°53'55"W	7860
Buckles Mountain	Kootenai	47°51'01"N	116°36'27"W	4737
Buckskin Mountain	Caribou	42°37'35"N	111°52'27"W	6440
Bud Lewis Hill	Twin Falls	42°10'18"N	114°57'47"W	5780
Buffalo Hump	Idaho	45°37'14"N	115°42'00"W	8938
Bugle Mountain	Boise	44°02'27"N	115°08'54"W	9193
Bull Camp Butte	Owyhee	42°00'16"N	116°40'36"W	5394
Bull Creek Point	Owyhee	42°06'08"N	115°52'26"W	6017
Bull Run Peak	Kootenai	47°31'10"N	116°26'45"W	3969
Bumblebee Peak	Shoshone	47°40'34"N	116°18'35"W	4746
Bunker Hill	Blaine	43°26'23"N	114°20'48"W	6178
Bunker Hill	Jefferson	43°54'30"N	112°17'36"W	4849
Burgdorf Summit	Idaho	45°19'01"N	115°35'38"W	8110
Burke Summit	Shoshone	47°31'02"N	115°42'38"W	6610
Burley Butte	Cassia	42°29'11"N	113°55'01"W	4644
Burnt Butte	Owyhee	42°13'59"N	115°17'46"W	5350
Burnt Cabin Summit	Kootenai	47°46'17"N	116°35'19"W	4004
Burnt Hill	Shoshone	47°49'15"N	115°59'59"W	4580
Burnt Knob	Idaho	45°42'09"N	114°59'27"W	8196
Burnt Knob	Idaho	45°28'06"N	115°16'55"W	7339
Burnt Strip Mountain	Idaho	45°49'54"N	114°39'10"W	7489
Burpee Mountain	Idaho	45°35'53"N	115°29'04"W	6848
Burton Peak	Boundary	48°45'15"N	116°29'52"W	6844
Bussard Mountain	Boundary	48°53'26"N	116°13'52"W	5968
Bussel Peak	Shoshone	47°09'08"N	116°07'47"W	4473
Buster Butte	Owyhee	42°21'14"N	115°50'40"W	5291
Butler Mountain	Bonner	48°07'50"N	116°31'28"W	4893
Butte, The	Bingham	43°22'07"N	112°06'04"W	4737
Buttercup Mountain	Blaine	43°30'56"N	114°34'34"W	9075
Butterfly Butte	Bonneville	43°33'36"N	112°19'32"W	5179
Buttes, The	Owyhee	42°23'41"N	116°57'28"W	5921
Button Butte	Clark	44°26'10"N	111°54'31"W	6781
Butts Creek Point	Idaho	45°21'43"N	114°44'12"W	7836
Butzien Butte	Idaho	45°28'52"N	115°54'10"W	6053
Buzzard Roost	Shoshone	46°59'16"N	115°40'50"W	4871
Cabin Creek Peak	Custer	44°24'28"N	114°53'49"W	9968
Cabin Mountain	Custer	43°51'18"N	113°46'19"W	11163

Mountain, Peak, or Summit	County	Latitude	Longitude	Elev.
Cabin Mountain	Idaho	45°23'50"N	114°56'43"W	6502
Cabin Peak	Shoshone	47°50'43"N	115°56'02"W	4900
Cabin Peak	Valley	44°43'36"N	115°37'40"W	8062
Cabin Point	Clearwater	46°35'46"N	115°29'04"W	4580
Cable Peak	Kootenai	47°35'56"N	117°00'17"W	4952
Cache Peak	Cassia	42°11'08"N	113°39'37"W	10339
Calamity Point	Bonneville	43°20'02"N	111°12'40"W	6502
Calder Mountain	Bonner	48°29'33"N	116°11'27"W	5699
Calder Point	Shoshone	47°14'19"N	116°12'26"W	4313
Calhoun Butte	Twin Falls	42°07'32"N	114°54'17"W	6530
Camas Butte	Jefferson	44°03'06"N	112°17'15"W	5138
Camel Hill	Idaho	46°20'54"N	115°33'08"W	4580
Camelback Mountain	Bannock	42°52'53"N	112°20'58"W	6586
Camelback Mountain	Shoshone	47°09'27"N	116°00'26"W	6578
Cameron Hill	Shoshone	47°21'33"N	116°05'22"W	5082
Camp Peak	Caribou	42°57'48"N	111°33'00"W	7729
Canfield Buttes	Kootenai	47°43'57"N	116°42'55"W	3900
Canida Peak	Boundary	48°59'59"N	116°38'57"W	5320
Cannon Ball Mountain	Idaho	45°19'53"N	116°28'04"W	7197
Cannonball Mountain	Camas	43°27'36"N	114°42'42"W	8330
Canyon Creek Butte	Madison	43°49'41"N	111°30'26"W	6121
Canyon Peak	Shoshone	47°00'28"N	115°37'19"W	4961
Cape Horn	Custer	44°21'16"N	115°06'53"W	6640
Cape Horn Mountain	Custer	44°24'07"N	115°14'25"W	9526
Cape Horn Peak	Bonner	47°59'44"N	116°31'15"W	4519
Cape Horn Summit	Custer	44°21'50"N	115°16'04"W	7306
Capital Hill	Shoshone	47°33'34"N	115°58'12"W	5318
Captain Butte	Owyhee	43°08'58"N	116°55'31"W	5891
Carbonate Hill	Adams	45°15'27"N	116°34'30"W	8150
Carbonate Mountain	Blaine	43°32'13"N	114°20'31"W	6530
Carey Dome	Idaho	45°24'07"N	115°54'14"W	7681
Carey Kipuka	Blaine	43°19'31"N	113°38'06"W	5378
Caribou Hill	Bonner	48°48'51"N	116°50'17"W	4592
Caribou Mountain	Bonneville	43°05'36"N	111°18'38"W	9803
Carill Peak	Kootenai	47°32'26"N	116°40'52"W	4098
Carpenter Mountain	Benewah	47°03'38"N	116°28'01"W	4660
Cascade Peak	Shoshone	47°03'27"N	115°08'55"W	6682
Cascade Point	Shoshone	47°03'26"N	115°10'34"W	6100
Casey Mountain	Adams	45°13'31"N	116°35'53"W	8753
Casey, Mount	Bonner	48°25'47"N	116°41'09"W	6706
Casner Mountain	Boise	43°45'50"N	115°56'57"W	5556
Castle Butte	Idaho	46°26'03"N	115°13'09"W	6659
Castle Butte	Owyhee	43°06'07"N	116°16'17"W	2710
Castle Peak	Clark	44°29'31"N	112°02'55"W	8124
Castle Peak	Custer	44°01'53"N	113°35'25"W	10924
Castle Rock	Boise	43°51'06"N	116°06'18"W	5415
Castle Rock	Bonneville	43°16'55"N	111°33'58"W	7006
Castle Rock	Custer	43°54'58"N	113°59'26"W	8781

Mountain, Peak, or Summit	County	Latitude	Longitude	Elev.
Castle Rock	Shoshone	46°59'16"N	116°02'28"W	5742
Castleford Butte	Twin Falls	42°30'42"N	115°01'40"W	4450
Castro Table	Owyhee	42°25'37"N	116°40'56"W	5085
Cat Creek Summit	Elmore	43°18'16"N	115°18'46"W	5601
Cataldo Mountain	Shoshone	47°35'42"N	116°19'45"W	4124
Cataract Peak	Kootenai	47°50'21"N	116°25'10"W	5039
Caterpillar Hill	Latah	46°52'33"N	116°22'14"W	3745
Cathedral Peak	Shoshone	47°56'39"N	116°08'27"W	4925
Cathedral Rock	Lemhi	45°32'19"N	114°26'41"W	7788
Cathedral Rock	Lemhi	45°08'09"N	114°32'46"W	9420
Catholic Butte	Owyhee	42°24'22"N	115°40'09"W	4790
Cavaney Hill	Owyhee	43°03'03"N	116°37'43"W	5291
Cave Point	Clearwater	46°42'36"N	115°32'00"W	4588
Cavieta Hill	Owyhee	42°20'53"N	117°01'04"W	5541
Cayuse Mountain	Idaho	45°44'31"N	114°38'10"W	6579
Cayuse Point	Elmore	43°42'09"N	115°11'56"W	8925
Cayuse Point	Lemhi	45°27'22"N	114°34'03"W	7586
Cedar Butte	Bannock	43°00'58"N	112°31'58"W	4478
Cedar Butte	Bingham	43°23'23"N	112°54'22"W	5828
Cedar Butte	Clark	44°05'10"N	112°22'08"W	5383
Cedar Butte	Shoshone	47°00'28"N	116°18'12"W	3460
Cedar Butte	Twin Falls	42°13'11"N	114°55'45"W	5785
Cedar Creek Peak	Cassia	42°26'50"N	113°03'17"W	7464
Cedar Hill	Bannock	42°19'41"N	111°57'48"W	6134
Cedar Hill	Twin Falls	42°25'15"N	114°17'56"W	4296
Cedar Knob	Idaho	46°15'15"N	115°34'00"W	4460
Cedar Knoll	Cassia	42°09'44"N	113°25'54"W	5690
Cedar Mountain	Bannock	42°33'02"N	112°19'02"W	6299
Cedar Mountain	Shoshone	47°21'15"N	115°50'30"W	5875
Center Mountain	Idaho	45°35'04"N	115°02'39"W	8260
Center Mountain	Lemhi	45°13'47"N	113°38'51"W	10362
Center Mountain	Valley	45°06'04"N	115°13'23"W	9323
Center Star Mountain	Idaho	45°48'18"N	115°33'28"W	5300
Chair Point	Idaho	45°28'16"N	116°13'43"W	6951
Chalk Hills	Owyhee	42°47'09"N	116°01'40"W	3042
Chamberlain Mountain	Clearwater	46°54'45"N	115°12'56"W	6612
Champion Point	Shoshone	47°24'44"N	115°47'07"W	6026
Character Peak	Shoshone	47°37'45"N	116°02'59"W	5048
Charles Butte	Benewah	47°07'17"N	116°44'31"W	4471
Chattin Hill	Elmore	43°01'42"N	116°03'34"W	2771
Chenoweth, Mount	Shoshone	47°12'54"N	115°37'57"W	5210
Cherry Butte	Latah	46°48'52"N	116°31'07"W	3763
Cherry Hill	Kootenai	47°41'17"N	116°45'20"W	2490
Chester Hill	Caribou	42°40'30"N	111°36'41"W	6520
Chicken Peak	Bonneville	43°36'35"N	111°22'05"W	8440
Chicken Peak	Idaho	45°17'35"N	115°23'46"W	8680
Chilco Mountain	Kootenai	47°53'30"N	116°31'45"W	5635
Chilcoot Peak	Valley	44°46'13"N	115°25'17"W	8998
Chilly Buttes	Custer	44°04'36"N	113°54'03"W	6884
Chimney Butte	Idaho	46°25'12"N	115°21'16"W	5642
Chimney Peak	Idaho	46°11'29"N	115°15'31"W	7660
Chimney Rock	Idaho	45°18'07"N	115°46'38"W	7681
Chimney Rock Middle Cairn	Boundary	48°37'10"N	116°41'50"W	7124
China Butte	Owyhee	42°56'51"N	116°43'46"W	7471
China Cap	Caribou	42°49'23"N	111°35'35"W	6778
China Hat	Caribou	42°48'30"N	111°36'09"W	7164
China Mountain	Idaho	45°14'07"N	115°36'35"W	7950
China Mountain	Twin Falls	42°03'21"N	114°51'25"W	7550
China Point	Idaho	45°56'04"N	115°46'49"W	5974
Chinamans Hat	Washington	44°32'36"N	117°02'03"W	6180
Chinese Peak	Bannock	42°50'56"N	112°21'44"W	6791
Chinook Mountain	Valley	44°43'57"N	115°21'06"W	9113
Christmas Mountain	Ada	43°16'57"N	116°12'40"W	3497
Churchill Mountain	Idaho	45°33'09"N	115°23'39"W	6700
Cinder Butte	Clark	44°18'37"N	111°55'07"W	6419
Cinder Butte	Jerome	42°39'37"N	114°04'44"W	4224
Cinder Cone Butte	Ada	43°13'10"N	115°59'32"W	3426
Cinnabar Mountain	Owyhee	42°58'27"N	116°39'23"W	8403
Cinnabar Peak	Valley	44°55'32"N	115°17'59"W	8680
Circular Butte	Jefferson	43°49'50"N	112°37'56"W	5064
Clarke Mountain	Clearwater	46°37'49"N	115°32'38"W	5285
Clarkia Peak	Shoshone	47°00'45"N	116°16'21"W	3480
Clay Bank Hills	Blaine	43°17'23"N	114°21'04"W	5242
Clay Butte	Jefferson	43°54'32"N	112°22'01"W	4898
Clear Creek Summit	Boise	44°14'01"N	115°29'33"W	7100
Clear Creek Summit	Boise	43°42'59"N	116°01'43"W	4940
Clear Creek Summit	Valley	44°31'18"N	115°47'04"W	6340
Cleveland Hill	Franklin	42°17'05"N	111°45'42"W	6478
Cliff Mountain	Adams	45°15'32"N	116°38'06"W	7384
Clifty Mountain	Boundary	48°36'51"N	116°13'09"W	6705
Cline Hill	Camas	43°23'08"N	114°41'23"W	5361
Cline Mountain	Valley	44°39'08"N	115°41'09"W	5900
Clover Butte	Owyhee	42°22'47"N	115°23'55"W	5172
Clover Knoll	Caribou	42°30'32"N	111°04'27"W	8442
Clover Mountain	Owyhee	42°44'06"N	116°25'05"W	6877
Club Point	Shoshone	47°01'34"N	115°24'48"W	6236
Coal Pit Butte	Cassia	42°15'02"N	114°11'15"W	7291
Cobb Peak	Blaine	43°43'52"N	114°07'32"W	11650
Cobblers Knob	Clearwater	46°54'09"N	115°51'50"W	3245
Coddington Peak	Shoshone	47°10'20"N	116°11'24"W	4743
Coeur d'Alene, Mount	Kootenai	47°34'53"N	116°41'24"W	4439
Coin Mountain	Valley	45°02'56"N	115°23'32"W	8994
Cold Mountain	Idaho	45°14'57"N	114°56'53"W	8084
Cold Springs Peak	Clearwater	46°47'11"N	115°18'12"W	6731
Collier Peak	Valley	44°28'09"N	116°06'45"W	7982
Collins Peak	Shoshone	46°57'32"N	115°29'05"W	6383

Mountains, Peaks, Summits

Mountain, Peak, or Summit	County	Latitude	Longitude	Elev.
Colt Mountain	Kootenai	47°49'31"N	116°28'51"W	4206
Concord Hill	Idaho	45°34'42"N	115°41'57"W	8333
Cone Peak	Idaho	46°11'40"N	114°53'43"W	3989
Congress Knob	Bonneville	43°20'34"N	111°15'29"W	7492
Conrad Peak	Shoshone	47°09'05"N	115°27'55"W	6441
Continental Mountain	Boundary	48°56'44"N	116°55'36"W	6677
Cony Peak	Boise	44°03'47"N	115°03'57"W	9606
Cook Knob	Valley	44°08'48"N	116°08'50"W	6507
Cook Mountain	Clearwater	46°35'21"N	115°18'10"W	6500
Cooks Peak	Boundary	48°42'31"N	116°33'22"W	5993
Coolin Mountain	Bonner	48°27'38"N	116°51'35"W	3475
Coolwater Mountain	Idaho	46°09'16"N	115°27'22"W	6929
Coonskin Butte	Owyhee	42°20'57"N	115°08'12"W	5310
Cooperation Point	Idaho	46°25'43"N	114°49'12"W	6785
Copper Basin Knob	Custer	43°45'21"N	113°50'53"W	10784
Copper Butte	Idaho	45°59'12"N	115°11'38"W	7178
Copper Mountain	Boundary	48°58'24"N	116°05'52"W	6196
Copper Mountain	Clark	44°10'26"N	112°49'50"W	10303
Copper Mountain	Custer	44°19'29"N	115°11'21"W	8895
Copper Mountain	Kootenai	47°38'52"N	116°28'48"W	5007
Copper Mountain	Lemhi	45°18'51"N	114°16'35"W	7093
Copper Point	Shoshone	47°02'31"N	115°21'56"W	5614
Copperhead Peak	Lemhi	45°18'14"N	113°43'41"W	10060
Corkscrew Mountain	Custer	44°31'48"N	114°22'57"W	9348
Corner Mountain	Washington	44°34'23"N	116°59'44"W	6308
Cornwall Point	Shoshone	47°05'05"N	116°04'25"W	5853
Corral Butte	Blaine	43°14'34"N	113°37'07"W	5052
Corral Creek Summit	Custer	43°54'21"N	113°47'39"W	8802
Corral Hill	Idaho	45°57'52"N	115°49'18"W	5968
Cottontail Point	Idaho	45°22'57"N	115°44'45"W	7665
Cottonwood Butte	Idaho	46°04'08"N	116°27'52"W	5730
Cottonwood Butte	Idaho	45°17'19"N	114°47'33"W	9349
Cottonwood Peak	Bannock	42°24'37"N	111°57'55"W	3500
Cottonwood Peak	Kootenai	47°32'05"N	116°38'22"W	3925
Cottonwood Point	Bonner	48°22'09"N	116°55'19"W	3081
Couch Summit	Camas	43°30'54"N	114°48'02"W	7194
Cougar Mountain	Valley	44°17'27"N	116°05'16"W	5460
Cougar Mountian	Idaho	45°51'50"N	115°49'20"W	5321
Cougar Peak	Bonner	48°14'42"N	116°13'11"W	6004
Cougar Peak	Valley	45°03'10"N	115°17'23"W	9120
Cougar Peak Loookout	Shoshone	47°40'58"N	116°12'34"W	5282
Cougar Rock	Clearwater	46°47'18"N	115°47'32"W	4200
Cougar Rock	Valley	44°35'53"N	115°43'10"W	7066
Coulter Summit	Boise	44°00'07"N	115°44'49"W	7220
Council Mountain	Valley	44°42'39"N	116°16'06"W	8126
Cove Mountain	Idaho	45°25'21"N	115°31'04"W	6100
Cove Peak	Idaho	46°07'55"N	114°33'52"W	7553
Coyote Butte	Ada	43°19'34"N	116°22'12"W	3051

Mountain, Peak, or Summit	County	Latitude	Longitude	Elev.
Coyote Butte	Blaine	43°14'37"N	113°24'51"W	5175
Coyote Butte	Butte	43°28'33"N	113°06'47"W	5153
Coyote Butte	Butte	43°24'14"N	113°31'08"W	5909
Coyote Butte	Shoshone	47°16'15"N	116°15'05"W	3012
Coyote Rock	Boise	43°48'21"N	116°05'59"W	6255
Crab Spring Butte	Owyhee	42°32'35"N	116°16'05"W	6296
Crabb Butte	Owyhee	42°53'10"N	115°46'14"W	2999
Craddock Peak	Shoshone	47°16'54"N	115°28'52"W	6327
Crag Peak	Shoshone	46°56'29"N	115°35'37"W	6879
Crags, The	Idaho	46°10'41"N	115°11'28"W	8021
Craig Mountain	Nez Perce	46°04'34"N	116°51'02"W	5290
Cramer, Mount	Custer	44°00'40"N	114°58'54"W	10716
Crane Hill	Idaho	46°03'28"N	115°48'06"W	3823
Crane Point	Benewah	47°02'38"N	116°48'12"W	3952
Crater Butte	Fremont	44°15'18"N	111°47'38"W	6503
Crater Butte	Lincoln	42°57'22"N	114°16'29"W	4132
Crater Mountain	Caribou	42°59'05"N	111°36'53"W	7066
Crater Peak	Shoshone	47°01'42"N	115°59'06"W	6444
Crater Peak	Valley	45°02'58"N	115°25'45"W	8860
Crescendo Peak	Shoshone	46°57'33"N	115°46'37"W	5890
Crescent Butte	Butte	43°24'35"N	113°30'08"W	5990
Crittenden Peak	Shoshone	47°20'03"N	115°32'59"W	6416
Croesus Peak	Custer	43°59'53"N	114°39'05"W	10288
Crofoot Point	Idaho	45°34'39"N	115°08'34"W	5780
Crooked River Point	Adams	44°51'32"N	116°41'37"W	6929
Crooked Summit	Boise	43°41'16"N	116°02'47"W	4740
Crown Point	Valley	44°32'32"N	116°02'50"W	5506
Crows Nest Butte	Owyhee	42°35'19"N	115°18'39"W	4261
Cruthers Butte	Butte	43°26'34"N	113°18'15"W	5517
Crystal Butte	Fremont	44°17'31"N	111°42'01"W	6974
Crystal Mountain	Idaho	45°17'21"N	115°53'32"W	7060
Crystal Peak	Shoshone	47°08'38"N	116°18'31"W	5427
Cuban Hill	Bonner	48°14'52"N	117°02'55"W	3435
Cuddy Mountain	Washington	44°47'04"N	116°47'26"W	7715
Culdesac Hill	Nez Perce	46°21'25"N	116°39'28"W	
Cuneo Point	Idaho	45°57'00"N	114°34'26"W	8285
Curren Mountain	Adams	45°12'09"N	116°30'14"W	7805
Custer Peak	Shoshone	47°32'02"N	115°50'10"W	6423
Cutoff Peak	Boundary	48°52'49"N	116°38'56"W	6844
Dads Hump	Cassia	42°05'27"N	114°09'39"W	7620
Dago Peak	Shoshone	47°31'43"N	115°56'53"W	5025
Dairy Mountain	Idaho	45°40'54"N	116°07'12"W	6480
Dalton Hill	Bonneville	43°21'24"N	111°36'13"W	6778
Danskin Peak	Elmore	43°24'47"N	115°39'31"W	6694
Dare Peak	Bingham	43°01'33"N	112°08'17"W	6429
Daugherty Hill	Shoshone	47°20'10"N	116°05'33"W	4827
Dave Lewis Peak	Valley	45°02'40"N	114°49'43"W	9252
Daveggio Knob	Shoshone	47°11'14"N	116°00'58"W	5882

Mountain, Peak, or Summit	County	Latitude	Longitude	Elev.
Davis Butte	Fremont	44°20'07"N	111°43'43"W	6974
Davis Mountain	Camas	43°12'50"N	114°54'43"W	6806
Dead Mule Peak	Idaho	45°15'01"N	115°15'03"W	8585
Dead Point	Idaho	45°34'44"N	116°08'46"W	7365
Deadhorse Mountain	Clearwater	46°44'11"N	115°37'40"W	4700
Deadtop Mountain	Idaho	45°40'27"N	114°50'16"W	7817
Deadwood Mountain	Idaho	45°46'49"N	115°31'19"W	5418
Deadwood Summit	Valley	44°32'51"N	115°33'28"W	6840
Deception Point	Clearwater	46°49'31"N	115°08'55"W	5462
Decker Peak	Custer	44°02'35"N	114°58'00"W	10704
Deep Creek Peak	Power	42°28'16"N	112°39'19"W	8748
Deer Heaven Mountain	Elmore	43°13'32"N	115°11'55"W	6409
Deer Mountain	Elmore	43°29'42"N	115°05'57"W	7931
Deer Point	Boise	43°45'18"N	116°05'53"W	7060
Degan Mountain	Lemhi	44°57'06"N	114°04'49"W	8748
DeLamar Mountain	Owyhee	43°00'41"N	116°49'52"W	6134
Della Mountain	Blaine	43°30'12"N	114°19'16"W	6772
Dellas Peak	Idaho	45°52'05"N	116°39'23"W	4290
Democrat Mountain	Clearwater	46°33'53"N	115°55'12"W	4788
Dennis Knob	Shoshone	47°17'25"N	115°59'08"W	4601
Dennis Mountain	Idaho	45°34'01"N	114°55'08"W	7720
Derr Point	Bonner	48°06'31"N	116°13'05"W	5220
Devil Creek Butte	Twin Falls	42°27'01"N	114°57'50"W	4415
Devil Peak	Shoshone	47°56'53"N	116°13'23"W	4615
Devils Bedstead, The	Blaine	43°47'56"N	114°09'05"W	11865
Devils Hill	Franklin	42°16'09"N	111°41'08"W	6691
Devils Point	Idaho	45°37'58"N	114°36'51"W	8217
Devils Pulpit	Shoshone	46°58'14"N	115°39'43"W	4015
Devils Throne	Idaho	45°18'35"N	116°33'15"W	9280
Devils Tooth	Idaho	45°20'39"N	116°32'19"W	7830
Dewey Hill	Valley	44°57'31"N	115°08'33"W	7895
Dewey Peak	Owyhee	42°55'36"N	116°47'59"W	6260
Diablo Mountain	Idaho	46°18'05"N	114°37'03"W	7461
Diamond Peak	Butte	44°08'32"N	113°04'59"W	12197
Diamond Peak	Caribou	42°47'42"N	111°11'04"W	8696
Diamond Point	Valley	45°09'09"N	115°11'24"W	8228
Diamond Rock	Valley	45°10'26"N	115°54'10"W	8460
Dickey Peak	Custer	44°13'46"N	113°52'57"W	11141
Dietrich Butte	Lincoln	42°56'46"N	114°12'49"W	4632
Dirty Head	Franklin	42°00'19"N	112°05'05"W	5240
Disgrace Butte	Idaho	45°54'31"N	115°10'56"W	6589
Dismal Mountain	Idaho	45°20'00"N	114°56'11"W	7104
Ditto Hill	Blaine	43°18'59"N	114°18'35"W	5361
Divide Peak	Shoshone	47°57'59"N	116°01'24"W	5205
Dixie Mountain	Valley	45°07'22"N	115°25'36"W	9070
Dodge Peak	Boundary	48°33'52"N	116°32'34"W	5027
Doe Point	Boise	43°45'18"N	116°05'42"W	7060
Dog Mountain	Elmore	43°31'09"N	115°20'42"W	7706

Mountain, Peak, or Summit	County	Latitude	Longitude	Elev.
Dollar Butte	Owyhee	42°22'58"N	116°19'26"W	5829
Dollar Mountain	Blaine	43°40'45"N	114°20'32"W	6578
Dollarhide Mountain	Blaine	43°36'33"N	114°40'51"W	9301
Dollarhide Summit	Blaine	43°35'38"N	114°40'57"W	8175
Dome Hill	Idaho	45°40'59"N	115°51'56"W	6828
Dome Mountain	Lemhi	45°14'02"N	114°31'17"W	9316
Dome, The	Owyhee	42°15'21"N	116°41'43"W	5131
Dominion Peak	Shoshone	47°22'02"N	115°34'39"W	6032
Dominion Point	Shoshone	47°09'12"N	115°31'03"W	5300
Donaldson Peak	Custer	44°03'53"N	113°41'50"W	12023
Dons Mountain	Shoshone	47°39'28"N	115°45'21"W	6002
Doris Butte	Clearwater	46°34'09"N	115°25'01"W	4753
Dorsey Butte	Elmore	43°05'09"N	116°03'15"W	3130
Dorsey Table	Owyhee	42°05'33"N	115°30'20"W	5630
Downey Peak	Shoshone	47°48'40"N	116°01'05"W	6010
Doyle Mountain	Owyhee	42°49'39"N	116°22'47"W	6170
Draney Peak	Caribou	42°44'39"N	111°09'32"W	9131
Drive Point	Shoshone	47°18'58"N	116°02'33"W	5043
Drummond Peak	Shoshone	47°31'37"N	116°15'16"W	3128
Drumond Peak	Shoshone	47°31'07"N	116°14'57"W	3091
Dry Buck Mountain	Boise	44°07'38"N	116°09'09"W	6226
Dry Hollow Mountain	Bannock	42°21'34"N	111°56'19"W	7220
Dry Point	Idaho	45°48'53"N	116°39'03"W	4042
Dryden Peak	Owyhee	43°19'06"N	116°52'27"W	5731
Duck Creek Point	Lemhi	44°54'41"N	114°29'47"W	9130
Duck Peak	Lemhi	44°54'53"N	114°29'03"W	8972
Dudley Peak	Kootenai	47°31'43"N	116°24'13"W	3983
Dugout Hill	Custer	43°56'41"N	113°43'37"W	7333
Dull Axe Mountain	Clearwater	46°41'21"N	115°46'32"W	5100
Duncecap Rock	Shoshone	46°59'32"N	115°47'28"W	6025
Dunn Peak	Shoshone	47°17'40"N	115°53'57"W	5607
Dusty Peak	Bonner	48°43'21"N	116°58'22"W	4857
Dutch Oven Ridge	Idaho	45°32'00"N	116°28'00"W	6200
Dutchler Mountain	Lemhi	45°26'35"N	114°14'24"W	7113
Dutchmans Hump	Lemhi	45°14'22"N	114°17'51"W	6560
Eagan Point	Valley	45°05'21"N	114°48'18"W	6743
Eagen, Mount	Bonner	48°16'13"N	116°18'01"W	5278
Eagle Mountain	Idaho	46°20'16"N	115°06'59"W	7427
Eagle Mountain	Lemhi	45°29'48"N	113°50'44"W	8266
Eagle Peak	Kootenai	47°26'49"N	116°29'47"W	5199
Eagle Point	Cassia	42°24'02"N	113°07'16"W	6550
Eagle Point	Clearwater	46°48'03"N	115°32'13"W	5709
Eagle Point	Idaho	45°51'32"N	114°52'34"W	5403
Eagle Point	Shoshone	47°13'09"N	115°31'13"W	5180
Eagle Rock	Idaho	45°59'44"N	114°48'56"W	5026
Eagle Rock	Power	42°42'28"N	112°57'33"W	4389
Eagles Nest	Lemhi	45°21'45"N	114°17'14"W	4851
Eagleson Summit	Boise	43°41'43"N	116°05'59"W	6089

Mountains, Peaks, Summits

Mountain, Peak, or Summit	County	Latitude	Longitude	Elev.
Eakin Point	Idaho	45°30'11"N	114°48'59"W	8115
Easley Peak	Custer	43°51'06"N	114°34'51"W	11108
East and West Peak	Lemhi	44°49'45"N	113°25'43"W	9924
East Butte	Bingham	43°30'05"N	112°39'42"W	6538
East Butte	Shoshone	47°01'13"N	115°47'03"W	6380
East Canfield Butte	Kootenai	47°43'56"N	116°42'32"W	3867
East Cathedral Peak	Shoshone	47°57'08"N	116°06'34"W	4100
East Dennis	Benewah	47°04'15"N	116°41'31"W	4626
East Elk Peak	Shoshone	47°05'02"N	116°15'35"W	5170
East Farnes Mountain	Madison	43°40'36"N	111°28'29"W	7300
East Fork Peak	Bonner	48°15'01"N	116°02'59"W	5987
East Gold Hill	Latah	46°58'37"N	116°46'42"W	4677
East Grouse Peak	Shoshone	47°30'23"N	115°46'51"W	6286
East Moscow Mountain	Latah	46°48'17"N	116°50'20"W	4721
East Mountain	Valley	44°26'33"N	115°52'09"W	7752
East Peak	Idaho	46°12'56"N	115°09'53"W	7852
East Sister	Clearwater	46°53'14"N	115°31'39"W	7043
East Sister Peak	Shoshone	47°09'40"N	115°37'55"W	6871
East Twin	Latah	46°48'43"N	116°54'15"W	4780
East Warrior Peak	Elmore	43°52'42"N	115°12'52"W	8758
Eccles Butte	Fremont	44°19'16"N	111°16'39"W	6525
Echo Peak	Kootenai	47°45'18"N	116°29'52"W	4409
Echols Mountain	Adams	45°09'38"N	116°31'58"W	8331
Edaho Mountain	Boise	44°00'11"N	115°08'48"W	9614
Eddy Peak	Boundary	48°38'06"N	116°43'01"W	6725
Eighteenmile Peak	Lemhi	44°26'52"N	112°59'41"W	11125
Eightmile Mountain	Boise	44°10'03"N	115°21'11"W	7871
Eighty Peak	Shoshone	47°53'19"N	115°56'47"W	6460
Eightyone, Point	Shoshone	47°16'50"N	115°32'06"W	6026
Eightyseven Mile Peak	Shoshone	47°49'38"N	115°51'04"W	5617
El Capitan	Blaine	43°56'28"N	114°55'59"W	9907
Eldridge, Mount	Valley	45°08'24"N	115°24'48"W	9207
Elephant Butte	Owyhee	43°27'35"N	116°51'05"W	3163
Elevator Mountain	Idaho	45°58'11"N	114°49'51"W	4108
Elizabeth Mountain	Clearwater	46°47'06"N	115°14'47"W	6464
Elk Butte	Cassia	42°12'58"N	114°16'31"W	7142
Elk Butte	Clearwater	46°50'28"N	116°07'00"W	5824
Elk Butte	Fremont	44°14'14"N	111°21'45"W	6372
Elk Butte	Idaho	45°30'52"N	115°47'09"W	6744
Elk Mountain	Blaine	43°33'14"N	113°59'33"W	8093
Elk Mountain	Clearwater	46°38'57"N	115°37'40"W	5299
Elk Mountain	Clearwater	46°38'43"N	115°36'21"W	5820
Elk Mountain	Custer	44°16'13"N	115°04'33"W	7925
Elk Mountain	Idaho	45°54'46"N	115°03'40"W	7826
Elk Mountain	Kootenai	47°35'50"N	116°36'32"W	4236
Elk Mountain	Lemhi	44°45'35"N	113°08'12"W	10194
Elk Peak	Boise	44°01'28"N	115°02'11"W	10582
Elk Summit	Idaho	45°52'47"N	115°32'22"W	6386

Mountain, Peak, or Summit	County	Latitude	Longitude	Elev.
Elkhorn Mountain	Bannock	42°21'36"N	112°19'03"W	8274
Elkhorn Peak	Bonneville	43°21'28"N	111°05'57"W	9940
Elkhorn Peak	Oneida	42°20'12"N	112°19'38"W	9095
Elkhorn, Mount	Oneida	42°22'32"N	112°29'44"W	6115
Elmira Peak	Bonner	48°29'33"N	116°25'46"W	3852
Elsie Peak	Shoshone	47°23'53"N	116°06'42"W	5257
Emerald Butte	Latah	47°00'53"N	116°23'13"W	4660
Emery Butte	Twin Falls	42°18'20"N	114°51'12"W	5030
Emida Peak	Benewah	47°06'44"N	116°32'54"W	4548
Emmett Mountain	Adams	45°11'55"N	116°37'03"W	8355
End Butte	Idaho	46°18'31"N	115°06'39"W	7303
Eneas Peak	Boundary	48°53'00"N	116°30'00"W	6570
Estes Mountain	Custer	44°26'45"N	114°42'39"W	9643
Evergreen Mountain	Shoshone	47°13'39"N	116°14'29"W	4450
Everly, Mount	Elmore	43°57'30"N	115°05'29"W	9852
Eyrie Peak	Oneida	42°18'19"N	112°56'55"W	6577
Fairview Knob	Benewah	47°04'35"N	116°44'15"W	3940
Falconberry Peak	Lemhi	44°44'57"N	114°45'44"W	9465
Fall Creek Point	Idaho	45°49'05"N	115°39'59"W	4929
Fall Point	Idaho	45°52'01"N	116°25'17"W	3450
Farber Point	Latah	46°58'39"N	116°54'01"W	3125
Farnes Mountain	Madison	43°40'07"N	111°29'53"W	7396
Farnham Peak	Boundary	48°51'00"N	116°29'49"W	7001
Farrow Mountain	Idaho	45°15'49"N	114°49'47"W	8992
Faset Peak	Bonner	47°56'24"N	116°20'21"W	5246
Federal Butte	Owyhee	43°08'28"N	116°37'23"W	4901
Fenn Mountain	Idaho	46°10'38"N	115°12'13"W	8021
Fernan Hill	Kootenai	47°41'26"N	116°43'09"W	2700
Ferry Butte	Bingham	43°06'46"N	112°29'33"W	4822
Fin Rock	Valley	44°26'32"N	115°36'17"W	6820
Fingers Butte	Butte	43°25'14"N	113°15'08"W	5624
Fir Grove Mountain	Camas	43°11'57"N	114°47'16"W	6236
Fire Mountain	Idaho	45°48'44"N	114°48'12"W	6880
Fish Butte	Idaho	46°19'53"N	115°23'40"W	6346
Fish Creek Point	Idaho	45°48'16"N	116°06'03"W	5185
Fisher Peak	Boundary	48°52'00"N	116°32'10"W	7580
Fishhook Peak	Shoshone	47°08'29"N	115°57'52"W	6531
Fissure Butte	Blaine	43°21'05"N	113°27'58"W	5877
Fitsum Peak	Valley	44°58'43"N	115°51'56"W	8583
Fitsum Summit	Valley	44°56'53"N	115°52'38"W	8394
Fitzgerald Peak	Benewah	47°21'06"N	116°15'18"W	4540
Five Lakes Butte	Shoshone	46°56'55"N	115°16'31"W	6713
Flag Knoll	Bonneville	43°16'56"N	111°28'56"W	7709
Flagstaff Butte	Ada	43°16'45"N	116°08'25"W	3390
Flash Peak	Shoshone	47°18'33"N	115°52'41"W	5685
Flat Mountain	Clearwater	46°44'52"N	115°16'25"W	6606
Flat Top	Custer	43°57'09"N	113°57'21"W	9409
Flat Top	Power	42°59'42"N	113°04'09"W	5138

Mountain, Peak, or Summit	County	Latitude	Longitude	Elev.
Flat Top Butte	Gooding	43°10'01"N	114°39'40"W	5557
Flat Top Butte	Jerome	42°43'57"N	114°24'54"W	4275
Flat Top Butte	Owyhee	43°30'11"N	116°56'37"W	2729
Flat Top Mountain	Elmore	43°55'04"N	115°08'02"W	9665
Flatiron	Bonneville	43°24'15"N	111°19'09"W	5820
Flatiron	Twin Falls	42°13'04"N	114°31'23"W	5243
Flatiron Butte	Owyhee	42°56'11"N	115°40'22"W	2829
Flatiron Butte	Owyhee	42°28'16"N	116°37'22"W	5378
Flatiron Hill	Power	42°43'37"N	112°33'53"W	5738
Flatiron Knobs	Shoshone	47°04'30"N	115°52'04"W	5490
Flatiron Mountain	Cassia	42°08'29"N	114°11'48"W	7809
Flatiron Mountain	Elmore	43°33'18"N	115°43'49"W	4105
Flatiron Mountain	Lemhi	44°27'40"N	113°31'31"W	11019
Flattop	Bonner	48°28'01"N	116°41'03"W	6465
Flattop Butte	Owyhee	43°14'17"N	116°56'29"W	5828
Flattop Mountain	Shoshone	47°14'47"N	115°19'23"W	6403
Fleck Summit	Camas	43°37'32"N	114°53'33"W	5982
Fleming Point	Bonner	48°05'02"N	116°20'40"W	3584
Flemming Point	Shoshone	47°14'24"N	115°52'01"W	4220
Fletcher Butte	Valley	44°11'37"N	116°11'40"W	5775
Flint Mesa	Elmore	42°51'55"N	115°12'44"W	3153
Flora Miller Hill	Kootenai	47°45'14"N	116°29'04"W	4582
Florida Mountain	Owyhee	43°01'00"N	116°45'18"W	7784
Fly Creek Point	Lemhi	44°40'53"N	114°34'09"W	8984
Fly Hill	Clearwater	46°51'17"N	115°14'25"W	6342
Fly Peak	Shoshone	47°04'58"N	115°24'21"W	5648
Flynn Butte	Latah	46°59'18"N	116°38'47"W	3604
Flytrap Butte	Idaho	46°24'50"N	115°02'58"W	6338
Fog Mountain	Idaho	46°05'36"N	115°14'08"W	6559
Fogg Butte	Fremont	44°17'47"N	111°40'54"W	6910
Fogg Hill	Bonneville	43°30'32"N	111°07'00"W	8933
Foolhen Mountain	Shoshone	47°22'07"N	115°55'33"W	3963
Forage Mountain	Shoshone	47°07'13"N	115°46'42"W	5785
Ford Rock	Kootenai	47°42'25"N	116°55'52"W	4274
Fort Hall Hill	Adams	44°49'21"N	116°25'44"W	3649
Fortune Point	Shoshone	47°07'01"N	115°31'20"W	5767
Foss Mountain	Butte	44°11'48"N	113°11'01"W	9190
Fossil Butte	Owyhee	43°06'31"N	116°26'52"W	3304
Four Bit Summit	Gem	44°27'42"N	116°14'36"W	6193
Fourth of July Peak	Bonneville	43°31'27"N	111°10'17"W	7496
Fourth Of July Summit	Kootenai	47°37'32"N	116°31'06"W	3500
Fox Butte	Clearwater	46°34'38"N	115°19'52"W	6334
Fox Peak	Blaine	43°43'31"N	114°30'49"W	9165
Fox Peak	Idaho	46°17'43"N	114°47'23"W	7279
Freds Mound	Cassia	42°11'04"N	114°11'23"W	7620
Free Use Point	Idaho	45°44'56"N	116°08'40"W	4980
Freeman Peak	Boise	43°57'05"N	115°42'25"W	8111
Freeman Peak	Idaho	46°08'51"N	114°47'01"W	7298
Freeman Peak	Lemhi	45°16'28"N	113°42'00"W	10279
Freezeout Hill	Gem	43°49'51"N	116°29'05"W	3209
Freezeout Mountain	Idaho	46°24'32"N	114°58'46"W	5956
Freize Knoll	Fremont	44°16'56"N	111°47'56"W	6535
French John Hill	Owyhee	43°24'45"N	116°52'35"W	4682
French Mountain	Clearwater	46°30'13"N	115°42'58"W	5327
Frenchman Butte	Idaho	46°18'58"N	115°32'23"W	5185
Friday Butte	Gem	44°17'22"N	116°15'05"W	4700
Frisco Peak	Idaho	46°12'31"N	115°06'24"W	6820
Fritz Peak	Clark	44°22'51"N	112°44'47"W	9738
Frog Peak	Idaho	46°20'53"N	114°27'26"W	8078
Frost Peak	Kootenai	47°28'49"N	116°20'33"W	5832
Fubar Peak	Shoshone	47°33'00"N	116°14'54"W	2800
Fulkerson Peak	Shoshone	47°35'45"N	115°51'32"W	5447
Fuller Peak	Cassia	42°13'53"N	114°09'48"W	7484
Gabes Peak	Gem	44°25'50"N	116°09'45"W	7655
Galbraith Hill	Bonneville	43°26'02"N	111°52'59"W	5880
Galena Peak	Blaine	43°53'23"N	114°36'10"W	11153
Galena Summit	Blaine	43°52'15"N	114°42'47"W	8990
Gallagher Peak	Boise	44°06'23"N	115°47'05"W	6073
Gallagher Peak	Clark	44°10'02"N	112°47'19"W	9825
Gant Mountain	Lemhi	45°14'03"N	114°22'18"W	8276
Garden Peak	Bingham	43°11'44"N	112°06'06"W	6357
Gardiner Peak	Idaho	45°58'12"N	114°45'57"W	6597
Garfield Mountain	Blaine	43°36'23"N	113°58'44"W	8627
Garns Mountain	Teton	43°39'48"N	111°19'55"W	9016
Gateway Peak	Idaho	46°15'50"N	114°39'32"W	6283
Gedney Butte	Owyhee	42°28'31"N	115°53'15"W	5045
Gedney Mountain	Idaho	46°09'00"N	115°15'31"W	7380
Geneva Summit	Bear Lake	42°19'58"N	111°08'17"W	7461
George, Mount	Idaho	45°54'58"N	114°37'32"W	7747
Gerber Butte	Fremont	43°58'57"N	112°02'14"W	5215
Gerdie Hill	Owyhee	43°01'46"N	116°38'14"W	
Germer Park	Custer	44°23'56"N	114°14'21"W	7684
Getaway Mountain	Owyhee	43°19'54"N	116°49'58"W	5641
Getaway Point	Shoshone	46°56'12"N	115°44'02"W	4891
Ghost Mountain	Idaho	46°09'27"N	115°20'47"W	6861
Giants Nose	Idaho	45°45'51"N	116°18'40"W	2981
Gibson Mountain	Bannock	42°46'25"N	112°27'11"W	6775
Gibson Point	Shoshone	47°22'51"N	115°51'17"W	5490
Gilman Butte	Blaine	43°27'20"N	114°22'02"W	6111
Gilmore Summit	Lemhi	44°25'47"N	113°13'46"W	7150
Gisborne Mountain	Bonner	48°20'53"N	116°44'41"W	5692
Glass Hill	Owyhee	43°04'13"N	116°46'30"W	6843
Glassford Peak	Custer	43°54'44"N	114°28'50"W	11602
Glens Peak	Elmore	43°57'00"N	115°00'38"W	10053
Glide Mountain	Custer	43°43'28"N	113°57'45"W	10256
Gnat Point	Shoshone	46°57'31"N	115°35'54"W	6540

Mountains, Peaks, Summits

Mountain, Peak, or Summit	County	Latitude	Longitude	Elev.
Goat Mountain	Boise	43°57'54"N	115°21'00"W	8835
Goat Mountain	Bonner	48°11'36"N	116°05'52"W	6380
Goat Mountain	Boundary	48°46'40"N	116°03'31"W	6641
Goat Mountain	Idaho	46°08'20"N	114°57'56"W	5288
Goat Mountain	Idaho	45°36'23"N	114°49'25"W	8600
Goat Mountain	Lemhi	45°12'48"N	114°36'46"W	9607
Goat Mountain	Lemhi	44°49'24"N	113°26'19"W	9943
Goat Mountain	Valley	45°03'12"N	115°20'56"W	9095
Goat Point	Idaho	45°22'34"N	114°50'10"W	7090
Goat Ridge	Idaho	45°23'07"N	116°36'26"W	4400
Goat Roost	Idaho	46°28'13"N	114°41'31"W	6783
Gobblers Knob	Clearwater	46°36'09"N	115°32'56"W	4100
Goblin Knob	Boundary	48°40'13"N	116°44'56"W	6606
Goblin, The	Idaho	45°19'08"N	116°31'26"W	8980
Goddard Point	Idaho	46°02'47"N	115°35'03"W	5630
Gold Butte	Clearwater	46°46'07"N	116°02'34"W	4490
Gold Crown Peak	Shoshone	47°01'28"N	115°05'32"W	7374
Gold Cup Mountain	Bonner	48°11'18"N	116°48'17"W	4561
Gold Hill	Bonner	48°14'24"N	116°29'44"W	4027
Gold Hill	Clearwater	46°29'13"N	115°51'33"W	3700
Gold Hill	Idaho	46°20'46"N	115°12'51"W	6449
Gold Hill	Latah	46°58'05"N	116°47'32"W	4661
Gold Hill	Latah	46°39'25"N	116°29'35"W	2920
Gold Hill	Nez Perce	46°04'59"N	116°55'43"W	3274
Gold Hill	Shoshone	47°25'51"N	115°49'27"W	6284
Gold Mountain	Bonner	48°13'10"N	116°28'54"W	4177
Gold Peak	Bonner	48°48'11"N	116°58'55"W	4534
Golden Gate Hill	Valley	44°56'49"N	115°28'55"W	6775
Goldstone Mountain	Lemhi	45°07'09"N	113°34'34"W	9909
Gooding Butte	Gooding	42°54'41"N	114°45'04"W	3841
Goose Creek Point	Idaho	45°49'14"N	116°07'11"W	5171
Goose Peak	Shoshone	47°33'46"N	115°50'34"W	6270
Gopher Knoll	Elmore	42°57'33"N	115°06'05"W	3136
Gorman Hill	Clearwater	46°42'15"N	115°03'22"W	5400
Gospel Hill	Idaho	45°37'48"N	115°57'20"W	8121
Gospel Hill	Shoshone	46°58'09"N	115°12'13"W	6457
Gospel Peak	Idaho	45°37'14"N	115°56'28"W	8345
Graham Mountain	Shoshone	47°36'12"N	116°08'05"W	5727
Graham Peak	Boise	43°58'54"N	115°17'45"W	7912
Graham Peak	Cassia	42°07'25"N	113°42'57"W	8867
Grand Mogul	Custer	44°04'48"N	114°57'32"W	9733
Grand Mountain	Elmore	43°51'29"N	115°25'23"W	7264
Grand View Peak	Twin Falls	42°15'19"N	114°17'56"W	7222
Grandfather Mountain	Shoshone	47°03'26"N	116°05'43"W	6306
Grandjean Peak	Boise	44°06'47"N	115°03'38"W	9105
Grandmother Mountain	Shoshone	47°02'52"N	116°04'20"W	6369
Granger Butte	Washington	44°26'38"N	116°26'26"W	3730
Granite Mountain	Adams	45°05'37"N	116°12'23"W	8478
Granite Mountain	Bonner	48°41'31"N	116°56'22"W	4780
Granite Mountain	Elmore	43°52'24"N	115°24'55"W	7084
Granite Mountain	Elmore	43°22'04"N	115°29'54"W	6818
Granite Mountain	Idaho	45°16'27"N	116°36'26"W	8015
Granite Mountain	Lemhi	45°33'00"N	113°58'49"W	6354
Granite Peak	Clearwater	46°42'57"N	114°42'44"W	7551
Granite Peak	Idaho	45°51'33"N	115°05'26"W	7232
Granite Peak	Shoshone	47°33'46"N	115°45'34"W	6815
Granite Peak	Shoshone	47°00'41"N	115°26'55"W	6481
Granite Peak	Valley	44°25'28"N	116°08'27"W	8273
Granite Point	Latah	46°48'52"N	116°52'47"W	4511
Grape Mountain	Elmore	43°36'30"N	115°46'12"W	5896
Grass Mountain	Boundary	48°56'47"N	116°51'37"W	6205
Grass Mountain	Idaho	45°23'07"N	114°56'56"W	6262
Grassy Cone	Butte	43°27'19"N	113°34'52"W	6315
Grassy Mountain	Benewah	47°24'10"N	116°33'56"W	4947
Grassy Mountain	Shoshone	47°47'23"N	116°12'16"W	5078
Grassy Point	Clearwater	46°54'33"N	115°38'28"W	6060
Grave Butte	Idaho	46°27'41"N	115°06'53"W	6190
Grave Meadow Peak	Idaho	45°57'46"N	115°04'26"W	7373
Grave Peak	Idaho	46°23'43"N	114°43'46"W	8282
Grave Point	Idaho	45°38'33"N	116°22'47"W	5630
Graves Peak	Shoshone	47°00'49"N	115°04'20"W	7235
Graveyard Point	Owyhee	43°34'17"N	117°01'18"W	2778
Grays Peak	Blaine	43°40'08"N	114°05'40"W	10563
Grays Peak	Shoshone	47°19'56"N	115°36'42"W	4944
Grays Peak	Valley	44°49'48"N	115°03'55"W	8806
Greeley Mountain	Valley	45°06'09"N	115°27'11"W	9233
Green Bonnet Mountain	Boundary	48°52'54"N	116°54'16"W	5991
Green Creek Point	Idaho	45°57'42"N	115°52'49"W	4980
Green Knob	Latah	46°34'24"N	116°47'56"W	2890
Green Monarch Mountain	Bonner	48°06'46"N	116°20'20"W	5082
Green Mountain	Benewah	47°05'41"N	116°25'12"W	3824
Green Mountain	Bonner	47°59'20"N	116°24'13"W	5148
Green Mountain	Caribou	42°35'53"N	111°14'24"W	8772
Green Mountain	Clearwater	46°50'14"N	116°02'28"W	5285
Green Mountain	Idaho	45°49'37"N	114°40'56"W	7330
Green Mountain	Idaho	45°46'16"N	115°04'33"W	7227
Green Mountain	Kootenai	47°52'39"N	116°35'44"W	4060
Green Mountain	Valley	44°49'32"N	115°53'12"W	8093
Greenside Butte	Idaho	46°18'00"N	115°13'07"W	6840
Grey Lock Peak	Elmore	43°50'20"N	115°05'07"W	9128
Greyhound Mountain	Custer	44°35'50"N	115°07'11"W	8995
Greylock Mountain	Elmore	43°50'39"N	115°05'45"W	9363
Greylock, Mount	Custer	44°25'12"N	114°41'47"W	9857
Greystone Butte	Idaho	46°23'38"N	115°04'47"W	6540
Grief Mountain	Bonner	48°18'10"N	116°20'50"W	4361
Griffin Butte	Blaine	43°43'48"N	114°25'08"W	8411

Mountain, Peak, or Summit	County	Latitude	Longitude	Elev.
Griffith Peak	Shoshone	47°54'38"N	116°16'23"W	4589
Grindstone Butte	Elmore	42°46'13"N	115°15'12"W	3494
Grizzly Hill	Lemhi	44°47'43"N	113°21'14"W	9290
Grizzly Mountain	Shoshone	47°42'47"N	116°05'36"W	5950
Grouse Butte	Elmore	43°33'15"N	115°09'06"W	7656
Grouse Creek Mountain	Custer	44°21'58"N	113°54'29"W	11085
Grouse Creek Peak	Custer	44°37'39"N	114°50'32"W	9454
Grouse Creek Point	Idaho	45°40'58"N	115°06'04"W	6516
Grouse Knoll	Valley	44°47'50"N	116°01'25"W	5531
Grouse Mountain	Bonner	48°29'51"N	116°15'54"W	5982
Grouse Mountain	Bonner	48°13'41"N	116°48'00"W	3775
Grouse Mountain	Bonner	48°10'38"N	116°27'41"W	4238
Grouse Peak	Custer	44°33'54"N	114°05'38"W	8464
Grouse Peak	Shoshone	47°30'20"N	115°48'31"W	6077
Guard Peak	Shoshone	47°48'10"N	116°00'30"W	6005
Guffey Butte	Owyhee	43°17'10"N	116°32'25"W	3130
Gunsight Peak	Elmore	43°41'52"N	115°04'14"W	9527
Gunsight Peak	Lemhi	44°35'37"N	113°31'14"W	10835
Haderlie Knoll	Caribou	42°58'20"N	111°02'39"W	6109
Half Cone	Butte	43°25'24"N	113°31'23"W	6055
Halfway Hill	Shoshone	47°14'17"N	115°35'07"W	4284
Halfway Peak	Shoshone	47°45'41"N	116°08'51"W	5254
Hall Mountain	Boundary	48°58'26"N	116°22'14"W	5650
Hamilton Mountain	Kootenai	47°52'23"N	116°20'19"W	5041
Hamilton Mountain	Shoshone	47°52'34"N	116°20'25"W	5056
Hammond Point	Shoshone	47°19'09"N	115°46'13"W	3884
Handwerk Peak	Blaine	43°45'33"N	114°09'30"W	10860
Hanover Mountain	Idaho	45°32'54"N	115°56'32"W	7966
Hansen Butte	Twin Falls	42°30'01"N	114°14'05"W	4455
Hard Butte	Idaho	45°15'38"N	116°12'32"W	8659
Hardpan Point	Shoshone	47°12'06"N	115°36'02"W	5150
Harrington Mountain	Idaho	45°32'14"N	114°55'46"W	8186
Harrington Peak	Bear Lake	42°34'58"N	111°20'17"W	8554
Harrington Peak	Cassia	42°14'01"N	114°12'11"W	7290
Harris Creek Summit	Boise	43°53'33"N	116°00'54"W	5202
Harrison Peak	Boundary	48°41'17"N	116°38'43"W	7292
Harrison, Mount	Cassia	42°18'53"N	113°39'29"W	9265
Hartley Peak	Power	42°24'29"N	112°56'57"W	7456
Harvey Mountain	Boundary	48°59'31"N	116°17'06"W	6402
Hat Butte	Canyon	43°24'29"N	116°35'47"W	3077
Hat, The	Owyhee	42°36'16"N	116°11'40"W	5690
Hatchery Butte	Fremont	44°16'07"N	111°21'51"W	6417
Hawks Peak	Bear Lake	42°31'44"N	111°16'44"W	9079
Hawley Mountain	Boise	43°59'33"N	116°01'48"W	7301
Hawley Mountain	Butte	44°06'19"N	113°20'22"W	9752
Hayden Peak	Owyhee	42°58'51"N	116°39'28"W	8403
Haystack	Washington	44°22'01"N	117°01'15"W	3884
Haystack Mountain	Bannock	42°42'53"N	112°06'52"W	9033
Haystack Mountain	Idaho	45°46'34"N	115°35'54"W	4980
Haystack Mountain	Idaho	45°42'50"N	116°21'56"W	4103
Haystack Mountain	Idaho	45°42'14"N	114°44'42"W	6875
Haystack Mountain	Lemhi	45°17'07"N	114°09'10"W	8800
Haystack Mountain	Shoshone	47°45'41"N	115°56'49"W	4615
Haystack Peak	Shoshone	47°31'36"N	116°08'04"W	3402
Hazelton Butte	Jerome	42°33'32"N	114°05'38"W	4401
He Devil	Idaho	45°19'26"N	116°32'54"W	9393
Heart Mountain	Bannock	42°18'59"N	111°55'32"W	6313
Heart Mountain	Clark	44°19'12"N	112°44'57"W	10422
Heart Peak	Shoshone	46°56'10"N	115°35'08"W	6870
Heavens Gate	Idaho	45°22'08"N	116°29'38"W	8429
Heinen, Mount	Boise	43°39'42"N	115°52'30"W	6336
Hells Half Acre Mountain	Idaho	45°38'45"N	114°37'39"W	8116
Hemingway Butte	Owyhee	43°19'27"N	116°38'43"W	2620
Hemlock Butte	Clearwater	46°55'00"N	116°09'12"W	5750
Hemlock Butte	Clearwater	46°28'25"N	115°37'40"W	6053
Hemlock Mountain	Kootenai	47°43'12"N	116°19'40"W	4932
Henderson Peak	Teton	43°39'13"N	111°15'16"W	8312
Henry Peak	Caribou	42°55'15"N	111°24'13"W	8319
Henry, Point	Shoshone	47°16'56"N	115°33'28"W	5195
Herd Peak	Custer	43°59'11"N	114°13'31"W	9860
Herds V	Owyhee	42°25'16"N	116°55'41"W	5852
Hershey Point	Idaho	45°18'45"N	116°07'21"W	8232
Heyburn Mountain	Custer	44°06'04"N	114°58'31"W	10220
Hidden Peak	Idaho	46°22'17"N	114°28'05"W	7826
Higgins Hump	Idaho	46°12'29"N	115°38'22"W	4475
High Point	Fremont	44°16'05"N	111°33'16"W	7281
Higham Peak	Bingham	43°08'28"N	112°04'57"W	6655
Highland Valley Summit	Ada	43°34'14"N	116°01'49"W	3769
Hill 36	Shoshone	47°17'32"N	116°00'29"W	4715
Hilo Peak	Shoshone	47°07'32"N	115°43'40"W	5762
Hitt Peak	Washington	44°35'39"N	116°55'29"W	7410
Hobo Hill	Shoshone	47°05'54"N	116°07'21"W	5072
Hog Cove Butte	Payette	44°07'03"N	116°33'16"W	3782
Hog Creek Butte	Washington	44°26'24"N	116°32'25"W	3411
Hogue Mountain	Boundary	48°58'58"N	116°11'43"W	4324
Holaki Knob	Benewah	47°03'12"N	116°45'21"W	3643
Hollister Mountain	Kootenai	47°51'41"N	116°40'03"W	4330
Homestead Hump	Shoshone	47°06'58"N	116°01'07"W	5680
Honey Jones Peak	Benewah	47°21'18"N	116°18'48"W	4100
Honey Mountain	Bonner	47°54'06"N	116°29'50"W	4580
Hoodoo Mountain	Bonner	48°04'44"N	116°57'09"W	5119
Hoodoo Mountain	Idaho	46°18'47"N	114°38'35"W	6908
Hoodoo Peak	Shoshone	47°03'18"N	115°40'48"W	4962
Horn Mountain	Bonner	48°08'10"N	116°48'34"W	2862
Horn, The	Idaho	46°05'23"N	115°49'41"W	3228
Hornet Point	Clearwater	46°51'43"N	115°11'58"W	5975

Mountains, Peaks, Summits

Mountain, Peak, or Summit	County	Latitude	Longitude	Elev.
Horse Butte	Cassia	42°30'10"N	113°23'35"W	4748
Horse Butte	Owyhee	42°25'10"N	115°14'01"W	4990
Horse Creek Butte	Lemhi	45°25'47"N	114°30'12"W	8351
Horse Heaven	Idaho	45°20'51"N	115°27'35"W	7523
Horse Heaven	Idaho	45°16'12"N	116°32'50"W	8220
Horse Heaven	Lemhi	45°15'46"N	114°32'43"W	8086
Horse Hill	Owyhee	42°44'03"N	115°48'16"W	3606
Horse Hill	Owyhee	42°03'33"N	115°16'33"W	5890
Horse Mountain	Adams	45°07'20"N	116°40'07"W	6887
Horse Mountain	Camas	43°15'13"N	114°28'06"W	5765
Horse Mountain	Clark	44°26'38"N	112°33'32"W	8827
Horse Mountain	Valley	45°07'46"N	114°52'43"W	8184
Horse Point	Idaho	45°58'41"N	115°19'19"W	5750
Horse Prairie Mountain	Lemhi	44°45'36"N	113°08'14"W	10194
Horse Ranch Mountain	Elmore	43°41'03"N	115°19'44"W	6788
Horsejaw Mountain	Idaho	45°40'36"N	114°42'14"W	6727
Horseshoe Peak	Shoshone	47°38'56"N	115°48'54"W	5734
Horstmann Peak	Custer	44°06'46"N	115°00'01"W	10470
Horton Peak	Custer	43°57'57"N	114°44'51"W	9896
Hot Springs Point	Idaho	46°28'37"N	114°50'19"W	5535
House Mountain	Elmore	43°26'00"N	115°29'12"W	7700
Howard Mountain	Bannock	42°51'53"N	112°30'45"W	5833
Howe Mountain	Bonner	48°11'25"N	116°13'07"W	3820
Howe Peak	Butte	43°42'48"N	113°05'58"W	8701
Hoyt Mountain	Shoshone	47°12'59"N	115°54'50"W	4925
Hub Butte	Twin Falls	42°25'53"N	114°28'52"W	4611
Huckleberry Butte	Clearwater	46°31'45"N	116°07'06"W	3729
Huckleberry Butte	Idaho	46°16'42"N	115°18'01"W	6710
Huckleberry Hill	Kootenai	47°50'59"N	116°57'04"W	4500
Huckleberry Mountain	Bonner	48°04'58"N	116°39'36"W	4224
Huckleberry Mountain	Kootenai	47°44'53"N	116°35'57"W	4865
Huckleberry Mountain	Latah	47°04'52"N	116°57'46"W	4060
Huckleberry Mountain	Shoshone	47°12'14"N	116°09'21"W	5662
Hudlow Mountain	Kootenai	47°48'57"N	116°40'32"W	3395
Hulliman Peak	Shoshone	47°49'29"N	115°55'10"W	5531
Hungry Creek Point	Idaho	45°21'23"N	114°51'45"W	7452
Hungry Point	Idaho	46°29'25"N	115°12'41"W	5770
Hunt Peak	Boundary	48°33'32"N	116°41'54"W	7058
Hunter Creek Summit	Custer	43°57'22"N	114°21'22"W	10060
Hunter Peak	Idaho	46°02'02"N	114°32'33"W	8472
Hyndman Peak	Blaine	43°44'57"N	114°07'48"W	12009
Hyram Butte	Bingham	43°03'24"N	112°21'57"W	4786
Ibex Peak	Cassia	42°06'05"N	114°04'49"W	7272
Ice Cave Knoll	Caribou	42°32'03"N	111°43'57"W	5620
Idaho Point	Shoshone	47°53'14"N	115°57'13"W	6505
Illinois Peak	Shoshone	47°01'35"N	115°04'16"W	7690
Independence Mountain	Cassia	42°11'23"N	113°40'06"W	
Independence, Mount	Cassia	42°11'51"N	113°40'27"W	9950
Indian Butte	Owyhee	42°45'28"N	115°04'15"W	3648
Indian Creek Butte	Clark	44°20'06"N	112°23'23"W	6959
Indian Dip	Shoshone	46°57'42"N	115°43'55"W	5714
Indian Grave Peak	Idaho	46°30'02"N	115°08'50"W	6446
Indian Head Mountain	Washington	44°16'32"N	117°08'59"W	3643
Indian Hill	Adams	45°16'00"N	116°21'56"W	5460
Indian Hill	Idaho	45°59'38"N	115°14'19"W	6881
Indian Hill	Idaho	45°47'25"N	114°32'35"W	8315
Indian Mountain	Adams	45°09'12"N	116°19'32"W	5708
Indian Mountain	Adams	44°35'58"N	116°14'54"W	7253
Indian Mountain	Bannock	42°44'45"N	112°20'44"W	7298
Indian Mountain	Idaho	45°21'35"N	116°17'44"W	7114
Indian Mountain	Kootenai	47°23'03"N	116°40'59"W	3499
Indian Peak	Idaho	45°57'01"N	115°06'01"W	7728
Indian Peak	Lemhi	45°29'27"N	114°06'27"W	7763
Indian Peak	Shoshone	47°04'59"N	115°19'34"W	5820
Indian Peak	Valley	44°55'00"N	115°33'52"W	7820
Indian Point	Lemhi	45°12'44"N	114°23'35"W	8453
Indian Trail Ridge	Idaho	45°24'11"N	116°32'00"W	6600
Inferno Cone	Butte	43°26'36"N	113°33'04"W	6181
Initial Peak	Kootenai	47°29'53"N	116°30'25"W	4060
Initial Point	Ada	43°22'20"N	116°23'35"W	3240
Invisible Mountain	Custer	43°58'00"N	113°30'50"W	11330
Irishmans Rock	Lemhi	45°32'31"N	114°11'01"W	7552
Iron Mountain	Boundary	48°33'56"N	116°12'13"W	6426
Iron Mountain	Camas	43°32'18"N	115°02'40"W	9694
Iron Mountain	Idaho	45°58'06"N	115°32'27"W	6815
Iron Mountain	Lemhi	44°50'42"N	114°07'13"W	7766
Iron Mountain	Washington	44°33'02"N	117°01'39"W	6492
Isabella Point	Clearwater	46°54'34"N	115°36'17"W	6073
Island, The	Owyhee	42°28'37"N	116°03'11"W	5665
Italian Peak	Boundary	48°57'22"N	116°37'37"W	6083
Italian Peak	Lemhi	44°21'37"N	112°51'20"W	10998
Jack Mountain	Idaho	45°37'37"N	115°22'51"W	6485
Jackass Butte	Owyhee	43°05'08"N	116°12'24"W	2852
Jackknife Peak	Kootenai	47°53'18"N	116°26'37"W	5242
Jackley Mountain	Adams	45°14'56"N	116°31'13"W	8747
Jackson Mountain	Clearwater	46°49'18"N	116°16'48"W	4685
Jackson Peak	Boise	44°04'41"N	115°24'47"W	8124
Jackson Peak	Boise	44°00'35"N	115°46'31"W	7344
Jacksons Knob	Shoshone	47°08'53"N	115°49'58"W	5700
Jakes Mountain	Bonner	48°07'46"N	116°15'30"W	4657
James Creek Summit	Elmore	43°45'25"N	115°14'45"W	7802
Janies Nipples	Power	42°42'06"N	113°02'52"W	4625
Jasper Mountain	Bonner	48°19'59"N	116°53'09"W	3940
Jay Peak	Boundary	48°30'51"N	116°20'31"W	4580
Jay Point	Idaho	46°29'54"N	114°44'44"W	5438
Jeanette Mountain	Idaho	46°17'46"N	114°34'05"W	7488

Mountain, Peak, or Summit	County	Latitude	Longitude	Elev.
Jefferson, Mount	Fremont	44°33'43"N	111°30'13"W	10203
Jericho Mountain	Clearwater	46°43'26"N	116°07'08"W	4325
Jerry Peak	Custer	44°03'37"N	114°06'24"W	10010
Jersey Mountain	Idaho	45°29'13"N	115°34'13"W	6867
Jeru Park	Boundary	48°31'23"N	116°40'25"W	6371
Jims Peak	Owyhee	42°45'57"N	116°58'00"W	6810
Joe Peak	Boundary	48°52'43"N	116°46'35"W	6748
Joe Point	Idaho	45°47'15"N	116°30'15"W	4601
Joes Butte	Gem	44°11'15"N	116°18'32"W	3420
Joes Mountain	Cassia	42°04'36"N	114°07'31"W	7700
John Day Mountain	Idaho	45°31'47"N	116°14'08"W	7424
John Lewis Mountain	Clearwater	46°41'39"N	115°56'44"W	4217
John Peak	Kootenai	47°41'43"N	116°24'04"W	4211
John Point	Benewah	47°10'53"N	116°41'38"W	3690
Johnagan Mountain	Clearwater	46°31'28"N	115°23'02"W	4387
Johnny Butte	Clearwater	46°35'16"N	115°25'38"W	4357
Johnny Long Mountain	Bonner	48°17'07"N	116°44'15"W	4668
Johns Mountain	Shoshone	47°39'33"N	115°46'01"W	6155
Johnson Butte	Idaho	45°28'59"N	115°49'53"W	5097
Johnson Butte	Lincoln	43°06'59"N	114°12'25"W	4867
Johnson Hill	Camas	43°17'13"N	114°56'02"W	5652
Johnson Hill	Camas	43°14'48"N	114°39'47"W	5678
Johnson Peak	Bonner	48°06'10"N	116°15'47"W	4478
Johnstone Peak	Blaine	43°43'02"N	114°13'55"W	9949
Joker Peak	Shoshone	46°57'59"N	115°43'40"W	5710
Jones Butte	Fremont	44°18'52"N	111°44'36"W	6722
Jordan, Mount	Custer	44°28'12"N	114°46'45"W	10063
Jug Rock	Shoshone	47°00'36"N	115°48'49"W	6580
Jughandle Mountain	Valley	44°50'07"N	115°57'23"W	8052
Jumbo Mountain	Camas	43°37'05"N	115°01'56"W	8216
Jumpoff Hill	Bonneville	43°18'23"N	111°39'06"W	6978
Jumpoff Peak	Butte	43°45'17"N	113°06'23"W	9045
Junction Peak	Lemhi	44°30'54"N	113°30'05"W	10608
Junction Peak	Shoshone	47°05'58"N	115°30'51"W	6288
June Grass Table	Owyhee	42°30'18"N	116°41'56"W	5400
Jungle Point	Clearwater	46°49'58"N	115°03'19"W	5210
Jungle Point	Idaho	45°44'14"N	116°03'36"W	5851
Juniper Butte	Owyhee	42°15'35"N	115°20'36"W	5475
Juniper Buttes	Fremont	44°01'51"N	111°53'07"W	5592
Juniper Mountain	Owyhee	42°26'39"N	116°51'10"W	6761
Jureano Mountain	Lemhi	45°12'14"N	114°14'33"W	8318
K Mountain	Lemhi	45°00'21"N	113°45'25"W	8063
Kaniksu Mountain	Boundary	48°59'35"N	116°55'02"W	5814
Katka Peak	Boundary	48°37'58"N	116°09'44"W	6208
Kellogg Peak	Shoshone	47°29'19"N	116°07'39"W	6297
Kelly Mountain	Blaine	43°29'33"N	114°27'59"W	8827
Kelly Mountain	Bonneville	43°37'25"N	111°36'17"W	6805
Kelly Mountain	Idaho	45°25'35"N	116°06'41"W	4810

Mountain, Peak, or Summit	County	Latitude	Longitude	Elev.
Kelly Mountain	Kootenai	47°42'22"N	116°38'30"W	4085
Kelly Mountain	Washington	44°22'01"N	117°02'30"W	4099
Kelly Pinnacle	Shoshone	47°21'54"N	115°37'43"W	5525
Kellys Sister	Clearwater	46°43'12"N	114°52'26"W	6822
Kennally Creek Summit	Valley	44°50'33"N	115°53'48"W	7940
Kent Peak	Blaine	43°53'45"N	114°24'15"W	11664
Kent Peak	Boundary	48°44'04"N	116°40'56"W	7243
Kents Peak	Oneida	42°19'07"N	112°18'46"W	8451
Keokee Mountain	Bonner	48°25'24"N	116°39'40"W	6448
Kepros Mountain	Ada	43°33'36"N	115°56'19"W	5428
Kern Butte	Kootenai	47°39'16"N	116°40'40"W	3200
Kettle Butte	Bonneville	43°35'34"N	112°22'42"W	5571
Keystone Mountain	Custer	44°25'55"N	114°20'38"W	9518
Killarney Mountain	Kootenai	47°35'24"N	116°32'49"W	3910
Kimama Butte	Minidoka	42°45'49"N	113°53'46"W	5074
King Hill	Adams	44°31'14"N	116°20'01"W	4299
King Mountain	Butte	43°46'42"N	113°16'23"W	10612
King Mountain	Lemhi	44°49'53"N	114°02'56"W	7956
Kings Crown	Elmore	43°02'31"N	115°13'52"W	3614
Kings Peak	Benewah	47°18'18"N	116°36'16"W	3500
Kings Point	Shoshone	47°38'26"N	115°57'02"W	4411
Kinport Peak	Bannock	42°48'28"N	112°29'14"W	7222
Kinzie Butte	Lincoln	43°05'17"N	114°19'46"W	4945
Kirby Mountain	Bonner	48°14'35"N	116°20'36"W	4078
Klootch Mountain	Boundary	48°43'12"N	116°45'22"W	6048
Knob Mountain	Clark	44°28'55"N	112°35'01"W	9762
Knob, The	Camas	43°13'01"N	115°00'22"W	6614
Knob, The	Camas	43°13'00"N	115°00'24"W	6614
Knob, The	Idaho	45°48'56"N	116°17'45"W	3460
Knoll	Clark	44°32'15"N	112°20'27"W	7464
Knolls, The	Cassia	42°21'31"N	113°53'21"W	4489
Kootenai Peak	Kootenai	47°25'12"N	116°27'13"W	5054
Kootenai Point	Boundary	48°42'08"N	116°29'26"W	5003
Krall Mountain	Elmore	43°34'01"N	115°41'50"W	5318
Krassell Knob	Valley	44°58'09"N	115°45'13"W	5839
Kuna Butte	Ada	43°26'44"N	116°26'48"W	3236
Lafoe Mountain	Bonner	48°21'33"N	116°03'32"W	6173
Laidlaw Butte	Blaine	43°12'58"N	113°39'53"W	5355
Lake Mountain	Boundary	48°51'31"N	116°41'28"W	6938
Lake Mountain	Lemhi	45°01'14"N	114°05'02"W	9274
Lake Mountain	Valley	44°43'55"N	115°24'23"W	8928
Lake Rock	Idaho	45°12'11"N	115°55'02"W	8001
Lakeview Mountain	Bonner	48°36'21"N	116°56'10"W	4074
Lamb Peak	Kootenai	47°27'21"N	116°41'40"W	3632
Lamb Peak	Shoshone	47°49'23"N	116°18'19"W	4598
Lambert Table	Owyhee	42°18'37"N	116°45'12"W	5624
Landmark Peak	Shoshone	47°05'25"N	115°41'27"W	6239
Langdon Point	Idaho	45°37'44"N	114°43'50"W	6347

Mountains, Peaks, Summits

Mountain, Peak, or Summit	County	Latitude	Longitude	Elev.
Langer Peak	Custer	44°29'25"N	115°07'15"W	9315
Larch Butte	Clearwater	46°32'22"N	115°31'36"W	5539
Larch Mountain	Kootenai	47°57'58"N	116°57'34"W	4624
Larch Mountain	Shoshone	48°01'01"N	116°16'22"W	6306
Larkins Peak	Shoshone	46°56'53"N	115°37'08"W	6661
Latour Baldy	Kootenai	47°28'12"N	116°20'45"W	6232
Latour Peak	Kootenai	47°26'02"N	116°21'39"W	6408
Lava Butte	Blaine	43°12'20"N	113°30'19"W	5079
Lava Butte	Idaho	45°17'10"N	116°07'22"W	8328
Lava Mountain	Elmore	43°35'00"N	115°33'11"W	7882
Lbex Peak	Cassia	42°06'05"N	114°04'49"W	7272
Lead Mountain	Owyhee	43°00'35"N	116°37'41"W	6535
Leadore Hill	Lemhi	44°35'18"N	113°22'51"W	8453
Lean-to Point	Clearwater	46°31'21"N	115°26'51"W	5078
Leatherman Peak	Custer	44°04'56"N	113°43'55"W	12228
Lee Peak	Custer	44°06'10"N	114°37'40"W	11342
Lehman Butte	Custer	43°57'47"N	113°50'07"W	7171
Leiberg Peak	Kootenai	47°46'26"N	116°20'20"W	4700
Lem Peak	Lemhi	44°46'50"N	113°51'56"W	10985
Lemhi Point	Idaho	45°26'03"N	115°23'48"W	7362
Lemonade Peak	Shoshone	47°23'04"N	116°11'10"W	5651
Leonia Knob	Boundary	48°37'30"N	116°05'04"W	3820
Leslie Butte	Custer	43°50'25"N	113°27'49"W	6478
Lewiston Hill	Nez Perce	46°27'49"N	116°58'33"W	2571
Liars Peak	Bonneville	43°37'09"N	111°21'43"W	8689
Liberal Mountain	Camas	43°28'39"N	114°41'07"W	8272
Liberty Butte	Benewah	47°09'58"N	117°00'12"W	3671
Liberty Butte	Canyon	43°29'55"N	116°45'11"W	2662
Lick Creek Point	Shoshone	47°09'59"N	115°51'30"W	4210
Lick Point	Idaho	45°57'37"N	115°27'57"W	5260
Lightning Creek Rocks	Boise	44°13'15"N	115°47'52"W	4829
Lightning Mountain	Bonner	48°17'21"N	116°05'54"W	6116
Lightning Peak	Bonner	48°24'41"N	116°20'02"W	3060
Lightning Peak	Valley	44°56'49"N	115°07'38"W	8528
Lightning Points	Clearwater	46°44'42"N	115°54'50"W	4206
Lime Mountain	Custer	43°50'48"N	113°44'50"W	11179
Limerock Mountain	Caribou	43°00'40"N	111°36'17"W	7475
Lincoln Peak	Bingham	43°05'10"N	112°03'37"W	6541
Lindsey Mountain	Bonner	48°14'25"N	116°48'15"W	3895
Lindstrom Peak	Benewah	47°14'49"N	116°33'11"W	4695
Line Point	Boundary	48°45'56"N	116°02'54"W	6673
Linfor Hill	Shoshone	47°36'05"N	116°11'40"W	4344
Lionhead	Fremont	44°43'43"N	111°19'23"W	9974
Lions Head, The	Boundary	48°47'11"N	116°42'55"W	7288
Little Bald Mountain	Latah	47°01'01"N	116°34'40"W	4866
Little Baldy	Idaho	45°56'12"N	115°40'57"W	5363
Little Baldy	Shoshone	47°41'36"N	115°50'23"W	5558
Little Baldy	Valley	44°41'22"N	115°25'54"W	8597
Little Baldy Mountain	Bonneville	43°25'26"N	111°13'01"W	8462
Little Blacktail Mountain	Bonner	48°05'40"N	116°33'10"W	4467
Little Blue Table	Owyhee	42°18'59"N	116°01'24"W	5900
Little Butte	Bingham	43°31'42"N	112°34'13"W	5533
Little Butte	Fremont	44°15'49"N	111°31'04"W	6796
Little Butte	Gem	43°56'06"N	116°27'41"W	3493
Little Buttes	Jefferson	43°43'18"N	111°56'33"W	4964
Little Copper Butte	Idaho	45°58'53"N	115°09'30"W	7181
Little Crater	Caribou	42°50'56"N	111°37'51"W	6540
Little Elk Mountain	Bonneville	43°12'52"N	111°19'16"W	8780
Little Fiddler	Elmore	43°29'00"N	115°43'12"W	5849
Little Grass Mountain	Bonner	48°45'36"N	117°01'26"W	5696
Little Grassy Butte	Fremont	43°59'16"N	112°05'23"W	5128
Little Grassy Hill	Owyhee	42°08'32"N	115°05'31"W	5858
Little Green Mountain	Clearwater	46°43'59"N	116°05'36"W	4741
Little Green Mountain	Idaho	45°46'50"N	115°05'30"W	6980
Little Guard Peak	Shoshone	47°47'54"N	116°00'18"W	6031
Little Haystack	Shoshone	47°46'44"N	115°56'56"W	3792
Little Horse Butte	Bonneville	43°23'28"N	111°42'14"W	6470
Little Joe Butte	Ada	43°09'43"N	116°00'32"W	3214
Little Joe Mountain	Shoshone	47°09'19"N	115°14'50"W	7052
Little Mountain	Franklin	42°10'07"N	111°57'02"W	5371
Little Mountain	Franklin	42°01'38"N	111°49'30"W	5228
Little Palisades Peak	Bonneville	43°26'18"N	111°05'02"W	9707
Little Pyramid	Idaho	45°46'44"N	114°48'28"W	6576
Little Round Top	Idaho	45°36'54"N	115°58'59"W	7495
Little Sand Mountain	Latah	46°55'52"N	116°39'04"W	3490
Little Sentinel Peak	Shoshone	47°50'01"N	116°02'21"W	6134
Little Sheep Peak	Adams	44°55'22"N	116°47'56"W	4631
Little Sheepeater Point	Idaho	45°23'38"N	115°23'55"W	8205
Little Sister Peak	Shoshone	47°11'30"N	115°40'07"W	6140
Little Snowy Top	Boundary	48°58'25"N	117°00'07"W	6829
Little Soldier Mountain	Custer	44°42'34"N	115°05'24"W	8813
Little Table Mountain	Lemhi	44°31'26"N	112°06'28"W	8680
Little Table Mountain	Power	42°37'38"N	112°58'24"W	5042
Little Tug	Twin Falls	42°00'42"N	114°50'26"W	6813
Little Weitas Butte	Idaho	46°26'52"N	115°25'44"W	5705
Little Wildhorse Butte	Blaine	43°16'46"N	113°00'09"W	5438
Liz Butte	Idaho	46°26'54"N	115°19'53"W	6098
Lizard Butte	Canyon	43°33'29"N	116°47'22"W	2634
Lizard Peak	Idaho	46°13'44"N	115°08'22"W	7104
Lochsa Peak	Idaho	46°18'37"N	115°20'02"W	4885
Lockman Butte	Elmore	43°12'51"N	115°43'03"W	3789
Lodge Point	Idaho	46°06'49"N	115°40'14"W	4761
Lodgepole Hump	Idaho	45°41'46"N	114°54'02"W	7246
Lodgepole Point	Idaho	45°22'03"N	115°09'53"W	6395
Loening, Mount	Custer	44°27'02"N	114°54'05"W	10012
Log Mountain	Valley	44°49'17"N	115°36'13"W	9179

Mountain, Peak, or Summit	County	Latitude	Longitude	Elev.
Logan Mountain	Valley	45°05'43"N	115°22'20"W	9002
Lone Butte	Clark	44°14'03"N	112°18'05"W	5406
Lone Fir Hill	Idaho	45°47'55"N	116°03'09"W	5450
Lone Jack Mountain	Latah	47°01'17"N	116°58'05"W	3853
Lone Knob	Idaho	46°20'20"N	115°16'06"W	5443
Lone Pine Peak	Custer	44°20'43"N	114°10'08"W	9658
Lone Tree	Adams	44°39'22"N	116°10'15"W	7836
Lone Tree Peak	Boundary	48°56'37"N	116°42'33"W	6732
Lonely Mountain	Idaho	45°44'57"N	114°51'30"W	6840
Long Butte	Owyhee	42°33'17"N	115°34'11"W	4130
Long Hike Peak	Shoshone	47°04'13"N	115°54'25"W	6216
Long Hike Rock	Shoshone	47°04'09"N	115°55'18"W	6498
Long Lake Point	Idaho	46°20'45"N	115°09'41"W	6888
Long Liz Point	Shoshone	47°20'21"N	115°40'48"W	4811
Long Mountain	Bonner	48°22'53"N	116°03'17"W	6102
Long Mountain	Bonner	48°04'36"N	116°42'26"W	4558
Long Mountain	Boundary	48°50'05"N	116°36'53"W	7265
Long Mountain	Lemhi	44°43'33"N	113°49'44"W	10696
Long Tom Mountain	Lemhi	45°20'53"N	114°34'37"W	8255
Looking Glass Butte	Idaho	45°28'02"N	116°03'35"W	6465
Lookout Butte	Fremont	44°12'54"N	111°26'13"W	6269
Lookout Butte	Idaho	46°02'46"N	115°39'28"W	5869
Lookout Butte	Owyhee	42°31'34"N	115°31'56"W	4401
Lookout Butte	Owyhee	42°07'51"N	116°31'57"W	5376
Lookout Mountain	Blaine	43°26'48"N	114°12'54"W	7539
Lookout Mountain	Boundary	48°46'22"N	116°46'06"W	6727
Lookout Mountain	Custer	44°11'34"N	114°45'28"W	9954
Lookout Mountain	Madison	43°38'09"N	111°34'01"W	6711
Lookout Mountain	Oneida	42°11'14"N	112°43'12"W	6080
Lookout Mountain	Shoshone	47°04'58"N	115°57'56"W	6789
Lookout Mountain	Valley	45°01'59"N	115°04'17"W	8680
Lookout Peak	Clearwater	46°35'56"N	115°12'56"W	6876
Lookout Peak	Valley	44°31'12"N	116°09'43"W	7813
Lookout Point	Camas	43°15'54"N	114°48'00"W	5365
Lookout Point	Clark	44°26'15"N	112°02'49"W	8716
Loon Creek Summit	Custer	44°27'53"N	114°44'01"W	8687
Lost Packer Peak	Idaho	45°28'48"N	114°47'37"W	8712
Lost Peak	Bonner	48°28'14"N	116°43'10"W	6285
Lost Peak	Shoshone	47°46'05"N	115°50'04"W	5940
Lower Shoepack Point	Shoshone	47°11'06"N	115°42'50"W	4780
Lucky Peak	Ada	43°36'19"N	116°03'37"W	5904
Lunch Peak	Bonner	48°22'30"N	116°11'35"W	6414
Lunde Peak	Clearwater	46°37'21"N	115°02'46"W	6367
Lupine Mountain	Custer	43°47'50"N	113°44'57"W	9554
Mabelle Hill	Blaine	43°23'07"N	114°17'47"W	5568
Mackay Peak	Custer	43°53'22"N	113°42'00"W	10273
Madden Butte	Canyon	43°29'05"N	116°32'58"W	2803
Magee Peak	Shoshone	47°52'57"N	116°17'42"W	4856

Mountain, Peak, or Summit	County	Latitude	Longitude	Elev.
Maggie Butte	Idaho	46°13'29"N	115°47'14"W	4355
Magruder Mountain	Idaho	45°42'25"N	114°49'20"W	7421
Mahogany Butte	Cassia	42°02'35"N	114°08'48"W	7200
Mahogany Butte	Clark	44°00'08"N	112°55'06"W	7006
Mahogany Hill	Custer	44°25'04"N	113°51'40"W	8468
Mahoney Butte	Blaine	43°35'49"N	114°27'12"W	7904
Malad Summit	Oneida	42°20'48"N	112°13'27"W	5615
Malamute Point	Shoshone	47°10'41"N	115°57'14"W	5045
Malin Point	Shoshone	47°15'22"N	115°33'15"W	5145
Mallard Peak	Shoshone	46°56'17"N	115°31'27"W	6870
Mammoth Mountain	Idaho	45°27'52"N	115°28'46"W	6600
Manning, Mount	Teton	43°42'10"N	111°17'09"W	7705
Maple Peak	Shoshone	47°37'58"N	115°45'33"W	5758
Marble Butte	Idaho	45°32'27"N	115°50'54"W	8079
Marble Mountain	Shoshone	47°09'03"N	116°01'53"W	6501
Marble Mountain	Valley	45°05'56"N	115°17'27"W	9128
Marcus Cook Peak	Shoshone	47°21'42"N	115°50'40"W	5875
Margaret, Mount	Latah	46°54'11"N	116°40'42"W	3765
Maria, Mount	Boise	43°57'26"N	116°11'13"W	3472
Marion, Mount	Bonner	48°16'28"N	116°18'43"W	5209
Marks Butte	Shoshone	47°01'33"N	116°03'11"W	6297
Marsh Hill	Benewah	47°03'11"N	116°51'26"W	3260
Marshall Butte	Owyhee	42°15'42"N	115°01'42"W	5508
Marshall Mountain	Idaho	45°22'28"N	115°50'44"W	8443
Marten Hill	Clearwater	46°33'24"N	115°08'35"W	5762
Marten Hill	Idaho	45°33'17"N	115°58'21"W	7479
Martin Mountain	Lemhi	44°50'07"N	114°40'58"W	9423
Mason Butte	Kootenai	47°23'00"N	116°51'45"W	3403
Mason Butte	Latah	46°38'01"N	116°21'35"W	3745
Mason Butte	Lewis	46°11'47"N	116°33'39"W	4639
Massacre Mountain	Custer	44°03'37"N	113°32'46"W	10924
Massacre Rocks	Power	42°40'26"N	112°59'08"W	4350
Mastodon Mountain	Shoshone	47°21'24"N	115°58'05"W	5905
Maternity Hill	Shoshone	47°15'36"N	116°07'21"W	3300
Mattingly Peak	Elmore	43°53'32"N	115°00'11"W	9921
Maud Mountain	Clark	44°13'01"N	112°37'51"W	6277
Maxim Ridge	Idaho	45°30'23"N	116°23'00"W	6052
Maxwell Point	Idaho	45°26'41"N	115°48'41"W	6190
May Mountain	Lemhi	44°41'12"N	113°49'56"W	10971
Mayfield Peak	Custer	44°31'40"N	114°44'19"W	9846
McCaleb, Mount	Custer	43°59'32"N	113°35'50"W	11592
McCartney Butte	Kootenai	47°22'21"N	116°55'42"W	3902
McCatron Ridge	Idaho	45°29'06"N	116°31'04"W	5000
McChord Butte	Washington	44°32'16"N	117°04'48"W	5748
McConnell Mountain	Idaho	46°21'31"N	114°54'54"W	7424
McDonald Peak	Blaine	43°56'47"N	114°54'20"W	10068
McDonald Peak	Shoshone	47°45'38"N	116°16'19"W	4576
McEleny Mountain	Lemhi	45°01'44"N	114°32'37"W	8938

Mountains, Peaks, Summits

Mountain, Peak, or Summit	County	Latitude	Longitude	Elev.
McElroy Butte	Canyon	43°23'53"N	116°33'13"W	2801
McGary Butte	Latah	46°46'50"N	116°21'27"W	4457
McGowan Butte	Kootenai	47°29'54"N	117°02'18"W	2950
McGown Peak	Custer	44°12'48"N	115°04'48"W	9860
McGuire, Mount	Lemhi	45°10'27"N	114°36'04"W	10082
McIntyre Hill	Canyon	43°43'19"N	116°36'57"W	2481
McKinney Butte	Gooding	43°01'13"N	114°49'08"W	3942
McLead Hill	Shoshone	47°33'26"N	116°12'33"W	3293
McLendon Butte	Idaho	46°18'28"N	115°25'12"W	5562
Meade Peak	Bear Lake	42°29'49"N	111°14'53"W	9957
Meadow Creek Mountain	Bingham	43°01'58"N	111°39'25"W	7425
Meadow Hill	Adams	44°55'52"N	116°12'25"W	5305
Meadow Peak	Custer	44°27'10"N	113°57'55"W	9099
Meadow Point	Clearwater	46°55'03"N	115°08'24"W	5536
Meadows Summit	Adams	44°56'37"N	116°12'05"W	5094
Medbury Hill	Elmore	42°57'46"N	115°31'49"W	2800
Medicine Mountain	Kootenai	47°28'43"N	116°36'02"W	2452
Menan Buttes	Madison	43°47'04"N	111°58'20"W	5227
Meridian Peak	Custer	43°56'45"N	114°19'07"W	10285
Merkley Mountain	Bear Lake	42°08'58"N	111°15'05"W	7422
Merriam Peak	Custer	44°03'08"N	114°34'49"W	10920
Mex Mountain	Idaho	46°19'38"N	115°35'59"W	5060
Mica Hill	Adams	44°39'11"N	116°14'21"W	5321
Mica Mountain	Latah	46°54'24"N	116°33'58"W	4900
Mica Peak	Kootenai	47°37'20"N	116°59'13"W	5241
Micky Point	Clearwater	46°48'07"N	115°44'14"W	5249
Middle Butte	Bingham	43°29'55"N	112°44'41"W	6391
Middle Butte	Idaho	46°18'03"N	115°28'11"W	5380
Middle Butte	Owyhee	42°10'56"N	115°22'10"W	5590
Middle Creek Butte	Clark	44°24'32"N	112°31'09"W	7966
Middle Fork Peak	Lemhi	44°57'43"N	114°39'19"W	9127
Middle Mountain	Adams	45°14'07"N	116°31'57"W	8437
Middle Mountain	Boundary	48°30'53"N	116°09'39"W	6220
Middle Mountain	Custer	43°45'24"N	113°43'57"W	8719
Middle Ridge	Idaho	45°23'16"N	116°31'19"W	6580
Middle Sister Peak	Shoshone	47°10'05"N	115°39'11"W	6898
Midget Peak	Shoshone	47°07'50"N	115°21'50"W	5765
Midnight Mountain	Bear Lake	42°16'13"N	111°34'18"W	9328
Midvale Hill	Washington	44°26'15"N	116°48'25"W	3650
Mill Creek Summit	Custer	44°28'19"N	114°29'18"W	8800
Mill Creek Summit	Gem	44°27'21"N	116°16'59"W	4780
Mill Mountain	Lemhi	44°41'03"N	113°39'24"W	10793
Miller Mountain	Boise	44°09'26"N	115°29'30"W	8088
Miller Peak	Custer	43°42'59"N	113°41'24"W	8610
Miller Peak	Shoshone	47°02'41"N	115°46'11"W	6060
Miller Water Table	Owyhee	42°39'01"N	115°44'08"W	3732
Mills, Mount	Custer	44°31'13"N	115°09'36"W	9185
Milner Butte	Cassia	42°27'28"N	114°01'06"W	4621

Mountain, Peak, or Summit	County	Latitude	Longitude	Elev.
Mineral Hill	Idaho	45°35'22"N	115°39'54"W	8500
Mineral Hill	Lemhi	44°44'01"N	113°24'09"W	6836
Mineral Mountain	Boise	43°59'59"N	115°53'44"W	5940
Mineral Mountain	Latah	47°04'01"N	116°52'55"W	4128
Miners Peak	Valley	44°52'07"N	115°46'22"W	7810
Minerva Peak	Bonner	48°04'09"N	116°23'28"W	4447
Mink Peak	Idaho	46°02'21"N	115°03'02"W	7054
Minkler Mountain	Kootenai	47°22'24"N	116°57'55"W	3398
Mirror Peak	Shoshone	47°16'51"N	115°35'06"W	4466
Mission Hill	Boundary	48°42'45"N	116°22'07"W	1968
Mission Mountain	Boundary	48°59'40"N	116°17'50"W	6206
Mission Mountain	Latah	47°04'03"N	116°56'34"W	4324
Moccasin Mountain	Idaho	45°25'32"N	115°33'03"W	6639
Moccasin Peak	Idaho	46°31'51"N	115°05'05"W	6759
Mockmer Butte	Idaho	45°50'42"N	116°24'18"W	3945
Moe Peak	Idaho	46°04'58"N	114°33'26"W	8106
Mogg Mountain	Lemhi	44°40'05"N	113°42'15"W	10573
Moh, Mount	Bannock	42°36'25"N	111°59'55"W	6245
Mollie Point	Idaho	45°49'05"N	116°22'50"W	4465
Mollies Nipple	Bingham	43°08'34"N	111°37'52"W	7093
Mollys Nipple	Power	42°31'11"N	112°51'42"W	5202
Monroe Butte	Clearwater	46°32'43"N	115°12'58"W	6522
Monroe Butte	Washington	44°31'14"N	117°01'38"W	6043
Montana Peak	Shoshone	47°04'33"N	115°43'12"W	5833
Montana Point	Shoshone	47°03'53"N	115°42'07"W	4925
Monte Verita	Boise	44°04'22"N	115°00'54"W	10140
Monument Butte	Fremont	44°18'09"N	111°43'32"W	6801
Monument Butte	Lincoln	43°09'03"N	113°57'20"W	4870
Monument Butte	Owyhee	42°25'06"N	115°42'09"W	4858
Monument Hill	Idaho	45°53'13"N	115°53'30"W	4500
Monument Mountain	Kootenai	47°43'32"N	116°26'25"W	5101
Monument Peak	Adams	45°13'37"N	116°33'25"W	8957
Monument Peak	Cassia	42°07'44"N	114°13'02"W	8050
Monument Peak	Lemhi	45°15'51"N	113°41'30"W	10323
Monument Peak	Teton	43°40'57"N	111°16'05"W	7655
Monumental Peak	Valley	44°27'51"N	115°42'40"W	7901
Monumental Summit	Valley	44°20'03"N	111°30'54"W	8920
Moon Hill	Latah	46°54'13"N	116°43'17"W	3145
Moon Peak	Shoshone	47°24'27"N	115°52'21"W	5681
Moonlight Mountain	Bannock	42°54'00"N	112°17'26"W	6639
Moonshine Mountain	Fremont	44°20'03"N	111°30'54"W	6930
Moonshine Peak	Power	42°35'42"N	112°39'52"W	7210
Moonstone Mountain	Blaine	43°21'50"N	114°27'41"W	2144
Moore Point	Valley	45°05'19"N	115°25'03"W	8971
Moose Butte	Idaho	45°40'58"N	115°29'05"W	7121
Moose Creek Butte	Fremont	44°11'58"N	111°11'08"W	7045
Moose Creek Buttes	Clearwater	46°44'51"N	115°10'49"W	6924
Moose Meadows Point	Idaho	45°19'16"N	115°08'08"W	6774

Mountain, Peak, or Summit	County	Latitude	Longitude	Elev.
Moose Mountain	Bonner	48°20'47"N	116°07'24"W	6543
Moose Mountain	Clearwater	46°46'47"N	115°11'02"W	6603
Morehead Mountain	Valley	44°33'53"N	115°21'31"W	8506
Mores Creek Summit	Boise	43°55'57"N	115°40'00"W	6117
Mores Mountain	Boise	43°47'29"N	116°05'23"W	7237
Morgan Creek Summit	Custer	44°50'30"N	114°15'56"W	7578
Morgan Mountain	Lemhi	45°31'14"N	113°50'00"W	8308
Mormon Hill	Blaine	43°37'49"N	113°57'20"W	
Mormon Mountain	Valley	45°00'56"N	114°52'48"W	9545
Morning Glory Peak	Lemhi	45°11'23"N	113°57'35"W	6460
Mosby Butte	Power	43°02'18"N	113°08'06"W	5498
Moscow Mountain	Latah	46°48'13"N	116°52'04"W	4983
Moses Butte	Shoshone	47°00'54"N	115°49'07"W	6630
Moses Mountain	Benewah	47°09'22"N	116°46'30"W	4949
Mosquito Lake Butte	Owyhee	42°13'13"N	115°20'11"W	5562
Mosquito Peak	Idaho	45°16'54"N	115°21'53"W	8774
Mosquito Peak	Shoshone	47°05'51"N	115°28'23"W	6298
Mother Lode Hill	Idaho	45°48'18"N	115°25'09"W	4500
Mountain View Peak	Twin Falls	42°13'54"N	114°18'29"W	7365
Moyer Peak	Lemhi	44°58'20"N	114°13'08"W	9085
Mozier Peak	Shoshone	47°21'53"N	115°47'36"W	5500
Mud Flat Hill	Owyhee	42°04'47"N	115°06'01"W	6014
Mule Butte	Blaine	43°04'31"N	113°21'30"W	6020
Mule Creek Point	Idaho	45°26'17"N	115°00'50"W	6961
Mule Hill	Valley	44°52'07"N	115°16'37"W	8356
Mulligan Hump	Shoshone	46°59'14"N	115°38'45"W	3940
Murphy Peak	Valley	44°53'27"N	115°14'55"W	9240
Murray Peak	Shoshone	47°39'38"N	115°49'44"W	5934
Myrtle Peak	Boundary	48°44'56"N	116°38'28"W	7122
N Mountain	Lemhi	45°26'34"N	113°51'49"W	8870
N P Hill	Shoshone	47°07'00"N	115°45'41"W	5690
Nahneke Mountain	Elmore	43°53'36"N	115°07'56"W	9582
Nakarna Mountain	Benewah	47°03'44"N	116°28'52"W	4995
Napoleon Hill	Lemhi	45°20'43"N	114°00'13"W	7433
Needle Butte	Jefferson	43°58'56"N	112°23'25"W	5025
Needle Peak	Bonneville	43°19'25"N	111°03'46"W	9220
Needle Peak	Shoshone	47°01'23"N	115°16'30"W	6643
Needles	Valley	44°44'05"N	115°49'37"W	8302
Negro Brown Hill	Benewah	47°19'40"N	116°40'52"W	2780
Negro Peak	Lemhi	44°29'01"N	113°21'37"W	10571
Nellie Mountain	Idaho	45°50'08"N	115°43'31"W	6180
Nelson Peak	Shoshone	47°16'19"N	115°43'25"W	5929
Nelson Point	Idaho	45°21'00"N	115°34'52"W	6902
Neversweat Peak	Shoshone	47°01'54"N	115°13'29"W	6665
New Butte	Blaine	43°12'18"N	113°20'27"W	5072
New York Summit	Owyhee	43°02'16"N	116°43'33"W	6676
Newman Peak	Camas	43°42'22"N	114°57'01"W	9603
Newport Hill	Bonner	48°14'01"N	117°06'58"W	4128

Mountain, Peak, or Summit	County	Latitude	Longitude	Elev.
Nez Perce Peak	Idaho	45°43'54"N	114°35'05"W	7531
Nick Peak	Valley	44°56'05"N	115°52'02"W	9064
Nick Wynn Mountain	Idaho	45°46'15"N	114°45'22"W	
Nickel Creek Table	Owyhee	42°34'37"N	116°45'59"W	5700
Nickelplate Mountain	Bonner	48°38'49"N	116°58'16"W	3888
Ninemile Knoll	Fremont	43°56'29"N	111°57'44"W	4984
Nipple Knob	Idaho	45°59'41"N	114°34'08"W	8505
Nipple Mountain	Idaho	45°42'55"N	115°36'26"W	7098
No Business Mountain	Adams	44°45'56"N	116°11'57"W	7340
North Blacktail Butte	Idaho	45°54'53"N	115°57'43"W	4660
North Butte	Idaho	45°31'08"N	115°51'18"W	7178
North Butte	Shoshone	47°02'18"N	115°47'57"W	6739
North Chapin Mountain	Cassia	42°30'16"N	113°10'17"W	5556
North Cone	Caribou	42°50'08"N	111°34'47"W	6591
North Laidlaw Butte	Blaine	43°21'04"N	113°31'43"W	5912
North Loon Mountain	Valley	45°06'46"N	115°52'18"W	9322
North Peak	Idaho	46°12'57"N	115°02'23"W	7025
North Pole	Idaho	45°38'51"N	115°40'59"W	8818
North Raker	Boise	43°59'44"N	115°06'17"W	9970
North Siwash	Shoshone	47°13'08"N	115°44'18"W	5290
North Snow Peak	Shoshone	47°57'36"N	116°17'10"W	4630
North Star Butte	Adams	45°09'19"N	116°27'50"W	7452
Norton Peak	Blaine	43°45'43"N	114°39'08"W	10336
Norton Ridge Peak	Valley	44°51'10"N	114°55'53"W	8252
No-see-um Butte	Idaho	46°23'34"N	115°16'15"W	5590
Notch Butte	Lincoln	42°53'06"N	114°25'02"W	4337
Notch Butte	Owyhee	42°41'12"N	115°19'24"W	3913
Notch Butte	Owyhee	42°40'04"N	115°10'22"W	3607
Nub, The	Clearwater	46°52'23"N	115°29'26"W	6924
Nugget Hill	Shoshone	47°13'23"N	115°33'00"W	4460
Nugget Point	Idaho	45°54'27"N	115°34'24"W	
Nut Hill	Idaho	45°33'17"N	115°31'10"W	6620
Obia Point	Idaho	46°21'41"N	115°25'57"W	4588
Observation Peak	Custer	44°10'16"N	115°06'29"W	9151
Ogre, The	Idaho	45°19'10"N	116°31'55"W	9210
O'Hara Point	Idaho	46°00'13"N	115°28'50"W	5661
Ola Summit	Gem	44°11'56"N	116°14'14"W	4930
Old Baldy Peak	Bonneville	43°07'10"N	111°12'58"W	8325
Old Baldy Peak	Franklin	42°08'57"N	112°04'33"W	8356
Old Hyndman Peak	Blaine	43°44'27"N	114°06'58"W	11775
Old Stormy	Idaho	46°30'18"N	114°22'58"W	8203
Old Timer Mountain	Idaho	45°25'47"N	116°30'21"W	6319
Old Tom Mountain	Bannock	42°37'26"N	112°19'10"W	8733
Oliver Peak	Teton	43°31'11"N	111°03'19"W	9004
One Man Butte	Idaho	45°18'13"N	115°10'48"W	7300
O'Neill Hill	Benewah	47°18'13"N	116°22'08"W	3391
Opal Mountain	Clark	44°23'22"N	112°05'10"W	7219
Oregon Butte	Idaho	45°31'18"N	115°40'06"W	8464

Mountains, Peaks, Summits

Mountain, Peak, or Summit	County	Latitude	Longitude	Elev.
Oro Mountain	Valley	44°33'24"N	115°50'27"W	7759
Orogrande Summit	Idaho	45°38'23"N	115°36'48"W	7266
Osborne Butte	Fremont	44°17'26"N	111°23'57"W	6301
Otter Butte	Idaho	46°01'59"N	115°10'51"W	6088
Otter Buttes	Owyhee	42°08'10"N	116°00'59"W	6296
Otter Peak	Shoshone	46°59'52"N	115°23'13"W	6370
Outlaw Point	Shoshone	47°09'16"N	115°53'47"W	4540
Outlet Mountain	Bonner	48°28'58"N	116°53'18"W	3540
Owinza Butte	Lincoln	42°52'12"N	114°03'30"W	4435
Oxford Peak	Franklin	42°16'11"N	112°05'50"W	9282
Packer Butte	Elmore	43°17'18"N	115°19'31"W	6020
Packer Butte	Elmore	43°16'55"N	115°18'37"W	5823
Packer John Mountain	Boise	44°12'59"N	116°03'53"W	7055
Packrat Peak	Custer	44°03'37"N	115°01'21"W	9600
Packsaddle Butte	Blaine	43°13'56"N	113°04'17"W	5256
Packsaddle Mountain	Bonner	48°01'51"N	116°21'18"W	6402
Paddy Flat Summit	Valley	44°47'13"N	115°59'44"W	5900
Padrick Ridge	Idaho	45°29'04"N	116°11'06"W	6200
Paisley Cone	Butte	43°27'02"N	113°32'53"W	6107
Palisades Peak	Bonneville	43°26'30"N	111°05'16"W	9778
Palmer Butte	Latah	47°03'55"N	117°00'34"W	3100
Paloma, Mount	Idaho	45°57'32"N	114°36'22"W	8371
Papoose Mountain	Shoshone	47°02'45"N	115°31'08"W	6090
Papoose Peak	Idaho	45°12'41"N	114°48'52"W	8989
Paps Mountain	Power	42°45'29"N	112°48'38"W	5124
Parachute Hill	Idaho	46°33'14"N	114°42'11"W	5616
Paradise Peak	Camas	43°44'19"N	114°50'54"W	9798
Paradise Point	Latah	46°48'59"N	116°56'13"W	4356
Paris Peak	Bear Lake	42°12'11"N	111°32'55"W	9575
Park Butte	Blaine	43°10'09"N	113°22'11"W	5054
Parker Mountain	Lemhi	45°28'17"N	114°39'49"W	8503
Parker Mountain	Lemhi	44°36'12"N	114°32'53"W	9151
Parker Peak	Boundary	48°52'26"N	116°35'12"W	7670
Parks Peak	Blaine	43°58'02"N	114°55'55"W	10208
Parks Peak	Valley	45°02'04"N	115°31'07"W	8833
Particular Knoll	Bonneville	43°13'25"N	111°30'54"W	6613
Pass Creek Summit	Butte	44°01'25"N	113°27'18"W	7637
Patrick Butte	Idaho	45°19'04"N	116°12'26"W	8841
Patrol Point	Idaho	46°03'09"N	114°52'46"W	5862
Patrol Point	Idaho	45°40'15"N	116°07'59"W	6106
Patrol Point	Idaho	45°25'56"N	114°55'51"W	6090
Patterson Peak	Blaine	43°31'54"N	114°13'27"W	8350
Paul Peak	Bonneville	43°17'07"N	111°05'24"W	7437
Paunch Mountain	Idaho	46°17'07"N	115°53'45"W	3216
Payette Peak	Boise	43°59'32"N	114°59'12"W	10211
Pearson Peak	Benewah	47°22'48"N	116°23'43"W	6136
Peck Mountain	Adams	44°51'44"N	116°36'39"W	5148
Peggy Peak	Shoshone	47°05'09"N	115°29'09"W	6568
Pegleg Mountain	Shoshone	47°09'07"N	115°18'52"W	5739
Pence Butte	Owyhee	42°45'55"N	115°34'41"W	3714
Pence Butte	Owyhee	42°05'48"N	115°17'11"W	5696
Pend Oreille, Mount	Bonner	48°24'56"N	116°10'31"W	6755
Penny Mountain	Blaine	43°41'14"N	114°21'40"W	6137
Pepperbox Hill	Adams	45°09'14"N	116°36'58"W	7800
Peterson Point	Benewah	47°10'33"N	116°33'05"W	4161
Petit Peak	Kootenai	47°27'20"N	116°33'20"W	3584
Petticoat Peak	Bannock	42°39'22"N	111°57'25"W	8033
Pettis Peak	Benewah	47°14'09"N	116°24'37"W	4235
Pewee Peak	Bonner	48°17'58"N	116°54'20"W	3611
Phantom Hill	Blaine	43°47'12"N	114°29'42"W	6540
Phelan Mountain	Lemhi	45°08'12"N	114°04'16"W	8905
Phi Kappa Mountain	Custer	43°48'53"N	114°12'19"W	10516
Phoebe Tip	Boundary	48°52'25"N	116°50'21"W	6620
Picket Mountain	Elmore	44°05'24"N	115°11'16"W	8123
Pickle Butte	Canyon	43°32'44"N	116°47'32"W	2319
Pickles Butte	Canyon	43°29'20"N	116°42'41"W	3083
Pierson Peak	Shoshone	47°53'00"N	116°15'13"W	4640
Pigeon Roost	Nez Perce	46°16'53"N	116°56'39"W	2340
Pigtail Butte	Twin Falls	42°16'30"N	114°55'27"W	5270
Pike Mountain	Cassia	42°11'24"N	114°16'19"W	7710
Pilgrim Mountain	Valley	44°28'17"N	115°33'35"W	8196
Pillar Butte	Power	42°53'10"N	113°12'54"W	5235
Pilot Knob	Clearwater	46°43'36"N	114°38'46"W	7262
Pilot Knob	Idaho	45°54'20"N	115°42'26"W	7135
Pilot Peak	Boise	43°57'38"N	115°41'10"W	8128
Pilot Peak	Valley	45°10'26"N	115°31'35"W	8057
Pilot Rock	Idaho	45°55'08"N	115°42'56"W	6952
Pinchot Butte	Shoshone	46°59'40"N	115°57'05"W	5994
Pinchot Mountain	Boise	44°01'01"N	115°06'12"W	9502
Pine Butte	Fremont	44°15'49"N	111°48'49"W	6435
Pine Knob	Cassia	42°24'47"N	113°38'35"W	5879
Pine Knob	Idaho	46°05'19"N	115°40'51"W	4961
Pine Mountain	Bingham	43°16'25"N	111°41'15"W	7279
Pine Mountain	Blaine	43°24'50"N	113°47'30"W	7332
Pine Summit	Custer	44°30'16"N	114°24'52"W	8020
Pine Tit	Cassia	42°02'19"N	114°12'30"W	7138
Pineapple Peak	Shoshone	47°05'43"N	115°36'35"W	6090
Piney Peak	Teton	43°37'48"N	111°21'28"W	9020
Pinnacles, The	Valley	45°01'50"N	115°20'42"W	9273
Pinyon Peak	Custer	44°34'12"N	114°54'58"W	994
Pioneer Butte	Gooding	42°59'35"N	115°03'04"W	3372
Pioneer Mountains	Custer	43°40'17"N	113°54'43"W	10493
Pistol Rock	Valley	44°47'21"N	115°23'24"W	9178
Pitz Mountain	Bonner	47°59'27"N	116°54'32"W	3213
Piute Butte	Owyhee	43°18'28"N	116°52'54"W	5978
Piute Butte	Owyhee	42°12'23"N	116°33'33"W	5272

Mountain, Peak, or Summit	County	Latitude	Longitude	Elev.
Placer Peak	Shoshone	47°24'58"N	115°54'25"W	5252
Player Butte	Twin Falls	42°00'33"N	114°49'03"W	6090
Pleasant Valley Table	Owyhee	42°34'37"N	116°49'15"W	5600
Plowboy Mountain	Bonner	48°46'17"N	116°56'53"W	5140
Plummer Butte	Benewah	47°18'47"N	116°52'03"W	4145
Plummer Peak	Elmore	43°56'46"N	115°04'32"W	9978
Pocono Hill	Shoshone	47°13'31"N	116°03'37"W	5053
Point Lookout	Bonneville	43°23'52"N	111°30'39"W	7460
Point of Rock	Lemhi	45°19'30"N	114°10'47"W	8368
Poison Butte	Owyhee	42°25'12"N	115°56'21"W	5519
Poison Butte	Owyhee	42°08'57"N	115°27'11"W	5676
Poison Peak	Lemhi	44°53'29"N	113°52'11"W	9364
Poison Timber Point	Adams	44°36'20"N	116°11'56"W	7212
Poker Peak	Bonneville	43°14'35"N	111°11'33"W	8439
Polaris Peak	Shoshone	47°28'48"N	116°02'21"W	5585
Pole Creek Breaks	Owyhee	42°29'45"N	117°01'42"W	5208
Pole Creek Top	Owyhee	43°22'04"N	117°01'20"W	5147
Pole Mountain	Idaho	45°41'20"N	114°37'36"W	6588
Pole Mountain	Shoshone	46°58'49"N	115°25'54"W	6601
Pollock Hill	Clearwater	46°45'59"N	115°00'15"W	5503
Pollock Mountain	Adams	45°11'22"N	116°24'44"W	8048
Pond Peak	Shoshone	47°50'55"N	116°03'01"W	6136
Pony Peak	Shoshone	47°35'11"N	115°51'01"W	5779
Pool Knob	Idaho	45°51'42"N	116°44'45"W	2900
Porphyry Peak	Custer	43°55'51"N	113°57'14"W	10012
Porphyry Peak	Idaho	46°31'08"N	115°14'36"W	6210
Porters Mountain	Idaho	45°43'44"N	115°27'29"W	6411
Portland Mountain	Lemhi	44°27'19"N	113°19'45"W	10820
Pot Hole Butte	Owyhee	42°42'45"N	115°30'46"W	3745
Pot Mountain	Clearwater	46°44'08"N	115°24'26"W	7139
Potaman Peak	Custer	44°13'14"N	114°23'45"W	9376
Potato Hill	Idaho	46°04'06"N	115°46'26"W	4090
Potato Hill	Idaho	45°18'32"N	116°34'42"W	8122
Potato Hill	Latah	46°48'55"N	116°33'27"W	4017
Potato Knob	Adams	44°32'52"N	116°15'40"W	5888
Potato Mountain	Custer	44°18'33"N	114°55'30"W	8163
Potlatch Hill	Kootenai	47°39'53"N	116°44'34"W	2626
Potter Butte	Blaine	43°13'56"N	113°33'09"W	5032
Power Mountain	Bonner	48°00'06"N	116°17'30"W	5394
Powers Butte	Canyon	43°25'37"N	116°31'26"W	2910
Prater Mountain	Bonner	48°18'06"N	116°48'14"W	4594
Pratt Butte	Butte	43°19'08"N	113°17'15"W	5434
Preacher Mountain	Idaho	45°25'42"N	116°20'46"W	4645
Preston Knob	Benewah	47°03'19"N	116°34'08"W	4140
Prichard Peak	Shoshone	47°39'49"N	115°55'36"W	4424
Prince Peak	Shoshone	47°18'32"N	115°59'34"W	5306
Proctor Mountain	Blaine	43°42'15"N	114°18'30"W	7798
Profile Peak	Valley	45°04'02"N	115°28'02"W	8965

Mountain, Peak, or Summit	County	Latitude	Longitude	Elev.
Prospect Mountain	Blaine	43°41'35"N	114°19'47"W	6758
Prospect Mountain	Shoshone	47°20'12"N	115°55'04"W	3895
Prospect Peak	Bonner	47°54'30"N	116°29'47"W	4820
Prospect Peak	Gem	43°50'48"N	116°18'11"W	4874
Prospect Peak	Latah	47°01'27"N	116°48'00"W	4138
Prospect Peak	Madison	43°39'20"N	111°22'29"W	8020
Prospect Point	Idaho	45°56'42"N	115°11'18"W	5940
Proux Mountain	Idaho	45°46'52"N	115°37'25"W	5380
Puddin Mountain	Lemhi	45°06'32"N	114°37'10"W	9684
Pulaski, Mount	Shoshone	47°27'40"N	115°57'58"W	5480
Pungo Mountain	Valley	44°47'07"N	115°02'05"W	8295
Purdy Mountain	Bonner	48°28'01"N	116°10'37"W	6062
Purple Butte	Blaine	43°15'46"N	113°26'02"W	5332
Putnam, Mount	Bannock	42°57'11"N	112°09'51"W	8810
Pyle Creek Point	Boise	44°08'54"N	115°52'45"W	5280
Pyramid Mountain	Idaho	45°46'26"N	114°50'22"W	7768
Pyramid Mountain	Idaho	45°17'15"N	116°34'15"W	8650
Pyramid Peak	Adams	45°10'53"N	116°33'51"W	8380
Pyramid Peak	Boundary	48°48'43"N	116°37'14"W	7355
Pyramid Peak	Custer	43°47'47"N	113°59'26"W	11628
Pyramid Peak	Idaho	45°38'00"N	115°55'55"W	8369
Pyramid Peak	Lemhi	45°27'19"N	113°47'04"W	9616
Pyramid Point	Valley	44°55'55"N	115°14'26"W	9021
Pyramid Point	Washington	44°46'55"N	116°37'32"W	6524
Quaking Asp Mountain	Oneida	42°15'04"N	112°51'57"W	7230
Quaking Aspen Butte	Butte	43°23'26"N	113°12'20"W	5864
Quarles Peak	Shoshone	47°17'57"N	115°31'43"W	6562
Quartz Mountain	Bonner	48°19'19"N	116°56'50"W	4091
Quartzite Butte	Idaho	45°30'54"N	115°44'33"W	8371
Quartzite Mountain	Lemhi	45°03'36"N	114°26'46"W	8724
Queen Mountain	Boundary	48°52'41"N	116°12'50"W	6112
Queens Crown	Blaine	43°17'24"N	113°59'16"W	5590
Quicksilver Mountain	Owyhee	42°56'52"N	116°37'56"W	8082
Rabbit Creek Summit	Boise	43°48'11"N	115°44'01"W	6460
Rabbit Mountain	Caribou	42°40'08"N	111°33'18"W	6642
Rabbit Point	Idaho	45°30'02"N	115°23'01"W	6335
Ragged Mountain	Kootenai	47°51'57"N	117°03'10"W	4893
Rainbow Peak	Valley	44°57'46"N	115°13'38"W	9325
Rainey Point	Idaho	45°36'10"N	115°08'49"W	5540
Rainy Day Point	Idaho	45°45'48"N	115°42'01"W	5959
Rainy Hill	Kootenai	47°28'33"N	116°35'14"W	2210
Rakers, The	Boise	43°59'40"N	115°06'15"W	9998
Ramey Hill	Valley	44°49'51"N	114°51'59"W	8120
Ramsey Mountain	Lemhi	44°52'23"N	113°34'50"W	8171
Ramshorn Mountain	Custer	44°25'31"N	114°21'23"W	9895
Randall Butte	Latah	46°52'24"N	117°01'13"W	3316
Ranger Peak	Idaho	46°30'08"N	114°23'56"W	8817
Rapid Peak	Valley	44°51'27"N	115°54'03"W	8264

Mountains, Peaks, Summits

Mountain, Peak, or Summit	County	Latitude	Longitude	Elev.
Raspberry Butte	Clearwater	46°35'53"N	115°07'48"W	6374
Rathdrum Mountain	Kootenai	47°50'49"N	116°55'35"W	5003
Rattle Mountain	Bonner	48°19'12"N	116°04'57"W	6079
Rattler Butte	Blaine	43°26'26"N	114°18'02"W	6265
Rattlesnake Butte	Blaine	43°15'15"N	113°15'08"W	5229
Rattlesnake Butte	Blaine	43°15'15"N	114°20'42"W	5365
Rattlesnake Butte	Owyhee	43°33'23"N	116°59'22"W	2630
Rattlesnake Butte	Power	42°45'06"N	113°09'36"W	4563
Rattlesnake Mountain	Elmore	43°37'24"N	115°34'47"W	8177
Rattlesnake Peak	Bannock	42°29'45"N	112°19'33"W	5915
Rattlesnake Peak	Idaho	45°12'48"N	115°29'02"W	7791
Rattlesnake Point	Clark	44°03'42"N	112°42'26"W	6648
Rattlesnake Point	Idaho	45°33'20"N	115°06'29"W	6002
Raumaker Butte	Jefferson	43°57'34"N	112°07'50"W	5039
Reas Peak	Fremont	44°32'09"N	111°30'44"W	9371
Red Butte	Bonneville	43°24'10"N	111°44'23"W	6203
Red Butte	Custer	44°38'18"N	114°19'52"W	8844
Red Butte	Madison	43°39'38"N	111°20'22"W	8108
Red Conglomerate Peaks	Clark	44°29'17"N	112°35'50"W	10250
Red Devil Mountain	Blaine	43°32'23"N	114°18'03"W	6603
Red Hill	Bannock	42°51'40"N	112°25'20"W	4788
Red Hill	Custer	44°28'09"N	114°04'12"W	7335
Red Horse Mountain	Kootenai	47°32'57"N	116°38'00"W	4398
Red Ives Peak	Shoshone	47°03'23"N	115°17'49"W	6368
Red Knoll	Oneida	42°07'15"N	112°09'26"W	6813
Red Mountain	Boise	44°15'04"N	115°24'17"W	
Red Mountain	Boise	42°28'30"N	111°07'33"W	8722
Red Mountain	Caribou	42°49'34"N	111°39'47"W	6817
Red Mountain	Custer	44°24'19"N	114°51'48"W	9387
Red Mountain	Elmore	43°32'08"N	115°24'49"W	6975
Red Mountain	Owyhee	42°58'07"N	116°35'33"W	6341
Red Mountain	Teton	43°37'47"N	111°16'18"W	8697
Red Peak	Bonneville	43°20'13"N	111°18'47"W	8740
Red Peak	Valley	44°51'41"N	115°11'18"W	9468
Red Rock Butte	Idaho	46°05'56"N	116°09'46"W	3000
Red Rock Mountain	Cassia	42°12'13"N	113°27'01"W	6382
Red Rock Mountain	Fremont	44°34'36"N	111°30'22"W	9512
Red Rock Peak	Lemhi	45°00'01"N	114°25'43"W	8191
Red Top	Boundary	48°53'50"N	116°43'29"W	6265
Redbird Mountain	Custer	43°50'12"N	113°43'58"W	11273
Reed Hill	Bonner	48°11'23"N	116°32'38"W	2270
Reed Mountain	Idaho	45°50'12"N	115°42'51"W	6290
Reeder Mountain	Bonner	48°40'06"N	116°59'40"W	4729
Reeds Baldy	Benewah	47°22'32"N	116°22'31"W	6153
Regan Butte	Gem	43°55'48"N	116°20'48"W	3310
Regan, Mount	Boise	44°09'36"N	115°03'37"W	10190
Renfro Peak	Shoshone	47°11'15"N	116°18'31"W	5301
Reserve Mountain	Custer	43°42'21"N	113°40'18"W	8304
Reservoir Mountain	Caribou	42°54'28"N	111°40'35"W	7143
Reward Peak	Custer	44°02'07"N	115°02'26"W	10074
Rhoda Point	Idaho	46°16'53"N	115°04'36"W	6882
Rhodes Peak	Clearwater	46°40'30"N	114°46'58"W	7930
Rice Peak	Valley	44°30'22"N	115°37'36"W	8696
Rich Hill	Kootenai	47°37'47"N	116°53'53"W	3290
Richard Butte	Butte	43°57'43"N	112°45'31"W	5161
Richardson Summit	Blaine	43°26'15"N	114°28'15"W	6110
Riley Butte	Washington	44°22'13"N	116°28'01"W	4137
Rimrock	Kootenai	47°51'31"N	116°43'55"W	2580
Ripley Butte	Fremont	44°21'13"N	111°19'32"W	6485
Rising Butte	Fremont	44°08'39"N	111°08'07"W	6686
Riverview Point	Clearwater	46°42'35"N	115°59'53"W	3115
Rizzi Table	Owyhee	42°03'21"N	115°43'15"W	5993
Robbers Roost	Caribou	42°57'18"N	111°30'33"W	6305
Roberts Mountain	Bonner	48°29'37"N	116°05'19"W	6675
Robinson Bar Peak	Custer	44°12'20"N	114°44'25"W	9900
Rochat Peak	Benewah	47°24'00"N	116°24'30"W	5643
Rock Butte	Butte	43°18'30"N	112°57'38"W	5436
Rock Corral Butte	Butte	43°18'11"N	113°00'01"W	5650
Rock Garden	Clearwater	46°36'37"N	114°58'48"W	7101
Rock Knoll	Bannock	42°46'25"N	112°29'37"W	7268
Rock Pillar	Idaho	45°59'46"N	115°03'58"W	7100
Rock Point	Idaho	46°07'28"N	114°34'48"W	7178
Rocky Butte	Blaine	43°27'21"N	114°22'31"W	5940
Rocky Knoll	Franklin	42°24'33"N	111°40'31"W	6180
Rocky Mountain	Lemhi	45°34'53"N	114°01'22"W	8690
Rocky Peak	Clearwater	46°42'45"N	114°37'27"W	6700
Rocky Peak	Franklin	42°15'28"N	111°47'01"W	6471
Rocky Peak	Idaho	45°56'15"N	115°06'17"W	7660
Rocky Peak	Lemhi	44°32'33"N	113°28'47"W	10555
Rocky Peak	Teton	43°35'02"N	111°12'36"W	7308
Rocky Peak	Valley	44°29'27"N	115°39'16"W	8393
Rocky Point	Idaho	46°39'40"N	114°19'55"W	7812
Rocky Point	Idaho	46°34'56"N	114°39'55"W	6280
Rocky Point	Latah	46°50'48"N	116°52'38"W	3737
Rocky Run Point	Shoshone	47°03'36"N	115°49'58"W	5060
Roland	Shoshone	47°23'09"N	115°40'51"W	5479
Roland Summit	Shoshone	47°23'12"N	115°39'55"W	4468
Roland Summit	Shoshone	47°23'11"N	115°38'50"W	5479
Roman Nose	Boundary	48°37'41"N	116°35'36"W	7260
Rookie Point	Custer	44°22'43"N	114°43'41"W	7297
Rooster Comb Peak	Owyhee	43°08'17"N	116°48'55"W	6336
Roothaan Mountain	Bonner	48°37'12"N	116°50'58"W	2985
Roothaan, Mount	Boundary	48°36'48"N	116°42'25"W	7326
Roots Knob	Idaho	45°20'03"N	114°59'20"W	7019
Ross Peak	Bonneville	43°29'15"N	111°16'40"W	7929
Ross Peak	Camas	43°43'30"N	115°00'53"W	9773

Mountain, Peak, or Summit	County	Latitude	Longitude	Elev.
Rough Mountain	Owyhee	42°45'21"N	116°22'13"W	7090
Rough Mountain	Owyhee	42°24'16"N	116°46'57"W	6300
Rough Mountain	Owyhee	42°02'11"N	115°55'41"W	6821
Round Knob	Idaho	45°28'57"N	116°25'35"W	6600
Round Knoll	Butte	43°28'24"N	113°28'16"W	5695
Round Mountain	Boundary	48°35'48"N	116°25'27"W	2851
Round Mountain	Cassia	42°01'55"N	113°13'59"W	5336
Round Mountain	Custer	43°47'17"N	113°43'10"W	8956
Round Mountain	Idaho	45°20'00"N	115°30'47"W	3840
Round Mountain	Kootenai	47°53'43"N	116°49'13"W	3456
Round Mountain	Kootenai	47°34'40"N	117°02'13"W	4060
Round Top	Benewah	47°23'34"N	116°30'29"W	5412
Round Top	Idaho	45°36'19"N	115°59'01"W	7842
Round Top	Shoshone	47°36'27"N	115°46'10"W	5045
Round Top Mountain	Bonner	48°16'48"N	116°14'54"W	6149
Round Top Mountain	Idaho	46°07'44"N	115°22'50"W	6807
Roundtop	Idaho	46°32'03"N	114°35'21"W	6560
Roundtop Mountain	Idaho	45°41'40"N	114°35'56"W	6634
Roundtop Mountain	Shoshone	47°06'40"N	115°49'56"W	6003
Ruby Mountain	Boundary	48°55'13"N	116°06'30"W	6256
Ruby Mountain	Idaho	45°13'32"N	115°51'07"W	7464
Ruby Point	Clearwater	46°44'52"N	115°05'08"W	4620
Ruby Point	Shoshone	46°58'09"N	115°21'18"W	5840
Ruffneck Peak	Custer	44°28'55"N	115°09'17"W	9407
Runaway Point	Idaho	45°20'37"N	114°54'13"W	7450
Runt Mountain	Shoshone	47°27'13"N	115°42'23"W	5540
Rupe Peak	Idaho	46°17'52"N	114°28'22"W	7651
Rush Creek Point	Valley	45°05'52"N	114°56'01"W	7223
Rush Peak	Washington	44°42'55"N	116°43'22"W	7634
Russell Mountain	Boundary	48°48'33"N	116°32'39"W	6818
Ruud Mountain	Blaine	43°42'21"N	114°19'49"W	6705
Ryan Peak	Custer	43°54'08"N	114°24'31"W	11714
Ryan Peak	Teton	43°42'25"N	111°20'38"W	8660
Sabe Mountain	Idaho	45°39'27"N	114°56'06"W	8245
Sable Hill	Idaho	45°51'29"N	115°11'32"W	6084
Saddle Butte	Butte	43°28'42"N	113°21'37"W	5602
Saddle Mountain	Boundary	48°57'02"N	116°46'12"W	6893
Saddle Mountain	Butte	43°56'15"N	112°57'14"W	10810
Saddle Mountain	Idaho	46°11'10"N	114°32'29"W	8258
Saddle, The	Bingham	43°14'23"N	111°42'30"W	6930
Sagebrush Hill	Washington	44°15'55"N	116°54'00"W	3335
Sagebrush Mountain	Lemhi	45°16'14"N	114°26'42"W	7147
Sailor Cap Butte	Owyhee	42°43'58"N	115°26'23"W	3590
Saint Joe Baldy	Benewah	47°21'49"N	116°24'39"W	5825
Saint Maries Peak	Benewah	47°18'52"N	116°32'52"W	2628
Saint Marys Nipple	Bonneville	43°26'45"N	112°29'30"W	5345
Sal Mountain	Lemhi	45°00'53"N	113°50'33"W	9592
Salamander Butte	Idaho	45°41'54"N	114°51'31"W	6054

Mountain, Peak, or Summit	County	Latitude	Longitude	Elev.
Salmon Butte	Owyhee	43°16'09"N	116°49'55"W	6100
Salmon Butte	Twin Falls	42°12'49"N	114°43'25"W	5389
Salmon Mountain	Idaho	45°37'02"N	114°50'13"W	8943
Salmon Point	Idaho	45°25'08"N	116°18'23"W	2940
Sampson, Mount	Idaho	45°19'43"N	116°26'40"W	6462
Sams Throne	Idaho	45°17'55"N	116°07'28"W	8283
Sand Butte	Lincoln	43°07'46"N	113°50'19"W	4974
Sand Mountain	Boundary	48°30'20"N	116°22'24"W	4490
Sand Mountain	Jefferson	43°58'04"N	111°58'41"W	5360
Sand Mountain	Latah	46°55'34"N	116°34'43"W	4980
Santa Peak	Benewah	47°09'30"N	116°21'47"W	4209
Saturday Mountain	Custer	44°16'43"N	114°29'45"W	8927
Savage Point	Valley	45°04'54"N	115°34'07"W	8119
Sawpit Hill	Valley	44°53'14"N	115°36'19"W	7578
Sawpit Peak	Owyhee	42°59'28"N	116°44'23"W	7810
Sawtell Peak	Fremont	44°33'42"N	111°26'37"W	9866
Sawtooth Peak	Shoshone	47°00'15"N	115°34'42"W	5895
Sawtooth Peak	Valley	44°59'47"N	115°55'33"W	8875
Scarface Mountain	Custer	44°41'52"N	115°00'13"W	7652
Schafer Peak	Bonner	48°06'36"N	116°22'07"W	5209
Schoolmarm Peak	Idaho	45°24'21"N	116°18'55"W	3571
Scotchman Number Two	Bonner	48°12'53"N	116°03'05"W	6989
Scotchman Peak	Bonner	48°11'20"N	116°04'51"W	7009
Scott Butte	Clark	44°04'05"N	112°47'33"W	6161
Scott Mountain	Boise	44°10'54"N	115°47'17"W	8215
Scott Peak	Lemhi	44°21'13"N	112°49'15"W	11393
Scott Table	Owyhee	41°59'04"N	115°41'12"W	6070
Scout Mountain	Bannock	42°40'41"N	112°20'41"W	8700
Scout, The	Boundary	48°52'17"N	116°03'00"W	6300
Scurvy Mountain	Clearwater	46°41'23"N	115°08'00"W	6691
Secret Cabin Butte	Twin Falls	42°20'58"N	114°51'44"W	4707
Sedgwick Peak	Bannock	42°30'56"N	111°55'23"W	9167
Sentinel Peak	Shoshone	47°50'25"N	116°02'21"W	6020
Sentinel, The	Butte	43°22'45"N	113°29'31"W	5812
Service Butte	Fremont	44°15'17"N	111°42'06"W	6225
Serviceberry Butte	Butte	43°18'43"N	113°10'38"W	5400
Sevenmile Point	Clearwater	46°34'08"N	115°30'11"W	4851
Sevy Peak	Custer	44°01'29"N	114°58'20"W	10485
Sexy Peak	Custer	44°01'29"N	114°58'22"W	10400
Shafer Butte	Boise	43°46'18"N	116°05'17"W	7582
Shale Butte	Lincoln	42°58'53"N	113°49'05"W	4578
Shale Mountain	Clearwater	46°42'50"N	114°47'15"W	7612
Shale Point	Bear Lake	42°24'15"N	111°10'05"W	7740
Shale Point	Shoshone	47°03'31"N	115°47'53"W	5743
Shares Snout	Owyhee	43°22'35"N	116°49'28"W	4837
Sharp Top	Benewah	47°21'22"N	116°32'49"W	3966
Shasta Butte	Kootenai	47°38'05"N	116°59'48"W	4877
Shattuck Butte	Bonneville	43°36'23"N	112°08'41"W	5163

Mountains, Peaks, Summits

Mountain, Peak, or Summit	County	Latitude	Longitude	Elev.
Shattuck Butte	Clearwater	46°50'13"N	116°13'36"W	4540
Shattuck Mountain	Idaho	46°17'32"N	114°28'00"W	8458
Shaw Mountain	Ada	43°36'19"N	116°03'37"W	5904
Shaw Mountain	Blaine	43°36'58"N	114°39'38"W	9650
Shay Hill	Benewah	47°19'21"N	116°39'26"W	2752
She Devil	Idaho	45°19'29"N	116°32'24"W	9280
Shearer Peak	Idaho	46°02'21"N	114°53'37"W	6226
Sheep Butte	Shoshone	47°19'14"N	116°01'47"W	5051
Sheep Creek Peak	Bonneville	43°22'08"N	111°07'39"W	9950
Sheep Dip Mountain	Oneida	41°59'52"N	112°09'42"W	7057
Sheep Hill	Idaho	46°11'14"N	115°25'41"W	5676
Sheep Hill	Idaho	45°35'17"N	115°05'16"W	8405
Sheep Horn Mountain	Valley	45°02'31"N	114°57'05"W	8256
Sheep Mountain	Bingham	43°11'55"N	111°40'43"W	7003
Sheep Mountain	Bonneville	43°23'23"N	111°09'51"W	9700
Sheep Mountain	Cassia	42°10'44"N	113°24'53"W	5764
Sheep Mountain	Clearwater	46°47'01"N	115°35'21"W	5708
Sheep Mountain	Custer	44°03'52"N	114°21'34"W	10910
Sheep Mountain	Custer	43°46'50"N	113°35'07"W	9649
Sheep Mountain	Elmore	43°41'52"N	115°29'55"W	8148
Sheep Mountain	Idaho	45°34'41"N	115°51'16"W	7821
Sheep Mountain	Idaho	45°20'44"N	116°16'47"W	7415
Sheep Mountain	Lemhi	45°28'49"N	113°45'34"W	9858
Sheep Mountain	Lemhi	44°23'26"N	113°17'11"W	10865
Sheep Peak	Adams	44°54'15"N	116°47'34"W	5257
Sheep Point	Camas	43°25'01"N	114°58'29"W	6987
Sheep Rock	Adams	45°11'30"N	116°40'13"W	6847
Sheep Trail Butte	Blaine	43°20'44"N	113°27'17"W	5614
Sheepeater Mountain	Idaho	45°23'24"N	115°20'46"W	8486
Sheepeater Point	Lemhi	45°23'18"N	114°21'29"W	7853
Sheepherder Monument Butte	Fremont	43°59'12"N	112°06'21"W	5075
Sheephorn Mountain	Lemhi	44°52'57"N	114°05'26"W	8159
Sheephorn Peak	Lemhi	44°30'37"N	113°24'21"W	10465
Shefoot Mountain	Shoshone	47°18'15"N	115°40'47"W	6349
Sheldon Peak	Lemhi	44°41'59"N	114°33'41"W	9278
Shell Rock Peak	Valley	44°48'58"N	115°34'48"W	8574
Shellrock Peak	Valley	44°57'24"N	114°56'10"W	9435
Shelly Mountain	Custer	43°49'36"N	113°43'34"W	11278
Shephard Peak	Boise	43°58'21"N	115°19'18"W	8833
Sherlock Peak	Shoshone	47°05'02"N	115°11'00"W	6404
Sherman Peak	Bear Lake	42°27'55"N	111°32'54"W	9682
Sherman Peak	Custer	44°31'56"N	114°38'23"W	9892
Sherman Peak	Idaho	46°24'55"N	115°19'23"W	6658
Sherwin Point	Latah	46°57'54"N	116°21'21"W	4074
Shin Point	Clearwater	46°42'43"N	115°35'11"W	4300
Shining Butte	Idaho	45°33'21"N	115°43'51"W	8348
Shoeffler Butte	Kootenai	47°22'55"N	116°47'09"W	3366
Shoepack Point	Shoshone	47°10'00"N	115°44'54"W	5300

Mountain, Peak, or Summit	County	Latitude	Longitude	Elev.
Short Point	Clearwater	46°54'31"N	115°00'59"W	6219
Shorty Peak	Boundary	48°56'25"N	116°41'07"W	6515
Shoshone Peak	Shoshone	47°54'06"N	116°02'07"W	5425
Sid Butte	Lincoln	42°50'53"N	113°57'26"W	4771
Signal Butte	Twin Falls	42°10'03"N	115°00'47"W	5913
Signal Peak	Clark	44°26'22"N	112°06'42"W	8534
Silent Cone	Butte	43°26'41"N	113°34'12"W	6357
Silver Butte	Clearwater	46°42'55"N	115°48'45"W	4680
Silver Creek Summit	Valley	44°24'49"N	115°45'34"W	6985
Silver Dollar Peak	Boundary	48°37'54"N	116°40'04"W	7181
Silver Dome	Idaho	45°53'24"N	115°46'01"W	6384
Silver Hill	Shoshone	47°26'58"N	116°07'07"W	5983
Silver Mountain	Boise	43°57'53"N	115°18'58"W	8573
Silver Mountain	Boundary	48°57'53"N	116°44'45"W	6555
Silver Peak	Blaine	43°50'44"N	114°32'13"W	11112
Simmons Peak	Shoshone	47°05'48"N	115°17'45"W	6648
Sinker Butte	Owyhee	43°13'54"N	116°23'39"W	3421
Sinker Creek Butte	Owyhee	43°11'02"N	116°24'16"W	3129
Sinker Mountain	Idaho	45°35'21"N	115°23'56"W	6420
Siwash Peak	Shoshone	47°12'19"N	115°44'29"W	5819
Sixmile Butte	Butte	43°30'45"N	113°15'11"W	5695
Sixmile Point	Valley	44°18'27"N	115°57'09"W	6290
Skeeter Point	Blaine	43°17'44"N	114°08'07"W	6325
Skeleton Butte	Jerome	42°35'07"N	114°13'55"W	4400
Skelly Hill	Bonneville	43°29'08"N	111°39'43"W	6162
Skillern Peak	Camas	43°41'15"N	114°48'54"W	8918
Skitwish Peak	Kootenai	47°42'05"N	116°29'04"W	5356
Skookum Butte	Idaho	46°39'53"N	114°23'39"W	7215
Skunk Creek Summit	Valley	44°22'49"N	115°55'42"W	7164
Sky Pilot Mountain	Idaho	46°26'13"N	114°22'03"W	8792
Slab Butte	Adams	45°05'33"N	116°08'25"W	8225
Slack Mountain	Owyhee	42°33'55"N	116°40'13"W	5855
Slacks Mountain	Owyhee	43°04'22"N	116°44'08"W	7339
Slate Mountain	Bannock	42°45'05"N	112°26'15"W	6980
Slate Peak	Shoshone	47°19'09"N	115°57'11"W	5429
Slate Point	Idaho	45°35'55"N	116°07'58"W	7291
Sleeping Deer Mountain	Lemhi	44°46'03"N	114°41'26"W	9881
Slide Mountain	Clark	44°32'58"N	111°51'24"W	9805
Sloans Point	Valley	44°44'43"N	115°55'45"W	6280
Smelter Butte	Blaine	43°34'45"N	113°54'36"W	6860
Smiley Mountain	Custer	43°41'59"N	113°48'35"W	11508
Smith Butte	Idaho	46°31'09"N	115°17'23"W	5384
Smith Knob	Idaho	45°17'17"N	115°30'52"W	6989
Smith Mountain	Adams	45°08'04"N	116°34'41"W	8005
Smith Mountain	Bonner	48°24'55"N	116°06'14"W	6510
Smith Peak	Boundary	48°50'03"N	116°40'10"W	7653
Smoky Dome	Camas	43°29'37"N	114°56'08"W	10095
Smoky Mountain	Cassia	42°03'34"N	113°40'46"W	7570

Mountain, Peak, or Summit	County	Latitude	Longitude	Elev.
Smoky Peak	Elmore	43°59'18"N	115°09'54"W	9294
Smoky Peak	Idaho	46°15'58"N	115°25'41"W	4557
Snake River Butte	Fremont	44°08'46"N	111°20'15"W	6494
Snow Cone	Butte	43°26'29"N	113°33'27"W	6030
Snow Creek Butte	Fremont	44°16'32"N	111°07'11"W	7679
Snow Peak	Bannock	42°45'05"N	112°08'10"W	9132
Snow Peak	Boundary	48°40'09"N	116°28'14"W	4547
Snow Peak	Shoshone	47°02'44"N	115°37'55"W	6760
Snowbank Mountain	Valley	44°27'05"N	116°07'40"W	8322
Snowshoe Butte	Clark	44°12'55"N	111°57'29"W	6001
Snowshoe Summit	Valley	44°39'41"N	115°26'54"W	8140
Snowslide Peak	Valley	45°05'13"N	115°14'32"W	9104
Snowslide Peak	Valley	44°58'37"N	115°55'57"W	8522
Snowslide Summit	Valley	44°58'47"N	115°55'35"W	8141
Snowstorm Peak	Shoshone	47°29'45"N	115°43'41"W	6362
Snowy Summit	Idaho	46°26'16"N	115°35'19"W	6000
Snowy Top	Boundary	48°59'32"N	116°59'09"W	7572
Snowyside Peak	Blaine	43°56'18"N	114°58'14"W	10651
Sob Point	Idaho	46°01'50"N	115°25'09"W	5618
Soda Peak	Bear Lake	42°31'38"N	111°37'28"W	8921
Soldier Cap	Owyhee	43°17'01"N	116°45'19"W	5430
Soloa Peak	Kootenai	47°24'12"N	116°38'41"W	3435
Solomon Mountain	Boundary	48°47'26"N	116°06'28"W	4201
Sommer Camp Mountain	Owyhee	43°01'08"N	116°58'09"W	5935
Sommers Butte	Madison	43°48'08"N	111°42'29"W	5381
Sonnickson Butte	Jerome	42°40'12"N	114°32'46"W	3741
Sourdough Peak	Idaho	45°43'38"N	115°48'53"W	6800
South Butte	Shoshone	47°01'13"N	115°48'25"W	6994
South Chapin Mountain	Cassia	42°27'25"N	113°10'20"W	5585
South Chilco Mountain	Kootenai	47°52'17"N	116°32'55"W	5665
South Hill	Bear Lake	42°23'39"N	111°23'18"W	7140
South Loon Mountain	Valley	45°06'01"N	115°52'49"W	9287
South Mountain	Gem	43°57'51"N	116°15'49"W	4688
South Putnam Mountain	Caribou	42°54'49"N	112°07'46"W	8949
South Raker	Boise	43°59'16"N	115°06'04"W	9620
South Tit	Cassia	42°04'47"N	114°09'43"W	7518
Southwest Butte	Idaho	45°30'27"N	116°12'04"W	7814
Spades Mountain	Kootenai	47°47'44"N	116°34'03"W	5038
Spain Hill	Owyhee	43°04'33"N	116°53'18"W	6181
Sparrow Point	Idaho	45°54'40"N	115°12'10"W	6100
Sparta Peak	Shoshone	47°39'49"N	115°58'58"W	3925
Spencer Butte	Owyhee	42°31'24"N	116°36'59"W	5481
Spion Kop	Shoshone	47°54'18"N	116°05'23"W	5540
Split Butte	Blaine	43°21'56"N	113°28'02"W	5652
Split Butte	Blaine	42°53'59"N	113°22'07"W	4795
Split Butte	Fremont	44°14'57"N	111°43'52"W	6252
Split Creek Point	Valley	45°06'29"N	115°46'27"W	6628
Split Rock	Owyhee	42°59'11"N	116°57'52"W	5695

Mountain, Peak, or Summit	County	Latitude	Longitude	Elev.
Split Top	Blaine	43°09'20"N	113°06'48"W	5563
Sponge Mountain	Idaho	46°20'37"N	115°07'01"W	7349
Spooky Butte	Shoshone	47°20'27"N	116°09'25"W	4910
Spot Mountain	Idaho	45°47'14"N	114°51'15"W	8024
Spotted Louis Point	Shoshone	47°02'29"N	115°39'23"W	5682
Spread Creek Point	Idaho	45°38'35"N	115°01'24"W	8190
Spring Butte	Owyhee	42°14'33"N	116°57'26"W	5566
Spring Hill	Clearwater	46°51'12"N	115°03'52"W	5557
Spring Hill	Custer	44°16'15"N	113°44'39"W	9449
Spring Mountain	Idaho	46°34'59"N	114°54'15"W	6715
Spring Mountain	Lemhi	44°21'18"N	113°16'29"W	10180
Sprout Mountain	Elmore	43°29'31"N	115°10'10"W	7059
Spruce Mountain	Bonner	47°59'07"N	116°20'18"W	5130
Spud Butte	Lincoln	43°10'15"N	113°49'26"W	5003
Spy Point	Clearwater	46°36'59"N	115°08'35"W	6455
Spyglass Peak	Shoshone	47°50'31"N	116°11'46"W	5318
Square Mountain	Blaine	43°23'12"N	114°30'10"W	5467
Square Mountain	Idaho	45°35'55"N	115°51'45"W	8020
Square Rock	Idaho	45°58'44"N	115°00'30"W	7300
Square Top	Idaho	45°28'23"N	114°40'36"W	8402
Square Top	Valley	44°45'31"N	115°47'49"W	8681
Squaw Butte	Gem	44°01'58"N	116°24'40"W	5896
Squaw Butte	Owyhee	43°15'47"N	116°53'03"W	6740
Squaw Hump	Benewah	47°07'21"N	116°50'31"W	3060
Squaw Peak	Lemhi	45°31'33"N	114°36'09"W	8660
Squaw Peak	Shoshone	47°05'41"N	115°32'27"W	6332
Squaw Point	Idaho	45°13'04"N	115°58'14"W	8286
Squawtit	Custer	44°10'40"N	113°36'36"W	9046
Squirrel Mountain	Teton	43°36'45"N	111°39'59"W	7780
Squirrel Mountain	Teton	43°36'45"N	111°13'58"W	7780
Standhope Peak	Custer	43°47'05"N	114°01'27"W	11878
Stanford Point	Latah	46°49'05"N	116°40'29"W	3010
Stanley Butte	Idaho	46°14'52"N	115°12'33"W	7362
Star Butte	Washington	44°22'32"N	116°41'26"W	4134
Star Mountain	Boundary	48°31'05"N	116°07'20"W	6107
Star Ranch Table	Owyhee	42°27'59"N	116°44'39"W	5312
Statue Hills	Owyhee	43°05'13"N	116°52'02"W	5874
Steakhouse Hill	Latah	46°48'02"N	116°58'47"W	2930
Steamboat	Twin Falls	42°00'14"N	114°50'11"W	7098
Steamboat Hill	Caribou	42°39'32"N	111°38'44"W	5892
Steamboat Peak	Shoshone	47°43'25"N	116°12'18"W	3329
Steel Mountain	Elmore	43°44'49"N	115°19'00"W	9730
Steep Hill	Idaho	45°39'34"N	114°30'24"W	7977
Stein Mountain	Boundary	48°59'36"N	116°28'48"W	2340
Stein Mountain	Lemhi	45°27'33"N	113°52'46"W	8555
Stevens Peak	Bingham	43°08'10"N	112°17'38"W	5372
Stevens Peak	Shoshone	47°25'32"N	115°46'17"W	6838
Stewart, Mount	Idaho	46°08'23"N	115°58'11"W	1960

Mountains, Peaks, Summits

Mountain, Peak, or Summit	County	Latitude	Longitude	Elev.
Stinson Hill	Blaine	43°26'05"N	113°51'39"W	6650
Stoddard Creek Point	Idaho	45°13'37"N	114°43'56"W	7410
Stone Johnny	Bonner	48°15'51"N	117°01'40"W	4648
Stony Butte	Shoshone	46°58'00"N	116°07'13"W	5690
Stony Point	Adams	44°46'31"N	116°36'44"W	6220
Stony Point	Clearwater	46°35'43"N	115°10'27"W	6204
Storm Mountain	Shoshone	47°18'51"N	115°50'39"W	5643
Storm Peak	Valley	45°07'56"N	115°55'06"W	9100
Stormy Hill	Owyhee	43°00'19"N	116°41'41"W	7800
Stormy Peak	Lemhi	45°21'03"N	114°12'40"W	8022
Stormy Point	Idaho	45°29'34"N	116°31'05"W	5929
Stouts Mountain	Bonneville	43°32'52"N	111°20'52"W	8620
Stratton Butte	Latah	47°03'51"N	117°02'25"W	3292
Strawberry Mountain	Bonner	48°27'54"N	116°12'53"W	5684
Stricker Butte	Twin Falls	42°28'11"N	114°21'33"W	4134
Stripe Mountain	Idaho	45°30'48"N	114°46'25"W	9001
Striped Peak	Shoshone	47°26'23"N	115°59'44"W	6316
Stubtoe Peak	Shoshone	47°01'28"N	115°42'02"W	5700
Stump Peak	Caribou	42°54'03"N	111°11'06"W	8603
Sturgill Peak	Washington	44°37'12"N	116°56'33"W	7589
Suburban Peak	Shoshone	47°48'42"N	116°08'09"W	4380
Sugar Bowl	Elmore	42°59'36"N	115°14'28"W	2908
Sugar Loaf	Lemhi	45°04'10"N	114°33'57"W	9045
Sugar Loaf	Owyhee	43°03'07"N	116°42'11"W	6345
Sugar Mountain	Valley	44°58'19"N	115°18'52"W	8778
Sugarloaf	Adams	44°38'15"N	116°18'50"W	5245
Sugarloaf	Idaho	45°56'32"N	115°56'46"W	2984
Sugarloaf	Owyhee	42°35'49"N	116°08'55"W	5952
Sugarloaf	Owyhee	42°01'21"N	115°54'42"W	7039
Sugarloaf	Twin Falls	42°21'24"N	114°21'39"W	5750
Sugarloaf	Twin Falls	42°10'41"N	114°55'15"W	5523
Sugarloaf	Valley	44°36'47"N	116°05'04"W	4968
Sugarloaf	Washington	44°20'18"N	116°43'43"W	4026
Sugarloaf Butte	Idaho	45°50'31"N	116°39'29"W	4430
Sugarloaf Butte	Idaho	45°27'03"N	115°39'21"W	3046
Sugarloaf Knob	Idaho	45°49'47"N	116°38'52"W	4347
Sugarloaf Mountain	Bingham	43°12'43"N	111°35'25"W	7140
Sugarloaf Mountain	Bonner	48°08'24"N	116°04'16"W	4205
Sugarloaf Rock	Boise	43°50'11"N	116°03'30"W	6364
Sulphur Peak	Caribou	42°38'32"N	111°25'39"W	8302
Sundance Mountain	Bonner	48°29'27"N	116°45'02"W	6298
Sunnyside Mountain	Bonner	48°17'46"N	116°23'40"W	2523
Sunset Butte	Twin Falls	42°33'39"N	114°52'57"W	4006
Sunset Butte	Twin Falls	42°33'38"N	114°52'56"W	4006
Sunset Cone	Butte	43°28'06"N	113°33'48"W	6410
Sunset Mountain	Benewah	47°01'36"N	116°26'56"W	5206
Sunset Mountain	Boise	43°53'55"N	115°38'45"W	7869
Sunset Peak	Butte	43°58'34"N	113°22'08"W	10693

Mountain, Peak, or Summit	County	Latitude	Longitude	Elev.
Sunset Peak	Shoshone	47°33'48"N	115°50'01"W	6424
Surveyors Peak	Shoshone	46°59'15"N	115°28'51"W	6517
Susies Nipple	Teton	43°50'01"N	111°19'28"W	6897
Swan Peak	Kootenai	47°30'23"N	116°38'24"W	4099
Swan Peak	Lemhi	45°00'52"N	114°11'31"W	8416
Swan Peak	Shoshone	47°06'48"N	116°08'46"W	4838
Swan Point	Clearwater	46°50'17"N	115°40'43"W	4780
Swanholm Peak	Elmore	43°53'34"N	115°19'30"W	8727
Swanty Peak	Cassia	42°04'02"N	114°09'29"W	7208
Swartz Meadow Point	Idaho	45°47'05"N	116°06'48"W	5220
Swede Peak	Blaine	43°38'25"N	113°59'32"W	8512
Sweeny Hill	Idaho	45°49'40"N	115°26'57"W	4140
Swensen Butte	Custer	44°02'54"N	113°58'50"W	7185
Swet Point	Idaho	45°31'21"N	114°49'43"W	8495
Swisher Mountain	Owyhee	43°02'46"N	116°59'14"W	6130
Switchback Hill	Clearwater	46°38'43"N	115°10'07"W	6431
Sydney Butte	Camas	43°32'06"N	114°47'28"W	8161
Table Butte	Jefferson	43°59'33"N	112°19'47"W	5235
Table Butte	Owyhee	42°32'08"N	115°50'35"W	4842
Table Legs Butte	Bingham	43°24'54"N	112°46'20"W	5385
Table Mountain	Bear Lake	42°29'35"N	111°04'18"W	8586
Table Mountain	Custer	44°44'40"N	114°12'36"W	8426
Table Mountain	Custer	44°26'59"N	114°00'29"W	9082
Table Mountain	Custer	43°49'35"N	113°38'16"W	7386
Table Mountain	Oneida	42°05'18"N	112°51'01"W	5436
Table Mountain	Power	42°35'28"N	112°58'22"W	5505
Table Rock	Ada	43°35'42"N	116°08'31"W	3658
Table Rock	Owyhee	43°09'11"N	116°56'09"W	5787
Tackobe Mountain	Elmore	43°57'09"N	115°10'13"W	9283
Talbot Hill	Kootenai	47°24'52"N	116°49'51"W	3070
Tamarack Mountain	Idaho	46°25'03"N	114°36'08"W	6132
Tanglefoot Point	Shoshone	47°19'45"N	115°41'47"W	5924
Tank Hill	Bonner	48°14'42"N	116°36'10"W	2220
Targhee Peak	Fremont	44°43'21"N	111°23'15"W	10300
Taylor Mountain	Bingham	43°18'23"N	111°55'13"W	7414
Taylor Mountain	Clark	44°33'41"N	111°40'54"W	9855
Taylor Mountain	Lemhi	44°53'06"N	114°12'55"W	9960
Taylor Mountain	Lemhi	44°11'15"N	113°28'39"W	8644
Taylor Peak	Shoshone	47°47'36"N	115°53'00"W	5585
Teakean Butte	Nez Perce	46°35'29"N	116°23'22"W	4140
Teakettle Butte	Butte	43°29'49"N	113°10'14"W	5418
Teapot Dome	Elmore	43°10'29"N	115°30'57"W	4713
Teapot Mountain	Valley	44°54'38"N	115°43'51"W	5518
Telephone Booth Hill	Shoshone	46°57'08"N	116°08'40"W	4330
Temple Mountain	Boundary	48°44'20"N	116°43'10"W	6815
Temple Peak	Teton	43°39'04"N	111°20'17"W	8219
Temple, The	Custer	44°00'43"N	114°59'30"W	9860
Tenmile Hill	Idaho	45°41'39"N	115°42'42"W	7241

Mountain, Peak, or Summit	County	Latitude	Longitude	Elev.
Tennessee Mountain	Owyhee	43°03'12"N	116°47'02"W	7004
Tepee Peak	Shoshone	47°44'20"N	116°17'30"W	4402
Tepee Summit	Shoshone	47°45'11"N	116°17'59"W	4087
Thatcher Hill	Franklin	42°24'02"N	111°51'08"W	7980
Thirteen Mountain	Idaho	45°31'45"N	114°45'42"W	8973
Thomas Hill	Shoshone	47°35'47"N	116°10'35"W	4951
Thompson Butte	Clearwater	46°47'44"N	115°42'00"W	3995
Thompson Peak	Boise	44°08'29"N	115°00'33"W	10751
Thompson Peak	Bonneville	43°28'49"N	111°08'43"W	9481
Thompson Point	Clearwater	46°51'08"N	115°42'42"W	3995
Thor Mountain	Shoshone	47°09'55"N	115°34'23"W	6384
Thorn Creek Butte	Boise	43°44'41"N	115°45'31"W	7515
Three Knolls	Bonneville	43°20'04"N	111°25'23"W	7300
Three Point Mountain	Elmore	43°30'23"N	115°56'00"W	5324
Three Prong Mountain	Idaho	45°46'22"N	114°55'33"W	8182
Three Sisters Peaks	Bonner	48°01'47"N	116°32'23"W	3980
Three Tree Butte	Latah	47°01'46"N	116°40'00"W	3646
Threemile Knoll	Caribou	42°42'30"N	111°34'32"W	6475
Threemile Knoll	Fremont	44°19'29"N	111°50'00"W	6900
Threemile Peak	Bear Lake	42°28'06"N	111°16'54"W	9215
Thumb, The	Clark	44°29'06"N	112°32'57"W	9787
Thunder Hill	Nez Perce	46°25'17"N	116°48'16"W	1060
Thunder Mountain	Cassia	42°10'18"N	113°36'46"W	7455
Thunder Mountain	Teton	43°38'56"N	111°18'07"W	8662
Thunder Mountain	Valley	44°57'01"N	115°07'57"W	8579
Thunderbolt Mountain	Valley	44°43'57"N	115°38'19"W	8652
Tiger Peak	Shoshone	47°32'11"N	115°49'50"W	6625
Timber Butte	Boise	44°02'47"N	116°14'42"W	4851
Timber Butte	Twin Falls	42°04'20"N	114°17'59"W	7027
Timber Creek Peak	Lemhi	44°29'47"N	113°24'20"W	10553
Timber Mountain	Boundary	48°31'57"N	116°06'13"W	6395
Timber Peak	Shoshone	47°01'19"N	115°23'15"W	5212
Timbered Dome	Butte	43°36'49"N	113°32'31"W	8356
Tims Peak	Owyhee	43°20'08"N	116°52'13"W	5250
Tin Can Hill	Shoshone	47°15'01"N	115°36'05"W	5065
Tin Cup Butte	Butte	43°28'06"N	113°12'14"W	5590
Tincup Hill	Lemhi	45°28'22"N	114°24'46"W	8234
Tincup Mountain	Caribou	43°00'49"N	111°13'27"W	8220
Toboggan Hill	Clearwater	46°40'04"N	114°59'03"W	6150
Tohobit Peak	Boise	44°05'30"N	115°01'16"W	10046
Tola Point	Bonner	48°29'49"N	117°00'38"W	3826
Tollgate Hill	Elmore	43°15'14"N	115°33'10"W	5260
Tom Beal Peak	Idaho	46°26'43"N	114°46'05"W	7568
Tomato Point	Idaho	45°25'08"N	115°42'51"W	6345
Tomer Butte	Latah	46°42'25"N	116°55'19"W	3466
Tongue, The	Owyhee	42°17'36"N	116°50'50"W	4921
Tool Cache	Washington	44°37'59"N	116°55'59"W	6220
Township Butte	Clearwater	46°45'37"N	115°53'14"W	4836

Mountain, Peak, or Summit	County	Latitude	Longitude	Elev.
Toy Mountain	Owyhee	42°51'24"N	116°30'21"W	6329
Trail Creek Summit	Custer	43°49'32"N	114°15'38"W	8140
Trail Creek Summit	Valley	44°18'11"N	115°50'33"W	5019
Trail Hill	Adams	44°56'24"N	116°43'17"W	4260
Trail Peak	Lemhi	44°20'11"N	113°15'04"W	10533
Trap Point	Clearwater	46°53'49"N	115°04'07"W	5748
Trapper Mountain	Valley	44°46'34"N	115°23'20"W	8982
Trapper Peak	Boundary	48°54'06"N	116°52'30"W	6187
Trapper Peak	Cassia	42°09'09"N	114°12'29"W	7892
Treasure Mountain	Kootenai	47°43'46"N	116°37'10"W	4605
Treasureton Hill	Franklin	42°18'22"N	111°52'03"W	6705
Tree Spring Mountain	Gem	44°20'37"N	116°16'24"W	5043
Trego Point	Shoshone	47°16'49"N	115°50'53"W	4757
Trestle Peak	Bonner	48°19'13"N	116°12'33"W	6320
Triangle Mountain	Idaho	45°35'10"N	116°27'12"W	3975
Triangle Peak	Shoshone	47°21'52"N	115°43'31"W	4485
Trimmed Tree Hill	Shoshone	47°05'52"N	115°38'51"W	6401
Trinity Mountain	Elmore	43°35'56"N	115°25'41"W	9451
Triplet Butte	Owyhee	42°01'25"N	115°38'42"W	6038
Tripod Peak	Valley	44°23'02"N	116°07'37"W	8086
Tripod Summit	Valley	44°17'25"N	116°08'04"W	5330
Trout Creek Mountain	Cassia	42°05'01"N	114°08'47"W	7860
Trout Peak	Bonner	48°19'45"N	116°16'24"W	5226
Trout Peak	Idaho	46°09'26"N	114°51'13"W	5420
Trout Point	Shoshone	47°22'16"N	116°12'45"W	4129
Tuanna Butte	Twin Falls	42°24'19"N	114°57'15"W	4601
Tubbs Hill	Kootenai	47°39'57"N	116°46'31"W	2530
Tules, The	Owyhee	42°12'34"N	116°30'14"W	5050
Tungsten Mountain	Boundary	48°54'41"N	116°16'35"W	5704
Turkey Head Butte	Gooding	43°03'20"N	114°40'21"W	3950
Turkey Point	Shoshone	47°18'25"N	115°39'20"W	6027
Turnbow Point	Latah	46°52'13"N	116°47'00"W	3410
Turnbull Butte	Blaine	43°15'40"N	113°34'43"W	5262
Turner Butte	Owyhee	42°22'53"N	116°08'25"W	6132
Turner Peak	Kootenai	47°33'43"N	116°43'56"W	3700
Turner Peak	Shoshone	47°16'21"N	115°38'33"W	5825
Turner Table	Owyhee	42°24'45"N	116°10'20"W	6003
Turntable Mountain	Owyhee	42°59'16"N	116°40'33"W	8122
Twentymile Butte	Elmore	42°48'25"N	115°08'30"W	3346
Twentymile Butte	Idaho	45°42'43"N	115°44'48"W	7265
Twentymile Peak	Boundary	48°33'27"N	116°17'02"W	5371
Twin Butte	Idaho	46°03'58"N	114°41'05"W	6878
Twin Butte	Owyhee	42°31'38"N	115°41'38"W	4308
Twin Buttes	Bingham	43°29'45"N	112°41'50"W	6572
Twin Buttes	Idaho	46°06'31"N	116°07'35"W	3745
Twin Buttes	Owyhee	42°44'30"N	115°44'36"W	3228
Twin Buttes	Owyhee	42°43'01"N	115°11'52"W	3597
Twin Buttes	Twin Falls	42°07'55"N	114°50'18"W	6780

Mountains, Peaks, Summits

Mountain, Peak, or Summit	County	Latitude	Longitude	Elev.
Twin Imps	Idaho	45°18'08"N	116°33'38"W	9005
Twin Knobs	Bannock	42°34'09"N	111°57'26"W	7768
Twin Oaks	Lincoln	43°08'37"N	114°33'39"W	5894
Twin Peak	Idaho	45°16'19"N	114°44'59"W	9258
Twin Peaks	Bonner	48°17'25"N	116°03'25"W	6262
Twin Peaks	Boundary	48°39'04"N	116°39'35"W	7599
Twin Peaks	Camas	43°12'55"N	114°51'34"W	6531
Twin Peaks	Custer	44°36'00"N	114°28'17"W	9190
Twin Peaks	Kootenai	47°33'57"N	116°58'35"W	4705
Twin Peaks	Lemhi	44°58'01"N	113°58'43"W	7104
Twin Peaks	Madison	43°37'59"N	111°28'33"W	7766
Twin Peaks	Owyhee	43°05'14"N	116°51'02"W	7073
Twin Peaks	Valley	44°51'26"N	115°56'51"W	7980
Twin Sisters	Adams	44°29'39"N	116°18'51"W	5105
Twin Sisters	Cassia	42°02'37"N	113°43'12"W	6838
Twitchells Knoll	Fremont	44°18'38"N	111°46'56"W	6735
Two Point Butte	Blaine	43°19'47"N	113°26'50"W	5566
Two Point Mountain	Camas	43°44'14"N	114°58'33"W	10124
Two Point Peak	Valley	44°59'18"N	114°54'06"W	9426
Two Tail Peak	Boundary	48°40'48"N	116°10'40"W	3949
Two Top, Mount	Fremont	44°37'30"N	111°15'30"W	8710
Twodot Peak	Shoshone	47°09'37"N	115°58'44"W	6508
Tyee Mountain	Boise	44°01'04"N	115°18'11"W	8753
Tyler Peak	Clark	43°59'19"N	112°57'53"W	10740
Tyson Peak	Benewah	47°06'03"N	116°31'08"W	4745
Ulm Peak	Shoshone	47°53'56"N	115°58'02"W	6444
Ulysses Mountain	Lemhi	45°25'47"N	114°06'38"W	7649
Umbrella Butte	Idaho	45°35'11"N	115°57'11"W	7929
Uranus Peak	Shoshone	47°42'49"N	116°01'20"W	4977
Van Horn Peak	Custer	44°46'47"N	114°19'28"W	9616
Van Meter Hill	Valley	44°59'03"N	115°30'21"W	8025
Van Point	Bonneville	43°18'16"N	111°09'39"W	6399
Vance Mountain	Idaho	45°54'43"N	114°29'34"W	8793
Vance Point	Idaho	45°54'00"N	114°33'28"W	6823
Vanderbilt Hill	Shoshone	46°56'49"N	115°07'50"W	5692
Vanity Summit	Custer	44°28'57"N	115°05'15"W	7813
Verita, Monte	Boise	44°04'30"N	115°02'30"W	9003
Vermilion Peak	Idaho	45°55'44"N	115°08'11"W	7575
Victor Peak	Valley	45°08'53"N	115°53'10"W	8718
Vinegar Hill	Valley	45°09'09"N	114°56'27"W	6020
Vitta Point	Idaho	45°33'23"N	115°18'21"W	5160
Wagon Butte	Blaine	43°13'21"N	113°47'58"W	4947
Wahoo Peak	Idaho	46°09'53"N	114°38'44"W	7682
Wakley Peak	Bannock	42°23'44"N	112°17'30"W	8801
Walde Mountain	Idaho	46°14'45"N	115°37'59"W	5221
Walkers Peak	Valley	44°39'28"N	115°21'08"W	9094
Wall Mountain	Boundary	48°50'48"N	116°14'15"W	5140
Wall Peak	Kootenai	47°36'50"N	116°24'11"W	4119
Wall Point	Idaho	45°57'54"N	115°51'00"W	5620
Wallow Mountain	Clearwater	46°50'38"N	115°23'35"W	5981
Walo Point	Shoshone	47°07'22"N	115°26'11"W	5855
Walters Butte	Canyon	43°20'48"N	116°33'05"W	2938
Wapi Park	Power	42°54'10"N	113°13'38"W	5148
Wapito Point	Clearwater	46°47'41"N	114°56'14"W	6396
War Eagle Mountain	Idaho	45°19'18"N	115°47'38"W	8200
War Eagle Mountain	Owyhee	43°00'25"N	116°42'13"W	8051
War Eagle Peak	Cassia	42°06'32"N	113°06'09"W	8716
Warbonnet Peak	Boise	44°04'44"N	115°00'33"W	10220
Ward Peak	Shoshone	47°15'59"N	115°22'55"W	7312
Wardner Peak	Shoshone	47°29'33"N	116°08'33"W	6198
Wards Butte	Lemhi	44°50'02"N	114°11'53"W	8731
Warm Lake Summit	Valley	44°38'42"N	115°35'04"W	7291
Warm River Butte	Fremont	44°10'28"N	111°14'06"W	6618
Warm Springs Point	Boise	43°48'01"N	115°55'51"W	6054
Warm Springs Point	Idaho	45°23'06"N	116°11'21"W	5839
Warners Mountain	Kootenai	47°22'47"N	116°57'58"W	3307
Warren Mountain	Custer	44°00'31"N	113°28'47"W	9469
Washington Peak	Custer	44°00'31"N	114°40'18"W	10519
Washout Point	Shoshone	47°07'01"N	115°13'26"W	5757
Wasset Peak	Shoshone	46°57'31"N	115°34'25"W	6384
Watchman, The	Butte	43°23'05"N	113°29'16"W	5858
Watchtower Peak	Idaho	45°49'46"N	114°30'39"W	8780
Watson Mountain	Bonner	48°40'06"N	116°54'57"W	3978
Watson Peak	Custer	44°07'51"N	114°41'05"W	10453
Watson Peak	Lemhi	44°53'19"N	113°50'20"W	8671
Watt Hill	Benewah	47°17'45"N	116°54'17"W	3090
Waugh Mountain	Idaho	45°29'39"N	114°47'30"W	8887
Weasel Point	Clearwater	46°36'42"N	114°53'57"W	5761
Weasel Point	Idaho	46°04'25"N	115°03'41"W	4546
Webber Peak	Lemhi	44°20'41"N	112°48'59"W	11223
Webster Butte	Madison	43°47'29"N	111°42'00"W	5373
Wedge Butte	Blaine	43°14'31"N	114°17'37"W	5607
Weitas Butte	Idaho	46°28'14"N	115°27'06"W	5967
Wells Summit	Camas	43°28'45"N	114°45'27"W	7080
West Butte	Fremont	44°00'10"N	111°54'03"W	6194
West Canfield Butte	Kootenai	47°44'02"N	116°43'33"W	4162
West Dennis	Benewah	47°05'33"N	116°43'06"W	4808
West Elk Peak	Shoshone	47°05'13"N	116°16'33"W	5054
West Emerald	Latah	47°00'33"N	116°24'56"W	4622
West Farnes Mountain	Madison	43°40'09"N	111°29'52"W	7396
West Fork Mountain	Boundary	48°49'54"N	116°44'33"W	6416
West Grouse Peak	Shoshone	47°29'58"N	115°50'15"W	6177
West Horse Point	Lemhi	45°25'29"N	114°45'55"W	7398
West Sister	Clearwater	46°52'25"N	115°33'47"W	6721
West Sister Peak	Shoshone	47°10'50"N	115°39'57"W	6781
West Twin	Latah	46°48'40"N	116°54'55"W	4533

Mountain, Peak, or Summit	County	Latitude	Longitude	Elev.
West Warrior Peak	Elmore	43°52'34"N	115°17'55"W	8930
West Willow Peak	Shoshone	47°26'25"N	115°47'00"W	6324
Weston Peak	Franklin	42°09'48"N	112°04'22"W	8165
Wheaton Mountain	Bonneville	43°36'27"N	111°26'53"W	7090
Wheeler Mountain	Idaho	45°46'13"N	115°27'25"W	5535
Whiskey Butte	Clearwater	46°34'56"N	115°58'41"W	4407
Whiskey Butte	Idaho	45°28'14"N	116°20'00"W	4449
Whiskey Butte	Minidoka	42°54'25"N	113°25'04"W	4633
Whiskey Mountain	Owyhee	43°09'52"N	116°48'02"W	6161
Whistling Peak	Shoshone	47°08'25"N	115°37'50"W	6003
White Bird Hill Summit	Idaho	45°50'46"N	116°13'30"W	4420
White Butte	Canyon	43°19'38"N	116°33'33"W	2590
White Cloud Peaks	Custer	44°05'50"N	114°37'37"W	11263
White Goat Mountain	Lemhi	44°43'10"N	114°24'22"W	9320
White Knob Mountains	Custer	43°55'56"N	114°01'52"W	9292
White Knob Mountains	Custer	43°48'00"N	113°42'00"W	11224
White Knob Summit	Custer	43°52'36"N	113°43'09"W	10529
White Mountain	Adams	45°11'01"N	116°35'24"W	8514
White Mountain	Boundary	48°31'13"N	116°31'50"W	5005
White Mountain	Custer	44°34'18"N	114°28'42"W	9870
White Mountain	Lemhi	44°44'30"N	114°25'04"W	9540
White Owl Butte	Madison	43°45'52"N	111°33'36"W	6323
White Peak	Shoshone	47°36'21"N	116°01'11"W	5228
White Quartz Mountain	Power	42°36'11"N	112°43'23"W	7777
White Rock	Shoshone	46°59'24"N	116°06'49"W	5444
White Rock Peak	Valley	44°47'12"N	115°43'48"W	7775
White Top Mountain	Idaho	45°43'00"N	114°57'15"W	7886
White Valley Mountain	Custer	44°37'17"N	114°26'52"W	9986
Whitehawk Mountain	Valley	44°17'15"N	115°31'49"W	8374
Whitehorse Butte	Owyhee	42°06'38"N	116°35'10"W	5469
Whitetail Butte	Bonner	48°23'53"N	116°51'25"W	3105
Whitetail Peak	Shoshone	48°00'35"N	116°09'24"W	5503
Whitetail Peak	Shoshone	47°10'40"N	115°23'43"W	4345
Wickiup Butte	Idaho	45°32'23"N	116°24'42"W	5534
Widow Mountain	Shoshone	47°03'37"N	115°56'50"W	6826
Wiessner, Mount	Kootenai	47°26'21"N	116°20'56"W	6187
Wigwams, The	Boundary	48°43'04"N	116°42'33"W	7033
Wilburn Butte	Washington	44°14'26"N	116°37'45"W	4048
Wild Buck Peak	Valley	44°23'18"N	115°39'49"W	7992
Wild Horse Butte	Idaho	45°40'55"N	116°25'58"W	5458
Wild Horse Butte	Owyhee	43°07'54"N	116°19'35"W	2776
Wild Horse Table	Owyhee	42°23'17"N	116°03'14"W	6100
Wild Mountain	Bannock	42°47'56"N	112°28'10"W	7220
Wilderness Peak	Franklin	42°02'26"N	111°38'46"W	9460
Wildhorse Butte	Butte	43°28'41"N	113°17'12"W	5714
Wildhorse Butte	Lincoln	43°02'57"N	113°46'49"W	4915
Willard, Mount	Bonner	48°26'12"N	116°10'29"W	6536
Williams Peak	Clearwater	46°39'22"N	114°47'55"W	7660
Williams Peak	Custer	44°09'09"N	115°00'18"W	10635
Williams Peak	Valley	45°01'07"N	115°39'43"W	6826
Willow Creek Peak	Cassia	42°04'05"N	114°08'05"W	7620
Willow Creek Summit	Custer	44°13'52"N	113°58'28"W	7161
Willow Peak	Benewah	47°02'25"N	116°25'13"W	4688
Willson Peak	Valley	45°03'12"N	115°26'45"W	8955
Wilson Butte	Jerome	42°47'06"N	114°12'48"W	4387
Wilson Mountain	Benewah	47°09'08"N	116°24'27"W	3780
Wilson Mountain	Lemhi	45°05'17"N	114°37'27"W	9570
Wilson Peak	Boise	43°58'01"N	115°44'59"W	7837
Wilson Peak	Gem	44°27'27"N	116°09'30"W	7865
Wilson Peak	Owyhee	43°18'16"N	116°44'37"W	5353
Wind Butte	Ada	43°14'11"N	116°07'00"W	3406
Windfall Peak	Shoshone	47°58'44"N	116°14'16"W	4605
Windy Devil	Custer	43°54'40"N	113°41'18"W	8204
Windy Peak	Shoshone	46°59'47"N	116°05'09"W	5723
Windy Point	Clearwater	46°51'45"N	116°07'38"W	5550
Windy Point	Idaho	45°48'36"N	116°34'07"W	4220
Windy Point	Owyhee	43°25'16"N	116°53'00"W	4365
Windy Point	Owyhee	43°14'23"N	116°41'39"W	5079
Winter Camp Butte	Owyhee	42°37'39"N	115°30'31"W	4419
Winter Camp Butte	Owyhee	42°37'38"N	115°30'30"W	4418
Wisdom Peak	Shoshone	47°01'42"N	115°07'43"W	6189
Wishard Peak	Shoshone	47°17'35"N	115°30'28"W	6403
Witter Point	Idaho	45°33'36"N	114°48'34"W	7988
Wolf Fang Peak	Valley	45°10'27"N	115°24'56"W	9007
Wolf Lodge Mountain	Kootenai	47°43'21"N	116°33'55"W	4735
Wolf Mountain	Boise	44°00'36"N	115°23'14"W	8876
Wolf Mountain	Caribou	42°40'36"N	111°22'52"W	7470
Wonderful Peak	Shoshone	47°24'32"N	115°45'35"W	5981
Wood Creek Mountain	Elmore	43°21'28"N	115°18'38"W	6351
Wood Hump	Idaho	45°33'06"N	114°40'12"W	7224
Woodall Mountain	Caribou	42°45'09"N	111°31'06"W	7822
Wooden Shoe Butte	Twin Falls	42°16'20"N	114°17'07"W	6793
Woodrat Mountain	Idaho	46°11'13"N	115°45'55"W	4983
Woods Peak	Custer	44°48'07"N	114°18'52"W	9730
Woodtick Summit	Lemhi	44°48'10"N	114°41'12"W	8863
Worley Mountain	Kootenai	47°23'22"N	116°53'29"W	3533
Wrangle Hill	Gooding	43°10'59"N	114°50'12"W	5668
Wylie Knob	Bonner	48°28'49"N	116°19'08"W	4382
Wylies Peak	Idaho	45°58'57"N	114°57'56"W	7799
Yandell Mountain	Bingham	43°05'54"N	112°09'16"W	6399
Yankee Peak	Shoshone	47°00'37"N	115°11'17"W	6721
Yellow Peak	Lemhi	44°31'22"N	113°31'05"W	10968
Zehm Hill	Kootenai	47°25'23"N	116°47'59"W	3190
Ziegler Mountain	Bingham	43°15'53"N	111°49'32"W	6915

Parks & Reserves

Park or Reserve	County	Latitude	Longitude	Elev.
Airport Park	Bingham	43°12'25"N	112°20'46"W	
Airway Park	Nez Perce	46°22'26"N	117°00'08"W	
Aldape Park	Ada	43°36'51"N	116°10'42"W	
Almo Park	Cassia	42°09'36"N	113°43'05"W	7985
Alpine Lake Wildlife Habitat Area	Kootenai	47°48'51"N	116°45'41"W	2440
Ammon Park	Bannock	42°53'01"N	112°25'50"W	
Anares Park	Bonneville	43°30'09"N	112°03'20"W	4725
Anderson Ranch Big Game Range	Elmore	43°24'57"N	115°19'34"W	4400
Ann Morrison Park	Ada	43°36'45"N	116°13'07"W	
Aspendale Wildlife Habitat Area	Latah	46°39'27"N	116°50'10"W	2600
Ayres Park	Ada	43°36'51"N	116°13'45"W	
Baker Creek Picnic Area	Blaine	43°47'05"N	114°32'58"W	
Baker Wildlife Habitat Area North	Cassia	42°25'11"N	114°02'24"W	4293
Balanced Rock County Park	Twin Falls	42°32'45"N	114°57'25"W	
Balanced Rock State Park	Twin Falls	42°32'52"N	114°57'27"W	3700
Barclay Bay Picnic Area	Ada	43°31'10"N	116°03'02"W	
Barry Park	Twin Falls	42°34'07"N	114°29'06"W	
Bear Lake National Wildlife Refuge	Bear Lake	42°09'48"N	111°18'42"W	
Beaver Dam Recreation Area	Clearwater	46°31'42"N	116°18'05"W	
Big Fir Picnic Area	Bannock	42°42'26"N	112°21'42"W	
Billingsley Creek Wildlife Management Area	Gooding	42°50'08"N	114°53'08"W	3000
Billingsley Creek Wildlife Management Area	Gooding	42°49'44"N	114°52'51"W	
Blackfoot River Park	Caribou	42°49'19"N	111°33'01"W	6150
Bogus Basin State Park	Boise	43°46'14"N	116°06'12"W	
Boise River Wildlife Management Area	Ada	43°34'32"N	115°43'56"W	
Bonneville Park	Bannock	42°52'41"N	112°25'35"W	
Bottle Lake Research Natural Area	Bonner	48°42'58"N	116°52'33"W	3000
Bottolfsen Park	Butte	43°37'56"N	113°18'16"W	
Bowden Park	Ada	43°35'42"N	116°13'14"W	
Bruneau Dunes State Park	Owyhee	42°53'41"N	115°41'50"W	
Buck Park	Idaho	45°25'54"N	116°33'41"W	5000
Buckhorn Wildlife Habitat Area	Cassia	42°20'44"N	114°00'42"W	4460
Bull Run Lake Access Area	Kootenai	47°31'47"N	116°28'52"W	2160
Burley Wildlife Habitat Area Southwest	Cassia	42°29'28"N	113°56'36"W	4390
Butte Wildlife Habitat Area North	Cassia	42°24'15"N	114°01'12"W	4235
Butte Wildlife Habitat Area South	Cassia	42°23'40"N	113°59'32"W	4347
C J Strike Wildlife Management Area	Elmore	42°55'22"N	115°54'29"W	
Caldwell Park	Bannock	42°52'20"N	112°26'22"W	
Camas National Wildlife Refuge	Jefferson	43°57'05"N	112°14'54"W	
Camas Prairie Cent. Marsh Wildlife Mngmt Area	Camas	43°17'22"N	115°01'03"W	5060
Camelsback Park	Ada	43°38'17"N	116°12'00"W	
Camozzi Park	Jerome	42°44'20"N	114°30'37"W	
Candlelight Park	Jerome	42°42'47"N	114°30'36"W	
Canyon Creek Recreational Natural Area	Bonner	48°21'10"N	116°45'01"W	
Capitol Avenue Green Belt	Bonneville	43°29'21"N	112°02'41"W	4682
Carey Creek Game Management Area	Bonner	48°08'48"N	116°51'14"W	2140
Carey Lake Wildlife Management Area	Blaine	43°19'25"N	113°55'50"W	4770
Carl Miller Park	Elmore	43°08'11"N	115°41'15"W	
Cascade Park	Twin Falls	42°34'30"N	114°26'45"W	3715
Cascade Park	Twin Falls	42°34'28"N	114°26'50"W	
Cascade Park	Valley	44°30'42"N	116°03'27"W	
Cedar Grove Picnic Area	Idaho	46°32'21"N	114°40'27"W	
Celebration County Park	Owyhee	43°17'57"N	116°31'19"W	
Centennial Waterfront Park	Jerome	42°36'02"N	114°28'08"W	
Central Park	Bonneville	43°30'07"N	112°01'20"W	
Central Park	Payette	44°04'28"N	116°55'55"W	
Chief Joseph War Historical Marker	Idaho	45°35'05"N	116°17'41"W	
City of Rocks National Reserve	Cassia	42°04'34"N	113°42'03"W	5970
Civitan Park	Bonneville	43°30'12"N	112°02'36"W	4710
Claytonia Pond Wildlife Habitat Area	Owyhee	43°34'08"N	116°49'24"W	2260
Clover Creek Wildlife Habitat Area	Elmore	43°00'38"N	115°06'48"W	3000
Coeur d'Alene Indian Reservation	Benewah	47°17'00"N	116°50'00"W	
Coeur d'Alene River Wildlife Management Area	Kootenai	47°22'14"N	116°41'19"W	2160
Corbin Park	Kootenai	47°42'15"N	116°59'22"W	
Craig Mountain Wildlife Management Area	Nez Perce	46°07'29"N	116°55'00"W	2300
Craters of the Moon National Monument	Butte	43°25'00"N	113°31'00"W	
Curley Park	Bonneville	43°29'22"N	112°01'35"W	
Cycle Wildlife Habitat Area	Cassia	42°30'39"N	113°55'47"W	4320
David Thompson State Wildlife Preserve	Bonner	48°12'14"N	116°16'16"W	
Deep Creek Picnic Area	Boundary	48°36'38"N	116°23'53"W	
Deer Creek Picnic Area	Blaine	43°31'44"N	114°30'07"W	
Deer Flat National Wildlife Refuge	Owyhee	43°18'29"N	116°34'36"W	
Deer Park	Camas	43°31'38"N	115°04'31"W	
Dierkes Lake Park	Twin Falls	42°35'43"N	114°23'17"W	
Dingle Swamp Wildlife Habitat Area	Bear Lake	42°17'02"N	111°21'25"W	5920
Discovery Picnic Area	Ada	43°31'32"N	116°03'49"W	
Dry Meadows Wildlife Habitat Area	Cassia	42°21'40"N	113°59'34"W	4405
Duck Valley Indian Reservation	Owyhee	42°04'39"N	116°11'15"W	
Duck Valley Indian Reservation	Owyhee	41°52'31"N	116°07'31"W	
Dworshak State Park	Clearwater	46°35'28"N	116°17'14"W	1840
Eagle Island State Park	Ada	43°41'13"N	116°23'02"W	2530
East Park	Cassia	42°32'20"N	113°47'16"W	
East Side Park	Elmore	43°07'53"N	115°40'45"W	3135
Eastman Park	Twin Falls	42°36'12"N	114°45'35"W	
Elba Park	Cassia	42°14'52"N	113°34'26"W	
Elm Grove Park	Ada	43°38'12"N	116°12'56"W	
Elm Street Park	Bonneville	43°29'20"N	112°01'59"W	4702
Esquire Acres Park	Bonneville	43°30'12"N	112°04'45"W	4728
Fairview Park	Ada	43°37'24"N	116°13'06"W	2676
Fairview Park	Bannock	42°53'07"N	112°26'36"W	
Farragut State Park	Kootenai	47°58'02"N	116°34'54"W	
Farragut Wildlife Management Area	Kootenai	47°58'19"N	116°32'26"W	2100
Filer Community Park	Twin Falls	42°34'03"N	114°36'15"W	
Fish Creek Recreation Area	Idaho	45°50'34"N	116°04'59"W	
Fort Boise Park	Ada	43°36'58"N	116°11'16"W	
Fort Boise Wildlife Management Area	Canyon	43°49'05"N	117°00'32"W	2190

Park or Reserve	County	Latitude	Longitude	Elev.
Fort Hall Historic Monument	Bannock	43°01'13"N	112°38'02"W	
Fort Hall Indian Reservation	Bingham	43°02'30"N	112°07'31"W	
Fort Henry Historic Monument	Fremont	43°55'29"N	111°46'15"W	
Fort Lemhi Monument	Lemhi	44°59'01"N	113°38'23"W	
Frank Church River of No Return	Lemhi	44°40'45"N	114°25'15"W	9030
Frank Church River Of No Return Wilderness	Idaho	45°25'00"N	115°51'59"W	
Fraser Park	Clearwater	46°23'17"N	116°02'49"W	
Freckleton Park	Bannock	42°52'42"N	112°27'13"W	
Freedom Park	Cassia	42°32'25"N	113°45'52"W	4175
Frontier Park	Twin Falls	42°34'48"N	114°27'57"W	
Gayle Forsyth Memorial Park	Jerome	42°43'56"N	114°30'14"W	
Gene Day Park	Shoshone	47°30'52"N	116°01'52"W	2630
Georgetown Summit Wildlife Mngmt Area	Bear Lake	42°33'41"N	111°24'22"W	6468
Gifford Spring Wildlife Habitat Area	Cassia	42°37'14"N	113°12'39"W	4300
Glenns Ferry Wildlife Habitat Area	Elmore	42°59'40"N	115°19'48"W	2800
Golden Valley Wildlife Habitat Areas	Cassia	42°20'50"N	113°59'59"W	4445
Gospel Hump Wilderness	Idaho	45°29'59"N	115°51'59"W	
Grange Park	Benewah	47°09'07"N	116°29'29"W	
Grays Lake National Wildlife Refuge	Bonneville	43°03'47"N	111°25'34"W	
Grimes Monument	Boise	44°01'21"N	115°50'37"W	
Hagerman City Park	Gooding	42°49'09"N	114°53'47"W	
Hagerman Fossil Beds National Monument	Twin Falls	42°47'30"N	114°56'48"W	
Hagerman Wildlife Management Area	Gooding	42°46'26"N	114°52'51"W	
Hagerman Wildlife Management Area	Gooding	42°46'22"N	114°52'48"W	2960
Hanging Trees Historical Marker	Clearwater	46°28'18"N	115°49'01"W	
Harmon Park	Twin Falls	42°33'13"N	114°27'14"W	
Harriman State Park	Fremont	44°20'10"N	111°27'38"W	6115
Harrison City Park	Kootenai	47°27'16"N	116°46'57"W	2320
Harrison Park	Twin Falls	42°34'10"N	114°28'06"W	3705
Hells Canyon National Recreation Area	Idaho	45°33'00"N	116°30'00"W	
Hells Canyon Park	Adams	45°02'44"N	116°47'47"W	1760
Hells Canyon Wilderness Area	Idaho	45°33'00"N	116°30'01"W	
Hells Gate State Recreation Area	Nez Perce	46°20'59"N	117°03'14"W	1200
Henry Stampede Park	Caribou	42°54'20"N	111°33'03"W	6135
Henrys Lake State Park	Fremont	44°37'12"N	111°22'23"W	6501
Heyburn State Park	Benewah	47°21'12"N	116°45'37"W	
Highland Park	Bonneville	43°30'23"N	112°02'15"W	
Historic Monument Site of First Ferry	Madison	43°49'37"N	111°54'18"W	
Hobo Botanical Area	Shoshone	47°05'02"N	116°06'58"W	4600
Horse Butte Wildlife Habitat Area	Cassia	42°30'37"N	113°23'49"W	4645
Horse Creek Administrative Research Area	Idaho	45°59'45"N	115°22'33"W	5916
Horsethief Reservoir State Park	Valley	44°30'47"N	115°54'49"W	5159
Hudspeth Cutoff Historical Marker	Oneida	42°15'47"N	112°45'43"W	5175
Hunt Girl Creek Research Natural Area	Boundary	48°31'54"N	116°09'46"W	6298
Hyland Park	Bannock	42°51'56"N	112°28'33"W	
Idaho National Engineering Laboratory	Butte	43°37'31"N	112°45'01"W	
Idaho Primitive Area	Lemhi	45°02'54"N	114°38'17"W	
Idahome Waterfowl Wildlife Habitat Areas	Cassia	42°26'31"N	113°21'39"W	4380
Idahome Wildlife Habitat Area	Cassia	42°25'09"N	113°24'02"W	4415
Indian Cove Wildlife Habitat Area	Owyhee	42°54'52"N	115°45'31"W	2620
Indian Massacre Historical Monument	Clark	44°06'53"N	112°52'55"W	
Indian Rocks State Park	Bannock	42°42'43"N	112°13'08"W	
Ivywild Park	Ada	43°34'54"N	116°11'20"W	2730
Jackson Lab Wildlife Habitat Area	Cassia	42°37'58"N	113°32'03"W	4190
J-C Park	Twin Falls	42°33'17"N	114°27'10"W	
Jensen Grove Park	Bingham	43°12'05"N	112°21'25"W	
Julia Davis Park	Ada	43°36'28"N	116°12'00"W	
Kaniksu Marsh Research Natural Area	Bonner	48°26'25"N	116°55'00"W	2525
Kelly Park	Caribou	42°39'56"N	111°34'42"W	5870
Kiwanis Park	Payette	44°04'06"N	116°56'10"W	
Kootenai Indian Reservation	Boundary	48°44'37"N	116°22'10"W	
Kootenai National Wildlife Refuge	Boundary	48°42'30"N	116°24'30"W	
Kube Park	Idaho	46°27'14"N	114°42'42"W	
Kurtz Park	Canyon	43°31'58"N	116°33'50"W	
Laird Park	Latah	46°56'35"N	116°38'56"W	
Lakeview Park	Canyon	43°32'06"N	116°32'55"W	
Land of the Yankee Fork State Park	Custer	44°30'27"N	114°10'52"W	
Langer Monument	Custer	44°27'56"N	115°06'11"W	7080
Lapwai State Game Farm	Nez Perce	46°22'42"N	116°47'38"W	
Lewis and Clark Monument	Lemhi	45°28'07"N	113°59'23"W	3910
Lewis-Clark Canoe Camp State Park	Clearwater	46°29'57"N	116°20'19"W	
Little Banks Island Wildlife Habitat Area	Payette	44°05'12"N	116°58'20"W	2140
Little Drops Recreation Area	Lincoln	42°58'47"N	114°25'13"W	
Locust Park	Owyhee	42°56'47"N	115°58'51"W	2435
Lucky Peak State Recreation Area	Ada	43°31'49"N	116°01'17"W	
Mae McEuen Playfield	Kootenai	47°40'18"N	116°46'42"W	2135
Magee Historic Site	Kootenai	47°51'20"N	116°15'05"W	
Malad Gorge State Park	Gooding	42°51'56"N	114°51'35"W	
Malad Gorge State Park	Gooding	42°51'28"N	114°52'41"W	
Mallard Larkins Pioneer Park	Shoshone	46°56'56"N	115°33'21"W	5800
Malta Wildlife Habitat Area North	Cassia	42°28'20"N	113°20'16"W	4370
Malta Wildlife Habitat Area South	Cassia	42°27'05"N	113°19'31"W	4380
Mann Creek State Park	Washington	44°23'57"N	116°54'20"W	3200
Marion No. One Wildlife Habitat Area East	Cassia	42°19'14"N	113°49'24"W	4425
Marion No. One Wildlife Habitat Area South	Cassia	42°18'20"N	113°51'02"W	4425
Marion No. Two Wildlife Habitat Area East	Cassia	42°18'45"N	113°49'47"W	4430
Marion No. Two Wildlife Habitat Area South	Cassia	42°18'19"N	113°50'06"W	4445
Mark Means Park	Nez Perce	46°24'56"N	116°58'31"W	
Market Lake Wildlife Management Area	Jefferson	43°46'54"N	112°08'46"W	
Mary L Gooding Memorial Park	Lincoln	42°56'17"N	114°24'33"W	
Mary Minerva McCroskey Memorial State Park	Latah	47°03'54"N	116°56'58"W	
Massacre Rocks State Park	Power	42°40'52"N	112°58'56"W	
McCauley Park	Ada	43°37'37"N	116°12'28"W	
McCowin Park	Bonneville	43°28'37"N	111°58'12"W	4720
McGinnis Park	Gooding	42°46'28"N	114°41'27"W	
Memorial Field	Kootenai	47°40'41"N	116°47'33"W	2230

Parks & Reserves

Park or Reserve	County	Latitude	Longitude	Elev.
Memorial Park	Bannock	42°51'52"N	112°27'27"W	
Memorial Park	Elmore	43°08'04"N	115°41'31"W	3145
Mica Bay County Campground	Kootenai	47°35'22"N	116°51'44"W	
Middle Fork Clearwater Wild and Scenic River	Idaho	46°30'18"N	114°47'38"W	3400
Middle Fork Salmon Wild And Scenic River	Valley	45°46'08"N	115°04'54"W	
Milner Butte Wildlife Habitat Areas	Cassia	42°27'42"N	114°01'31"W	4510
Milner Historical Recreation Area	Jerome	42°31'24"N	114°00'11"W	
Milner Recreation Area	Cassia	42°31'31"N	113°56'21"W	4185
Minidoka Forest State Bird Sanctuary	Cassia	42°09'29"N	113°07'29"W	
Minidoka National Wildlife Refuge	Cassia	42°40'13"N	113°23'19"W	
Montpelier Wildlife Management Area	Bear Lake	42°20'19"N	111°15'05"W	7400
Moose Meadow Creek Research Natural Area	Idaho	45°38'00"N	115°29'30"W	7425
Morton Slough Game Management Area	Bonner	48°12'09"N	116°41'31"W	2120
Mowry State Park	Kootenai	47°27'45"N	116°51'24"W	2200
Mud Lake Wildlife Management Area	Jefferson	43°53'58"N	112°23'27"W	
Mullen Trail Park Historical Monument	Benewah	47°19'08"N	116°34'58"W	2200
Murtaugh Lake County Park	Twin Falls	42°27'44"N	114°10'05"W	
Myrtle Creek Game Preserve	Boundary	48°43'45"N	116°33'08"W	
N O P Park	Bannock	42°53'51"N	112°27'36"W	
Nelson Game Enclosure	Caribou	42°37'07"N	111°41'28"W	6080
Neptune Park	Minidoka	42°36'38"N	113°40'24"W	
Nez Perce Indian Reservation	Lewis	46°17'30"N	116°22'30"W	
Nez Perce National Hist. Park - Spalding Area	Nez Perce	46°27'03"N	116°48'55"W	
Nez Perce National Historical Park	Idaho	46°12'34"N	116°00'20"W	
Nez Perce National Historical Park	Idaho	45°47'38"N	116°16'38"W	2200
Nez Perce National Historical Park	Nez Perce	46°27'03"N	116°48'55"W	
Nezperce Indian War Historical Monument	Idaho	45°47'35"N	116°16'16"W	
Niagara Springs Wildlife Management Area	Gooding	42°40'15"N	114°42'45"W	
Niagara Springs Wildlife Management Area	Gooding	42°40'03"N	114°41'25"W	3025
Noble Park	Ada	43°36'41"N	116°11'30"W	
North Beach State Park	Bear Lake	42°01'40"N	111°15'02"W	6393
North Park	Kootenai	47°43'46"N	116°56'47"W	2218
North Rock Wildlife Habitat Area	Cassia	42°16'23"N	113°51'24"W	4520
North Shore Game Management Area	Bonner	48°10'35"N	116°57'02"W	2185
North Shore Picnic Ground	Custer	44°08'20"N	114°55'28"W	
Oakley Wildlife Habitat Area East	Cassia	42°13'47"N	113°52'12"W	4700
Oden Bay Game Management Area	Bonner	48°19'07"N	116°26'25"W	
O'Hara Creek Research Natural Area	Idaho	45°59'30"N	115°31'45"W	6040
Old Mission State Park	Kootenai	47°32'57"N	116°21'36"W	
Oregon Trail Historical Marker	Elmore	43°11'56"N	115°33'16"W	
Pack River Flats Rifuge	Bonner	48°19'55"N	116°22'50"W	
Packer Johns Cabin State Park	Adams	44°57'34"N	116°13'31"W	
Palisades Cr. Winter Range Wildlife Habitat Area	Bonneville	43°23'39"N	111°15'12"W	5600
Palisades Winter Range	Bonneville	43°20'08"N	111°10'58"W	6600
Panhandle Wildlife Mangement Area	Bonner	48°08'38"N	116°12'18"W	2095
Paradise Wildlife Habitat Area	Latah	46°42'26"N	116°56'22"W	2850
Payette River Wildlife Management Area	Gem	43°52'21"N	116°33'08"W	2320
Payette River Wildlife Management Area	Payette	43°59'17"N	116°47'43"W	2205
Pend Oreille Wildlife Management Area	Bonner	48°18'00"N	116°22'40"W	
Pend Oreille Wildlife Management Area:				
Carey Unit	Bonner	48°08'52"N	116°51'21"W	2060
Hoodoo Creek Unit	Bonner	48°09'26"N	116°45'22"W	2060
Mallard Bay Unit	Bonner	48°14'11"N	116°41'10"W	2060
Morton Slough Unit	Bonner	48°12'21"N	116°41'25"W	2060
Oden Bay Unit	Bonner	48°18'38"N	116°27'12"W	2060
Pack River Unit	Bonner	48°19'11"N	116°22'53"W	2062
Priest River Unit	Bonner	48°10'46"N	116°53'05"W	2077
Riley Creek Unit	Bonner	48°09'39"N	116°46'44"W	2085
Piemeisel Wildlife Habitat Area	Cassia	42°30'20"N	113°53'35"W	4337
Pine Street Park	Bonner	48°16'25"N	116°33'35"W	2096
Pioneer Women Historical Monument	Franklin	42°08'27"N	111°54'37"W	
Pocatello Elk Refuge	Bannock	42°48'36"N	112°28'53"W	6700
Pocatello Game Preserve	Bannock	42°46'34"N	112°28'00"W	
Poitevin Park	Bonneville	43°29'10"N	112°02'21"W	4695
Ponderosa State Park	Valley	44°56'42"N	116°04'24"W	
Portneuf Wildlife Management Area	Bannock	42°42'17"N	112°12'02"W	4900
Power Line Wildlife Habitat Area	Cassia	42°23'20"N	113°55'19"W	4327
Priest Lake State Forest	Boundary	48°34'00"N	116°46'30"W	
Priest River Park	Bonner	48°10'45"N	116°53'20"W	
Railroad Park	Elmore	43°07'54"N	115°41'46"W	3145
Rainey Park	Bannock	42°51'30"N	112°26'39"W	
Rapid National Wild River	Adams	45°12'10"N	116°27'59"W	5400
Raymond Park	Bannock	42°52'11"N	112°27'53"W	
Rhinehart Park	Bonneville	43°30'17"N	112°03'54"W	4755
Richard Aguirre Park	Elmore	43°08'24"N	115°41'41"W	3170
Riley Creek Recreation Area	Bonner	48°09'40"N	116°46'13"W	
River Park	Kootenai	47°42'07"N	116°56'50"W	2185
Riverfront Park	Lewis	46°13'45"N	116°00'57"W	1160
Roadside Park	Latah	47°02'27"N	116°52'17"W	2830
Rock Creek Park	Twin Falls	42°34'14"N	114°30'25"W	
Rock Wildlife Habitat Area	Cassia	42°15'40"N	113°51'35"W	4557
Rollandet Park	Bonneville	43°28'40"N	112°02'35"W	4695
Rose Lake Wildlife Habitat Area	Kootenai	47°32'58"N	116°29'30"W	2140
Rosedale Memorial Park	Payette	44°05'26"N	116°54'40"W	
Ross Park	Bannock	42°50'59"N	112°25'18"W	
Round Lake State Park	Bonner	48°09'40"N	116°38'09"W	
Rowell Marsh Wildlife Habitat Area	Canyon	43°44'50"N	117°00'22"W	2200
Russ Freeman Park	Bonneville	43°30'53"N	112°03'04"W	
Sacajawea Monument	Lemhi	44°58'45"N	113°38'19"W	
Saint Joe Wild and Scenic River	Shoshone	47°03'41"N	115°21'08"W	
Saint Maries Wildlife Management Area	Benewah	47°15'47"N	116°33'01"W	3600
Salmon River Breaks Primitive Area	Idaho	45°32'37"N	114°53'02"W	
Salmon Wild and Scenic River	Idaho	45°30'01"N	115°00'01"W	
Samowen Park	Bonner	48°13'12"N	116°17'25"W	2100
Sand Hollow Wildlife Habitat Area	Canyon	43°46'47"N	116°37'37"W	2540
Scardino Park	Bannock	42°54'07"N	112°26'27"W	

Park or Reserve	County	Latitude	Longitude	Elev.
Shamrock Park	Bonneville	43°28'10"N	112°00'32"W	4710
Shoshone Falls Park	Twin Falls	42°35'33"N	114°24'20"W	
Shoshone Park	Shoshone	47°27'55"N	115°43'38"W	
Sikes Act Wildlife Habitat Area	Blaine	42°46'45"N	113°25'45"W	4360
Smiths Ferry Wildlife Habitat Area	Valley	44°16'42"N	116°04'25"W	4520
Snake National Wild River	Idaho	45°34'01"N	116°29'59"W	
Snake River Bird of Prey Natural Area	Owyhee	43°10'19"N	116°19'44"W	
Snake River Birds of Prey Area	Owyhee	42°52'31"N	115°37'31"W	
Snow Peak Wildlife Management Area	Shoshone	47°02'40"N	115°38'52"W	5600
Soldier Mountain Game Preserve	Elmore	43°19'05"N	115°14'59"W	
South Fork Payette River Game Preserve	Boise	44°08'47"N	115°16'02"W	
Sportsman Park	Kootenai	47°48'42"N	116°41'36"W	
Sportsmens Park	Kootenai	47°48'42"N	116°41'40"W	
Springfield Bird Haven	Bingham	43°04'44"N	112°41'25"W	
Star Lane Wildlife Habitat Area	Gem	43°50'38"N	116°33'11"W	2340
Sterling Wildlife Management Area	Bingham	42°58'22"N	112°46'56"W	
Stock Park	Bear Lake	42°19'24"N	111°17'46"W	
Stuart Park	Bannock	42°55'38"N	112°28'27"W	4480
Sunrise Park	Twin Falls	42°34'03"N	114°26'58"W	
Sunset Park	Ada	43°38'30"N	116°13'45"W	2670
Tautphaus Park	Bonneville	43°28'23"N	112°02'18"W	
Ted Trueblood Wildlife Habitat Area	Elmore	43°00'39"N	116°06'12"W	2400
Tepee Creek Research Natural Area	Bonner	48°43'34"N	116°52'17"W	3000
Tex Creek Wildlife Management Area	Bonneville	43°28'05"N	111°40'40"W	6100
Thorne Creek Wildlife Habitat Area	Gooding	43°06'05"N	114°36'57"W	4450
Three Devils Picnic Area	Idaho	46°08'08"N	115°38'43"W	
Three Island Crossing State Park	Elmore	42°56'17"N	115°19'07"W	
Three Island State Park	Elmore	42°55'59"N	115°18'54"W	2740
Tom Beal Park	Idaho	46°26'28"N	114°43'56"W	
Triangle Park	Bonner	48°15'57"N	116°32'14"W	2082
Trueblood Wildlife Area	Elmore	43°00'37"N	116°05'56"W	
United States Sheep Experiment Station	Clark	44°17'04"N	112°08'02"W	5700
University of Idaho Experimental Forest	Latah	46°47'54"N	116°48'42"W	
Upper Fishook Research Natural Area	Shoshone	47°07'08"N	115°51'28"W	4881
Van Tassel Wildlife Habitat Area North	Cassia	42°23'14"N	113°56'44"W	4352
Van Tassel Wildlife Habitat Area South	Cassia	42°22'37"N	113°56'15"W	4362
Veterans Memorial State Park	Ada	43°38'16"N	116°14'08"W	2654
Vollmer Park	Nez Perce	46°24'51"N	117°00'52"W	
Ward Memorial State Park	Canyon	43°40'30"N	116°36'34"W	
Webb Brothers Wildlife Habitat Area	Cassia	42°34'09"N	113°15'23"W	4300
West Park	Cassia	42°32'11"N	113°47'52"W	
West Shoshone Park	Shoshone	47°32'32"N	116°13'38"W	2211
William Craig Historical Monument	Nez Perce	46°21'27"N	116°46'07"W	
Winchester State Park	Lewis	46°14'05"N	116°37'13"W	
Winstead Park	Ada	43°37'41"N	116°15'18"W	2704
Wood River Deer Migration Corridor	Blaine	43°19'45"N	114°22'59"W	4880
Yale Wildlife Habitat Area East	Cassia	42°37'27"N	113°10'35"W	4300
Yale Wildlife Habitat Area West	Cassia	42°37'27"N	113°12'08"W	4300

Pass, Gap, or Saddle	County	Latitude	Longitude	Elev.
12 Mile Saddle	Idaho	46°30'35"N	115°09'13"W	5537
Antelope Pass	Custer	43°46'18"N	113°45'50"W	8934
Archer Saddle	Idaho	45°55'06"N	114°56'28"W	
Avery Creek Nocelly Gulch Saddle	Shoshone	47°40'59"N	115°54'22"W	3750
Avery Saddle	Shoshone	47°41'23"N	115°53'37"W	4125
Bannack Pass	Clark	44°28'42"N	112°47'08"W	
Banner Summit	Boise	44°18'23"N	115°13'51"W	7020
Bannock Pass	Lemhi	44°48'50"N	113°16'14"W	
Basin-Elba Pass	Cassia	42°14'41"N	113°42'24"W	7106
Bean Creek Saddle	Idaho	45°29'47"N	116°21'56"W	
Bear Creek Pass	Idaho	46°06'59"N	114°29'33"W	
Bear Creek Pass	Idaho	46°26'07"N	114°22'35"W	
Bear Saddle	Adams	45°08'16"N	116°31'09"W	
Bear Saddle	Washington	44°36'08"N	116°59'31"W	6100
Bear Trap Saddle	Valley	45°04'47"N	114°59'41"W	
Beartrap Saddle	Idaho	45°50'14"N	116°01'24"W	
Beartrap Saddle	Idaho	45°22'35"N	115°49'34"W	
Beauty Saddle	Kootenai	47°33'48"N	116°37'16"W	3485
Beaver Dam Saddle	Idaho	46°25'28"N	115°33'36"W	5302
Beaver Pass	Bonner	48°44'08"N	116°57'40"W	
Beaver Saddle	Idaho	46°27'34"N	115°36'52"W	
Beaverdam Pass	Cassia	42°05'15"N	114°06'10"W	
Beaverland Pass	Butte	43°44'55"N	113°16'53"W	
Beers Spur	Franklin	42°08'48"N	111°57'46"W	4670
Bell Saddle	Adams	44°51'54"N	116°11'30"W	
Bennetts Pass	Bannock	42°59'38"N	112°03'07"W	6297
Benton Saddle	Washington	44°38'52"N	116°58'54"W	5740
Big Canyon Saddle	Idaho	45°41'23"N	116°25'25"W	
Big Creek Summit	Valley	44°37'33"N	115°47'50"W	6577
Big Fog Saddle	Idaho	46°06'50"N	115°12'11"W	
Big Hole Pass	Lemhi	45°32'58"N	113°49'13"W	
Big Horse Basin Gap	Owyhee	42°42'57"N	116°04'05"W	
Bingo Saddle	Clearwater	46°47'27"N	115°44'53"W	
Bishop Saddle	Shoshone	47°58'42"N	116°17'25"W	
Black Bear Saddle	Clearwater	46°46'28"N	115°47'10"W	4850
Black Canyon Saddle	Shoshone	47°42'03"N	116°13'51"W	4307
Blackburn Saddle	Idaho	45°12'20"N	114°58'48"W	
Blackmare Summit	Valley	44°48'19"N	115°47'43"W	7900
Blacktail Pass	Teton	43°39'15"N	111°22'22"W	
Blodgett Pass	Idaho	46°16'47"N	114°26'50"W	
Boardman Pass	Camas	43°30'39"N	114°58'24"W	
Bootjack Pass	Fremont	44°36'04"N	111°23'05"W	
Boulder Pass	Idaho	46°19'18"N	115°12'30"W	
Bradley Summit	Blaine	43°20'46"N	114°03'36"W	5410
Breezy Saddle	Shoshone	47°06'17"N	115°53'11"W	
Brown Creek Saddle	Shoshone	47°43'12"N	116°02'39"W	
Browns Rock Saddle	Clearwater	46°42'31"N	115°42'56"W	
Brush Creek Saddle	Idaho	45°16'30"N	116°38'20"W	5500

Passes, Gaps, Saddles

Pass, Gap, or Saddle	County	Latitude	Longitude	Elev.
Buck Creek Saddle	Idaho	45°32'16"N	114°57'43"W	
Buck Saddle	Idaho	45°14'56"N	115°35'57"W	
Buck Saddle	Valley	44°49'30"N	116°11'53"W	
Buckskin Saddle	Bonner	48°02'03"N	116°10'58"W	4519
Bullion Pass	Shoshone	47°24'26"N	115°40'34"W	5449
Butte Creek Saddle	Clearwater	46°47'44"N	115°43'40"W	4576
Cabinet Pass	Boundary	48°53'20"N	116°58'22"W	
Cache Saddle	Clearwater	46°41'50"N	114°45'14"W	7105
Canada Saddle	Idaho	45°23'59"N	115°48'09"W	
Canuck Pass	Boundary	48°55'07"N	116°03'32"W	
Canyon Gap	Bonner	48°17'09"N	116°43'43"W	
Caribou Pass	Boundary	48°40'13"N	116°32'40"W	
Cascade Saddle	Kootenai	47°48'55"N	116°25'20"W	
Cedar Saddle	Idaho	45°55'29"N	114°40'29"W	5650
Cedar Saddle	Kootenai	47°52'53"N	116°37'15"W	3753
Center Ridge Saddle	Idaho	45°37'04"N	116°22'12"W	
Chamook Saddle	Idaho	46°21'05"N	115°41'26"W	
Chatfield Saddle	Kootenai	47°32'15"N	116°36'31"W	3248
Chief Joseph Pass	Lemhi	45°41'06"N	113°55'58"W	7264
Chilcoot Pass	Valley	44°45'58"N	115°25'25"W	
Chimney Saddle	Idaho	45°17'40"N	116°38'57"W	4700
Clarks Cut	Caribou	43°00'24"N	111°29'24"W	
Cloochman Saddle	Valley	45°12'09"N	115°59'44"W	
Coffee Can Saddle	Idaho	45°26'51"N	116°04'35"W	
Cold Spring Summit	Adams	44°47'06"N	116°18'22"W	6218
Cold Springs Saddle	Adams	45°11'01"N	116°25'17"W	
Cold Springs Saddle	Idaho	45°42'40"N	116°02'30"W	5600
Cold Storage Saddle	Idaho	46°31'20"N	114°48'45"W	4364
Cooks Pass	Boundary	48°42'30"N	116°33'23"W	
Cooper Pass	Shoshone	47°31'30"N	115°42'18"W	5802
Coopers Saddle	Clearwater	46°53'59"N	115°56'32"W	2780
Cougar Creek Summit	Valley	44°48'40"N	115°47'46"W	7820
Cougar Saddle	Valley	44°50'02"N	115°49'10"W	
Cow Creek Saddle	Clearwater	46°33'56"N	115°57'08"W	3580
Cow Creek Saddle	Idaho	45°32'54"N	116°24'34"W	
Curran Saddle	Kootenai	47°38'29"N	116°27'21"W	3970
Cutoff Saddle	Adams	44°47'05"N	116°18'20"W	5615
Danish Pass	Franklin	42°04'57"N	111°35'21"W	
Daves Pass	Cassia	42°04'54"N	114°02'38"W	
Davies Pass	Shoshone	47°04'59"N	116°09'16"W	
Dead Man Saddle	Idaho	45°35'43"N	115°04'35"W	
Deadhorse Saddle	Clearwater	46°44'16"N	115°37'54"W	
Deadhorse Saddle	Idaho	45°47'35"N	116°25'03"W	
Deadman Pass	Lemhi	44°48'13"N	113°12'06"W	
Deadman Pass	Lemhi	44°48'13"N	113°12'06"W	
Deep Saddle	Idaho	46°31'32"N	114°55'41"W	
Deep Saddle	Idaho	46°24'56"N	115°24'40"W	5027
Deep Saddle	Idaho	45°27'53"N	116°13'08"W	5955

Pass, Gap, or Saddle	County	Latitude	Longitude	Elev.
Deer Creek Pass	Butte	44°04'44"N	113°20'29"W	
Deer Creek Pass	Valley	44°24'25"N	115°31'37"W	
Deerhorn Pass	Cassia	42°07'12"N	114°11'55"W	
Dennis Saddle	Clearwater	46°53'08"N	115°49'55"W	4150
Devils Elbow	Washington	44°19'50"N	116°55'47"W	2380
Devils Gap	Clark	44°09'19"N	112°38'11"W	
Devils Gate	Caribou	42°53'52"N	111°49'47"W	6780
Dixie Summit	Elmore	43°17'32"N	115°26'47"W	5220
Dixie Summit	Idaho	45°36'52"N	115°26'48"W	6204
Dobson Pass	Shoshone	47°32'13"N	115°53'08"W	4235
Dodson Pass	Gem	44°20'07"N	116°20'20"W	
Double W Divide	Bingham	43°17'08"N	111°52'23"W	7071
Doublespring Pass	Custer	44°13'25"N	113°50'23"W	8318
Drake Saddle	Idaho	45°59'33"N	115°07'39"W	
Dry Buck Summit	Valley	44°09'07"N	116°10'01"W	5890
Dry Saddle	Idaho	45°40'39"N	114°59'31"W	7863
Eagle Pass	Power	42°32'49"N	112°27'49"W	
East Saddle	Clearwater	46°40'52"N	115°04'43"W	4616
Eighty Day Saddle	Shoshone	47°41'37"N	116°07'02"W	4020
Elk Summit	Valley	45°09'02"N	115°25'21"W	8670
Fall Creek Saddle	Valley	44°57'20"N	115°58'28"W	
Fan Creek Saddle	Idaho	46°14'46"N	115°39'14"W	
Fernan Saddle	Kootenai	47°44'09"N	116°36'17"W	4061
Finns Saddle	Clearwater	46°53'35"N	115°49'03"W	3990
Fire Camp Saddle	Idaho	45°26'23"N	116°28'23"W	
Firebox Summit	Lemhi	44°25'19"N	113°21'16"W	9018
Fish Butte Saddle	Idaho	46°19'36"N	115°24'29"W	
Fish Lake Saddle	Idaho	46°19'30"N	115°05'47"W	
Fisher Creek Saddle	Idaho	45°09'26"N	116°06'20"W	
Five Fingers Saddle	Kootenai	47°44'45"N	116°31'14"W	3500
Flat Creek Saddle	Shoshone	47°46'34"N	116°10'27"W	4598
Fog Mountain Saddle	Idaho	46°04'42"N	115°15'41"W	
Freeman Pass	Caribou	42°39'37"N	111°10'59"W	
Freezeout Saddle	Shoshone	47°01'34"N	116°02'49"W	
French Saddle	Clearwater	46°29'59"N	115°41'43"W	
Friday Pass	Idaho	46°23'05"N	114°43'48"W	
Game Warden Saddle	Idaho	45°19'40"N	116°26'54"W	
Garden Creek Gap	Bannock	42°34'24"N	112°18'41"W	
Georgetown Summit	Bear Lake	42°31'30"N	111°24'52"W	
Glidden Pass	Shoshone	47°32'01"N	115°42'49"W	
Glover Saddle	Idaho	46°08'52"N	115°22'30"W	
Golden Gate	Bonneville	43°18'27"N	111°22'14"W	7580
Goldman Cut	Valley	45°08'45"N	115°23'10"W	
Goldstone Pass	Lemhi	45°08'36"N	113°33'59"W	
Government Trail Pass	Teton	43°37'58"N	111°15'34"W	
Granite Pass	Cassia	41°59'32"N	113°51'26"W	
Granite Pass	Cassia	41°59'32"N	113°51'00"W	7000
Granite Pass	Clearwater	46°39'33"N	114°37'40"W	

Pass, Gap, or Saddle	County	Latitude	Longitude	Elev.
Grave Creek Saddle	Idaho	45°27'12"N	116°24'40"W	
Green Canyon Pass	Fremont	44°20'30"N	111°32'10"W	6927
Green Pass	Bear Lake	42°16'52"N	111°34'41"W	8139
Green Saddle	Idaho	46°24'58"N	115°27'18"W	
Green Saddle	Idaho	45°30'53"N	115°46'10"W	
Grimes Creek Pass	Boise	43°59'08"N	115°49'22"W	4980
Grimes Pass	Boise	44°01'11"N	115°50'34"W	
Griner Saddle	Clearwater	46°44'53"N	115°55'18"W	3180
Grizzly Saddle	Idaho	46°05'47"N	114°57'14"W	
Gunsight, The	Custer	44°07'45"N	114°36'20"W	
Half Moon Pass	Owyhee	42°45'50"N	116°21'20"W	
Hamby Saddle	Idaho	45°59'53"N	115°39'06"W	
Hanson Saddle	Clearwater	46°49'23"N	115°43'02"W	4180
Happy Fork Gap	Bonner	48°17'59"N	116°42'11"W	
Harrington Saddle	Idaho	45°32'17"N	114°54'55"W	7091
Haystack Saddle	Idaho	45°42'54"N	114°46'17"W	
Haystack Saddle	Shoshone	47°45'38"N	115°56'19"W	3628
Hells Half Acre Saddle	Idaho	45°40'36"N	114°36'41"W	
High Pass	Elmore	43°56'46"N	115°10'27"W	
Hobo Pass	Shoshone	47°04'47"N	116°08'19"W	
Holbrook Saddle	Adams	45°14'14"N	116°31'27"W	
Hollow Top	Blaine	43°19'10"N	113°35'04"W	5377
Hoodoo Pass	Shoshone	46°58'32"N	115°01'34"W	
Horse Creek Pass	Lemhi	45°30'11"N	114°23'38"W	7305
Horse Heaven Saddle	Idaho	45°38'14"N	114°54'34"W	
Horseheaven Pass	Custer	44°14'03"N	113°45'15"W	
Hudlow Saddle	Kootenai	47°50'21"N	116°33'36"W	
Hyde Saddle	Owyhee	42°54'40"N	116°32'12"W	
Independence Saddle	Kootenai	47°51'42"N	116°25'57"W	4486
Inkom Pass	Bannock	42°48'10"N	112°08'21"W	
Inman Pass	Caribou	42°51'16"N	112°07'55"W	
Jack Creek Summit	Idaho	45°39'13"N	115°20'40"W	5646
Jensen Pass	Oneida	42°20'53"N	112°30'42"W	
Jesse Pass	Idaho	46°09'54"N	115°12'58"W	
Jim Brown Pass	Power	42°35'44"N	112°42'07"W	
Joes Gap	Adams	45°11'18"N	116°35'11"W	
Joes Gap	Bear Lake	42°22'15"N	111°16'10"W	
John Saddle	Kootenai	47°41'38"N	116°25'30"W	3790
Johnson Saddle	Bonner	48°03'43"N	116°16'20"W	4770
Johnson Saddle	Idaho	45°29'12"N	115°50'05"W	
Johnstone Pass	Blaine	43°43'30"N	114°04'26"W	
Jordan Saddle	Shoshone	47°54'37"N	116°00'51"W	4000
Katka Pass	Boundary	48°37'56"N	116°09'55"W	
Kauffman Saddle	Clearwater	46°49'42"N	115°43'03"W	
Kellogg Saddle	Shoshone	47°25'13"N	116°07'02"W	
Kelly Pass	Boundary	48°31'24"N	116°15'11"W	
Kim Creek Saddle	Idaho	45°40'23"N	114°46'21"W	
Kings Pass	Shoshone	47°37'23"N	115°54'34"W	3319

Pass, Gap, or Saddle	County	Latitude	Longitude	Elev.
Laverne Saddle	Shoshone	47°41'02"N	116°18'08"W	3980
Leatherman Pass	Custer	44°05'02"N	113°44'31"W	
Leiberg Saddle	Kootenai	47°46'40"N	116°19'05"W	
Lemhi Pass	Bingham	43°17'28"N	112°25'26"W	4564
Lemhi Pass	Lemhi	44°58'28"N	113°26'40"W	
Lemon Creek Saddle	Idaho	45°29'34"N	115°31'08"W	
Lick Creek Summit	Valley	45°02'11"N	115°55'54"W	
Lightning Creek Saddle	Idaho	45°27'08"N	116°16'02"W	
Little Freezeout	Gem	43°50'05"N	116°37'15"W	
Little Half Moon Pass	Owyhee	42°46'09"N	116°19'26"W	
Little Horse Basin Gap	Owyhee	42°43'11"N	116°05'11"W	
Little Slate Creek Saddle	Idaho	45°27'54"N	116°06'31"W	5465
Lockwood Saddle	Adams	45°07'53"N	116°39'04"W	
Lolo Pass	Benewah	47°13'11"N	116°50'07"W	3373
Lolo Pass	Clearwater	46°38'07"N	114°34'44"W	5235
Lone Pine Pass	Clark	44°26'23"N	112°06'08"W	
Long Canyon Pass	Boundary	48°46'55"N	116°39'32"W	
Lookout Pass	Shoshone	47°27'23"N	115°41'46"W	4725
Lost Hat Saddle	Clearwater	46°27'17"N	115°41'37"W	4380
Lost Horse Pass	Idaho	46°09'55"N	114°30'07"W	6616
Lost Trail Pass	Lemhi	45°41'38"N	113°56'50"W	
Lovell Saddle	Clearwater	46°49'20"N	115°46'28"W	4501
Low Pass	Washington	44°17'35"N	116°20'49"W	4468
Low Saddle	Idaho	45°29'54"N	116°30'24"W	
Lyman Pass	Cassia	42°04'52"N	113°46'14"W	6196
Lynes Saddle	Adams	45°05'34"N	116°42'33"W	
Magruder Saddle	Idaho	45°41'10"N	114°48'55"W	
Malad Pass	Oneida	42°06'08"N	112°25'25"W	
Marie Saddle	Kootenai	47°39'21"N	116°28'13"W	4174
Mason Saddle	Kootenai	47°38'41"N	116°29'47"W	4395
Meadows Divide, The	Oneida	42°10'24"N	112°46'10"W	
Mertary Creek Saddle	Idaho	45°43'53"N	115°58'22"W	4854
Meyers Saddle	Kootenai	47°40'42"N	116°36'59"W	2740
Mica Saddle	Adams	44°36'07"N	116°14'25"W	
Monida Pass	Clark	44°33'31"N	112°18'19"W	
Moon Pass	Shoshone	47°25'03"N	115°51'40"W	4946
Moon Saddle	Idaho	46°32'10"N	115°00'59"W	
Moon Saddle	Shoshone	47°35'59"N	116°02'35"W	4660
Morris Saddle	Clearwater	46°53'42"N	115°45'56"W	4443
Moses Pass	Bingham	43°01'37"N	111°38'30"W	6730
Motthorn Saddle	Idaho	45°40'36"N	116°24'13"W	
Muldoon Summit	Blaine	43°28'58"N	114°07'10"W	6498
Mullan Pass	Shoshone	47°27'51"N	115°38'16"W	
Mush Saddle	Clearwater	46°45'12"N	115°21'58"W	5378
Narrows, The	Bingham	43°16'32"N	111°57'47"W	5300
Narrows, The	Caribou	42°47'53"N	111°21'20"W	
Narrows, The	Cassia	42°03'50"N	113°29'03"W	5004
Narrows, The	Idaho	45°25'04"N	116°28'49"W	

Passes, Gaps, Saddles

Pass, Gap, or Saddle	County	Latitude	Longitude	Elev.
Needles Summit	Valley	44°45'19"N	115°49'06"W	7900
Nez Perce Pass	Idaho	45°43'00"N	114°30'04"W	6587
No Business Saddle	Valley	44°45'02"N	116°11'14"W	
North Fitsum Summit	Valley	44°59'03"N	115°53'04"W	7540
O'Hara Saddle	Idaho	45°57'08"N	115°30'58"W	
Oxbow Saddle	Adams	45°14'08"N	116°35'57"W	
P K Pass	Blaine	43°41'53"N	114°04'40"W	
Packbox Pass	Idaho	46°28'03"N	114°23'09"W	
Papoose Saddle	Idaho	46°35'12"N	114°43'55"W	5647
Papoose Saddle	Idaho	45°24'19"N	116°25'19"W	
Parker Pass	Benewah	47°20'13"N	116°40'12"W	
Pete Creek Divide	Lemhi	44°32'02"N	112°04'41"W	7640
Pickett Creek Saddle	Owyhee	42°57'41"N	116°39'01"W	
Pierce Divide	Clearwater	46°27'35"N	115°50'36"W	
Pine Creek Pass	Teton	43°34'18"N	111°12'52"W	
Pine Gap	Bear Lake	42°06'04"N	111°09'27"W	
Piney Pass	Teton	43°39'09"N	111°20'22"W	
Pinnacle Pass	Cassia	42°02'22"N	113°42'39"W	6260
Pipe Saddle	Idaho	45°28'06"N	116°24'07"W	
Pittsburg Saddle	Idaho	45°39'51"N	116°23'34"W	
Poe Saddle	Idaho	45°49'42"N	116°14'45"W	
Porcupine Pass	Clark	44°28'57"N	112°07'11"W	7062
Porcupine Pass	Shoshone	47°50'49"N	115°53'20"W	5069
Profile Gap	Valley	45°03'27"N	115°24'54"W	
Pueblo Summit	Idaho	45°12'15"N	115°20'03"W	
Purgatory Saddle	Adams	45°11'26"N	116°34'11"W	
Pyramid Pass	Boundary	48°48'52"N	116°36'57"W	
Railroad Pass	Valley	44°33'18"N	115°45'40"W	
Railroad Saddle	Adams	45°04'19"N	116°28'07"W	
Rainbow Saddle	Valley	44°58'58"N	115°30'33"W	
Rape Saddle	Idaho	45°50'21"N	116°15'20"W	4389
Raynolds Pass	Fremont	44°42'41"N	111°28'04"W	6836
Reas Pass	Fremont	44°34'04"N	111°11'20"W	
Reas Pass	Fremont	44°33'58"N	111°11'27"W	6940
Red Rock Pass	Bannock	42°21'23"N	112°02'33"W	
Red Rock Pass	Fremont	44°35'56"N	111°31'22"W	
Remenclau Saddle	Lemhi	45°05'23"N	114°27'52"W	
Remount	Idaho	46°08'22"N	115°24'46"W	6503
Riley Saddle	Kootenai	47°45'58"N	116°15'16"W	
Riverview Saddle	Idaho	45°40'08"N	116°21'40"W	
Ruby Pass	Boundary	48°38'43"N	116°33'02"W	
Sabe Saddle	Idaho	45°41'01"N	114°56'50"W	
Saddle, The	Bonneville	43°21'31"N	111°45'10"W	6020
Saddle, The	Owyhee	42°16'18"N	116°00'03"W	
Sage Creek Saddle	Kootenai	47°51'48"N	116°34'44"W	
Saint Paul Pass	Shoshone	47°23'24"N	115°38'50"W	5200
Saint Regis Pass	Shoshone	47°27'02"N	115°43'17"W	
Savage Pass	Idaho	46°27'57"N	114°37'53"W	6168
Saw Pit Saddle	Idaho	45°31'32"N	116°29'42"W	
Scott Saddle	Idaho	45°27'07"N	115°59'28"W	
Secesh Summit	Valley	45°11'38"N	115°58'10"W	
Sheep Mountain Saddle	Clearwater	46°45'42"N	115°37'09"W	
Sherman Saddle	Idaho	46°25'04"N	115°21'58"W	4737
Shuck Saddle	Idaho	45°45'11"N	116°23'07"W	
Skookum Saddle	Kootenai	47°42'21"N	116°25'56"W	4020
Skull Saddle	Shoshone	47°44'13"N	116°18'30"W	3900
Slate Creek Saddle	Shoshone	47°24'55"N	115°53'38"W	4767
Sleepy Hollow	Idaho	45°23'03"N	115°21'44"W	
Sleepy Saddle	Idaho	45°21'33"N	115°34'25"W	
Smith Saddle	Idaho	46°11'05"N	115°42'30"W	
Smith Saddle	Idaho	45°17'02"N	115°31'57"W	6039
Solitare Saddle	Kootenai	47°52'07"N	116°27'30"W	
South Fork Buckhorn Summit	Valley	44°50'39"N	115°50'11"W	7100
Squaw Pass	Fremont	44°36'22"N	111°31'24"W	7262
Squaw Saddle	Idaho	46°31'51"N	114°51'09"W	
Squaw Saddle	Idaho	45°26'49"N	116°12'28"W	5683
Stevens Saddle	Adams	45°15'02"N	116°34'24"W	
Stines Pass	Cassia	42°08'53"N	113°42'53"W	
Stormy Pass	Idaho	46°30'20"N	114°21'41"W	
Studebaker Saddle	Idaho	45°25'18"N	115°57'00"W	
Stull Saddle	Shoshone	47°44'36"N	116°18'07"W	3800
Suicide Pass	Caribou	42°54'58"N	111°09'49"W	
Summit Springs Pass	Oneida	42°17'48"N	112°56'09"W	
Swan Saddle	Kootenai	47°31'18"N	116°37'37"W	3000
Swanson Saddle	Clearwater	46°46'10"N	115°36'25"W	4534
Swaps Pass	Caribou	42°56'12"N	111°50'56"W	6690
Sylvan Saddle	Clearwater	46°30'54"N	115°33'49"W	
Tamarack Saddle	Idaho	45°51'00"N	115°11'10"W	
Tamarack Saddle	Shoshone	47°33'24"N	115°49'29"W	5543
Tanner Pass	Bingham	43°08'25"N	111°47'13"W	7010
Targhee Pass	Fremont	44°40'29"N	111°16'30"W	
Taylor Saddle	Shoshone	47°46'57"N	115°51'23"W	
Telephone Saddle	Washington	44°38'50"N	116°56'16"W	5830
Tenmile Pass	Caribou	42°45'44"N	111°45'33"W	
Thompson Pass	Shoshone	47°34'38"N	115°42'59"W	
Tie Lavin Saddle	Kootenai	47°46'08"N	116°21'50"W	
Timber Creek Pass	Lemhi	44°27'27"N	113°26'43"W	
Township Saddle	Clearwater	46°45'57"N	115°49'15"W	4238
Toy Pass	Owyhee	42°54'15"N	116°32'46"W	
Trap Creek Narrows	Custer	44°18'58"N	115°05'01"W	
Treasure Saddle	Kootenai	47°43'12"N	116°37'43"W	3725
Tribulation Saddle	Idaho	46°18'47"N	115°02'14"W	
Trout Creek Pass	Cassia	42°05'33"N	114°10'12"W	
Twentymile Pass	Boundary	48°33'54"N	116°15'11"W	
Twin Creek Saddle	Shoshone	47°07'13"N	115°47'22"W	
Vance Creek Saddle	Idaho	45°40'29"N	114°36'00"W	

Pass, Gap, or Saddle	County	Latitude	Longitude	Elev.
Wahoo Pass	Idaho	46°09'41"N	114°30'51"W	
Walker Saddle	Kootenai	47°48'26"N	116°27'56"W	
Warm Springs Pass	Idaho	46°21'42"N	114°47'43"W	
Warm Springs Saddle	Idaho	45°16'05"N	116°11'19"W	
Warren Summit	Idaho	45°13'48"N	115°37'45"W	
Weber Saddle	Bonner	47°53'52"N	116°25'11"W	4598
White Creek Saddle	Shoshone	47°36'36"N	116°00'15"W	4350
White Pass	Bear Lake	42°29'05"N	111°16'11"W	
Wild Horse Divide	Bannock	42°45'30"N	112°29'19"W	6526
Wildhorse Saddle	Idaho	45°18'05"N	116°24'00"W	
Williams Creek Summit	Lemhi	45°05'37"N	114°05'13"W	
Wilson Pass	Caribou	42°59'37"N	111°36'08"W	6615
Windfall Pass	Benewah	47°14'26"N	116°50'11"W	3559
Windy Devil	Custer	44°04'14"N	114°35'42"W	10020
Windy Gap	Elmore	43°19'15"N	115°24'47"W	5170
Windy Pass	Bannock	42°33'54"N	111°57'17"W	
Windy Saddle	Idaho	46°22'23"N	115°32'31"W	
Windy Saddle	Idaho	45°52'59"N	115°02'29"W	
Wolf Lodge Saddle	Kootenai	47°42'23"N	116°30'25"W	4710
Woods Creek Pass	Lemhi	45°31'50"N	114°26'08"W	7190

Point, Penninsula, or Cape	County	Latitude	Longitude	Elev.
Anderson Point	Bonner	48°14'57"N	116°22'31"W	
Arrow Point	Kootenai	47°37'56"N	116°46'04"W	
Bear Point	Idaho	45°33'06"N	115°02'43"W	7974
Beaver Point	Idaho	45°44'54"N	114°46'07"W	
Beedle Point	Kootenai	47°23'21"N	116°45'18"W	
Black Rock	Kootenai	47°31'06"N	116°50'54"W	
Blackwell Point	Kootenai	47°58'08"N	116°32'16"W	
Bliss Point	Gooding	42°58'27"N	114°55'53"W	
Bottle Bay Point	Bonner	48°15'20"N	116°26'55"W	
Bronco Point	Power	42°56'11"N	112°41'22"W	
Browns Point	Kootenai	47°24'48"N	116°45'58"W	
Canoe Point	Bonner	48°43'16"N	116°50'11"W	
Cape Horn	Bonner	48°36'51"N	116°52'10"W	2540
Castle Rock	Clearwater	46°39'07"N	115°31'37"W	
Cellar Point	Kootenai	47°37'18"N	116°47'12"W	
Chicken Point	Caribou	42°56'49"N	111°38'11"W	6120
Chicken Point	Kootenai	47°46'44"N	116°41'05"W	
Chippy Point	Kootenai	47°25'45"N	116°46'41"W	2210
Clark Point	Kootenai	47°45'25"N	116°43'27"W	
Conkling Point	Kootenai	47°24'37"N	116°45'32"W	
Contest Point	Bonner	48°16'14"N	116°30'01"W	
Cooper Point	Idaho	45°49'44"N	114°33'19"W	8194
Crappie Point	Owyhee	42°07'01"N	114°44'35"W	
Crown Point	Idaho	45°37'17"N	115°10'48"W	
Deadman Point	Bonner	48°07'37"N	116°19'37"W	
Delcardo Bay	Kootenai	47°34'54"N	116°49'09"W	
Donavons Point	Kootenai	47°39'52"N	116°48'49"W	
Driftwood Point	Kootenai	47°35'28"N	116°47'54"W	
Dukes	Owyhee	42°18'21"N	116°57'07"W	
East Point	Kootenai	47°28'40"N	116°51'18"W	
English Point	Kootenai	47°46'08"N	116°42'47"W	
Eureka Point	Valley	44°28'54"N	115°46'09"W	7225
Evernade Point	Kootenai	47°46'29"N	116°42'25"W	
Fenstermaker Point	Power	42°51'18"N	112°50'43"W	
Fick Point	Idaho	45°51'58"N	116°22'57"W	
Fuzzy Peak	Shoshone	47°10'50"N	115°34'10"W	
Gasser Point	Kootenai	47°27'44"N	116°51'23"W	
Gotts Point	Canyon	43°32'37"N	116°38'20"W	
Graham Point	Bonner	47°58'14"N	116°27'37"W	2066
Granite Point	Bonner	48°05'59"N	116°25'52"W	
Grouse Mountain Point	Bonner	48°09'29"N	116°26'33"W	
Halfmoon Point	Boise	43°56'28"N	115°52'09"W	
Harer Point	Bear Lake	42°11'04"N	111°10'59"W	6666
Harlow Point	Kootenai	47°27'31"N	116°49'00"W	
Harts Island	Kootenai	47°52'48"N	116°52'06"W	
Harvey Point	Owyhee	42°32'20"N	116°05'19"W	
Hawkins Point	Bonner	48°16'59"N	116°22'40"W	
Hess Point	Bonner	48°31'24"N	116°51'04"W	

Points, Penninsulas, Capes

Point, Penninsula, or Cape	County	Latitude	Longitude	Elev.
Hida Point	Idaho	45°31'22"N	115°13'03"W	6758
Hill Pasture Point	Owyhee	42°33'37"N	116°01'26"W	
Hope Point	Bonner	48°13'33"N	116°18'23"W	
Horn, Cape	Bonner	47°59'32"N	116°29'44"W	
Horn, Cape	Idaho	45°30'32"N	115°34'33"W	
Independence Point	Kootenai	47°40'26"N	116°47'11"W	2130
Indian Point	Bonner	48°07'49"N	116°22'03"W	
Indian Point	Elmore	42°56'26"N	115°33'38"W	
Indian Rock	Bonner	48°34'52"N	116°53'53"W	
J-P Point	Owyhee	42°27'34"N	115°36'50"W	
Kinney Point	Valley	44°38'23"N	115°40'22"W	
Kootenai Point	Bonner	48°17'39"N	116°28'49"W	2075
Lees Point	Kootenai	47°46'26"N	116°40'47"W	
Little Deer Point	Boise	43°44'38"N	116°06'34"W	6575
Little Point	Owyhee	42°27'26"N	116°38'39"W	
Long Point	Bonner	48°10'26"N	116°23'56"W	
Lookout Point	Owyhee	42°18'37"N	115°38'12"W	
Lower Point	Boise	43°44'16"N	116°07'16"W	
Lucks Point	Valley	44°56'40"N	116°03'38"W	
Luds Point	Twin Falls	42°06'48"N	114°44'52"W	
Lyndale Landing	Kootenai	47°36'28"N	116°51'00"W	2160
Maiden Rock	Bonner	48°05'58"N	116°30'20"W	
Martin Peak	Shoshone	46°58'22"N	115°32'16"W	
McDonald Point	Kootenai	47°33'18"N	116°49'04"W	
Mineral Point	Bonner	48°10'32"N	116°23'26"W	
Mush Point	Clearwater	46°45'39"N	115°23'23"W	
Myrtle Point	Blaine	43°16'00"N	114°22'40"W	
North Cape	Kootenai	47°38'41"N	116°47'33"W	
Nugget Point	Shoshone	47°11'49"N	115°37'02"W	
Paradise Point	Valley	44°57'57"N	116°03'38"W	
Picard Point	Bonner	48°12'59"N	116°21'24"W	
Picnic Point	Valley	44°39'12"N	115°40'10"W	
Pilot Rock Sandpoint	Kootenai	47°32'28"N	116°49'28"W	
Pinto Point	Bonner	48°36'15"N	116°50'43"W	
Pleasant Valley V	Owyhee	42°35'25"N	116°54'09"W	
Plumbago Point	Bonner	48°29'49"N	116°50'17"W	2540
Plummer Peninsula	Benewah	47°21'43"N	116°46'31"W	
Ponder Point	Bonner	48°18'10"N	116°30'34"W	
Porcupine Point	Valley	44°57'37"N	116°04'58"W	5130
Preceptor Point	Clearwater	46°51'31"N	115°33'43"W	
Prospect Point	Custer	44°13'11"N	114°45'16"W	9867
Rattlesnake Point	Nez Perce	46°30'34"N	116°33'31"W	
Reynolds Point	Kootenai	47°27'08"N	116°50'20"W	
Rockford Point	Kootenai	47°29'17"N	116°52'42"W	
Rocky Point	Bonner	48°32'22"N	116°50'15"W	
Rocky Point	Bonner	48°15'09"N	116°35'19"W	
Rocky Point	Idaho	45°29'57"N	115°07'46"W	7056
Rose Lewis Point	Idaho	45°48'16"N	116°25'00"W	

Point, Penninsula, or Cape	County	Latitude	Longitude	Elev.
Sabe Vista Point	Idaho	45°39'09"N	114°56'54"W	
Saddle Point	Owyhee	42°07'27"N	114°44'13"W	
Sheepherder Point	Bonner	48°11'40"N	116°17'12"W	
Sourdough Point	Bonner	48°15'26"N	116°27'55"W	
Spokane Point	Kootenai	47°26'47"N	116°48'14"W	
Steamboat Rock	Kootenai	47°57'25"N	116°31'47"W	2060
Stevens Point	Boise	44°09'53"N	115°37'43"W	
Stevens Point	Kootenai	47°38'35"N	116°47'08"W	
Swede Point	Kootenai	47°36'48"N	116°47'55"W	
Threemile Point	Kootenai	47°37'44"N	116°46'57"W	
Timber Point	Boise	44°04'15"N	115°42'10"W	
Tripod Point	Bonner	48°43'07"N	116°51'10"W	
University Point	Kootenai	47°30'13"N	116°52'02"W	2160
Valhalla Point	Kootenai	47°36'31"N	116°48'38"W	
Wickahoney Point	Owyhee	42°24'02"N	116°01'29"W	6081
Willies Point	Owyhee	42°32'54"N	116°06'28"W	
Willow Point	Idaho	46°22'52"N	115°23'52"W	
Windy Peak	Bonneville	43°23'34"N	111°10'40"W	8701
Windy Point	Boise	44°04'25"N	115°43'11"W	
Windy Point	Bonner	48°07'39"N	116°23'12"W	
Windy Point	Kootenai	47°45'08"N	116°42'18"W	
Yellowstone Point	Kootenai	47°45'48"N	116°41'55"W	

Rapid	County	Latitude	Longitude	Elev.
Bargamin Rapids	Idaho	45°33'59"N	115°11'58"W	2880
Bell Rapids (historical)	Gooding	42°47'26"N	114°56'25"W	
Big Hill Rapids	Idaho	46°08'19"N	115°41'11"W	
Big Mallard Rapids	Idaho	45°32'18"N	115°16'08"W	
Big Sulphur Rapids	Idaho	45°46'54"N	116°38'19"W	
Blackhawk Rapid	Idaho	45°37'18"N	116°17'55"W	1540
Blackhawk Rapids	Idaho	45°37'40"N	116°18'06"W	
Boulder Rapids	Idaho	45°46'48"N	116°38'51"W	
Buffalo Rapids	Nez Perce	46°11'01"N	116°56'23"W	
Caldron Linn	Twin Falls	42°29'44"N	114°07'44"W	
Canapener Rapids	Idaho	45°27'02"N	115°45'35"W	
Captain John Rapids	Nez Perce	46°08'28"N	116°56'00"W	
Captain Lewis Rapids	Nez Perce	46°05'55"N	116°57'51"W	
Carey Falls	Idaho	45°27'25"N	115°56'15"W	
Cattle Rapids	Adams	45°00'35"N	116°50'45"W	
China Rapids	Idaho	45°36'18"N	116°27'48"W	
Chipmunk Rapids	Bonner	48°25'23"N	116°54'28"W	
Chittam Rapids	Idaho	45°27'30"N	115°53'10"W	
Cliff Mountain Rapids	Adams	45°15'30"N	116°41'35"W	1485
Cliffside Rapids	Lemhi	45°13'11"N	114°40'48"W	3760
Cochran Rapids	Nez Perce	45°57'58"N	116°53'34"W	860
Coffee Pot Rapids	Fremont	44°29'45"N	111°23'41"W	
Copper Creek Rapids	Idaho	45°45'11"N	116°32'59"W	
Cottonwood Rapids	Idaho	45°41'52"N	116°32'10"W	1090
Davis Creek Rapids	Idaho	45°39'34"N	116°30'00"W	
Deadhorse Rapids	Valley	44°53'41"N	115°29'38"W	
Deadman Rapids	Idaho	45°22'42"N	115°30'25"W	2160
Deep Creek Rapids	Idaho	45°46'49"N	116°39'14"W	
Devils Teeth Rapids	Idaho	45°26'09"N	114°53'32"W	
Divide Creek Rapids	Idaho	45°49'34"N	116°44'13"W	
Dolman Rapids	Twin Falls	42°46'14"N	114°54'58"W	
Double Drop Rapids	Idaho	46°07'12"N	114°56'58"W	
Dried Meat Rapids	Idaho	45°27'57"N	115°49'55"W	
Dug Creek Rapids	Idaho	45°47'26"N	116°40'30"W	
Durham Rapids	Idaho	45°37'08"N	116°27'58"W	
Dutch Oven Rapids	Lemhi	45°19'53"N	114°21'52"W	
Elkhorn Rapids	Idaho	45°30'59"N	115°19'32"W	
Fiddle Creek Rapids	Idaho	45°29'28"N	116°19'07"W	
Five Pine Rapids	Idaho	45°46'47"N	116°35'55"W	
Fivemile Rapids	Idaho	45°33'27"N	115°13'38"W	
Frenchy Rapids	Nez Perce	45°53'25"N	116°50'15"W	
Granite Creek Rapids	Idaho	45°21'25"N	116°38'38"W	1920
Growler Rapids	Idaho	45°30'01"N	115°19'46"W	
Gunbarrel Rapids	Lemhi	45°23'25"N	114°42'58"W	3480
Hancock Rapids	Idaho	45°31'17"N	115°05'29"W	
Hancock Rapids	Lemhi	45°15'36"N	114°38'40"W	
Haystack Rapids	Valley	44°58'29"N	114°44'04"W	3620
Highrange Rapids	Idaho	45°44'13"N	116°32'10"W	1060
Imnaha Rapids	Idaho	45°49'06"N	116°45'53"W	
Indian Rapids	Idaho	46°08'29"N	115°41'59"W	
Indian Riffles	Custer	44°15'22"N	114°41'22"W	
Jack Creek Rapids	Lemhi	45°00'08"N	114°42'42"W	
Kanaka Rapids	Gooding	42°39'54"N	114°48'08"W	
Kerrs Rapids	Adams	45°01'58"N	116°50'05"W	
Lantz Rapids	Idaho	45°25'06"N	114°52'19"W	
Limekiln Rapids	Nez Perce	46°04'34"N	116°57'54"W	
Limekiln Rapids	Nez Perce	46°04'34"N	116°57'52"W	
Little Devils Teeth Rapids	Idaho	45°27'48"N	114°56'40"W	
Little Growler Rapids	Idaho	45°27'17"N	115°46'15"W	
Little Mallard Rapids	Idaho	45°32'01"N	115°18'01"W	
Long Tom Rapids	Lemhi	45°18'02"N	114°34'44"W	
Lower Bernard Creek Rapids	Idaho	45°24'18"N	116°36'30"W	1340
Lower Cache Creek Rapids	Nez Perce	45°59'26"N	116°54'38"W	
Lower Dug Bar Rapids	Idaho	45°48'48"N	116°41'36"W	
Lower Kirby Rapids	Idaho	45°35'37"N	116°28'17"W	
Lower Pittsburg Rapids	Idaho	45°38'14"N	116°28'54"W	
Ludwig Rapids	Idaho	45°23'36"N	115°29'28"W	
McDuff Rapids	Nez Perce	45°59'51"N	116°54'59"W	
Middle Kirby Rapids	Idaho	45°35'04"N	116°28'57"W	
Mountain Sheep Rapids	Idaho	45°50'14"N	116°47'17"W	
Negro Head Rapids	Nez Perce	46°02'39"N	116°56'13"W	
Negro Head Rapids	Nez Perce	46°02'38"N	116°56'08"W	
Ninefoot Rapids (historical)	Elmore	42°59'20"N	115°55'01"W	
Ouzel Rapids	Idaho	45°14'05"N	114°40'13"W	
Pine Bar Rapids	Idaho	45°53'22"N	116°19'51"W	
Pine Creek Rapids	Lemhi	45°21'49"N	114°18'00"W	
Pistol Creek Rapids	Valley	44°43'49"N	115°09'00"W	4820
Pleasant Valley Rapids	Idaho	45°38'57"N	116°29'11"W	
Porcupine Rapids	Lemhi	45°07'45"N	114°43'35"W	
Rainier Rapids	Idaho	45°23'47"N	114°49'09"W	
Redside Rapids	Lemhi	45°09'06"N	114°43'50"W	
Robinson Gulch Rapids	Idaho	45°46'58"N	116°39'54"W	
Rocky Point Rapids	Idaho	45°17'30"N	116°40'16"W	1435
Roland Bar Rapids	Idaho	45°45'53"N	116°34'17"W	
Ruby Rapids	Idaho	45°24'16"N	116°11'36"W	
Sheep Creek Rapids	Idaho	45°28'10"N	115°48'30"W	
Shovel Creek Rapids	Nez Perce	46°01'15"N	116°55'36"W	
Shovel Creek Rapids	Nez Perce	46°01'14"N	116°55'32"W	
Ski Jump Rapid	Valley	44°44'17"N	115°00'49"W	
Snow Hole Rapids	Idaho	46°00'32"N	116°36'18"W	
Split Rock Rapids	Idaho	45°33'21"N	115°14'17"W	
Sturgill Rapids	Washington	44°40'50"N	117°04'39"W	
T-Bone Creek Rapids	Idaho	45°27'07"N	115°46'01"W	
Tenmile Rapids	Nez Perce	46°18'26"N	117°00'26"W	
Upper Bernard Creek Rapids	Idaho	45°24'07"N	116°36'44"W	1335
Upper Cache Creek Rapids	Nez Perce	45°58'54"N	116°53'53"W	

Rapids

Rapid	County	Latitude	Longitude	Elev.
Upper Kirby Rapids	Idaho	45°34'55"N	116°28'53"W	
Upper Pleasant Valley Rapids	Idaho	45°38'32"N	116°28'58"W	
Vinegar Creek Rapids	Idaho	45°27'33"N	115°54'20"W	
Wapshilla Rapids	Nez Perce	45°56'21"N	116°45'18"W	
Warm Springs Rapids	Idaho	45°49'16"N	116°42'00"W	
Whitehorse Rapids	Idaho	45°49'36"N	116°42'44"W	
Whitehorse Rapids	Valley	44°47'07"N	115°32'00"W	
Wild Goose Rapids	Nez Perce	46°02'17"N	116°55'49"W	
Wild Sheep Rapids	Idaho	45°20'00"N	116°39'32"W	2060
Wildhorse Rapids	Washington	44°51'17"N	116°53'47"W	
Willow Creek Rapids	Idaho	45°29'33"N	116°33'13"W	1250
Wolf Creek Rapids	Idaho	45°46'15"N	116°34'52"W	
Zigzag Rapids	Idaho	45°49'34"N	116°43'14"W	

Rapid	County	Latitude	Longitude	Elev.

Ridge or Slope	County	Latitude	Longitude	Elev.
Abercombie Ridge	Idaho	45°49'10"N	116°28'42"W	4415
Aldermand Ridge	Clearwater	46°40'51"N	116°17'25"W	8700
Allen Ridge	Shoshone	47°12'52"N	115°40'19"W	
American Ridge	Latah	46°39'55"N	116°43'17"W	
Angel Ridge	Nez Perce	46°28'42"N	116°29'40"W	2928
Antelope Ridge	Clark	44°26'33"N	111°49'32"W	
Antelope Ridge	Owyhee	42°43'55"N	116°31'06"W	
Antimony Ridge	Valley	44°55'20"N	115°26'58"W	7722
Antler Ridge	Lemhi	44°52'02"N	114°29'33"W	
Apache Ridge	Boundary	48°35'42"N	116°35'23"W	
Argument Ridge	Madison	43°39'40"N	111°27'22"W	
Arlington Ridge	Idaho	45°28'29"N	115°41'04"W	
Arrison Ridge	Idaho	45°36'55"N	116°11'12"W	3620
Austin Ridge	Idaho	46°21'12"N	115°40'08"W	
Avalanche Ridge	Clearwater	46°54'29"N	115°29'12"W	
Banner Ridge	Boise	44°02'18"N	115°31'02"W	
Barren Ridge	Idaho	46°17'23"N	114°59'57"W	
Battle Creek Ridge	Idaho	46°12'41"N	114°37'35"W	7132
Battle Ridge	Idaho	46°05'26"N	115°56'26"W	2315
Bear Grass Ridge	Idaho	45°37'00"N	115°46'07"W	7909
Beartrap Ridge	Lemhi	45°25'08"N	114°23'53"W	
Beaver Ridge	Benewah	47°12'33"N	116°24'01"W	
Beaver Ridge	Bonneville	43°17'33"N	111°29'33"W	7591
Beaver Ridge	Idaho	46°32'55"N	114°32'50"W	6284
Bedstead Ridge	Owyhee	42°24'13"N	116°53'42"W	
Bentz Ridge	Idaho	45°47'17"N	116°10'44"W	4326
Biddle Ridge	Idaho	45°48'34"N	116°24'10"W	
Big Baldy Ridge	Valley	44°48'11"N	115°18'29"W	8787
Big Bear Ridge	Latah	46°43'14"N	116°37'03"W	
Big Bend Ridge	Fremont	44°10'31"N	111°25'48"W	
Big Creek Ridge	Valley	45°05'15"N	115°03'57"W	8330
Big Grassy Ridge	Fremont	44°02'39"N	111°58'00"W	
Big Point	Owyhee	42°23'51"N	116°36'37"W	
Big Ridge	Owyhee	42°28'12"N	116°56'21"W	6204
Big Tree	Owyhee	42°22'00"N	116°49'55"W	
Bimetallic Ridge	Bonner	48°07'58"N	116°30'08"W	
Bingman Ridge	Idaho	45°57'09"N	116°28'26"W	
Birch Ridge	Clearwater	46°52'02"N	115°09'28"W	6101
Birch Ridge	Kootenai	47°48'32"N	116°58'35"W	3163
Black Ridge	Idaho	45°31'24"N	114°59'54"W	
Blackrock Ridge	Kootenai	47°28'05"N	116°28'33"W	
Blacktail Ridge	Idaho	46°12'31"N	115°06'02"W	
Blacktail Ridge	Idaho	45°54'24"N	115°57'25"W	4754
Blackwell Hump	Shoshone	47°10'48"N	116°18'33"W	5304
Blanchard Ridge	Teton	43°33'45"N	111°09'38"W	
Blue Bunch Ridge	Adams	41°54'50"N	104°01'56"W	
Blue Bunch Ridge	Valley	44°28'16"N	115°16'38"W	8608
Blue Ridge	Bingham	43°11'19"N	111°52'19"W	

Ridge or Slope	County	Latitude	Longitude	Elev.
Blue Ridge	Owyhee	42°36'50"N	115°06'00"W	
Blue Star Ridge	Shoshone	47°32'17"N	116°12'47"W	
Bluff Divide	Shoshone	47°07'40"N	115°33'46"W	
Bobby Anderson Ridge	Shoshone	47°30'53"N	116°13'32"W	
Bobtail Ridge	Shoshone	47°43'38"N	115°50'53"W	
Boise Ridge	Boise	43°50'33"N	116°01'56"W	6983
Boquet Ridge	Teton	43°38'07"N	111°13'02"W	
Bouffard Ridge	Idaho	45°41'44"N	116°10'55"W	
Boulder Divide	Shoshone	47°12'24"N	115°56'27"W	4470
Boundary Ridge	Bear Lake	42°04'33"N	111°03'33"W	7650
Brown Creek Ridge	Clearwater	46°19'18"N	115°49'12"W	
Browns Ridge	Shoshone	47°34'25"N	116°00'12"W	
Bruin Ridge	Clearwater	46°45'43"N	114°53'02"W	5684
Brushy Ridge	Idaho	45°32'38"N	116°12'20"W	
Buck Ridge	Clearwater	46°35'09"N	115°20'53"W	5746
Buckhorn Ridge	Boundary	48°52'36"N	116°02'55"W	
Buckskin Morgan Ridge	Madison	43°38'52"N	111°36'22"W	
Buckskin Ridge	Twin Falls	42°13'32"N	114°18'17"W	
Buffalo Ridge	Custer	44°24'31"N	114°24'54"W	
Bugle Ridge	Boundary	48°51'08"N	116°47'25"W	
Bull Creek Ridge	Idaho	45°29'19"N	115°44'31"W	
Bull Elk Ridge	Idaho	45°40'26"N	115°47'47"W	6862
Bulldog Ridge	Valley	44°15'38"N	115°51'25"W	
Burns Ridge	Boise	44°02'10"N	115°41'10"W	6810
Burnt Ridge	Elmore	42°59'06"N	115°06'21"W	
Burnt Ridge	Latah	46°42'54"N	116°43'01"W	
Butcherknife Ridge	Lemhi	45°32'22"N	114°05'47"W	
Camel Ridge	Idaho	45°32'02"N	116°26'41"W	4675
Camels Prairie	Boundary	48°32'30"N	116°45'22"W	
Camp Howard Ridge	Idaho	45°43'01"N	116°24'24"W	5725
Canyon Creek Ridge	Idaho	45°21'43"N	114°49'41"W	7036
Caribou Ridge	Boundary	48°38'54"N	116°29'19"W	
Cascade Ridge	Boundary	48°45'45"N	116°30'37"W	
Cascade Spur	Kootenai	47°49'19"N	116°22'44"W	
Cedar Mountain	Shoshone	47°21'01"N	115°50'29"W	
Cedar Ridge	Idaho	46°19'17"N	114°41'03"W	6931
Cedar Ridge	Power	42°41'51"N	113°01'07"W	
Cedar Ridge	Power	46°39'07"N	116°29'36"W	
Cedar Ridge	Power	42°29'19"N	112°53'47"W	5612
Cemetery Ridge	Shoshone	47°22'20"N	116°02'45"W	5591
Center Ridge	Idaho	45°47'10"N	116°24'13"W	
Center Ridge	Idaho	45°36'25"N	116°20'19"W	
Center Ridge	Idaho	45°13'21"N	116°04'10"W	
Center Ridge	Lemhi	44°26'40"N	113°12'20"W	7576
Central Ridge	Lewis	46°24'33"N	116°20'25"W	
Chamook Ridge	Idaho	46°20'56"N	115°42'03"W	
Character Ridge	Shoshone	47°36'52"N	116°03'27"W	4980
Charters Mountain	Boise	44°01'58"N	116°00'53"W	6525

Ridges & Slopes

Ridge or Slope	County	Latitude	Longitude	Elev.
Chicken Ridge	Bonneville	43°13'38"N	111°14'09"W	
Chicken Ridge	Clark	44°12'07"N	111°55'09"W	
China Point	Owyhee	43°01'18"N	116°44'43"W	
China Point Ridge	Idaho	45°57'31"N	115°46'23"W	
Clarks Ridge	Idaho	45°28'00"N	116°30'02"W	5200
Coeur d'Alene Saint Joe Divide	Shoshone	47°26'03"N	115°50'52"W	
Cold Mountain Ridge	Idaho	45°13'22"N	114°57'26"W	8138
Cold Spring Ridge	Adams	44°34'02"N	116°12'38"W	6229
Columbia Ridge	Idaho	45°36'30"N	115°33'54"W	7387
Columbia River Great Basin Divide	Caribou	42°41'11"N	111°27'11"W	
Combination Ridge	Owyhee	42°43'51"N	116°45'23"W	6713
Commissary Ridge	Bonneville	43°17'14"N	111°25'13"W	
Connor Ridge	Cassia	42°18'50"N	113°34'11"W	8228
Coolwater Ridge	Idaho	46°08'14"N	115°31'13"W	
Copper Ridge	Boundary	48°57'06"N	116°05'04"W	
Copper Ridge	Idaho	45°58'05"N	115°13'06"W	6000
Copper Ridge	Shoshone	47°01'30"N	115°24'56"W	
Corey Ridge	Idaho	45°30'38"N	114°59'36"W	
Corral Creek Ridge	Caribou	42°49'57"N	111°44'33"W	6840
Corrigan Ridge	Shoshone	47°31'19"N	116°12'09"W	
Cotton Ridge	Cassia	42°12'02"N	114°13'38"W	
Cottonwood Ridge	Cassia	42°16'06"N	114°01'04"W	
Cowcatcher Ridge	Blaine	43°28'59"N	114°15'05"W	
Coyote Ridge	Bonneville	43°17'00"N	111°48'44"W	6658
Coyote Ridge	Idaho	45°43'52"N	116°25'36"W	5481
Cream Ridge	Nez Perce	46°32'42"N	116°30'38"W	
Crooked Ridge	Kootenai	47°51'07"N	116°25'29"W	
Crooked Ridge	Shoshone	47°08'19"N	115°13'30"W	6411
Crystal Springs Grade	Twin Falls	42°39'03"N	114°39'02"W	
Dairy Ridge	Bear Lake	42°08'47"N	111°11'02"W	7021
Dam Ridge	Shoshone	47°24'08"N	115°58'35"W	
Dan Lee Ridge	Clearwater	46°27'02"N	115°42'47"W	5005
Dan Ridge	Idaho	46°28'07"N	114°27'16"W	
Danby Ridge	Kootenai	47°50'56"N	116°58'50"W	4302
Dawson Ridge	Boundary	48°46'56"N	116°12'45"W	
Dead Indian Ridge	Washington	44°18'34"N	117°10'12"W	4281
Deadhorse Ridge	Bonneville	43°20'46"N	111°22'05"W	
Deadline Ridge	Twin Falls	42°02'30"N	114°17'20"W	7547
Deadman Ridge	Valley	44°59'24"N	115°39'15"W	6352
Deadwood Ridge	Boise	46°47'12"N	115°07'57"W	
Deadwood Ridge	Boise	44°08'01"N	115°37'46"W	7862
Deep Creek Ridge	Lemhi	45°05'09"N	114°06'22"W	8575
Deer Ridge	Boundary	48°51'51"N	116°06'45"W	
Deer Ridge	Clearwater	46°39'45"N	114°52'58"W	6528
Deerfoot Ridge	Kootenai	47°48'35"N	116°36'57"W	4135
Defeat Ridge	Owyhee	42°18'27"N	116°49'18"W	
Delmage Ridge	Idaho	45°43'12"N	116°08'39"W	
Delyle Ridge	Bonner	48°03'12"N	116°08'51"W	
Dennis Ridge	Idaho	45°33'28"N	114°52'38"W	
Dent Ridge	Idaho	45°54'58"N	115°18'17"W	
Dentist Parlor Ridge	Idaho	45°45'27"N	116°07'41"W	4781
Diamond Ridge	Valley	45°09'52"N	115°55'18"W	8618
Dickshooter Ridge	Owyhee	42°19'04"N	116°32'36"W	5366
Dinosaur Ridge	Blaine	43°14'23"N	114°20'07"W	
Divide Number Three	Idaho	46°11'26"N	114°28'09"W	8417
Dogtown	Idaho	45°22'51"N	116°15'10"W	6059
Driscoll Ridge	Latah	46°41'07"N	116°46'09"W	2750
Drollinger Ridge	Owyhee	43°03'28"N	116°39'12"W	
Drumlummen Ridge	Idaho	45°31'36"N	115°37'28"W	8129
Dry Diggins Ridge	Idaho	45°22'27"N	116°33'10"W	
Dry Ridge	Bear Lake	42°31'44"N	111°16'44"W	
Dry Ridge	Caribou	42°40'11"N	111°13'54"W	
Dry Ridge	Latah	46°44'49"N	116°37'40"W	
Duncan Ridge	Blaine	43°45'03"N	114°08'40"W	11491
Eakin Ridge	Idaho	45°27'38"N	114°51'12"W	
Eightyeight Ridge	Idaho	45°12'47"N	115°12'48"W	
Eisenhour Ridge	Kootenai	47°46'32"N	116°59'53"W	
Elbow Ridge	Shoshone	47°00'21"N	115°19'50"W	
Eldorado Ridge	Idaho	46°17'24"N	115°40'56"W	
Elk Point	Fremont	44°05'01"N	111°06'26"W	
Elk Ridge	Bonner	48°02'18"N	116°03'50"W	
Elk Ridge	Camas	43°29'10"N	114°46'32"W	
Elk Ridge	Clearwater	46°40'42"N	116°12'26"W	2775
Elmberry Ridge	Clearwater	46°47'14"N	116°00'44"W	
Ericson Ridge	Idaho	45°53'27"N	115°27'33"W	
Eureka Ridge	Clearwater	46°31'20"N	116°15'37"W	
Farnham Ridge	Boundary	48°51'16"N	116°30'59"W	
Fawn Ridge	Idaho	45°26'24"N	114°49'54"W	
Fern Ridge	Clearwater	46°52'19"N	115°35'57"W	
Fishfin Ridge	Lemhi	45°08'46"N	114°35'18"W	
Fishing Ridge	Lemhi	45°08'50"N	114°33'28"W	
Fix Ridge	Latah	46°36'40"N	116°46'28"W	
Flat Ridge	Power	42°20'47"N	112°46'24"W	
Flatiron Ridge	Idaho	45°52'58"N	115°22'54"W	
Flatiron Ridge	Owyhee	43°11'03"N	116°34'25"W	
Fluorspar Ridge	Lemhi	44°51'13"N	114°30'06"W	7115
Folsom Ridge	Kootenai	47°39'29"N	116°37'41"W	
Fort Hall Ridge	Adams	44°49'04"N	116°22'19"W	
Fort Simons Ridge	Nez Perce	46°10'18"N	116°50'30"W	
Four Bit Spur	Gem	44°28'06"N	116°14'12"W	
Fourth Of July Ridge	Bonneville	43°18'38"N	111°24'27"W	7830
Free Use Ridge	Idaho	45°45'05"N	116°11'40"W	4200
Freeman Ridge	Caribou	42°38'42"N	111°11'10"W	8701
Freezeout Ridge	Shoshone	47°01'00"N	116°01'24"W	6266
Friday Ridge	Idaho	46°22'06"N	114°44'05"W	
Frye Point	Nez Perce	46°01'40"N	116°47'30"W	

Ridge or Slope	County	Latitude	Longitude	Elev.
Gabe Ridge	Idaho	45°41'11"N	116°10'43"W	
Gant Ridge	Lemhi	45°11'53"N	114°25'01"W	
Glover Ridge	Idaho	46°07'16"N	115°19'34"W	
Goat Heaven Peaks	Idaho	46°15'02"N	114°37'54"W	
Goat Ridge	Idaho	45°58'00"N	114°54'48"W	6562
Gold Ridge	Bonner	48°27'05"N	116°20'25"W	
Gold Ridge	Shoshone	47°14'07"N	116°17'55"W	
Gold Ridge	Valley	44°57'44"N	115°11'05"W	7402
Goldbug Ridge	Lemhi	44°57'38"N	113°53'35"W	9010
Golden Ridge	Lewis	46°18'28"N	116°37'03"W	3800
Goose Ridge	Clearwater	46°52'56"N	115°01'00"W	
Graham Ridge	Shoshone	47°36'07"N	116°07'38"W	
Grassy Ridge	Adams	45°09'46"N	116°41'40"W	
Grassy Ridge	Owyhee	42°09'42"N	116°49'15"W	
Green Monarch Ridge	Bonner	48°06'38"N	116°18'48"W	
Green Ridge	Idaho	45°55'38"N	115°20'50"W	
Green Ridge	Idaho	45°50'09"N	114°50'23"W	6600
Greyhound Ridge	Custer	44°35'18"N	115°08'06"W	8870
Grice Ridge	Shoshone	46°57'22"N	116°04'30"W	4673
Grizzly Ridge	Shoshone	47°45'30"N	116°08'49"W	
Grouse Ridge	Idaho	45°52'51"N	114°55'42"W	
Grouse Ridge	Idaho	45°49'59"N	116°25'13"W	
Hagen Ridge	Idaho	45°34'15"N	115°53'05"W	7100
Haley Ridge	Adams	45°12'17"N	116°40'21"W	
Hanson Ridge	Clearwater	46°43'42"N	114°53'18"W	6081
Harrington Ridge	Idaho	45°31'39"N	114°52'25"W	
Harris Ridge	Idaho	46°11'40"N	115°52'47"W	
Hellroaring Ridge	Bonner	48°28'37"N	116°38'35"W	
Hemlock Ridge	Clearwater	46°31'50"N	115°33'07"W	5490
Henderson Point	Owyhee	42°54'31"N	116°32'31"W	
Henderson Ridge	Lemhi	45°31'50"N	114°09'40"W	8111
Hewed Log Ridge	Idaho	45°51'10"N	116°19'01"W	
Hida Ridge	Idaho	45°30'53"N	115°12'43"W	6967
Hidden Creek Ridge	Idaho	46°22'20"N	114°32'33"W	
Highline Ridge	Idaho	46°00'13"N	115°07'46"W	
Highline Ridge	Idaho	45°24'56"N	115°16'49"W	
Highrange Ridge	Idaho	45°43'38"N	116°27'33"W	5148
Hogback Ridge	Kootenai	47°30'31"N	116°36'02"W	2810
Hogback Ridge	Valley	45°07'47"N	115°19'01"W	
Hogsback Ridge	Bear Lake	42°02'33"N	111°28'14"W	
Home Ridge	Bonneville	43°19'58"N	111°24'16"W	
Hoodoo Ridge	Shoshone	47°04'14"N	115°39'56"W	5543
Hornet Ridge	Adams	44°46'15"N	116°29'25"W	
Horse Heaven Ridge	Idaho	45°19'17"N	115°25'29"W	
Horse Race Ridge	Owyhee	43°17'25"N	116°46'45"W	
Horse Ridge	Clark	44°13'52"N	112°45'33"W	8172
Horse Ridge	Idaho	45°58'17"N	115°20'23"W	5791
Horton Ridge	Boundary	48°36'28"N	116°44'05"W	

Ridge or Slope	County	Latitude	Longitude	Elev.
Huckleberry Ridge	Fremont	44°07'59"N	111°12'42"W	
Hudson Ridge	Cassia	42°09'48"N	114°09'09"W	
Huffman Ridge	Latah	46°42'22"N	116°34'25"W	2780
Hughes Ridge	Boundary	48°51'14"N	116°59'09"W	
Hungarian Ridge	Boise	43°49'41"N	115°35'34"W	6183
Hungry Ridge	Idaho	45°45'33"N	115°57'51"W	
Idaho Ridge	Teton	43°44'13"N	111°18'27"W	
Incline, The	Shoshone	47°06'07"N	116°12'20"W	
Independence Ridge	Clearwater	46°46'13"N	115°05'01"W	4594
Indian Henry Ridge	Clearwater	46°51'13"N	115°21'09"W	5463
Indian Ridge	Idaho	45°48'53"N	114°41'27"W	6666
Indian Ridge	Owyhee	42°42'40"N	115°03'23"W	3580
Indian Ridge	Valley	44°56'37"N	115°38'54"W	
Indian Ridge	Valley	44°55'21"N	115°34'37"W	7872
Jackass Ridge	Lemhi	45°17'08"N	114°01'07"W	
Jackson Ridge	Idaho	45°33'38"N	115°54'28"W	7474
Jeru Ridge	Boundary	48°30'17"N	116°37'52"W	
Johnagan Ridge	Clearwater	46°32'31"N	115°19'37"W	
Johnson Grade	Gooding	42°49'36"N	114°52'22"W	3150
Johnson Ridge	Idaho	45°33'52"N	116°21'38"W	5494
Junction Ridge	Shoshone	47°07'53"N	115°29'47"W	
Justice Grade	Gooding	42°50'21"N	114°53'16"W	
Katka Ridge	Boundary	48°39'09"N	116°09'30"W	6208
Keating Ridge	Idaho	45°27'49"N	116°09'23"W	4555
Kerlee Ridge	Idaho	45°49'26"N	116°25'41"W	
Kidder Ridge	Idaho	46°11'12"N	115°56'40"W	
Kingston Ridge	Shoshone	47°32'54"N	116°14'37"W	
Kirkham Ridge	Boise	44°05'28"N	115°32'55"W	5773
Knife Edge Ridge	Idaho	46°11'44"N	115°29'35"W	4854
Lake Ridge	Bear Lake	42°09'28"N	111°15'05"W	7382
Laundry Ridge	Clearwater	46°46'43"N	115°02'48"W	4841
Lava Ridge	Butte	43°56'02"N	112°48'31"W	
Lava Ridge	Idaho	45°17'10"N	116°07'22"W	
Leacock Point	Lemhi	45°07'32"N	114°14'23"W	
Lean-To Ridge	Clearwater	46°30'54"N	115°27'43"W	4677
Leg Bone Ridge	Shoshone	47°08'13"N	115°21'04"W	5920
Lemhi Ridge	Bingham	43°17'10"N	112°25'56"W	
Lightning Ridge	Boise	44°13'23"N	115°50'04"W	
Lightning Ridge	Bonneville	43°19'49"N	111°23'16"W	7837
Lightning Ridge	Idaho	45°27'22"N	116°29'28"W	
Lightning Ridge	Shoshone	47°03'18"N	115°34'05"W	
Lions Head Ridge	Boundary	48°46'35"N	116°41'30"W	
Little Bear Ridge	Latah	46°42'00"N	116°41'11"W	2734
Little Deer Creek Ridge	Lemhi	45°08'03"N	114°17'16"W	
Little Grassy Ridge	Jefferson	43°52'36"N	112°03'06"W	
Little Gray Ridge	Caribou	42°59'01"N	111°27'55"W	
Little Moose Ridge	Clearwater	46°44'37"N	114°58'52"W	
Log Ridge	Idaho	46°12'19"N	114°58'12"W	

Ridges & Slopes

Ridge or Slope	County	Latitude	Longitude	Elev.
Lone Pine Ridge	Bonneville	43°27'44"N	111°40'28"W	6530
Lone Pine Ridge	Bonneville	43°15'44"N	111°24'59"W	
Long Creek Ridge	Clearwater	46°55'26"N	115°04'26"W	
Long Gulch Ridge	Adams	44°52'06"N	116°32'56"W	
Long Ridge	Caribou	42°42'49"N	111°40'06"W	
Long Ridge	Owyhee	43°00'47"N	116°53'17"W	
Long Ridge	Teton	43°42'04"N	111°15'32"W	6640
Long Tom Ridge	Lemhi	45°24'32"N	114°30'52"W	7991
Lookout Mountain Ridge	Valley	44°58'57"N	115°05'34"W	7958
Lookout Ridge	Kootenai	47°46'31"N	116°22'52"W	
Lost Creek Ridge	Shoshone	47°43'47"N	115°53'16"W	
Lost Ridge	Clearwater	46°54'33"N	115°23'59"W	5154
Lower Cream Ridge	Nez Perce	46°32'43"N	116°30'39"W	
Lucky Ridge	Idaho	45°24'44"N	114°46'52"W	
Lunde Ridge	Clearwater	46°36'18"N	114°58'02"W	
Magruder Ridge	Idaho	45°43'09"N	114°48'43"W	
Mahogany Ridge	Bear Lake	42°06'52"N	111°14'19"W	
Mahogany Ridge	Bonneville	43°20'58"N	111°27'57"W	6852
Mahogany Ridge	Bonneville	43°21'48"N	111°26'22"W	7211
Mains Ridge	Custer	44°19'49"N	115°16'19"W	8584
Mammoth Ridge	Owyhee	42°56'55"N	116°41'58"W	
Mann Creek Ridge	Idaho	45°21'37"N	115°36'34"W	5906
Maple Lake Ridge	Idaho	46°16'48"N	114°45'06"W	7220
Matteson Ridge	Idaho	45°50'55"N	115°08'06"W	
Maude Ridge	Shoshone	47°31'27"N	116°12'54"W	
McCalla Ridge	Idaho	45°21'41"N	115°05'38"W	
McClinery Ridge	Idaho	45°22'57"N	116°27'09"W	
McCormack Ridge	Nez Perce	46°18'30"N	116°46'28"W	
McCormick Ridge	Boundary	48°35'28"N	116°40'42"W	
McGinty Ridge	Boundary	48°36'24"N	116°09'26"W	
Meadow Ridge	Valley	44°50'15"N	115°22'43"W	8712
Meeker Ridge	Idaho	46°08'41"N	115°00'24"W	4184
Mica Ridge	Adams	44°38'09"N	116°11'23"W	
Middle Ridge	Idaho	46°28'04"N	115°24'50"W	5375
Middle Ridge	Lemhi	44°28'18"N	113°13'17"W	
Mineral Ridge	Kootenai	47°36'33"N	116°38'36"W	3550
Minerva Ridge	Bonner	48°03'50"N	116°23'24"W	
Missouri Ridge	Valley	44°59'47"N	115°21'53"W	
Montana Ridge	Shoshone	47°05'23"N	115°42'59"W	5818
Montgomery Ridge	Shoshone	47°33'52"N	116°03'12"W	3640
Monumental Buttes	Shoshone	47°01'36"N	115°48'19"W	
Moose Ridge	Idaho	46°03'13"N	114°57'15"W	7017
Morgan Ridge	Blaine	43°43'09"N	114°17'12"W	8532
Morgan Ridge	Franklin	42°04'25"N	111°41'45"W	6270
Morrison Ridge	Idaho	45°21'52"N	116°26'37"W	
Moscow Bar Ridge	Clearwater	46°47'00"N	115°23'50"W	5274
Mosquito Ridge	Idaho	45°13'57"N	115°22'58"W	
Moughmer Ridge	Idaho	45°58'07"N	116°24'41"W	

Ridge or Slope	County	Latitude	Longitude	Elev.
Mud Springs Ridge	Idaho	45°34'57"N	116°21'18"W	5368
Muldoon Ridge	Blaine	43°35'30"N	113°53'40"W	8124
Napoleon Ridge	Lemhi	45°21'58"N	114°00'24"W	
Never Again Ridge	Clearwater	46°39'38"N	115°06'57"W	
Niagara Springs Grade	Gooding	42°40'12"N	114°41'26"W	
North Cottonwood Ridge	Twin Falls	42°16'18"N	114°24'35"W	
Norton Ridge	Valley	44°51'10"N	114°55'55"W	8252
Norwegian Ridge	Latah	46°39'38"N	116°22'19"W	
Nugget Ridge	Shoshone	47°10'55"N	115°37'57"W	
Old Man Ridge	Caribou	42°58'08"N	111°05'03"W	6785
Oreana Ridge	Lemhi	45°27'28"N	114°30'20"W	7451
Osier Ridge	Clearwater	46°49'50"N	115°01'04"W	6065
Otter Ridge	Idaho	45°23'58"N	114°51'39"W	
Outlet Ridge	Bingham	43°12'52"N	111°35'08"W	
Oxford Ridge	Franklin	42°13'41"N	112°04'09"W	8385
Parachute Ridge	Idaho	45°51'53"N	114°53'29"W	
Paradise Ridge	Latah	46°40'40"N	116°58'33"W	3702
Pasture Ridge	Idaho	45°41'48"N	114°41'57"W	6287
Patrol Ridge	Idaho	45°53'04"N	114°57'13"W	6508
Patterson Ridge	Teton	43°38'46"N	111°14'35"W	
Peep a Day Ridge	Bonner	48°03'42"N	116°19'12"W	
Pelican Ridge	Caribou	42°58'23"N	111°34'13"W	7618
Pelke Divide	Bonner	48°28'10"N	117°04'00"W	
Pepper Creek Ridge	Lemhi	45°03'37"N	114°08'57"W	
Pete Creek Divide	Clark	44°31'58"N	112°04'41"W	
Peters Point	Owyhee	43°04'19"N	116°42'41"W	
Peterson Hill	Bonneville	43°24'11"N	111°51'37"W	6215
Pettibone Ridge	Idaho	46°05'45"N	114°43'32"W	5298
Phantom Ridge	Idaho	45°23'27"N	114°50'37"W	5858
Phelan Ridge	Lemhi	45°10'01"N	114°05'40"W	8025
Phillips Ridge	Idaho	45°32'39"N	116°23'25"W	
Pine Creek Ridge	Lemhi	45°19'08"N	114°10'28"W	
Pine Spring Ridge	Bear Lake	42°08'01"N	111°13'52"W	7878
Pinnacle Ridge	Idaho	45°46'39"N	116°08'38"W	
Pinnacle Ridge	Shoshone	47°09'22"N	115°36'11"W	
Pistol Creek Ridge	Valley	44°39'34"N	115°20'23"W	
Pleasant Ridge	Adams	44°50'08"N	116°31'01"W	
Point Siam	Shoshone	47°09'39"N	115°29'18"W	
Pollock Ridge	Clearwater	46°46'06"N	115°00'42"W	5448
Polly Ridge	Nez Perce	46°15'37"N	116°44'25"W	4035
Pony Ridge	Elmore	43°41'03"N	115°12'48"W	
Porphyry Ridge	Lemhi	45°02'55"N	114°22'56"W	
Pot Mountain Ridge	Clearwater	46°49'10"N	115°17'41"W	6472
Potlatch Ridge	Nez Perce	46°36'12"N	116°32'55"W	
Potlatch Ridge	Nez Perce	46°36'01"N	116°28'43"W	
Potosi Ridge	Owyhee	43°00'12"N	116°44'37"W	
Poulsen Ridge	Bear Lake	42°22'17"N	111°27'17"W	
Price Ridge	Idaho	45°55'19"N	116°27'34"W	

Ridge or Slope	County	Latitude	Longitude	Elev.
Prospect Ridge	Idaho	45°30'39"N	114°56'48"W	4926
Prospector Ridge	Shoshone	47°12'08"N	115°38'58"W	5570
Purser Ridge	Owyhee	43°06'11"N	117°00'58"W	
Quartz Ridge	Idaho	45°52'24"N	115°52'59"W	
Quartz Ridge	Shoshone	47°13'25"N	115°26'59"W	5787
Quartzite Ridge	Idaho	45°32'15"N	115°42'25"W	8202
Radcliff Ridge	Idaho	45°58'36"N	115°37'33"W	
Railroad Ridge	Custer	44°08'25"N	114°33'23"W	9648
Rainbold Ridge	Idaho	45°47'08"N	115°48'09"W	
Rainbow Ridge	Valley	45°00'58"N	115°32'30"W	8681
Rainbow Ridge	Valley	44°59'27"N	115°31'00"W	
Ramey Ridge	Idaho	45°14'12"N	115°15'18"W	
Rams Horn Ridge	Cassia	42°18'58"N	114°11'07"W	
Rasmussen Ridge	Caribou	42°51'06"N	111°21'01"W	7487
Rattlesnake Ridge	Idaho	45°56'26"N	116°40'57"W	
Red Hog	Kootenai	47°34'24"N	116°52'08"W	3805
Red Horse Ridge	Idaho	45°49'17"N	115°23'09"W	4932
Red Pine Ridge	Caribou	42°26'23"N	111°37'07"W	7503
Red Ridge	Bonneville	43°19'47"N	111°18'12"W	8648
Red Ridge	Custer	44°05'38"N	114°30'54"W	9070
Red Ridge	Idaho	45°47'41"N	116°08'58"W	
Relay Ridge	Teton	43°42'35"N	111°20'52"W	
Roan Ridge	Lemhi	45°27'29"N	114°37'39"W	8024
Rocky Ridge	Bonneville	43°27'38"N	111°14'53"W	7525
Rocky Ridge	Idaho	46°26'46"N	115°30'40"W	6540
Rose Ridge	Owyhee	42°44'42"N	116°44'33"W	6672
Ruby Ridge	Boundary	48°54'53"N	116°05'59"W	
Ruby Ridge	Boundary	48°37'29"N	116°28'58"W	5635
Runaway Ridge	Idaho	45°19'28"N	114°54'46"W	
Russell Ridge	Boundary	48°48'54"N	116°30'43"W	
Saddle Blanket Ridge	Idaho	45°55'25"N	114°55'39"W	7200
Saddle Horse Ridge	Idaho	45°46'15"N	116°25'31"W	
Saddle Ridge	Idaho	46°02'25"N	115°28'45"W	
Saint Joe Divide	Shoshone	47°23'43"N	116°21'48"W	6063
Salamander Ridge	Idaho	45°40'06"N	114°51'42"W	6647
Salmon Ridge	Clearwater	46°52'46"N	115°40'03"W	
Sand Ridge	Bonner	48°28'33"N	116°22'43"W	
Saturday Ridge	Idaho	46°22'12"N	114°46'01"W	7010
Savage Ridge	Idaho	46°27'27"N	114°35'15"W	6916
Sawmill Ridge	Caribou	42°56'54"N	111°50'56"W	7050
Sawyer Ridge	Idaho	45°41'24"N	115°57'14"W	
Schmid Ridge	Caribou	42°41'25"N	111°19'07"W	7550
Scofield Divide	Clearwater	46°39'17"N	115°44'39"W	
Serrate Ridge	Custer	44°02'47"N	114°35'20"W	1068
Sheep Drive Ridge	Idaho	45°53'40"N	115°16'54"W	5708
Sheep Ridge	Idaho	45°51'03"N	115°59'17"W	
Sheep Ridge	Payette	44°03'57"N	116°40'22"W	3074
Sheepeater Ridge	Idaho	45°20'56"N	115°21'23"W	7969

Ridge or Slope	County	Latitude	Longitude	Elev.
Shell Ridge	Shoshone	47°39'44"N	116°17'35"W	
Shellrock Ridge	Valley	44°56'35"N	115°00'50"W	8441
Sheridan Ridge	Clark	44°27'07"N	111°42'47"W	
Sherman Ridge	Bonner	48°05'55"N	116°22'51"W	
Silver Ridge	Idaho	45°51'03"N	115°46'37"W	
Silver Spur Ridge	Idaho	45°34'01"N	115°34'52"W	
Simmons Ridge	Idaho	45°51'35"N	115°16'49"W	6150
Sixmile Ridge	Valley	44°56'50"N	115°46'02"W	5891
Sixtytwo Ridge	Idaho	46°09'59"N	115°02'37"W	
Skitwish Ridge	Kootenai	47°40'28"N	116°30'37"W	
Skoug Ridge	Idaho	45°24'46"N	115°45'12"W	
Slate Ridge	Boundary	48°33'48"N	116°09'05"W	
Sliderock Ridge	Custer	44°39'40"N	114°55'38"W	8404
Sly Ridge	Idaho	45°41'59"N	116°11'05"W	
Smith Ridge	Idaho	45°31'03"N	114°58'21"W	6209
Smith Ridge	Kootenai	47°26'08"N	116°31'13"W	3608
Snow Ridge	Boundary	48°40'23"N	116°30'42"W	
Sourdough Ridge	Clearwater	46°48'47"N	116°01'26"W	4125
South Fork Ridge	Shoshone	47°33'38"N	116°11'10"W	
Split Creek Ridge	Idaho	46°14'09"N	115°22'02"W	
Spring Ridge	Camas	43°29'01"N	115°01'57"W	
Stewart Creek Ridge	Shoshone	47°50'35"N	116°21'44"W	
Storm Ridge	Idaho	45°35'54"N	114°37'38"W	7282
Strychnine Ridge	Latah	46°58'14"N	116°36'43"W	
Suicide Point	Idaho	45°33'04"N	116°31'29"W	1400
Summers Point	Adams	44°50'14"N	116°44'40"W	7716
Sunny Slope Ridge	Kootenai	47°22'59"N	116°50'47"W	3276
Sunset Ridge	Butte	43°27'53"N	113°12'24"W	
Surveyors Ridge	Shoshone	47°00'17"N	115°37'52"W	
Swamp Ridge	Clearwater	46°48'02"N	115°01'02"W	5060
Swede Ridge	Idaho	45°53'43"N	115°53'21"W	
Swisher Ridge	Owyhee	43°02'31"N	116°56'27"W	
Table Lands	Idaho	45°36'05"N	116°23'13"W	
Tag Creek Ridge	Idaho	45°23'46"N	114°53'31"W	5406
Tahoe Ridge	Idaho	46°07'12"N	115°52'53"W	
Tamarack Ridge	Clearwater	46°33'19"N	115°35'28"W	
Tamarack Ridge	Idaho	45°47'50"N	115°58'38"W	
Tamarack Ridge	Latah	46°41'32"N	116°25'34"W	4104
Telephone Ridge	Idaho	45°35'09"N	116°06'10"W	
Tenderfoot Ridge	Kootenai	47°48'13"N	116°35'04"W	
Tenmile Ridge	Boise	44°05'46"N	115°17'33"W	8705
Tenmile Ridge	Idaho	45°40'03"N	115°41'04"W	
Tennison Ridge	Washington	44°19'33"N	116°29'46"W	
Texas Ridge	Latah	46°43'40"N	116°31'19"W	
Third Fork Ridge	Gem	44°25'35"N	116°14'35"W	6040
Three Prong Ridge	Idaho	45°46'49"N	114°55'11"W	8109
Three Sisters	Shoshone	47°10'17"N	115°39'19"W	
Thurmon Ridge	Fremont	44°22'24"N	111°29'54"W	

Ridges & Slopes

Ridge or Slope	County	Latitude	Longitude	Elev.
Tick Ridge	Clearwater	46°44'32"N	116°11'19"W	
Timber Ridge	Valley	44°53'17"N	116°04'24"W	
Toboggan Ridge	Clearwater	46°39'26"N	114°56'44"W	6091
Toms Ridge	Bonner	48°05'06"N	116°18'33"W	4518
Trapper Ridge	Lemhi	45°11'39"N	114°16'35"W	
Trestle Ridge	Bonner	48°20'56"N	116°14'17"W	
Trout Creek Ridge	Idaho	45°27'02"N	115°18'34"W	7500
Tussel Ridge	Adams	45°09'26"N	116°40'42"W	
Twin Creek Ridge	Teton	43°40'44"N	111°15'10"W	
Twin Lakes Ridge	Valley	44°53'57"N	115°59'07"W	
Twin Springs Ridge	Owyhee	42°28'09"N	116°52'33"W	
Tygee Ridge	Caribou	42°48'26"N	111°03'37"W	
Tylers Ridge	Shoshone	47°08'48"N	116°13'52"W	5154
Tyndall Ridge	Valley	44°32'30"N	115°32'05"W	
Uranus Ridge	Shoshone	47°43'08"N	116°01'19"W	
Vader Grade	Gooding	42°46'53"N	114°51'34"W	
Van Ridge	Idaho	45°28'54"N	116°07'54"W	
Viola Ridge	Latah	46°53'41"N	117°00'52"W	
Wapshilla Ridge	Nez Perce	45°56'55"N	116°48'25"W	4566
Ward Ridge	Kootenai	47°32'22"N	116°33'04"W	3148
Warnock Ridge	Idaho	45°29'05"N	115°54'00"W	
Waugh Ridge	Lemhi	45°28'02"N	114°47'25"W	
Webb Ridge	Nez Perce	46°17'13"N	116°47'44"W	2819
Weitas Ridge	Idaho	46°28'29"N	115°21'00"W	5168
Wendover Ridge	Idaho	46°32'23"N	114°48'59"W	
White Bird Hill	Idaho	45°50'22"N	116°14'39"W	4613
White Bird Ridge	Idaho	45°46'42"N	116°32'01"W	
White Bird Ridge	Idaho	45°19'43"N	116°23'33"W	4566
White Monument	Adams	45°09'16"N	116°38'11"W	7623
White Quartz Ridge	Bonner	48°05'51"N	116°22'09"W	
Wild Horse Ridge	Idaho	45°40'47"N	116°27'49"W	4291
Williams Creek Ridge	Idaho	45°43'55"N	115°37'36"W	6035
Willow Ridge	Washington	44°16'21"N	116°26'49"W	4694
Willow Ridge	Washington	44°12'22"N	116°26'55"W	5280
Wilson Ridge	Bingham	43°01'30"N	111°38'15"W	
Wind Ridge	Blaine	43°18'45"N	114°19'36"W	
Windy Ridge	Adams	45°02'09"N	116°46'48"W	
Windy Ridge	Caribou	42°55'48"N	111°52'02"W	7250
Windy Ridge	Clearwater	46°33'06"N	115°13'23"W	
Windy Ridge	Idaho	45°49'20"N	116°34'42"W	
Windy Ridge	Kootenai	47°44'30"N	116°33'15"W	
Windy Ridge	Madison	43°39'19"N	111°32'48"W	
Witter Ridge	Idaho	45°34'37"N	114°45'53"W	
Wolf Fangs	Valley	45°10'33"N	115°24'28"W	
Wounded Doe Ridge	Idaho	46°20'54"N	114°58'20"W	
Wylies Ridge	Idaho	45°59'56"N	114°55'54"W	
Yellowbanks Ridge	Kootenai	47°45'58"N	116°38'56"W	
Yew Ridge	Idaho	45°55'36"N	115°50'41"W	5108

Ridge or Slope	County	Latitude	Longitude	Elev.

Rock or Pillar	County	Latitude	Longitude	Elev.
Balanced Rock	Twin Falls	42°32'54"N	114°57'24"W	
Bath Rock	Cassia	42°04'32"N	113°43'17"W	
Bell Rock	Blaine	43°36'10"N	113°58'57"W	
Bertha Rock	Jefferson	43°38'10"N	111°40'25"W	5070
Black Pole	Valley	44°57'15"N	115°03'29"W	
Blanche Rock	Jefferson	43°38'19"N	111°39'14"W	5650
Bread Loaf	Cassia	42°05'17"N	113°43'41"W	6980
Browns Rock	Clearwater	46°42'47"N	115°42'39"W	
Camel Rock	Cassia	42°10'47"N	113°46'12"W	
Camel Rock	Cassia	42°06'19"N	113°41'27"W	
Camp Rock	Cassia	42°04'26"N	113°41'24"W	
Castle Peak	Custer	44°02'23"N	114°35'06"W	11815
Castle Rock	Blaine	43°38'10"N	114°34'31"W	
Castle Rock	Cassia	42°07'28"N	113°40'15"W	
Castle Rock	Lemhi	44°47'43"N	114°27'42"W	
Castle Rock	Teton	43°39'15"N	111°18'59"W	
Cathedral Rocks	Elmore	43°25'12"N	115°36'58"W	
Cave Rock	Butte	43°34'05"N	113°38'58"W	
Chateau Rock	Clearwater	46°40'44"N	115°30'05"W	
City Of Rocks	Cassia	42°05'15"N	113°42'00"W	6267
City Of Rocks	Gooding	43°08'21"N	114°46'35"W	
Crane Rock	Owyhee	42°56'12"N	115°35'12"W	
Devils Chair	Idaho	46°31'03"N	115°04'43"W	6422
Diamond Rock	Idaho	45°55'24"N	114°57'02"W	
Eagle Rock	Valley	44°52'47"N	115°38'23"W	
Eagletail Rock	Power	42°45'59"N	112°37'15"W	
Elephant Rock	Cassia	42°04'18"N	113°42'39"W	6285
Elizabeths Finger	Clearwater	46°47'07"N	115°13'24"W	
Farm Rock	Cassia	42°06'39"N	113°41'22"W	6310
Finger Of Fate	Custer	44°01'35"N	114°57'40"W	9775
Finger Rock	Cassia	42°06'08"N	113°44'31"W	
Garrett Rock	Custer	44°13'42"N	113°36'02"W	
Gold Fork Rock	Valley	44°40'13"N	115°47'49"W	8165
Horse Collar Rock	Blaine	43°26'22"N	114°15'54"W	
Indian Point	Valley	45°00'48"N	115°42'58"W	
Indian Postoffice	Idaho	46°32'46"N	114°59'12"W	
Kellys Finger	Clearwater	46°41'14"N	114°50'17"W	7454
Kellys Thumb	Clearwater	46°42'44"N	114°56'21"W	5430
Landmark Rock	Valley	44°37'35"N	115°35'35"W	8433
Little City Of Rocks	Gooding	43°07'09"N	114°41'16"W	
Lone Rock	Cassia	42°07'22"N	113°41'31"W	6372
Loon Creek Point	Custer	44°47'09"N	114°49'19"W	7070
Map Rock	Canyon	43°25'16"N	116°42'11"W	
Marshall Peak	Camas	43°46'22"N	114°53'21"W	9766
Needles, The	Butte	43°36'08"N	113°40'09"W	
Old Warriors Face	Idaho	45°42'05"N	114°38'19"W	
Parking Lot Rock	Cassia	42°05'08"N	113°43'18"W	6830
Patterson Peak	Custer	44°02'56"N	114°37'04"W	10872
Peace Rock	Valley	44°18'46"N	115°44'48"W	8139
Pinnacle	Cassia	42°07'55"N	114°11'04"W	7643
Register Rock	Cassia	42°04'02"N	113°41'48"W	
Register Rock	Power	42°39'09"N	113°00'58"W	4225
Rock Of Ages	Custer	44°09'30"N	113°37'04"W	
Rye Grass Pinnacle	Custer	44°41'59"N	114°15'25"W	
Skull Rock	Elmore	43°19'44"N	115°18'33"W	
Sneak Point	Clearwater	46°48'18"N	115°28'16"W	
Spion Kop Rock	Shoshone	47°53'15"N	116°05'36"W	5271
Split Rock	Clark	44°14'24"N	111°51'43"W	
Stack Rock	Boise	43°45'08"N	116°09'48"W	5895
Standing Rock	Franklin	42°06'26"N	112°05'20"W	
Steamboat Rock	Shoshone	47°39'50"N	116°09'09"W	
Suicide Rock	Valley	44°59'31"N	115°09'28"W	6367
Summit Rock	Custer	44°29'36"N	114°29'46"W	9333
Table Rock	Bonneville	43°36'02"N	111°34'41"W	5500
Table Rock Cross	Ada	43°35'56"N	116°08'48"W	
Toad Rock	Kootenai	47°35'49"N	116°49'18"W	
Train Rock	Cassia	42°06'32"N	113°41'40"W	
Treasure Rock	Cassia	42°04'27"N	113°41'43"W	
Turtle Rock	Cassia	42°04'33"N	113°42'35"W	6407
Twentymile Rock	Bonneville	43°33'19"N	112°26'27"W	
Whiskey Rock	Bonner	48°03'02"N	116°27'28"W	
White Cloud Peaks	Custer	44°03'31"N	114°36'06"W	10857
White Rock	Bingham	43°05'08"N	112°12'51"W	5590
White Rocks	Valley	44°46'22"N	115°50'19"W	

Springs

Spring	County	Latitude	Longitude	Elev.
330 Spring	Cassia	42°18'48"N	114°15'41"W	
Abbott Springs	Elmore	43°25'35"N	115°19'06"W	
Ace Spring	Owyhee	42°48'15"N	116°22'31"W	
Aitken Spring	Idaho	45°19'45"N	116°25'34"W	
Albert Spring	Owyhee	42°24'01"N	116°36'07"W	
Albertson Spring	Lemhi	45°09'05"N	113°39'27"W	
Alder Spring	Idaho	45°19'59"N	116°28'44"W	
Alhands Spring	Lemhi	44°27'24"N	113°13'56"W	
Alibi Spring	Owyhee	42°55'31"N	116°30'45"W	
Alkali Spring	Custer	44°17'36"N	114°17'14"W	
Alkali Spring	Owyhee	43°26'15"N	116°51'06"W	
Allhands Spring	Lemhi	44°27'25"N	113°14'02"W	6980
Alpine Hot Springs	Bonneville	43°14'13"N	111°06'30"W	5540
Alpine Spring	Franklin	42°22'04"N	111°35'56"W	
Alto Spring	Owyhee	42°28'35"N	116°52'40"W	
Anderson Spring	Oneida	42°02'41"N	112°59'18"W	
Andrews Spring	Lemhi	44°54'49"N	113°47'39"W	
Angleworm Spring	Owyhee	42°28'48"N	116°09'57"W	
Antelope Spring	Cassia	42°23'27"N	113°05'55"W	
Antelope Spring	Cassia	42°18'47"N	114°14'40"W	
Antelope Spring	Clark	44°13'41"N	112°36'30"W	
Antelope Spring	Owyhee	42°57'26"N	116°24'55"W	
Antelope Spring	Owyhee	42°47'28"N	116°35'10"W	
Antelope Spring	Owyhee	42°29'36"N	116°35'33"W	
Antelope Spring	Owyhee	42°26'11"N	116°02'20"W	
Antelope Spring	Owyhee	42°19'17"N	116°53'38"W	
Antelope Spring	Owyhee	42°19'03"N	116°38'08"W	
Antelope Spring	Owyhee	42°03'37"N	116°20'10"W	
Antelope Spring	Twin Falls	42°13'21"N	114°47'06"W	
Antler Spring	Blaine	43°16'08"N	114°06'24"W	5130
Arkansas Spring	Twin Falls	42°17'00"N	114°27'19"W	
Arling Hot Spring	Valley	44°38'26"N	116°02'42"W	4843
Arrow Spring	Owyhee	43°10'00"N	116°55'09"W	
Arrowhead Spring	Blaine	43°15'37"N	114°07'50"W	5290
Arrowhead Spring	Cassia	42°17'41"N	113°41'10"W	
Arrowhead Spring	Washington	44°40'15"N	116°59'08"W	5460
Arrowhead Spring Number Two	Blaine	43°15'51"N	114°09'53"W	5195
Ashbury Spring	Twin Falls	42°08'18"N	114°21'27"W	
Aspen Spring	Adams	45°06'23"N	116°29'50"W	6840
Aspen Spring	Bear Lake	42°31'55"N	111°32'30"W	7170
Aspen Spring	Blaine	43°24'30"N	114°07'38"W	
Aspen Spring	Cassia	42°10'40"N	114°04'50"W	
Aspen Spring	Franklin	42°10'28"N	111°38'17"W	
Aspen Spring	Franklin	42°01'42"N	111°41'19"W	
Atlanta Hot Springs	Elmore	43°48'38"N	115°06'46"W	
Aunt Mellys Spring	Valley	44°42'31"N	115°41'41"W	4980
Avandale Spring	Owyhee	42°25'55"N	116°29'52"W	
Babbington Spring Number 2	Owyhee	43°12'17"N	116°45'32"W	
Baby Creek Spring	Custer	44°10'26"N	113°35'13"W	
Bache Spring	Lemhi	44°47'12"N	113°18'37"W	5238
Badger Gulch Spring	Cassia	42°06'10"N	114°09'29"W	
Badger Hole Spring	Oneida	42°10'03"N	112°47'38"W	
Badger Mountain Spring	Cassia	42°07'42"N	114°07'51"W	
Bailey Creek Spring	Bear Lake	42°33'35"N	111°35'58"W	6740
Bainbridge Spring	Elmore	43°25'12"N	115°20'33"W	
Baker Spring	Franklin	42°03'31"N	111°41'32"W	
Bald Hill Spring Number 1	Twin Falls	42°10'49"N	114°20'21"W	
Bald Hill Spring Number 2	Twin Falls	42°11'21"N	114°20'19"W	
Bald Mountain Spring	Clark	44°06'10"N	112°58'49"W	
Bald Mountain Spring	Owyhee	42°42'36"N	116°20'58"W	
Balsam Spring	Bear Lake	42°16'37"N	111°32'33"W	
Balsam Spring	Franklin	42°22'57"N	111°36'29"W	
Banbury Springs	Gooding	42°41'25"N	114°49'10"W	
Bancroft Springs	Elmore	42°56'03"N	115°09'17"W	
Banks Spring	Owyhee	42°24'33"N	116°08'00"W	
Bannock Spring	Power	42°35'45"N	112°42'25"W	
Bannon Spring	Blaine	43°25'50"N	114°18'11"W	
Barbour Spring	Twin Falls	42°01'21"N	114°51'51"W	
Bare Hill Spring	Owyhee	43°04'03"N	116°53'03"W	
Bare Spring	Lemhi	45°33'13"N	114°07'53"W	
Barnes Spring	Cassia	42°09'57"N	113°10'16"W	6110
Barney Hot Springs	Custer	44°16'08"N	113°26'55"W	
Barrel Spring	Blaine	43°20'15"N	114°25'50"W	
Barrel Spring	Owyhee	42°57'07"N	116°47'24"W	
Barrel Spring	Valley	44°47'40"N	115°30'27"W	
Barrel Springs	Blaine	43°25'11"N	114°10'58"W	
Barrel Water Hole	Camas	43°12'19"N	114°33'27"W	
Barth Hot Springs	Idaho	45°30'44"N	115°02'34"W	
Barton Spring	Idaho	45°22'56"N	116°29'16"W	7800
Basin Patch Spring	Cassia	42°15'18"N	114°08'21"W	
Basin Spring	Owyhee	42°30'11"N	116°32'49"W	
Basin Spring	Twin Falls	42°15'28"N	114°23'18"W	
Basque Spring	Boise	44°06'12"N	115°50'26"W	
Basque Spring	Owyhee	42°03'23"N	115°04'22"W	6180
Basque Spring	Twin Falls	42°03'03"N	114°19'53"W	
Bat Hot Spring (historical)	Owyhee	42°48'00"N	115°43'36"W	
Bat Spring	Owyhee	42°33'56"N	116°39'44"W	
Batiste Springs	Power	42°54'54"N	112°31'17"W	
Battle Creek Crossing Waterhole Number 1	Owyhee	42°23'42"N	116°23'03"W	
Battle Creek Crossing Waterhole Number 2	Owyhee	42°22'51"N	116°24'15"W	
Bear Camp Spring	Lemhi	45°22'11"N	114°34'17"W	
Bear Gulch Spring	Cassia	42°13'41"N	114°11'32"W	
Bear Spring	Fremont	44°10'58"N	111°12'50"W	
Bear Spring	Shoshone	47°13'48"N	115°23'35"W	
Bear Springs	Idaho	45°47'39"N	115°51'48"W	
Bear Wallow Spring	Custer	44°14'27"N	114°11'32"W	

Spring	County	Latitude	Longitude	Elev.
Bear Wallow Spring	Franklin	42°23'58"N	111°36'32"W	
Bear Wallow Spring	Fremont	44°14'57"N	111°33'16"W	
Bear Wallow Spring	Lemhi	45°32'44"N	114°06'39"W	
Beartrap Spring	Lemhi	45°23'50"N	114°26'29"W	
Beatty Spring	Twin Falls	42°04'15"N	114°17'24"W	
Beaver Creek Springs	Owyhee	42°24'45"N	116°46'29"W	
Beaverdam Pass Spring	Cassia	42°05'02"N	114°06'08"W	
Bed Spring	Cassia	42°14'53"N	114°12'49"W	
Bed Springs	Shoshone	47°15'29"N	115°20'01"W	
Bedke Spring	Cassia	42°15'28"N	113°45'15"W	
Bedstead Spring	Owyhee	42°23'36"N	116°55'35"W	
Beer Can Spring	Elmore	43°12'50"N	115°20'33"W	5900
Beer Mug Spring	Twin Falls	42°05'15"N	114°17'01"W	
Bell Mare Spring	Elmore	43°03'30"N	115°08'14"W	
Bell Spring	Caribou	42°38'41"N	111°14'34"W	
Bench Spring	Oneida	42°03'14"N	112°59'26"W	4792
Bennett Spring	Cassia	42°19'37"N	113°35'58"W	
Bennetts Spring	Bear Lake	42°11'41"N	111°14'03"W	
Benton Spring	Bonner	48°20'59"N	116°45'48"W	4840
Bergguist Spring	Franklin	42°08'14"N	111°38'30"W	
Berneathy Spring	Elmore	43°20'27"N	115°36'14"W	
Between The Fields Spring	Owyhee	42°31'41"N	116°58'12"W	
Bex Spring	Teton	43°44'50"N	111°22'31"W	
Bickel Spring	Gooding	42°45'29"N	114°51'13"W	
Big Boggy Spring	Owyhee	42°38'14"N	116°46'52"W	
Big Buck Spring	Butte	43°37'06"N	113°42'59"W	
Big Canyon Spring	Bear Lake	42°33'54"N	111°23'06"W	
Big Horse Spring	Bannock	42°58'57"N	112°19'56"W	
Big Malad Spring	Oneida	42°13'18"N	112°21'51"W	
Big Mike Spring	Gooding	43°11'23"N	114°45'59"W	
Big Pipe Spring	Cassia	42°12'41"N	114°01'15"W	
Big Spring	Bannock	42°59'16"N	112°07'19"W	
Big Spring	Bannock	42°58'44"N	112°21'29"W	
Big Spring	Bear Lake	42°31'03"N	111°19'33"W	
Big Spring	Butte	43°49'58"N	113°19'17"W	
Big Spring	Camas	43°14'08"N	114°55'20"W	
Big Spring	Caribou	42°46'03"N	112°05'58"W	
Big Spring	Caribou	42°41'41"N	111°39'55"W	
Big Spring	Clark	44°27'26"N	112°41'46"W	
Big Spring	Custer	43°39'32"N	113°41'21"W	
Big Spring	Twin Falls	42°05'13"N	114°58'06"W	
Big Springs	Gooding	42°49'13"N	114°51'57"W	
Big Springs	Owyhee	42°59'11"N	116°55'01"W	
Big Springs	Owyhee	42°34'33"N	116°26'01"W	
Big Tamarack Spring	Adams	45°02'26"N	116°36'50"W	5432
Big Willlow Spring	Owyhee	42°28'50"N	116°59'47"W	
Big Willow Spring	Bingham	43°01'58"N	112°05'28"W	5980
Bill George Spring	Washington	44°24'01"N	116°24'00"W	

Spring	County	Latitude	Longitude	Elev.
Billingsley Spring	Blaine	43°25'12"N	113°55'21"W	6035
Bills Spring	Owyhee	42°30'51"N	116°48'52"W	
Birch Spring	Cassia	42°09'39"N	114°02'56"W	
Birch Spring	Oneida	42°15'56"N	112°15'19"W	
Birch Springs	Custer	44°07'54"N	113°50'13"W	
Bishop Spring	Custer	44°23'52"N	114°13'37"W	
Bishop Spring	Shoshone	47°58'11"N	116°17'32"W	
Bissell Spring	Oneida	42°19'48"N	112°15'07"W	
Bitterroot Springs	Shoshone	47°27'30"N	115°43'09"W	
Bitton Spring	Caribou	42°28'34"N	111°41'39"W	
Black Canyon Spring	Twin Falls	42°08'46"N	114°49'43"W	
Black Horse Spring	Owyhee	43°06'35"N	116°40'39"W	
Black Pine Spring	Cassia	42°06'00"N	113°05'09"W	
Black Pine Spring	Oneida	42°03'26"N	112°58'51"W	
Black Sand Spring	Cassia	42°10'45"N	113°30'34"W	
Black Spring	Fremont	44°20'11"N	111°15'50"W	
Blackhawk Spring	Elmore	43°11'04"N	115°14'48"W	5790
Blackstock Spring	Owyhee	43°18'47"N	116°56'10"W	
Blind Spring	Bannock	42°42'38"N	112°24'03"W	
Blind Spring	Power	42°45'12"N	112°43'58"W	
Blind Springs	Butte	44°07'33"N	113°11'15"W	
Bloom Spring	Shoshone	47°44'44"N	115°49'06"W	4835
Blue Bunch Spring	Adams	44°50'16"N	116°15'30"W	6334
Blue Canyon Spring	Clark	44°11'00"N	112°47'59"W	
Blue Clay Pit Spring	Owyhee	43°06'46"N	116°57'05"W	
Blue Creek Spring	Clark	44°10'11"N	112°33'08"W	
Blue Gulch Spring	Idaho	45°20'16"N	116°26'45"W	
Blue Mud Spring	Twin Falls	42°11'09"N	114°21'59"W	
Blue Pond Spring	Bear Lake	42°06'18"N	111°29'42"W	6477
Blue Spring	Fremont	44°12'40"N	111°36'12"W	
Blue Spring	Idaho	45°50'18"N	116°35'03"W	
Blue Springs	Gooding	42°42'40"N	114°49'47"W	
Boatman Spring	Clark	44°31'38"N	112°10'20"W	
Boche Spring	Lemhi	44°47'25"N	113°18'14"W	8000
Boche Spring	Lemhi	44°47'24"N	113°18'02"W	
Boggy Spring	Owyhee	42°27'16"N	116°47'23"W	
Boghole Spring	Owyhee	43°13'14"N	116°49'15"W	
Bohannon Spring	Lemhi	45°10'26"N	113°41'09"W	
Boiling Springs	Valley	44°21'51"N	115°51'24"W	4260
Bone Spring	Twin Falls	42°14'24"N	114°23'51"W	
Boni Spring	Owyhee	42°32'30"N	116°46'25"W	
Bonneville Hot Springs	Boise	44°09'25"N	115°18'48"W	
Books Spring	Caribou	42°34'48"N	111°08'40"W	
Boosinger Spring	Cassia	42°09'37"N	113°09'47"W	6060
Boquet Springs	Teton	43°37'55"N	111°12'10"W	
Bostetter Pasture Spring	Cassia	42°10'06"N	114°09'50"W	
Bostetter Spring	Cassia	42°09'52"N	114°09'17"W	
Bottle Spring	Owyhee	42°31'52"N	116°42'48"W	

Springs

Spring	County	Latitude	Longitude	Elev.
Boulder Spring	Clark	44°14'17"N	112°44'34"W	
Boundary Spring	Caribou	42°26'36"N	111°39'21"W	
Bourbon Spring	Elmore	43°10'55"N	115°13'43"W	5580
Bowers Spring	Cassia	42°18'14"N	114°08'00"W	
Box Canyon Spring	Twin Falls	42°02'14"N	114°19'05"W	
Box Spring	Gem	44°02'58"N	116°19'49"W	
Box Spring	Lemhi	45°29'04"N	114°01'31"W	
Box Spring	Twin Falls	42°09'41"N	114°48'50"W	
Box Spring	Washington	44°42'11"N	116°58'47"W	
Box Springs	Lemhi	44°17'18"N	113°08'43"W	
Bradford Spring	Payette	44°06'03"N	116°38'25"W	
Bradshaw Spring	Custer	44°19'20"N	114°12'14"W	
Breazeale Spring	Lemhi	44°22'56"N	113°03'36"W	
Briar Spring	Owyhee	43°10'37"N	116°35'59"W	
Bridal Wreath Springs	Gooding	42°45'10"N	114°50'53"W	3000
Bridger Spring	Cassia	42°28'38"N	113°31'39"W	
Briggs Spring	Cassia	42°14'17"N	114°09'33"W	7138
Broken Tank Spring	Camas	43°17'19"N	114°55'41"W	
Bronco Spring	Bingham	43°04'40"N	112°04'02"W	5955
Brownell Spring	Camas	43°16'22"N	114°56'59"W	5085
Browns Spring	Idaho	46°03'59"N	115°38'40"W	
Browns Spring	Twin Falls	42°11'00"N	114°23'03"W	
Bruce Young Spring	Cassia	42°25'23"N	113°05'29"W	
Brunzell Spring	Owyhee	43°07'51"N	116°39'07"W	
Buck Gulch Spring	Owyhee	42°45'49"N	116°29'48"W	
Buck Spring	Cassia	42°04'09"N	114°09'53"W	
Buck Spring	Madison	43°40'20"N	111°34'46"W	
Buck Spring	Madison	43°39'25"N	111°21'41"W	
Buck Spring	Owyhee	42°23'16"N	116°53'20"W	
Buck Spring	Power	42°20'45"N	112°47'41"W	
Buck Springs	Lemhi	44°08'30"N	113°29'55"W	
Buck Trough Spring	Twin Falls	42°10'26"N	114°23'10"W	
Buckaroo Spring	Owyhee	42°35'05"N	116°13'34"W	
Buckaroo Spring	Owyhee	42°29'43"N	116°47'22"W	
Buckbrush Spring	Cassia	42°09'26"N	114°06'44"W	
Buckbrush Spring	Elmore	43°02'17"N	115°11'13"W	
Bucket Spring	Owyhee	43°04'58"N	116°53'12"W	
Buckhorn Hot Spring	Valley	44°53'08"N	115°50'16"W	
Buckhorn Spring	Cassia	42°20'31"N	114°05'39"W	
Buckhorn Spring	Idaho	45°30'41"N	116°25'21"W	
Buckhorn Spring	Madison	43°38'12"N	111°36'12"W	
Buckroo Spring	Owyhee	42°49'35"N	116°33'15"W	
Buckskin Spring	Bingham	43°05'40"N	112°14'33"W	
Buckskin Spring	Owyhee	42°24'15"N	116°47'53"W	
Bud Spring	Twin Falls	42°09'44"N	114°21'03"W	
Budge Spring	Franklin	42°10'58"N	111°37'51"W	
Buggy Spring	Fremont	44°00'16"N	111°05'51"W	
Bull Basin Spring	Owyhee	42°21'11"N	116°54'59"W	
Bull Creek Hot Springs	Valley	44°25'47"N	115°45'43"W	
Bull Spring	Cassia	42°14'37"N	113°41'55"W	
Bull Spring	Gooding	43°10'53"N	114°43'22"W	
Bull Spring	Lemhi	44°54'03"N	113°46'52"W	
Bull Spring	Owyhee	42°25'14"N	116°58'18"W	
Bull Spring	Owyhee	42°12'43"N	115°58'41"W	
Bull Spring	Power	42°31'24"N	112°26'51"W	
Bull Spring	Twin Falls	42°04'47"N	114°29'07"W	
Bull Springs	Idaho	45°38'17"N	116°22'51"W	
Bull Summit Waterhole	Lemhi	44°53'47"N	113°46'10"W	
Bullet Spring	Elmore	43°04'58"N	115°13'35"W	
Burbank Spring	Franklin	42°06'12"N	111°40'51"W	
Burchertt Spring	Caribou	42°39'13"N	111°25'00"W	
Burn Spring	Cassia	42°03'22"N	114°12'21"W	
Burnt Cabin Spring	Twin Falls	42°03'06"N	114°51'55"W	
Burnt Spring	Bear Lake	42°30'33"N	111°31'24"W	7500
Butch Spring	Bannock	42°27'02"N	112°16'06"W	
Butler Spring	Bingham	43°11'08"N	111°47'14"W	6419
Butler Spring	Cassia	42°22'59"N	113°09'47"W	
Buttars Spring	Cassia	42°09'33"N	114°14'18"W	6727
Buzzard Spring	Twin Falls	42°12'00"N	114°22'07"W	
C C C Spring	Owyhee	42°26'53"N	116°57'40"W	
Cabin Spring	Cassia	42°11'30"N	114°10'26"W	
Cabin Spring	Idaho	45°32'04"N	114°56'15"W	
Cabin Spring Number Two	Cassia	42°11'17"N	114°09'36"W	
Cabin Springs	Power	42°36'18"N	112°41'09"W	
Calder Spring	Cassia	42°30'14"N	113°02'25"W	
Calhoun Spring	Blaine	43°16'46"N	114°21'17"W	
Camp Spring	Bear Lake	42°32'31"N	111°32'54"W	6780
Camp Spring	Owyhee	42°29'59"N	116°42'41"W	
Camp Spring	Twin Falls	42°10'17"N	114°20'09"W	
Camp Springs	Lemhi	44°07'37"N	113°29'14"W	
Camp Thomas Spring	Idaho	45°54'49"N	116°39'33"W	
Camp, Spring	Idaho	45°21'19"N	115°28'18"W	
Campbell Spring	Oneida	42°18'41"N	112°10'09"W	
Cannon Ball Spring	Idaho	45°19'39"N	116°27'55"W	
Carlson Spring	Cassia	42°02'21"N	114°08'20"W	
Carter Spring	Owyhee	42°22'18"N	116°44'43"W	
Casey Spring	Adams	45°14'07"N	116°35'58"W	
Castlehead Spring	Owyhee	42°25'41"N	116°48'03"W	
Cat Spring	Owyhee	42°12'30"N	116°01'53"W	
Cavannah Spring	Owyhee	42°13'25"N	115°57'45"W	
Cavieta Spring	Owyhee	42°21'08"N	117°00'18"W	5369
Ccc Spring	Owyhee	42°36'28"N	116°32'07"W	
Cedar Birch Spring	Cassia	42°10'02"N	114°00'32"W	
Cedar Hollow Spring	Cassia	42°17'30"N	113°44'37"W	
Cedar Spring	Bingham	43°03'45"N	112°15'36"W	
Cedar Spring	Elmore	43°01'36"N	115°15'22"W	

Spring	County	Latitude	Longitude	Elev.
Cedar Tree Spring	Gooding	43°05'53"N	115°04'11"W	
Chad Spring	Twin Falls	42°13'06"N	114°21'00"W	
Chain Spring	Idaho	45°16'23"N	116°27'03"W	
Chalk Spring	Cassia	42°05'56"N	114°03'11"W	
Chalk Spring	Elmore	43°01'26"N	115°08'06"W	
Challis Hot Springs	Custer	44°31'21"N	114°10'25"W	
Chandler Spring	Caribou	42°43'10"N	111°42'45"W	6281
Chandler Spring	Clark	44°11'09"N	112°43'22"W	
Charity Spring	Owyhee	42°53'54"N	116°32'59"W	
Charlie Brackett Spring	Twin Falls	42°07'54"N	114°50'53"W	
Chattanooga Hot Spring	Elmore	43°48'47"N	115°06'59"W	
Chatterton Spring	Bannock	42°33'56"N	112°06'33"W	
Chatterton Spring	Franklin	42°01'55"N	111°42'43"W	
Cherry Spring	Bannock	42°18'54"N	112°09'22"W	
Cherry Spring	Lemhi	44°57'27"N	113°41'05"W	
Cherry Spring	Twin Falls	42°00'48"N	114°18'15"W	
Cherry Spring	Washington	44°12'21"N	116°22'26"W	
Chet Rowe Spring	Lemhi	45°08'43"N	113°41'48"W	
Chicken Spring	Camas	43°15'15"N	114°32'39"W	
Chicken Spring	Caribou	42°46'56"N	111°14'16"W	
Chicken Spring	Cassia	42°10'13"N	114°10'33"W	
Chicken Spring	Idaho	45°17'24"N	115°23'41"W	
Chicken Spring	Owyhee	42°00'09"N	116°00'04"W	
Chimney Spring	Owyhee	42°47'41"N	116°41'19"W	
China Spring	Idaho	45°34'24"N	116°20'31"W	
China Spring	Lemhi	45°00'46"N	114°05'26"W	
Chipmunk Spring	Adams	44°56'35"N	116°36'55"W	
Chokecherry Spring	Blaine	43°23'32"N	114°03'38"W	6800
Chokecherry Spring	Cassia	42°10'00"N	114°04'48"W	
Chokecherry Spring	Owyhee	42°41'48"N	116°16'57"W	
Christenson Spring	Bannock	42°32'23"N	112°06'16"W	
Chubb Springs	Caribou	42°59'45"N	111°30'18"W	
Chubby Spring	Owyhee	43°03'47"N	116°55'14"W	
City Spring	Custer	43°54'03"N	113°37'41"W	
Clark Spring	Caribou	42°42'29"N	111°57'52"W	5990
Clark Spring	Twin Falls	42°09'57"N	114°55'44"W	
Clay Bank Spring	Blaine	43°18'26"N	114°22'50"W	
Cliff Spring	Twin Falls	42°08'03"N	114°21'47"W	
Clifton Spring	Adams	44°56'08"N	116°42'36"W	
Clint Palmer Spring	Cassia	42°27'08"N	113°06'35"W	5045
Clover Spring	Owyhee	42°44'56"N	116°26'47"W	
Clyde Spring	Idaho	45°21'53"N	116°25'44"W	
Coal Banks Spring	Cassia	42°04'14"N	113°59'06"W	
Coal Kiln Spring	Lemhi	44°18'37"N	113°10'43"W	
Coal Mine Gulch Spring	Lemhi	45°07'39"N	113°42'29"W	
Coal Pit Spring	Cassia	42°14'25"N	114°10'48"W	
Coal Spring	Owyhee	43°14'47"N	116°50'44"W	
Coby Spring	Bear Lake	42°05'07"N	111°26'38"W	6920
Coffee Spring	Washington	44°37'23"N	116°58'10"W	6590
Cold Spring	Adams	44°56'48"N	116°26'14"W	4880
Cold Spring	Bingham	43°07'12"N	112°03'39"W	
Cold Spring	Bonneville	43°36'50"N	111°20'01"W	
Cold Spring	Butte	43°35'58"N	113°37'05"W	
Cold Spring	Camas	43°28'18"N	115°03'36"W	
Cold Spring	Camas	43°18'00"N	114°25'43"W	
Cold Spring	Caribou	42°55'34"N	111°54'46"W	6360
Cold Spring	Caribou	42°48'40"N	112°07'10"W	
Cold Spring	Cassia	42°19'00"N	114°08'00"W	
Cold Spring	Cassia	42°16'05"N	113°00'04"W	
Cold Spring	Cassia	42°15'05"N	114°13'36"W	
Cold Spring	Cassia	42°15'00"N	113°42'03"W	
Cold Spring	Elmore	43°28'52"N	115°41'58"W	
Cold Spring	Franklin	42°13'36"N	111°47'51"W	
Cold Spring	Fremont	44°35'37"N	111°22'34"W	
Cold Spring	Fremont	44°11'29"N	111°33'54"W	
Cold Spring	Gooding	43°01'58"N	114°58'27"W	3148
Cold Spring	Lemhi	44°33'52"N	113°20'39"W	
Cold Spring	Madison	43°38'57"N	111°33'38"W	
Cold Spring	Owyhee	43°07'40"N	116°50'36"W	
Cold Spring	Owyhee	43°04'25"N	116°52'18"W	
Cold Springs	Adams	45°10'58"N	116°25'06"W	
Cold Springs	Idaho	45°49'30"N	114°40'36"W	
Cold Springs	Owyhee	42°14'32"N	116°03'56"W	
Coldwater Spring	Cassia	42°07'13"N	114°09'23"W	
Colgate Warm Springs	Idaho	46°27'56"N	114°56'22"W	
Commerford Springs	Elmore	42°59'54"N	115°29'58"W	
Conner Spring	Madison	43°38'27"N	111°32'23"W	
Conners Spring	Owyhee	42°34'13"N	116°07'33"W	
Connors Spring	Owyhee	42°54'37"N	116°58'42"W	
Cook Spring	Elmore	43°25'28"N	115°19'56"W	
Co-op Spring	Oneida	42°07'15"N	112°37'01"W	
Cora Spring	Idaho	45°18'00"N	116°24'36"W	
Corbett Spring Number 1	Lemhi	44°52'55"N	114°00'42"W	
Corduroy Spring	Bear Lake	42°11'08"N	111°31'51"W	
Corduroy Spring	Caribou	42°54'32"N	111°52'34"W	6060
Corker Springs	Elmore	43°05'44"N	115°14'42"W	
Corner Spring	Franklin	42°19'15"N	111°40'45"W	
Corral Spring	Bannock	42°41'58"N	112°24'39"W	
Corral Spring	Cassia	42°04'11"N	113°05'06"W	6200
Corral Spring	Power	42°37'38"N	112°41'28"W	
Corta Spring	Owyhee	42°41'57"N	116°59'28"W	
Cotton Spring	Cassia	42°12'15"N	114°13'19"W	
Cottonwood Ridge Spring	Twin Falls	42°16'54"N	114°24'44"W	
Cottonwood Spring	Cassia	42°08'38"N	113°09'15"W	
Cottonwood Spring	Idaho	45°19'16"N	116°27'23"W	
Cottonwood Spring	Owyhee	42°25'35"N	116°54'18"W	

Springs

Spring	County	Latitude	Longitude	Elev.
Cottonwood Spring	Washington	44°39'11"N	116°59'14"W	
Cottonwood Springs	Elmore	43°20'31"N	115°37'59"W	
Cottrel Spring	Bannock	42°28'45"N	112°17'10"W	
Cougar Spring	Owyhee	42°33'30"N	116°39'16"W	
Cougar Spring	Owyhee	42°13'16"N	115°33'37"W	
Counting Chute Spring	Cassia	42°10'18"N	114°08'33"W	7314
Cove Spring	Bingham	43°12'12"N	112°03'09"W	
Cove Spring	Blaine	43°23'44"N	114°08'38"W	
Covered Springs	Bingham	43°06'05"N	112°03'37"W	
Cow Camp Spring	Clark	44°26'43"N	112°11'54"W	
Cow Gulch Spring	Cassia	42°22'00"N	113°29'48"W	
Cow Heaven Spring	Camas	43°12'18"N	114°48'46"W	6060
Cow Spring	Bingham	43°02'13"N	112°07'41"W	
Cow Spring	Cassia	42°14'37"N	114°13'20"W	
Cow Spring	Owyhee	42°31'23"N	116°33'21"W	
Cow Spring	Twin Falls	42°14'13"N	114°25'26"W	
Cowboy Spring	Bingham	43°05'35"N	112°13'14"W	
Cowboy Spring	Cassia	42°12'50"N	113°59'18"W	
Cowley Spring	Oneida	42°27'49"N	112°24'15"W	
Cow-S Spring	Cassia	42°14'35"N	114°13'17"W	
Cox Hot Springs	Valley	44°47'06"N	114°51'18"W	
Coyote Hill Spring	Gooding	43°06'38"N	114°56'52"W	4570
Coyote Pup Spring	Owyhee	42°13'07"N	115°54'08"W	
Coyote Spring	Bingham	43°04'30"N	112°12'13"W	
Coyote Spring	Butte	44°02'07"N	113°08'14"W	
Coyote Spring	Cassia	42°17'12"N	114°09'41"W	
Coyote Spring	Cassia	42°12'21"N	114°09'40"W	
Coyote Spring	Custer	44°03'06"N	113°47'08"W	
Coyote Spring	Elmore	43°07'46"N	115°23'13"W	3926
Coyote Spring	Gooding	43°08'44"N	114°45'54"W	
Coyote Spring	Idaho	45°14'33"N	114°50'31"W	
Coyote Spring	Lemhi	45°26'43"N	114°11'55"W	
Coyote Spring	Owyhee	43°24'49"N	116°47'58"W	
Coyote Spring	Twin Falls	42°43'26"N	114°56'47"W	3338
Coyote Spring	Twin Falls	42°17'56"N	114°22'01"W	
Crab Spring	Owyhee	42°33'05"N	116°17'09"W	
Craig Spring	Owyhee	42°21'56"N	116°48'54"W	
Crandall Spring	Teton	43°39'47"N	111°03'10"W	
Cranny Spring	Cassia	42°13'33"N	114°07'28"W	
Craster Spring	Elmore	43°12'33"N	115°22'20"W	
Crest Spring	Owyhee	43°14'05"N	116°50'34"W	
Crib Spring	Bear Lake	42°08'02"N	111°27'09"W	
Crib Spring	Franklin	42°12'46"N	111°37'50"W	
Crib Spring	Lemhi	45°08'35"N	114°02'03"W	
Crockett Spring	Twin Falls	42°18'29"N	114°20'22"W	
Croney Hollow Spring	Franklin	42°16'45"N	111°40'21"W	
Crossroads Spring	Owyhee	42°54'55"N	116°57'25"W	
Crows Nest Spring	Owyhee	43°10'00"N	116°49'11"W	
Crystal Spring	Bear Lake	42°20'52"N	111°27'04"W	
Crystal Spring	Cassia	42°05'35"N	114°09'55"W	
Crystal Spring	Valley	45°08'47"N	115°23'27"W	
Crystal Springs	Bannock	42°39'50"N	112°06'40"W	
Crystal Springs	Gooding	42°39'38"N	114°38'29"W	
Cub Basin Spring	Franklin	42°06'26"N	111°36'48"W	
Cub Spring	Twin Falls	42°12'40"N	114°23'18"W	
Curly Jack Spring	Bannock	42°19'28"N	112°10'18"W	
Cy Springs	Valley	44°27'20"N	115°16'26"W	
D H Spring	Oneida	42°04'57"N	112°23'23"W	
Dairy Spring	Cassia	42°30'21"N	113°01'53"W	
Dairy Springs	Caribou	42°56'07"N	111°48'49"W	
Damon Springs	Owyhee	42°02'57"N	116°01'18"W	
Dan Spring	Twin Falls	42°11'59"N	114°35'10"W	
Daugherty Spring	Custer	44°29'02"N	114°19'38"W	
Dave Shea Spring	Owyhee	42°40'55"N	116°59'17"W	
Daves Pass Spring	Cassia	42°05'07"N	114°02'33"W	
Daveys Spring	Elmore	43°26'45"N	115°23'58"W	
Davis Spring	Idaho	45°35'03"N	116°22'56"W	
Davis Spring	Owyhee	42°24'54"N	116°45'51"W	
Davis Springs	Power	42°45'55"N	112°53'28"W	
De Lamar Spring	Owyhee	42°58'57"N	116°45'40"W	
Dead Cow Spring	Bannock	42°47'36"N	112°30'57"W	
Dead Horse Spring	Gooding	43°03'09"N	115°01'54"W	
Dead Horse Spring	Owyhee	42°53'12"N	116°57'07"W	
Dead Horse Spring	Owyhee	42°01'48"N	115°34'41"W	
Deadeye Spring	Cassia	42°19'41"N	114°14'59"W	
Deadwater Spring	Lemhi	45°23'16"N	114°03'09"W	
Deaf Joe Spring	Owyhee	43°00'11"N	116°50'03"W	
Dean Spring	Adams	44°54'20"N	116°34'21"W	4623
Dean Spring	Custer	44°07'05"N	114°02'37"W	
Deep Creek Spring	Owyhee	42°21'22"N	116°38'39"W	
Deer House Spring	Oneida	42°04'23"N	112°10'02"W	
Deer Mountain Spring	Blaine	43°22'20"N	114°05'11"W	6030
Deer Spring	Camas	43°12'44"N	114°58'15"W	5700
Deer Spring	Twin Falls	42°13'19"N	114°19'56"W	
Deer Water	Owyhee	42°41'47"N	115°46'32"W	
Deerhorn Spring	Blaine	43°16'26"N	114°06'32"W	5620
Denton Spring	Washington	44°16'48"N	116°24'16"W	
Devils Hole	Washington	44°21'21"N	117°09'47"W	
Diablo Spring	Idaho	46°18'23"N	114°37'40"W	
Diamond Creek Spring	Owyhee	43°06'05"N	116°36'18"W	
Diamond Flat Spring	Twin Falls	42°07'30"N	114°19'45"W	
Dishpan Spring	Lemhi	45°27'23"N	114°27'22"W	
Dishrag Spring	Lemhi	44°55'49"N	114°11'43"W	
Ditch Spring	Blaine	43°14'35"N	114°06'22"W	
Dixey Spring	Bingham	43°05'55"N	112°12'36"W	
Dock Spring	Franklin	42°13'29"N	111°40'32"W	

Spring	County	Latitude	Longitude	Elev.
Doe Spring	Elmore	43°10'48"N	115°11'22"W	5410
Dord Spring	Bonneville	43°37'35"N	111°35'09"W	
Doublespring	Custer	44°19'13"N	113°47'15"W	6873
Dougal Waterhole	Owyhee	42°19'08"N	116°47'54"W	
Dougherty Springs	Owyhee	42°37'37"N	116°57'48"W	
Douglas Spring	Custer	44°10'07"N	114°02'34"W	
Dove Spring	Twin Falls	42°06'00"N	114°53'30"W	
Doyle Mountain Spring	Owyhee	42°49'53"N	116°22'27"W	
Dragroad Spring	Lemhi	44°55'37"N	113°34'18"W	
Drexall Spring	Shoshone	47°51'08"N	116°11'22"W	
Driscol Spring	Owyhee	42°59'20"N	116°38'35"W	
Drollinger Spring	Owyhee	43°02'25"N	116°38'52"W	
Dry Canyon Spring	Bonneville	43°34'22"N	111°21'58"W	
Dry Creek Spring	Oneida	42°04'29"N	112°08'05"W	
Dry Creek Spring Number 1	Custer	44°10'01"N	113°33'00"W	
Dry Creek Spring Number 2	Custer	44°09'42"N	113°33'11"W	
Dry Flat Spring	Cassia	42°13'36"N	114°14'47"W	
Dry Fork Spring	Cassia	42°16'59"N	114°13'08"W	
Dry Fork Spring	Cassia	42°10'14"N	114°02'59"W	
Dry Fork Swanty Creek Spring	Cassia	42°03'56"N	114°12'14"W	
Dry Gulch Spring	Cassia	42°03'24"N	114°03'04"W	
Dry Hole Spring	Owyhee	42°09'42"N	115°54'47"W	
Dry Hollow Spring	Bannock	42°21'39"N	111°56'58"W	
Dry Hollow Spring	Power	42°36'13"N	112°40'32"W	
Dry Lake Spring	Owyhee	42°30'41"N	116°39'35"W	
Dry Pine Spring	Oneida	42°04'46"N	112°22'41"W	
Duck Spring	Twin Falls	42°13'47"N	114°32'57"W	
Dug Spring	Owyhee	42°25'45"N	116°45'40"W	
Dugout Spring	Custer	43°56'20"N	113°43'00"W	
Dukes V Spring	Owyhee	42°18'50"N	116°56'49"W	
Dummer Spring	Cassia	42°09'20"N	114°04'50"W	
Duncan Creek Spring	Owyhee	42°27'22"N	116°03'19"W	
Dune Spring	Bear Lake	42°05'19"N	111°26'38"W	6900
Dutch Oven Spring	Adams	44°40'31"N	116°20'58"W	4440
Dutch Oven Spring	Valley	45°00'53"N	115°39'28"W	
Dutcher Spring	Owyhee	42°32'33"N	117°00'03"W	4890
Dynamite Spring	Bingham	43°13'17"N	111°41'44"W	6315
Eagle Spring	Blaine	43°21'52"N	114°24'12"W	
Eagle Spring	Owyhee	42°45'46"N	116°21'59"W	
Eagle Spring	Twin Falls	42°07'54"N	114°18'35"W	
East Farnes Mountain Spring	Madison	43°40'43"N	111°28'48"W	
East Kurtz Spring	Oneida	42°17'17"N	112°49'10"W	
East Line Spring	Cassia	42°15'10"N	114°13'47"W	
East Spring	Owyhee	42°24'55"N	116°48'52"W	
East Springs	Jefferson	43°45'45"N	112°06'37"W	
Echo Spring	Elmore	43°06'26"N	115°06'03"W	
Echo Springs	Benewah	47°18'18"N	116°41'22"W	
Eckersell Spring	Lemhi	44°58'56"N	113°31'56"W	

Spring	County	Latitude	Longitude	Elev.
Edie Spring	Clark	44°27'43"N	112°31'42"W	8480
Eds Spring	Owyhee	42°00'25"N	115°57'52"W	
Electric Spring	Twin Falls	42°14'19"N	114°21'08"W	
Elk Basin Spring	Cassia	42°01'46"N	114°11'28"W	
Elk Butte Spring	Cassia	42°12'06"N	114°16'46"W	
Elk Corral Spring	Twin Falls	42°11'53"N	114°17'24"W	
Elk Creek Hot Spring	Camas	43°25'24"N	114°37'39"W	
Elk Horn Spring	Idaho	45°35'01"N	116°26'09"W	
Elk Spring	Blaine	43°24'17"N	114°07'53"W	
Elk Spring	Idaho	45°13'29"N	115°23'13"W	
Elk Spring	Twin Falls	42°09'11"N	114°17'08"W	
Elkhorn Warm Spring	Madison	43°39'36"N	111°42'55"W	
Ellis Spring	Fremont	44°12'52"N	111°33'52"W	
Ellisons Spring	Jerome	42°38'24"N	114°34'25"W	3000
Emerson Spring	Twin Falls	42°10'59"N	114°32'08"W	
Emery Canyon Spring	Cassia	42°05'18"N	113°44'02"W	
Emigration Spring	Bear Lake	42°21'41"N	111°33'24"W	
Empey Spring	Bingham	43°15'27"N	111°40'24"W	6330
Erma Spring	Camas	43°13'12"N	114°52'34"W	5580
Ether Park Spring	Cassia	42°23'18"N	113°03'43"W	5945
Eusebio Spring	Camas	43°13'39"N	114°37'14"W	
Evans Lick Spring	Owyhee	42°44'54"N	116°30'38"W	
Eyrie Spring	Oneida	42°18'22"N	112°56'30"W	
F S Spring	Cassia	42°08'55"N	114°15'11"W	
Fallert Springs	Butte	44°00'24"N	113°05'01"W	
Fallini Spring	Custer	43°49'56"N	113°36'30"W	
Falls Hot Springs	Owyhee	42°45'43"N	115°44'18"W	2685
Fawn Spring	Twin Falls	42°10'37"N	114°19'05"W	
Fenn Cabin Spring	Idaho	45°44'39"N	116°06'12"W	
Fenwick Spring	Owyhee	42°27'46"N	116°56'23"W	
Fir Point Spring	Adams	44°51'42"N	116°28'07"W	3950
First Basin Spring	Lemhi	44°58'22"N	113°43'15"W	
Fish Creek Spring	Fremont	44°09'39"N	111°11'05"W	
Fish Dam Spring	Idaho	45°33'14"N	116°21'12"W	
Fisher Spring	Cassia	42°08'01"N	113°09'39"W	
Fisher Spring	Power	42°43'00"N	112°40'48"W	
Five Spring	Owyhee	42°58'24"N	116°57'53"W	
Flag Spring	Washington	44°11'08"N	116°36'15"W	
Flatiron Spring	Cassia	42°08'02"N	114°11'36"W	
Florence Spring	Gooding	42°50'35"N	114°52'40"W	3158
Florence Spring	Owyhee	42°50'05"N	116°38'14"W	
Formation Spring	Caribou	42°41'54"N	111°32'34"W	6155
Formation Springs	Cassia	42°03'42"N	113°06'48"W	6760
Fort Hall Spring	Bannock	42°45'42"N	112°21'30"W	
Fourbit Spring	Adams	44°28'03"N	116°14'12"W	5920
Fourmile Spring	Cassia	42°12'16"N	113°59'56"W	
Fourth of July Spring	Idaho	45°44'22"N	116°27'50"W	
Fourth of July Spring	Twin Falls	42°13'35"N	114°25'13"W	

Springs

Spring	County	Latitude	Longitude	Elev.
Fowler Springs	Butte	44°03'45"N	113°08'13"W	
Fox Spring	Bear Lake	42°22'54"N	111°13'40"W	
Frahm Spring	Cassia	42°09'02"N	114°16'12"W	
Franks Hollow Spring	Cassia	42°11'50"N	113°30'18"W	
Franks Spring	Cassia	42°17'01"N	114°07'35"W	
Freds Spring	Idaho	45°48'20"N	115°53'02"W	
Freezeout Spring	Bonner	48°01'55"N	116°18'35"W	
Freezeout Spring	Shoshone	46°59'44"N	116°01'36"W	
Freighter Spring	Custer	44°10'17"N	113°51'19"W	
French Spring	Washington	44°13'32"N	116°27'17"W	
Frenchman Spring	Elmore	43°12'30"N	115°34'28"W	
Frenchmans Spring	Butte	43°24'53"N	113°02'42"W	
Frog Spring	Idaho	45°14'22"N	115°01'36"W	
Frog Spring (historical)	Clark	44°30'16"N	111°44'34"W	
Frying Pan Spring	Owyhee	42°32'38"N	116°44'19"W	
Frying Pan Spring	Owyhee	42°25'08"N	116°25'05"W	
Garden Spring	Power	42°20'22"N	112°47'08"W	
Gardner Spring	Owyhee	42°00'41"N	115°33'37"W	
Garner Springs	Fremont	44°34'30"N	111°17'43"W	
Gaston Beatty Spring	Caribou	42°57'45"N	111°35'45"W	6490
Gibbs Spring	Oneida	42°01'24"N	112°21'43"W	
Gifford Spring	Power	42°38'26"N	113°11'28"W	
Giusti Spring	Owyhee	42°41'34"N	116°34'57"W	
Glen Canyon Spring	Oneida	42°12'52"N	112°56'50"W	
Goat Spring	Elmore	43°27'45"N	115°20'47"W	
Goat Springs	Twin Falls	42°17'21"N	114°27'42"W	
Gold Fork Hot Spring	Valley	44°40'32"N	115°56'35"W	
Gold Spring	Bingham	43°12'34"N	111°39'50"W	6354
Goodhart Spring	Bear Lake	42°35'51"N	111°14'43"W	8246
Goodheart Spring	Caribou	42°41'06"N	111°18'09"W	6630
Goodman Gulch Spring	Owyhee	42°47'06"N	116°26'31"W	
Goose Creek Spring	Owyhee	42°55'00"N	116°56'05"W	
Gooseberry Spring	Custer	44°28'37"N	114°06'36"W	
Gooseberry Spring	Owyhee	42°44'38"N	116°35'38"W	
Gossi Spring	Custer	44°14'55"N	114°08'39"W	
Grace Spring	Owyhee	43°05'00"N	116°57'00"W	
Graham Hollow Spring	Franklin	42°13'40"N	111°39'06"W	
Graham Spring	Teton	43°44'30"N	111°23'42"W	
Granite Spring	Custer	43°49'30"N	113°31'53"W	
Granite Spring	Elmore	43°23'40"N	115°32'39"W	
Granite Spring	Idaho	45°43'29"N	115°07'40"W	
Granite Spring	Owyhee	43°11'45"N	116°48'48"W	
Granite Spring	Owyhee	42°02'01"N	115°04'18"W	6610
Granite Springs	Owyhee	42°25'35"N	116°52'50"W	
Grass Mountain Spring	Idaho	45°23'28"N	114°56'32"W	
Grasshopper Spring	Owyhee	42°34'07"N	116°54'14"W	
Grassy Spring	Twin Falls	42°18'15"N	114°27'35"W	
Gravel Spring	Cassia	42°07'17"N	114°11'36"W	
Graves Spring	Custer	44°20'51"N	113°34'44"W	
Gray Stud Spring	Custer	44°15'53"N	114°14'39"W	
Green Canyon Spring	Bear Lake	42°05'19"N	111°29'39"W	
Green Pass Spring	Bear Lake	42°17'22"N	111°35'03"W	
Greenfield Spring	Washington	44°26'30"N	116°47'37"W	
Greyhound Pass Spring	Caribou	42°31'36"N	111°13'12"W	
Gridley Spring	Camas	43°13'14"N	114°49'13"W	5180
Griffith Spring	Owyhee	43°11'05"N	116°39'00"W	
Griffith Springs	Fremont	44°12'13"N	111°36'29"W	
Grim Springs	Caribou	42°47'32"N	111°56'43"W	5310
Grindstone Spring	Lemhi	45°29'07"N	114°06'15"W	
Grizzly Spring	Lemhi	45°31'26"N	114°07'36"W	
Ground Hog Spring	Elmore	43°08'06"N	115°11'53"W	
Grouse Creek Spring	Twin Falls	42°06'31"N	114°17'40"W	
Grouse Spring	Adams	44°51'49"N	116°36'00"W	
Grouse Spring	Cassia	42°08'54"N	114°12'16"W	
Guyer Hot Springs	Blaine	43°41'00"N	114°24'35"W	
Gwin Spring	Gooding	43°08'31"N	114°38'35"W	
Hailey Hot Springs	Blaine	43°30'20"N	114°21'17"W	
Hale Spring	Twin Falls	41°59'56"N	114°35'05"W	
Half Moon Spring	Owyhee	42°46'02"N	116°21'29"W	
Halfway Spring	Twin Falls	42°12'26"N	114°22'10"W	
Halogeton Spring	Gooding	43°00'56"N	114°58'27"W	3250
Hamilton Springs	Custer	43°59'59"N	113°51'55"W	
Hands Spring	Teton	43°36'51"N	111°13'24"W	
Hansen Spring	Oneida	42°15'07"N	112°49'38"W	
Hardiman Spring	Owyhee	42°49'24"N	116°41'09"W	
Harrington Spring	Adams	44°54'11"N	116°35'54"W	
Harris Spring	Caribou	42°28'46"N	111°41'26"W	
Harrison Spring	Fremont	44°28'34"N	111°24'36"W	
Hash Spring	Camas	43°14'14"N	114°40'54"W	
Hat Spring	Elmore	43°03'26"N	115°06'45"W	
Hawkins Spring	Bannock	42°31'52"N	112°24'54"W	
Hawks Nest Spring	Owyhee	42°16'42"N	116°13'07"W	
Hawley Spring	Madison	43°39'27"N	111°34'09"W	
Hawley Warm Spring	Madison	43°39'13"N	111°42'03"W	
Head of the Ditch Springs	Owyhee	42°21'12"N	116°55'19"W	
Heart Mountain Spring	Bannock	42°19'23"N	111°56'35"W	
Heath Spring	Cassia	42°09'11"N	113°39'22"W	6331
Heath Spring	Oneida	42°17'18"N	112°14'04"W	
Heaton Spring	Bingham	43°04'47"N	111°50'30"W	6400
Heifer Spring	Owyhee	42°09'29"N	116°02'02"W	
Heifer Springs	Owyhee	42°14'24"N	115°58'49"W	
Heise Hot Springs	Jefferson	43°38'37"N	111°41'05"W	5090
Hells Kitchen Spring	Franklin	42°19'14"N	111°35'48"W	
Hemlock Spring	Shoshone	47°01'06"N	116°00'09"W	
Henderson Springs	Owyhee	42°54'33"N	116°32'41"W	
Henley Spring	Owyhee	42°32'23"N	116°34'58"W	

Spring	County	Latitude	Longitude	Elev.
Henry Moore Spring	Idaho	45°23'42"N	115°29'28"W	
Hess Spring	Caribou	42°35'56"N	111°15'31"W	
Hibner Spring	Caribou	42°53'07"N	111°55'58"W	5460
Hicks Springs	Owyhee	42°30'19"N	116°04'09"W	
Hidden Rock Spring	Cassia	42°06'06"N	113°07'42"W	
Higgins Spring	Owyhee	42°00'52"N	115°07'04"W	6210
High Spring	Elmore	43°09'53"N	115°26'52"W	
High Spring	Idaho	46°10'55"N	115°02'11"W	
High Spring	Lemhi	44°54'29"N	113°34'33"W	
Higley Spring	Oneida	42°02'10"N	112°59'33"W	
Hill Spring	Caribou	42°38'30"N	111°10'34"W	
Hillside Spring	Owyhee	42°34'29"N	116°40'43"W	
Hillyard Spring	Franklin	42°07'33"N	111°38'02"W	
Hinton Spring	Twin Falls	42°09'57"N	114°18'14"W	
Hite Spring	Idaho	45°24'13"N	116°24'36"W	
Hobby Spring	Gooding	43°11'34"N	114°53'48"W	5700
Hodges Spring	Lemhi	44°54'28"N	113°43'49"W	
Hog Creek Spring	Elmore	43°05'50"N	115°08'10"W	
Hole in the Wall Spring	Cassia	42°05'36"N	114°04'11"W	
Holmes Spring	Owyhee	43°21'18"N	117°01'00"W	
Homestead Spring	Adams	44°42'07"N	116°35'13"W	4030
Homestead Spring	Owyhee	42°30'42"N	116°42'45"W	
Hoodoo Spring	Shoshone	47°06'04"N	115°38'31"W	
Hoolie Springs	Clark	44°19'03"N	112°20'49"W	5810
Hooper Spring	Caribou	42°40'45"N	111°36'17"W	
Hoopgobel Spring	Bear Lake	42°21'30"N	111°24'25"W	
Hoops Spring	Twin Falls	42°19'37"N	114°27'48"W	
Hopper Gulch Spring	Twin Falls	42°13'34"N	114°26'07"W	
Horn Spring	Camas	43°13'08"N	114°38'47"W	
Horner Springs	Idaho	45°35'00"N	116°22'00"W	
Hornet Spring	Caribou	42°50'12"N	112°03'19"W	
Horse Basin Spring	Franklin	42°12'49"N	111°39'30"W	
Horse Basin Spring	Twin Falls	42°17'48"N	114°26'25"W	
Horse Basin Spring Number 1	Custer	44°10'53"N	114°04'05"W	
Horse Basin Spring Number 2	Custer	44°12'19"N	114°03'42"W	
Horse Creek Hot Springs	Lemhi	45°30'12"N	114°27'47"W	
Horse Spring	Adams	44°52'30"N	116°36'52"W	
Horse Spring	Custer	44°28'56"N	113°58'27"W	
Horse Spring	Twin Falls	42°01'25"N	114°18'18"W	
Horse Springs	Twin Falls	42°07'13"N	114°57'07"W	
Horsefly Spring	Fremont	44°11'33"N	111°09'50"W	
Horsefly Spring	Lemhi	45°23'28"N	114°22'13"W	
Horsehead Spring	Owyhee	42°45'52"N	116°20'10"W	
Horsehead Spring	Owyhee	42°29'01"N	116°59'10"W	
Horsemint Spring	Fremont	44°31'33"N	111°31'43"W	
Horseshoe Spring	Cassia	42°01'08"N	113°59'02"W	
Hospital Hot Spring	Valley	44°50'10"N	114°47'26"W	
Hot Spring	Boise	44°48'35"N	115°49'06"W	

Spring	County	Latitude	Longitude	Elev.
Hot Spring	Camas	43°22'59"N	114°55'53"W	
Hot Spring	Cassia	42°10'22"N	113°51'39"W	5322
Hot Spring	Gooding	43°10'13"N	114°47'48"W	5097
Hot Spring (historical)	Lemhi	44°43'11"N	114°00'58"W	4720
Hot Springs	Adams	45°02'21"N	116°17'32"W	
Hot Springs	Adams	44°40'11"N	116°18'18"W	4480
Hot Springs	Boise	44°09'11"N	115°59'36"W	
Hot Springs	Boise	43°40'14"N	115°42'02"W	
Hot Springs	Cassia	42°19'58"N	113°08'41"W	
Hot Springs	Custer	43°59'25"N	114°48'01"W	
Hot Springs	Elmore	43°09'19"N	115°31'00"W	
Hot Springs	Idaho	45°16'34"N	115°54'46"W	
Hot Springs	Jefferson	43°38'39"N	111°41'11"W	
Hot Springs	Lemhi	45°18'23"N	114°20'16"W	
Hot Springs	Lemhi	44°53'45"N	114°33'52"W	
Hot Springs	Valley	44°15'10"N	115°53'32"W	
Hot Sulphur Springs	Twin Falls	42°41'31"N	114°51'33"W	
Howard Spring	Fremont	44°40'08"N	111°17'11"W	7030
Howard Spring	Idaho	45°21'50"N	116°28'57"W	
Howell Creek Spring	Cassia	42°19'36"N	113°36'19"W	
Howell Spring	Custer	44°03'53"N	114°03'52"W	
Howerton Spring	Nez Perce	46°16'54"N	116°44'29"W	2830
Huckleberry Spring	Bear Lake	42°17'26"N	111°32'27"W	
Hudson Ridge Spring	Cassia	42°10'35"N	114°08'22"W	
Huffman Springs	Oneida	42°11'59"N	112°45'41"W	
Humberg Spring	Bear Lake	42°24'31"N	111°36'05"W	
Humphrey Spring	Cassia	42°03'38"N	114°13'59"W	
Hutch Springs	Owyhee	42°37'08"N	116°19'12"W	
Hutchey Spring	Cassia	42°11'27"N	113°08'26"W	6430
Ibex Spring	Owyhee	42°54'27"N	116°57'22"W	
Ice Springs	Elmore	43°29'03"N	115°23'49"W	
Icicle Spring	Idaho	46°09'35"N	115°35'26"W	
Idaho Spring	Owyhee	42°23'01"N	117°00'59"W	5310
Indian Camp Spring	Cassia	42°03'12"N	114°16'15"W	
Indian Hot Springs	Owyhee	42°20'18"N	115°39'11"W	
Indian Meadows Spring	Owyhee	42°41'01"N	116°46'19"W	
Indian Spring	Caribou	42°52'12"N	111°40'21"W	6270
Indian Spring	Cassia	42°25'25"N	113°03'11"W	
Indian Spring	Cassia	42°16'10"N	114°11'28"W	
Indian Spring	Franklin	42°09'29"N	111°37'49"W	
Indian Spring	Lemhi	45°31'29"N	114°05'34"W	
Indian Spring	Twin Falls	42°16'18"N	114°21'33"W	
Indian Springs	Adams	45°08'49"N	116°28'06"W	
Indian Springs	Idaho	45°25'38"N	116°25'37"W	
Indian Springs	Power	42°43'29"N	112°52'23"W	
Indian Writing Water Hole	Gooding	43°09'28"N	115°01'56"W	
Irish Bay Springs	Lemhi	44°46'04"N	113°22'08"W	
Iron Spring	Custer	44°14'07"N	113°30'28"W	

Springs

Spring	County	Latitude	Longitude	Elev.
Iron Spring	Elmore	43°22'38"N	115°42'43"W	
Iron Springs	Custer	44°16'49"N	113°28'57"W	
Italian Spring	Blaine	43°29'11"N	114°08'11"W	
Jack Spring (historical)	Elmore	43°00'38"N	116°02'21"W	
Jackass Spring	Owyhee	42°32'06"N	116°41'25"W	
Jackson Mill Spring	Fremont	44°14'46"N	111°35'51"W	
Japanese Spring	Twin Falls	42°03'36"N	114°18'22"W	
Jarvis Spring	Bear Lake	42°11'28"N	111°28'56"W	
Jarvis Spring	Cassia	42°10'01"N	114°11'11"W	
Jay Creek Spring	Cassia	42°00'54"N	114°09'07"W	
Jean Heazle Spring	Owyhee	43°00'13"N	116°50'52"W	
Jeffs Spring	Cassia	42°20'19"N	114°13'42"W	
Jenkins Hollow Spring	Oneida	42°01'35"N	112°08'27"W	
Jensen Spring	Bear Lake	42°19'10"N	111°24'18"W	
Jenson Cabin Spring	Custer	44°22'16"N	114°11'38"W	
Jenson Spring	Custer	43°48'11"N	113°31'16"W	
Jerry Johnson Hot Springs	Idaho	46°27'46"N	114°52'18"W	
Jim Lee Spring	Owyhee	42°05'17"N	114°49'45"W	
Jim Sage Spring	Cassia	42°07'19"N	113°30'40"W	
Jimmy Cappel Spring	Owyhee	42°30'39"N	116°00'43"W	
Joe Spring	Idaho	45°38'48"N	116°21'59"W	
Joes Spring	Twin Falls	42°14'50"N	114°18'28"W	
John Spring	Twin Falls	42°03'12"N	114°28'37"W	
Johnston Spring	Custer	44°14'25"N	114°11'19"W	
Johnstone Spring	Owyhee	43°21'36"N	116°56'06"W	
Jones Spring	Cassia	42°08'03"N	114°14'27"W	
Jordan Spring	Shoshone	47°58'07"N	116°03'01"W	
Josie Spring	Caribou	42°41'49"N	111°55'55"W	
Jug Spring	Shoshone	47°00'38"N	115°49'34"W	
Junction Spring	Caribou	42°41'58"N	111°25'04"W	
Junction Spring	Twin Falls	42°09'20"N	114°22'35"W	
Juniper Spring	Owyhee	43°15'22"N	116°58'41"W	
Juniper Spring	Owyhee	42°23'45"N	116°55'08"W	
Kackley Spring	Caribou	42°32'00"N	111°47'28"W	
Kane Spring	Cassia	42°11'38"N	113°26'47"W	
Kane Spring	Owyhee	43°11'13"N	116°39'30"W	
Keg Hollow Spring	Cassia	42°08'35"N	113°31'03"W	
Keg Spring	Clark	44°32'05"N	111°37'05"W	
Keg Spring	Lemhi	44°16'00"N	113°04'40"W	
Kelly Spring	Madison	43°38'35"N	111°37'10"W	
Kessler Creek Spring	Idaho	45°27'55"N	116°24'31"W	
Kettle Spring	Owyhee	42°20'28"N	116°43'56"W	
Kidd Spring	Cassia	42°05'46"N	114°12'09"W	
King Canyon Spring	Caribou	42°36'06"N	111°40'33"W	6540
King Creek Spring	Caribou	42°49'32"N	112°02'56"W	
King Spring	Butte	43°43'14"N	113°13'26"W	
Kirkham Hot Springs	Boise	44°04'18"N	115°32'31"W	
Kitty Spring	Clark	44°13'47"N	112°48'16"W	
Kiyi Spring	Owyhee	43°01'08"N	116°53'50"W	
Kossman Seep	Oneida	42°18'47"N	112°56'53"W	
Kurtz Spring	Oneida	42°17'56"N	112°49'48"W	
Kwiskwis Hot Spring	Valley	44°49'53"N	115°12'56"W	
Kyle Spring	Butte	43°55'01"N	112°53'29"W	
Lake Fork Spring	Cassia	42°23'17"N	113°02'04"W	
Lamb Spring	Twin Falls	42°09'45"N	114°21'56"W	
Lamont Spring	Adams	45°16'03"N	116°40'55"W	1720
Land Ranch Spring	Cassia	42°18'31"N	113°48'04"W	4644
Land Spring	Cassia	42°20'03"N	113°38'01"W	
Lansing Springs	Elmore	43°12'04"N	115°18'58"W	
Last Chance Spring	Adams	45°00'44"N	116°11'19"W	
Last Chance Spring	Owyhee	41°59'56"N	115°43'30"W	
Last Chance Spring	Owyhee	41°59'54"N	115°43'34"W	6020
Last Chance Springs	Blaine	43°24'05"N	114°00'40"W	5150
Latham Spring	Fremont	44°27'37"N	111°08'39"W	
Latty Hot Spring	Elmore	43°06'55"N	115°18'19"W	
Lava Spring	Twin Falls	42°14'20"N	114°20'17"W	
Layton Spring	Cassia	42°10'17"N	114°06'41"W	
Leahy Spring	Camas	43°16'22"N	114°55'53"W	
Leavitt Spring	Custer	43°56'48"N	113°48'21"W	
Ledge Rock Spring	Cassia	42°12'14"N	114°12'47"W	
Leduck Canyon Spring	Blaine	43°16'56"N	114°08'05"W	5800
Left Fork Fish Haven Spring	Bear Lake	42°02'03"N	111°28'44"W	
Lemonade Spring	Owyhee	43°18'30"N	116°50'59"W	
Leslie Springs	Boundary	48°45'46"N	116°17'22"W	
Lester Creek Springs	Elmore	43°26'40"N	115°21'27"W	
Lidy Hot Springs	Clark	44°08'43"N	112°33'14"W	
Light Spring	Elmore	43°26'31"N	115°22'25"W	
Lightfoot Hot Springs	Camas	43°36'18"N	114°57'01"W	
Lilly Spring	Twin Falls	42°08'33"N	114°53'22"W	
Lime Spring	Twin Falls	42°06'44"N	114°19'17"W	
Lime Spring	Washington	44°41'17"N	116°54'57"W	
Lincoln Peak Spring	Bingham	43°05'13"N	112°04'15"W	5930
Line Spring	Twin Falls	41°59'57"N	114°18'23"W	6700
Lineham Spring	Owyhee	42°53'07"N	116°32'33"W	
Little Bear Spring	Adams	45°07'12"N	116°36'19"W	6443
Little Cottonwood Spring	Cassia	42°12'11"N	114°05'04"W	
Little Cottonwood Springs	Butte	43°29'34"N	113°36'07"W	
Little Fork Spring	Cassia	42°10'25"N	114°15'47"W	
Little Half Moon Spring	Owyhee	42°46'15"N	116°19'37"W	
Little Horse Spring	Bannock	42°59'16"N	112°20'41"W	
Little Jackass Spring	Owyhee	42°32'14"N	116°41'44"W	
Little Joe Spring	Bear Lake	42°31'37"N	111°36'46"W	7700
Little Kane Spring	Owyhee	43°12'22"N	116°40'00"W	
Little Malad Spring	Oneida	42°22'42"N	112°27'15"W	
Little Piney Spring	Cassia	42°05'04"N	114°13'00"W	
Little Rock Spring	Oneida	42°08'21"N	112°46'34"W	

Spring	County	Latitude	Longitude	Elev.
Little Spring	Bear Lake	42°06'34"N	111°28'59"W	
Little Squaw Creek Spring	Cassia	42°09'45"N	114°06'35"W	
Little Tigert Spring	Owyhee	42°38'24"N	116°10'51"W	
Little Willow Spring	Cassia	42°00'49"N	114°11'49"W	
Little Willow Spring	Owyhee	42°26'30"N	116°59'23"W	
Litz Spring	Franklin	42°06'33"N	111°38'57"W	
Lloyd Spring	Cassia	42°11'37"N	113°36'02"W	
LMB Spring	Custer	44°14'37"N	114°21'07"W	
Lodgepole Springs	Valley	44°18'53"N	115°49'35"W	
Log Spring	Bear Lake	42°20'12"N	111°33'40"W	
Log Spring	Franklin	42°12'13"N	111°37'45"W	
Log Trough Spring	Idaho	45°18'51"N	116°28'17"W	
Logged Up Springs	Idaho	45°32'20"N	116°22'34"W	
Logger Spring	Cassia	42°08'43"N	113°43'52"W	8102
Lone Cedar Spring	Cassia	42°07'02"N	114°01'15"W	
Lone Pine Spring	Bingham	43°01'58"N	112°10'42"W	
Lone Pine Spring	Caribou	42°40'45"N	111°12'00"W	
Lone Pine Spring	Cassia	42°06'07"N	114°06'03"W	
Lone Pine Spring	Clark	44°24'17"N	112°21'58"W	8120
Lone Pine Spring	Custer	44°22'24"N	114°10'10"W	
Lone Pine Spring	Twin Falls	42°11'26"N	114°23'09"W	
Lone Tree Spring	Elmore	43°06'56"N	115°24'56"W	3522
Lone Tree Spring	Owyhee	42°45'52"N	116°39'08"W	
Lone Tree Spring	Owyhee	42°23'16"N	116°54'57"W	
Lone Willow Spring	Owyhee	42°01'14"N	115°42'22"W	
Lonetree Spring	Caribou	42°42'06"N	111°19'00"W	
Long Creek Spring	Custer	44°10'30"N	113°34'20"W	
Long Tom Troughs	Owyhee	42°38'45"N	116°30'38"W	
Long Valley Spring	Bingham	43°14'22"N	111°43'13"W	6410
Long Valley Spring	Owyhee	42°43'26"N	116°38'12"W	
Lonigan Springs	Oneida	42°18'05"N	112°47'27"W	
Looney Spring	Washington	44°30'39"N	116°48'07"W	
Lost Hat Spring	Cassia	42°15'15"N	114°12'26"W	
Lost Spring	Franklin	42°22'47"N	111°37'18"W	
Lost Spring	Madison	43°44'00"N	111°24'44"W	
Lost Spring	Twin Falls	42°09'29"N	114°21'29"W	
Lost Tunnel Spring	Cassia	42°04'55"N	113°04'27"W	6340
Lost Valley Springs	Owyhee	42°25'56"N	116°18'54"W	
Lower Birch Spring	Owyhee	42°51'53"N	116°22'05"W	
Lower Bradshaw Spring	Custer	44°17'59"N	114°14'25"W	
Lower Corral Creek Spring	Gem	44°00'11"N	116°26'35"W	
Lower Crystal Spring	Clark	44°11'41"N	112°42'10"W	
Lower Flat Spring	Owyhee	43°04'48"N	116°52'58"W	
Lower Heglar Spring	Cassia	42°26'19"N	113°01'16"W	
Lower Hogpen Spring	Owyhee	42°51'46"N	116°23'13"W	
Lower Spring	Bingham	43°14'46"N	111°42'39"W	6555
Lower Spring	Oneida	42°09'59"N	112°29'52"W	
Lower Station Spring	Oneida	42°18'52"N	112°14'52"W	

Spring	County	Latitude	Longitude	Elev.
Lufkin Spring	Bonneville	43°36'53"N	111°29'11"W	
Lunch Spring	Twin Falls	42°16'35"N	114°26'08"W	
Lupe Springs	Power	42°33'03"N	112°55'08"W	
Lyda Spring	Idaho	45°52'24"N	116°26'47"W	
Lyle Spring	Fremont	44°18'01"N	111°32'39"W	
Mabey Spring	Cassia	42°06'49"N	114°06'08"W	
Mabey Spring Number 2	Cassia	42°07'25"N	114°06'18"W	
Macks Creek Spring	Owyhee	43°11'15"N	116°49'37"W	
Magic Hot Springs	Owyhee	42°00'29"N	114°30'11"W	
Magpie Spring	Cassia	42°05'53"N	114°01'22"W	
Magpie Spring	Custer	44°43'32"N	114°04'27"W	5150
Magpie Spring	Lemhi	44°16'34"N	113°07'16"W	
Magpie Spring Number 2	Lemhi	45°09'25"N	113°40'35"W	
Magpie Springs	Lemhi	44°14'22"N	113°16'22"W	
Mahogany Basin Spring	Bear Lake	42°18'44"N	111°31'26"W	
Mahogany Spring	Owyhee	42°30'33"N	116°58'53"W	
Main Canyon Spring	Franklin	42°23'01"N	111°37'14"W	
Malm Spring	Custer	44°20'29"N	114°13'36"W	
Mammath Spring	Caribou	42°42'49"N	111°37'22"W	5975
Mandolin Spring	Blaine	43°39'58"N	113°58'17"W	
Maple Canyon Spring	Bear Lake	42°04'53"N	111°28'38"W	
Maple Grove Hot Springs	Franklin	42°18'29"N	111°42'24"W	4900
Maple Hollow Spring	Oneida	42°01'07"N	112°20'14"W	
Maple Spring	Franklin	42°04'22"N	111°40'08"W	
Maple Spring	Oneida	42°18'09"N	112°15'01"W	
Marafio Spring	Custer	44°17'08"N	114°14'54"W	
Marco Spring	Custer	44°09'42"N	114°22'14"W	
Marlin Spring	Lemhi	45°30'13"N	114°15'28"W	
Marmaduke Spring	Owyhee	42°36'46"N	116°35'11"W	
Marsh Spring	Camas	43°14'11"N	114°25'19"W	
Marten Hot Springs	Idaho	46°00'20"N	115°01'15"W	
Martin Spring	Bingham	43°03'08"N	112°11'29"W	5740
Martin Spring	Blaine	43°18'27"N	114°20'38"W	
Martin Spring	Idaho	45°31'32"N	116°13'55"W	7120
Martin Spring	Owyhee	42°03'33"N	115°58'08"W	
Martindale Spring	Cassia	42°14'42"N	114°12'44"W	
Mauldin Springs	Twin Falls	42°10'48"N	114°55'08"W	
McClenden Spring	Cassia	42°20'12"N	113°24'25"W	
McConnell Spring	Twin Falls	42°13'37"N	114°21'00"W	
McCoullock Spring	Bonneville	43°35'07"N	111°21'40"W	
McCoy Spring	Lemhi	44°16'45"N	113°08'34"W	
McDonald Spring	Shoshone	47°45'32"N	116°16'02"W	
McFarney Spring	Elmore	43°25'36"N	115°19'34"W	
McGuire Spring	Elmore	43°20'30"N	115°36'55"W	
Meadow Brook Spring	Oneida	42°11'16"N	112°45'07"W	
Meadow Spring	Caribou	42°38'23"N	111°22'47"W	
Meadow Spring	Cassia	42°03'38"N	114°16'28"W	
Meadow Spring	Owyhee	42°31'58"N	116°46'56"W	

Springs

Spring	County	Latitude	Longitude	Elev.
Meadow Springs	Twin Falls	42°01'34"N	114°54'22"W	
Metcalf Spring	Camas	43°18'22"N	114°23'56"W	
Mickelson Spring	Caribou	42°26'46"N	111°37'32"W	
Middle Ridge Spring	Cassia	42°02'58"N	113°06'21"W	6660
Middle Spring	Custer	44°11'46"N	113°34'58"W	
Middle Water Hole	Lincoln	43°11'17"N	114°33'07"W	
Midnight Spring	Caribou	42°53'58"N	111°21'49"W	
Miles Canyon Spring	Bear Lake	42°16'09"N	111°29'50"W	
Milk Spring	Owyhee	43°06'39"N	116°39'32"W	
Milkshake Spring	Caribou	42°43'27"N	111°59'00"W	6155
Mill Canyon Spring	Cassia	42°25'52"N	113°00'50"W	5950
Miller Spring	Adams	45°05'50"N	116°28'51"W	6580
Millward Spring	Caribou	42°52'52"N	111°51'33"W	6000
Mine Dump Spring	Cassia	42°26'01"N	113°05'25"W	
Mineral Springs	Bonneville	43°25'24"N	111°24'51"W	
Miners Creek Spring	Clark	44°27'19"N	112°10'54"W	
Mink Creek Spring	Franklin	42°17'54"N	111°35'35"W	
Minks Drink	Owyhee	42°30'54"N	116°48'12"W	
Mission Spring (historical)	Latah	46°48'37"N	116°59'30"W	2930
Moffett Springs	Custer	44°15'37"N	113°26'14"W	
Monahan Spring	Owyhee	43°03'14"N	116°56'55"W	
Montana Springs	Shoshone	47°06'26"N	115°40'40"W	
Montrose Spring	Elmore	43°19'32"N	115°34'17"W	
Moonshine Spring	Cassia	42°23'40"N	113°01'45"W	
Moonshine Spring	Owyhee	42°32'34"N	116°36'24"W	
Moonshine Spring	Owyhee	42°26'31"N	116°48'46"W	
Moonshine Spring	Twin Falls	42°19'23"N	114°26'41"W	
Moorcastle Springs	Owyhee	42°10'15"N	116°02'07"W	
Moore Spring	Gem	44°19'11"N	116°21'30"W	
Moores Spring	Elmore	43°22'12"N	115°14'00"W	
Morris Canyon Spring	Bannock	42°46'33"N	112°18'14"W	
Morrison Ridge Spring	Idaho	45°21'53"N	116°27'36"W	
Mortenson Spring	Cassia	42°11'34"N	113°08'00"W	6590
Mosquito Springs	Idaho	45°15'53"N	115°22'48"W	
Moss Spring	Fremont	44°07'34"N	111°05'18"W	
Mossman Spring	Washington	44°12'37"N	116°25'13"W	
Mounment Springs	Owyhee	42°06'06"N	114°51'35"W	
Mountain Spring	Custer	43°46'14"N	113°35'29"W	
Mountain Spring	Custer	43°44'52"N	113°36'01"W	
Mountain View Spring	Twin Falls	42°14'08"N	114°18'01"W	
Mowers Spring	Power	42°39'40"N	113°00'23"W	
Mud Flat Spring	Lemhi	45°01'27"N	113°35'08"W	
Mud Flat Spring	Owyhee	42°36'18"N	116°32'59"W	
Mud Lick	Lemhi	45°11'28"N	114°26'22"W	
Mud Spring	Bannock	42°48'21"N	112°26'11"W	
Mud Spring	Bingham	43°06'09"N	112°01'48"W	6115
Mud Spring	Bonneville	43°31'04"N	111°38'53"W	
Mud Spring	Butte	44°09'51"N	113°11'03"W	

Spring	County	Latitude	Longitude	Elev.
Mud Spring	Camas	43°12'50"N	114°40'16"W	
Mud Spring	Caribou	42°25'04"N	111°39'14"W	
Mud Spring	Cassia	42°23'52"N	113°04'36"W	
Mud Spring	Cassia	42°20'42"N	114°08'21"W	
Mud Spring	Cassia	42°15'17"N	114°05'53"W	
Mud Spring	Cassia	42°11'09"N	113°06'48"W	
Mud Spring	Custer	44°12'46"N	113°44'05"W	
Mud Spring	Elmore	43°24'47"N	115°44'01"W	
Mud Spring	Lemhi	44°08'15"N	113°28'49"W	
Mud Spring	Oneida	42°08'44"N	112°27'53"W	
Mud Spring	Owyhee	43°07'56"N	116°58'02"W	
Mud Spring	Power	42°32'44"N	112°27'04"W	
Mud Spring	Twin Falls	42°16'45"N	114°21'38"W	
Mud Springs	Blaine	43°27'49"N	113°58'09"W	5710
Mud Springs	Elmore	43°06'29"N	115°15'21"W	
Mud Springs	Idaho	45°34'17"N	116°20'04"W	
Mud Springs	Lewis	46°12'32"N	116°35'07"W	
Mud Springs	Madison	43°43'17"N	111°31'32"W	
Mud Springs	Owyhee	42°56'16"N	116°56'30"W	
Muddy Spring	Twin Falls	42°01'36"N	114°18'22"W	
Mule Creek Springs	Twin Falls	42°01'43"N	114°36'18"W	6000
Mule Spring	Custer	44°14'02"N	113°48'41"W	
Muleshoe Springs	Lemhi	45°32'37"N	114°14'32"W	8582
Mullin Spring	Idaho	45°18'39"N	116°24'00"W	
Munn Spring	Bannock	42°45'58"N	112°21'17"W	
Murdicks Spring	Idaho	45°53'58"N	116°22'21"W	
N W Spring	Cassia	42°00'58"N	114°05'47"W	
Napo Spring	Lemhi	44°53'12"N	113°35'34"W	
Nat-Soo-Pah Warm Spring	Twin Falls	42°20'45"N	114°30'31"W	
Nelson Spring	Caribou	42°36'45"N	111°40'00"W	4220
Nelson Spring	Twin Falls	42°11'44"N	114°25'34"W	
Neville Spring	Teton	43°49'41"N	111°19'34"W	
Nez Perce Spring	Lemhi	44°31'19"N	113°21'05"W	
Niagara Springs	Gooding	42°39'57"N	114°40'24"W	
Nibbs Spring	Cassia	42°18'57"N	113°27'06"W	
Nichols Spring	Owyhee	43°01'07"N	116°56'00"W	
Nickel Creek Spring	Owyhee	42°32'51"N	116°46'31"W	
Nieber Spring	Bear Lake	42°17'25"N	111°30'59"W	
Ninety Percent Spring	Caribou	42°41'04"N	111°37'54"W	5780
Noon Creek Spring	Owyhee	42°41'31"N	116°48'41"W	
North Antelope Spring	Fremont	44°18'36"N	111°36'17"W	
North Kurtz Spring	Oneida	42°18'24"N	112°51'28"W	
North Queens Crown Springs	Blaine	43°18'12"N	113°58'57"W	5260
North Water Spring	Cassia	42°12'18"N	114°03'33"W	
North Willow Springs	Twin Falls	42°18'35"N	114°21'10"W	
Northeast Spring	Twin Falls	42°10'09"N	114°23'01"W	
Northside Spring	Camas	43°19'09"N	114°27'01"W	
O E Spring	Caribou	42°50'21"N	111°02'37"W	7000

Spring	County	Latitude	Longitude	Elev.
Oe Spring	Caribou	42°50'29"N	111°02'36"W	
Officer Spring	Power	42°42'03"N	112°36'07"W	
Old Canyon Spring	Oneida	42°17'04"N	112°15'41"W	
Oliver Spring	Cassia	42°16'05"N	114°08'15"W	
Opal Spring	Custer	44°28'46"N	114°01'15"W	
Orangeburg Spring	Cassia	42°00'20"N	114°04'17"W	
Oreana Spring	Lemhi	45°27'53"N	114°29'04"W	
Orin Spring	Twin Falls	42°09'01"N	114°20'46"W	
Orman Spring	Caribou	42°48'42"N	111°59'47"W	5490
Orr Springs	Elmore	43°25'21"N	115°19'20"W	
Osborne Springs	Fremont	44°18'32"N	111°24'27"W	
Otter Springs	Fremont	44°09'18"N	111°12'46"W	
Otter Springs	Owyhee	42°08'15"N	116°01'18"W	
Outlaw Spring	Bannock	42°50'37"N	112°30'55"W	
Overstreet Spring	Payette	44°08'15"N	116°42'34"W	
Owens Corral Spring	Cassia	42°06'10"N	114°01'44"W	
Owens Spring	Clark	44°27'15"N	112°11'42"W	
Owl Creek Hot Springs	Lemhi	45°20'40"N	114°27'47"W	
Oxbow Spring	Adams	45°15'05"N	116°36'52"W	
Pack Hook Spring	Owyhee	42°56'43"N	116°55'53"W	
Packer Spring	Boise	44°11'41"N	115°54'26"W	
Packsaddle Spring	Teton	43°43'26"N	111°21'20"W	
Palmer Spring	Cassia	42°26'31"N	113°05'24"W	
Palmer Springs	Clark	44°27'06"N	112°08'54"W	
Paradise Spring	Teton	43°38'58"N	111°11'40"W	
Paris Spring	Bear Lake	42°12'33"N	111°29'53"W	
Parke Spring	Cassia	42°23'11"N	113°28'38"W	6180
Parke Spring	Oneida	42°17'14"N	112°56'24"W	
Parker Spring	Elmore	43°13'28"N	115°38'42"W	
Parker Spring	Owyhee	42°04'22"N	115°58'29"W	
Parker Spring	Twin Falls	42°09'25"N	114°23'26"W	
Parsons Spring	Idaho	46°11'43"N	115°02'18"W	
Pasco Spring	Blaine	43°16'58"N	114°08'48"W	5900
Pass Creek Spring	Clark	44°28'57"N	112°06'54"W	
Paul Jones Spring	Franklin	42°06'25"N	111°40'00"W	
Peck Mountain Spring	Adams	44°51'48"N	116°37'00"W	4770
Pecks Canyon Spring	Custer	44°07'11"N	114°02'01"W	
Pedro Spring	Twin Falls	42°09'27"N	114°18'17"W	
Pence Hot Spring	Owyhee	42°47'53"N	115°43'11"W	
Peppermint Spring	Owyhee	42°33'11"N	116°56'32"W	5387
Peter Ready Spring	Idaho	45°40'48"N	116°03'11"W	
Peterson Hollow Spring	Franklin	42°13'59"N	111°40'32"W	
Peterson Spring	Oneida	42°16'48"N	112°47'03"W	
Pettit Spring	Oneida	42°17'04"N	112°34'12"W	
Pevo Spring	Power	42°43'24"N	112°34'23"W	
Philips Spring	Custer	44°21'52"N	114°09'18"W	
Picabo Town Spring	Blaine	43°17'05"N	114°05'02"W	4920
Picketts Lower Trough Spring	Cassia	42°07'30"N	114°10'13"W	

Spring	County	Latitude	Longitude	Elev.
Picketts Upper Trough Spring	Cassia	42°07'40"N	114°11'37"W	
Picnic Spring	Oneida	42°03'57"N	112°11'01"W	
Pig Foot Spring	Idaho	45°43'27"N	116°03'34"W	
Pike Mountain Spring	Cassia	42°10'44"N	114°16'24"W	
Pilgrim Spring	Elmore	42°53'28"N	115°07'44"W	
Pincock Hot Springs	Madison	43°47'29"N	111°26'08"W	
Pine Basin Spring	Twin Falls	42°16'29"N	114°18'19"W	
Pine Corral Spring	Franklin	42°15'18"N	112°04'45"W	
Pine Flat Hot Springs	Boise	44°03'45"N	115°41'07"W	
Pine Grove Spring	Cassia	42°21'01"N	113°04'29"W	5850
Pine Grove Spring	Cassia	42°21'01"N	113°04'30"W	6287
Pine Spring	Bannock	42°55'48"N	112°17'26"W	
Pine Spring	Bear Lake	42°09'21"N	111°12'41"W	
Pine Spring	Franklin	42°12'06"N	111°39'45"W	
Pine Spring	Oneida	42°19'13"N	112°10'37"W	
Pine Springs	Oneida	42°17'12"N	112°55'19"W	
Pineapple Spring	Shoshone	47°05'39"N	115°36'20"W	
Pinkston Spring	Owyhee	42°00'37"N	115°04'00"W	6950
Pipeline Spring	Franklin	42°19'35"N	111°39'07"W	
Pleasant Spring	Cassia	42°05'54"N	114°15'06"W	
Pleasantview Warm Springs	Oneida	42°09'19"N	112°20'53"W	
Poacher Spring	Owyhee	43°01'57"N	117°00'41"W	
Pocket Spring	Cassia	42°12'53"N	114°10'32"W	
Pofferman Spring	Camas	43°17'14"N	114°24'36"W	
Point of Rocks Spring	Owyhee	43°07'59"N	116°36'39"W	
Point Spring	Cassia	42°14'39"N	113°12'17"W	
Point Spring	Cassia	42°00'06"N	114°07'03"W	
Poison Gulch Spring	Lemhi	45°00'29"N	113°34'08"W	
Poison Gulch Spring	Owyhee	42°46'50"N	116°19'33"W	
Poison Spring	Custer	44°12'11"N	113°53'14"W	
Poison Spring	Lemhi	44°33'08"N	113°17'52"W	
Poison Spring	Oneida	42°11'59"N	112°04'34"W	
Poison Springs	Custer	44°12'28"N	113°40'35"W	
Pole Camp Spring	Twin Falls	42°15'10"N	114°21'55"W	
Pole Springs	Bannock	42°45'15"N	112°29'34"W	
Pollard Spring	Cassia	42°04'03"N	113°05'29"W	6360
Pond Spring	Twin Falls	42°08'45"N	114°21'31"W	
Porcupine Spring	Cassia	42°19'39"N	114°11'20"W	
Porcupine Spring	Cassia	42°10'06"N	114°16'01"W	
Porcupine Spring	Lemhi	45°29'26"N	114°12'45"W	
Porcupine Spring	Lemhi	44°58'01"N	113°52'20"W	
Port Spring	Twin Falls	42°07'13"N	114°17'14"W	
Portland Gulch Spring	Owyhee	42°46'01"N	116°27'07"W	
Post Hollow Spring	Franklin	42°22'19"N	111°38'32"W	
Pot Hole Spring Number 2	Twin Falls	42°02'40"N	114°18'03"W	
Potholes, The	Owyhee	42°29'22"N	116°11'18"W	
Potter Spring	Cassia	42°24'09"N	113°29'34"W	5338
Poulton Spring	Cassia	42°06'25"N	113°56'42"W	

Springs

Spring	County	Latitude	Longitude	Elev.
Preis Hot Spring	Camas	43°34'34"N	114°49'49"W	
Prince Albert Spring	Elmore	43°07'47"N	115°20'15"W	4300
Prince Albert Spring (historical)	Valley	44°37'26"N	115°48'07"W	
Pritchert Spring	Caribou	42°34'08"N	111°18'06"W	
Purcell Spring	Lemhi	44°33'12"N	113°21'09"W	
Qeedup Spring	Bingham	43°01'35"N	112°00'38"W	
Quaking Aspen Spring	Cassia	42°22'41"N	113°05'08"W	
Quaking Aspen Springs	Idaho	45°26'05"N	115°02'51"W	
Quaking Asphalt Spring	Oneida	42°13'56"N	112°49'56"W	
Quartz Spring	Idaho	45°20'09"N	115°26'17"W	
Queedup Spring	Bingham	43°01'34"N	112°01'28"W	5735
Rabbit Spring	Owyhee	42°03'43"N	114°40'12"W	
Rabbit Springs	Owyhee	43°02'37"N	116°08'30"W	
Radarmacher Spring	Camas	43°12'13"N	114°51'07"W	5775
Railroad Spring	Caribou	42°42'38"N	111°55'09"W	
Rainbow Spring	Valley	44°58'51"N	115°37'01"W	
Rat Spring	Owyhee	42°42'19"N	116°28'44"W	
Rattlesnake Spring	Cassia	42°02'00"N	114°16'13"W	
Rattlesnake Spring	Custer	44°26'08"N	113°54'04"W	
Rattlesnake Spring	Elmore	43°12'47"N	115°32'59"W	
Rattlesnake Spring	Gooding	43°09'41"N	114°36'17"W	
Rattlesnake Spring	Idaho	45°33'14"N	115°05'57"W	
Rattlesnake Spring	Idaho	45°24'27"N	116°01'10"W	
Rattlesnake Spring	Idaho	45°16'40"N	116°23'51"W	
Rattlesnake Spring	Oneida	42°16'09"N	112°08'28"W	
Rattlesnake Spring	Owyhee	42°20'53"N	116°49'16"W	
Rattlesnake Springs	Elmore	42°59'11"N	115°44'29"W	
Rattlesnake Springs	Owyhee	42°02'11"N	114°34'45"W	
Rawhide Spring	Bingham	43°04'57"N	112°14'41"W	
Raymer Spring	Payette	44°00'57"N	116°31'59"W	
Raymond Spring	Owyhee	43°04'13"N	116°53'33"W	
Reager Waterhole	Gooding	43°04'23"N	114°54'38"W	3950
Red Bluff Spring	Cassia	42°07'01"N	114°11'14"W	
Red Ives Spring	Shoshone	47°03'04"N	115°17'35"W	
Red Knoll Spring	Oneida	42°07'26"N	112°09'06"W	
Red Pine Spring	Bear Lake	42°31'06"N	111°21'45"W	
Red Pine Spring	Bear Lake	42°02'15"N	111°29'59"W	
Red Rock Spring	Bingham	43°06'05"N	112°01'12"W	6095
Red Rock Spring	Blaine	43°21'58"N	114°01'51"W	5615
Red Rock Spring	Cassia	42°12'41"N	113°27'24"W	
Red Rock Spring	Idaho	45°40'20"N	116°29'25"W	
Reed Spring	Cassia	42°02'00"N	113°36'43"W	
Reservation Spring	Owyhee	42°50'04"N	116°26'21"W	
Reynolds Springs	Twin Falls	42°11'44"N	114°52'02"W	
Rice Canyon Spring	Cassia	42°03'57"N	113°07'27"W	7240
Rice Creek Spring	Twin Falls	42°15'10"N	114°28'21"W	
Rice Spring	Cassia	42°17'14"N	113°26'42"W	4965
Rice Spring	Cassia	42°17'08"N	113°35'11"W	

Spring	County	Latitude	Longitude	Elev.
Richardson Spring	Custer	43°43'37"N	113°35'19"W	
Rickenbacker Spring	Twin Falls	42°12'29"N	114°17'32"W	
Rigger Spring	Lemhi	44°58'33"N	113°42'42"W	
Right Hand Fork Spring	Cassia	42°04'04"N	114°13'30"W	
Rim Rock Spring	Cassia	42°07'30"N	114°09'59"W	
Rim Spring	Owyhee	43°09'18"N	116°56'52"W	
Rimrock Spring	Twin Falls	42°17'50"N	114°27'03"W	
Rings Springs	Twin Falls	42°42'26"N	114°53'07"W	3290
Rizzi Spring	Owyhee	42°06'22"N	115°49'24"W	
Roan Spring	Owyhee	43°14'29"N	116°40'02"W	
Roaring Spring	Owyhee	42°22'53"N	116°50'26"W	
Robb Springs	Valley	44°31'43"N	115°51'38"W	
Roberts Spring	Nez Perce	46°03'07"N	116°50'43"W	
Robertson Spring	Lemhi	44°26'32"N	113°13'38"W	
Rock Cabin Spring	Owyhee	42°01'19"N	114°34'37"W	
Rock Spring	Bear Lake	42°14'10"N	111°32'16"W	
Rock Spring	Bingham	43°11'19"N	111°41'17"W	7070
Rock Spring	Bonner	47°58'50"N	116°39'12"W	2218
Rock Spring	Cassia	42°17'05"N	113°35'40"W	7164
Rock Spring	Cassia	42°04'51"N	114°01'32"W	
Rock Spring	Clark	44°32'07"N	112°00'11"W	
Rock Spring	Custer	44°21'20"N	113°47'47"W	
Rock Spring	Owyhee	42°41'35"N	116°29'41"W	
Rock Spring	Owyhee	42°34'27"N	116°18'48"W	
Rock Spring	Owyhee	42°25'12"N	116°49'46"W	
Rock Spring	Owyhee	42°20'23"N	117°00'06"W	5350
Rock Spring	Power	42°32'07"N	112°25'34"W	
Rock Springs	Oneida	42°15'27"N	112°47'39"W	
Rock Springs	Power	42°36'01"N	112°34'48"W	
Rocky Bench Spring	Franklin	42°08'53"N	111°39'54"W	
Rodeo Spring	Cassia	42°06'38"N	114°05'20"W	
Rodeo Spring	Owyhee	42°17'52"N	116°12'31"W	
Rodger Spring	Idaho	45°21'49"N	116°29'13"W	
Rogerson Spring	Twin Falls	42°12'18"N	114°35'41"W	
Roll Inn Spring	Owyhee	42°18'14"N	115°56'01"W	
Rose Briar Spring	Owyhee	42°42'29"N	116°17'48"W	
Rose Creek Spring	Owyhee	42°42'02"N	116°42'25"W	
Rose Spring	Oneida	42°01'36"N	112°59'26"W	
Roseberry Springs	Blaine	43°21'56"N	114°24'41"W	
Rosebrior Spring	Custer	44°19'11"N	114°11'51"W	
Rosie Spring	Caribou	42°56'31"N	111°32'30"W	6275
Ross Spring	Owyhee	42°22'05"N	116°54'51"W	
Round Mountain Spring	Cassia	42°01'37"N	113°14'17"W	
Roy Farnes Spring	Bonneville	43°33'21"N	111°21'13"W	
Roystone Hot Springs	Gem	43°57'16"N	116°21'19"W	
Rueger Springs	Power	42°46'15"N	112°52'50"W	
Rustican Spring	Elmore	43°06'32"N	115°17'43"W	3795
Rye Grass Spring	Custer	44°14'33"N	114°14'16"W	

Spring	County	Latitude	Longitude	Elev.
Ryegrass Spring	Owyhee	43°14'52"N	116°43'47"W	
Ryegrass Spring	Owyhee	42°16'29"N	116°41'14"W	
Sacajawea Hot Springs	Custer	44°09'37"N	115°10'38"W	5000
Saddle Spring	Caribou	42°38'02"N	111°22'38"W	
Saddle Spring	Idaho	45°20'11"N	116°37'19"W	
Saddle Spring	Lemhi	45°30'49"N	114°15'09"W	
Saddlehorse Spring	Owyhee	42°23'10"N	117°00'09"W	5470
Sadducee Spring	Bear Lake	42°03'05"N	111°27'36"W	
Sage Hen Spring	Cassia	42°06'14"N	114°02'44"W	
Sage Hen Springs	Owyhee	42°10'37"N	116°00'51"W	
Sage Hen Springs	Twin Falls	42°06'30"N	114°53'17"W	
Sagebrush Spring	Blaine	43°14'52"N	114°13'20"W	4820
Sagebrush Spring	Lemhi	45°16'01"N	114°26'41"W	
Sagebrush Spring	Lemhi	44°15'23"N	113°06'11"W	
Sagehen Spring	Owyhee	42°30'15"N	116°42'50"W	
Sagehen Spring	Twin Falls	42°17'39"N	114°18'16"W	
Sagehen Springs	Owyhee	42°22'06"N	116°58'58"W	
Sagehen Springs	Power	42°19'56"N	112°46'51"W	
Sager Spring	Oneida	42°17'35"N	112°54'17"W	
Sago Spring	Bear Lake	42°20'55"N	111°25'14"W	
Saint Charles Spring	Bear Lake	42°06'36"N	111°27'52"W	
Salmon Hot Spring	Lemhi	45°05'39"N	113°50'11"W	
Salt Lick Spring	Cassia	42°13'02"N	113°44'24"W	5910
Salyer Spring	Oneida	42°14'08"N	112°47'43"W	
Sam Noble Spring	Owyhee	42°37'29"N	116°31'43"W	
Sand Spring	Cassia	42°02'28"N	113°57'51"W	
Sand Spring	Owyhee	42°04'07"N	114°39'18"W	
Sand Springs	Gooding	42°43'38"N	114°50'26"W	2900
Sandy Spring	Washington	44°19'33"N	116°46'20"W	2670
Sanford Spring	Cassia	42°12'15"N	113°44'21"W	6462
Sanitary Spring	Twin Falls	42°08'19"N	114°20'51"W	
Savage Hollow Spring	Cassia	42°09'36"N	113°31'50"W	
Sawmill Gulch Spring	Lemhi	45°09'04"N	113°40'51"W	
Sawmill Spring	Franklin	42°05'41"N	111°39'42"W	
Sawmill Spring	Power	42°38'20"N	112°43'00"W	
School House Spring	Bannock	42°40'51"N	111°58'23"W	6570
Scott Spring	Owyhee	42°27'03"N	116°55'44"W	
Scratching Post Spring	Custer	44°14'32"N	113°55'58"W	
Scrub Spring	Elmore	43°11'48"N	115°11'55"W	5740
Secret Spring	Oneida	42°17'12"N	112°17'20"W	
Section 9 Spring	Cassia	42°03'03"N	114°00'07"W	
Section Spring	Twin Falls	42°12'05"N	114°22'50"W	
Section Twentyseven Spring	Owyhee	42°27'28"N	116°55'12"W	
Seep Spring	Washington	44°40'44"N	116°53'31"W	5000
Self Help Spring	Franklin	42°07'14"N	111°41'34"W	
Severe Spring	Cassia	42°11'34"N	114°05'28"W	
Shady Spring	Blaine	43°24'23"N	114°07'39"W	
Shamrock Spring	Clark	44°12'37"N	112°43'40"W	

Spring	County	Latitude	Longitude	Elev.
Sharkey Hot Spring	Lemhi	45°00'48"N	113°36'22"W	
Sharp Springs	Fremont	44°12'40"N	111°36'23"W	
Shaws Spring	Valley	44°52'23"N	116°02'03"W	
Sheef Spring	Bonner	47°59'09"N	116°38'20"W	2280
Sheep Creek Spring	Oneida	42°14'42"N	112°32'00"W	
Sheep Spring	Cassia	42°05'37"N	114°06'09"W	
Sheep Spring	Idaho	45°46'16"N	116°23'21"W	
Sheep Spring	Idaho	45°35'17"N	115°05'40"W	
Sheep Spring	Owyhee	42°56'13"N	116°58'41"W	
Sheep Spring	Owyhee	42°19'48"N	116°39'20"W	
Sheep Spring	Payette	44°04'47"N	116°37'42"W	
Sheep Spring	Twin Falls	42°05'59"N	114°19'29"W	
Sheep Springs	Bonner	47°59'00"N	116°46'55"W	
Sheepeater Hot Springs	Valley	44°37'41"N	115°11'49"W	
Sheepherder Spring	Owyhee	43°02'52"N	116°43'49"W	
Sheephorn Spring	Lemhi	44°52'55"N	114°06'09"W	
Shellworth Spring	Adams	44°49'00"N	116°12'14"W	
Sherma Spring	Bear Lake	42°05'56"N	111°26'43"W	6720
Sherman Springs	Teton	43°34'08"N	111°06'24"W	
Shirley G Spring	Lincoln	42°53'17"N	114°03'55"W	
Shirley Spring	Blaine	43°22'09"N	114°28'33"W	
Shoofly Springs	Owyhee	42°19'54"N	116°18'08"W	
Short Creek Spring Number 1	Custer	44°11'46"N	113°36'41"W	
Short Creek Spring Number 2	Custer	44°11'13"N	113°36'33"W	
Short Creek Spring Number 3	Custer	44°10'25"N	113°36'24"W	
Shoshone Spring	Twin Falls	42°12'45"N	114°20'43"W	
Shower Bath Springs	Custer	44°37'38"N	114°36'04"W	
Shrives Spring	Bannock	42°28'06"N	112°16'56"W	
Sibley Spring	Cassia	42°18'55"N	113°31'55"W	
Silver Creek Plunge	Valley	44°19'47"N	115°48'05"W	4850
Silver Hills Spring	Cassia	42°05'37"N	113°04'10"W	7040
Simms Spring	Camas	43°14'17"N	114°52'00"W	5270
Simpkins Spring	Elmore	42°59'02"N	115°28'10"W	
Sink Spring	Franklin	42°19'41"N	111°40'06"W	
Siver Springs	Valley	44°18'47"N	115°48'05"W	4850
Sixmile Spring	Cassia	42°07'35"N	113°09'31"W	
Skamfer Spring	Owyhee	42°20'56"N	116°56'54"W	
Skillern Hot Springs	Camas	43°38'49"N	114°48'47"W	
Skull Spring	Custer	44°07'32"N	114°14'49"W	
Skull Spring	Owyhee	43°21'39"N	116°58'56"W	
Skunk Spring	Idaho	45°12'13"N	115°36'03"W	
Slate Creek Hot Spring	Custer	44°10'16"N	114°37'28"W	
Slate Rock Spring	Franklin	42°09'40"N	111°40'00"W	
Slaughterhouse Spring	Lemhi	44°26'55"N	113°13'39"W	
Slaughterhouse Springs	Custer	43°51'53"N	113°34'37"W	
Slide Spring	Cassia	42°06'35"N	113°51'18"W	
Slide Spring	Owyhee	42°47'56"N	116°24'22"W	
Smith Spring	Lemhi	44°57'34"N	113°43'13"W	

Springs

Spring	County	Latitude	Longitude	Elev.
Smoky Spring	Adams	45°15'30"N	116°26'25"W	
Snipe Spring	Bannock	42°58'26"N	112°20'44"W	
Snow Creek Spring	Owyhee	42°43'08"N	116°19'03"W	
Snowmaker Spring	Owyhee	42°21'18"N	116°57'33"W	
Snyder Springs	Custer	44°15'14"N	114°37'33"W	
Soapstone Spring	Lemhi	44°45'34"N	113°18'45"W	
Soda Spring	Owyhee	43°02'07"N	116°53'56"W	
Soldier Spring	Idaho	45°19'29"N	115°25'28"W	
Soldier Spring	Twin Falls	42°14'29"N	114°26'10"W	
Sommercamp Spring	Owyhee	43°06'46"N	116°55'38"W	
Sorenson Spring	Butte	43°38'17"N	113°29'30"W	
Sorrel Spring	Custer	44°15'17"N	114°10'11"W	
South Canyon Spring	Oneida	42°19'27"N	112°17'04"W	
South Cottonwood Spring	Cassia	42°10'56"N	114°04'36"W	
South Fisher Spring	Cassia	42°07'34"N	113°10'37"W	5980
South Fork Fall Creek Spring	Cassia	42°22'02"N	113°03'52"W	6287
South Fork Fall Creek Spring	Cassia	42°22'00"N	113°04'07"W	6060
South Fork Kelsaw Spring	Cassia	42°05'02"N	113°08'36"W	6740
South Fork Spring	Bannock	42°28'08"N	112°19'36"W	
South Fork Spring	Power	42°19'49"N	112°59'01"W	
South Heglar Spring	Cassia	42°23'25"N	113°03'11"W	5995
South Spring	Owyhee	42°57'05"N	116°58'45"W	
South Willow Spring	Twin Falls	42°10'37"N	114°22'01"W	
Southeast Spring	Twin Falls	42°09'58"N	114°23'32"W	
Spar Mountain Spring	Custer	44°14'07"N	114°07'12"W	
Sparks Spring	Cassia	41°59'52"N	113°43'34"W	
Spring Gulch Spring	Custer	44°08'06"N	114°13'18"W	
Spring Ranch Spring	Owyhee	43°04'21"N	116°32'06"W	
Sproat Spring	Canyon	43°21'17"N	116°34'10"W	
Sprout Spring	Gem	44°19'43"N	116°13'24"W	5060
Spud Patch Spring	Blaine	43°17'22"N	114°15'43"W	
Squaw Creek Spring	Custer	44°11'11"N	113°37'27"W	
Squaw Joe Spring	Twin Falls	42°18'49"N	114°28'20"W	
Squaw Spring	Owyhee	42°32'09"N	116°15'17"W	
Squaw Spring	Owyhee	42°08'57"N	116°02'43"W	
Squaw Spring	Shoshone	47°00'23"N	116°04'51"W	
Squaw Springs	Butte	44°07'25"N	113°23'49"W	
Squaw Springs	Franklin	42°07'07"N	111°55'45"W	
Squawberry Spring	Idaho	46°26'36"N	115°35'24"W	
Stageroad Spring	Owyhee	42°47'48"N	116°34'01"W	
Stanford Spring	Owyhee	43°06'34"N	116°57'20"W	
Stanford Spring	Owyhee	42°26'58"N	116°50'48"W	
Stanley Hot Springs	Idaho	46°18'59"N	115°15'28"W	
Star Ranch Spring	Owyhee	42°26'48"N	116°46'45"W	
Starkey Hot Springs	Valley	44°51'22"N	116°26'38"W	3200
State 40 Spring	Owyhee	42°45'35"N	116°27'26"W	
Station Spring	Oneida	42°18'38"N	112°16'13"W	
Station Spring	Owyhee	42°42'21"N	116°30'19"W	
Steamboat Spring	Caribou	42°39'27"N	111°38'37"W	5720
Steele Spring	Fremont	44°12'25"N	111°36'18"W	
Steep Spring	Bear Lake	42°30'57"N	111°33'48"W	7270
Sterpernell Spring	Elmore	43°11'17"N	115°12'03"W	5635
Stinking Spring	Bonneville	43°36'44"N	111°37'26"W	
Stinking Spring	Clark	44°14'23"N	112°32'34"W	
Stockade Spring	Adams	45°03'23"N	116°40'55"W	5075
Stone Spring	Oneida	42°03'54"N	112°59'05"W	
Stoneman Creek Spring	Owyhee	42°33'49"N	116°45'03"W	
Stoneman Spring	Owyhee	42°36'58"N	116°46'25"W	
Stove Gulch Spring	Twin Falls	42°12'39"N	114°21'37"W	
Stove Spring	Owyhee	43°11'36"N	116°48'13"W	
Stovepipe Spring	Custer	44°15'46"N	114°42'21"W	
Stratton Spring	Twin Falls	42°12'43"N	114°26'22"W	
Strawberry Spring	Adams	44°42'21"N	116°34'53"W	4220
Strawberry Spring	Franklin	42°18'42"N	111°37'23"W	
Strawberry Spring	Twin Falls	42°00'35"N	114°57'43"W	
Strode Spring	Blaine	43°18'44"N	113°58'34"W	5020
Stuart Hot Springs	Idaho	46°08'17"N	115°05'23"W	
Stub Creek Spring	Idaho	45°21'14"N	114°45'13"W	
Study Spring	Power	42°41'33"N	112°40'21"W	
Sublett Troughs	Oneida	42°16'20"N	112°53'28"W	
Success Spring	Owyhee	43°07'47"N	116°36'04"W	
Sugarloaf Spring	Twin Falls	42°22'17"N	114°21'19"W	
Sullivan Hot Springs	Custer	44°15'17"N	114°26'35"W	
Sullivan Spring	Bonner	48°04'57"N	116°25'04"W	
Sullivan Spring	Owyhee	43°03'24"N	116°53'59"W	
Sulphur Spring	Caribou	42°38'43"N	111°30'18"W	
Summit Spring	Custer	44°17'18"N	114°06'55"W	
Summit Spring	Owyhee	42°42'45"N	116°22'05"W	
Summit Spring	Twin Falls	42°12'11"N	114°18'10"W	
Summit Spring Number One	Owyhee	42°55'57"N	116°57'56"W	
Summit Spring Number Two	Owyhee	42°56'04"N	116°58'02"W	
Summit Springs	Oneida	42°17'53"N	112°56'19"W	
Summit Springs	Owyhee	42°23'06"N	116°55'37"W	
Summit Springs	Shoshone	47°13'59"N	115°18'55"W	
Sunbeam Hot Springs	Custer	44°16'05"N	114°44'50"W	
Sunflower Hot Springs	Valley	44°43'46"N	114°59'34"W	4353
Sunnyside Spring	Oneida	42°16'23"N	112°54'25"W	
Sunset Spring	Blaine	43°23'24"N	114°03'12"W	6855
Surprise Spring	Owyhee	42°38'26"N	116°46'45"W	
Swamp Spring	Owyhee	42°37'29"N	116°47'01"W	
Swanty Spring	Cassia	42°04'53"N	114°12'21"W	
Sweat Springs	Clark	44°22'54"N	112°07'32"W	
Swede Spring	Cassia	42°05'00"N	114°05'05"W	
Sweet Anise Spring	Idaho	45°31'02"N	115°51'31"W	
Swisher Spring	Owyhee	42°25'38"N	116°44'24"W	
T L Spring	Owyhee	42°32'47"N	116°40'10"W	

Spring	County	Latitude	Longitude	Elev.
Table Spring	Idaho	45°22'57"N	114°52'42"W	
Tandem Spring	Owyhee	43°09'19"N	116°38'19"W	
Tank Spring	Twin Falls	42°09'42"N	114°48'24"W	
Tate Spring	Elmore	43°25'56"N	115°21'41"W	
Taylor Spring	Boise	44°05'12"N	115°08'22"W	
Taylor Spring	Cassia	42°06'09"N	113°43'45"W	
Taylor Spring	Oneida	42°16'02"N	112°33'16"W	
Teacup Spring	Owyhee	42°30'38"N	116°45'31"W	
Teakettle Spring	Owyhee	42°30'20"N	116°45'36"W	
Teapot Spring	Elmore	43°10'33"N	115°29'05"W	4070
Techick Spring	Butte	43°36'47"N	113°29'42"W	
Teds Spring	Owyhee	42°00'30"N	115°57'53"W	6985
Telephone Pole Spring	Lemhi	45°24'57"N	114°22'45"W	
Telephone Spring	Adams	44°34'46"N	116°19'47"W	4580
Telephone Spring	Cassia	42°09'35"N	114°13'07"W	
Tent Spring	Cassia	42°04'29"N	114°05'29"W	
Tepee Springs	Idaho	45°12'32"N	116°16'15"W	
Terrells Spring	Cassia	42°14'10"N	114°08'25"W	
Terteling Springs	Ada	43°40'36"N	116°12'28"W	
Texas Basin Spring	Owyhee	43°13'00"N	116°57'46"W	
Thimbleberry Spring	Franklin	42°24'14"N	111°38'28"W	
Third Spring	Custer	44°39'18"N	114°17'44"W	6780
Thistle Spring	Owyhee	42°36'23"N	116°31'53"W	
Thomas Creek Spring	Owyhee	42°28'12"N	116°51'59"W	
Thomas Davis Springs	Oneida	42°04'37"N	112°21'40"W	
Thomas Spring	Franklin	42°08'45"N	111°38'28"W	
Thomas Spring	Idaho	45°33'42"N	116°22'27"W	
Thompson Spring	Bear Lake	42°32'56"N	111°17'33"W	
Thorn Spring	Idaho	45°53'52"N	116°38'52"W	
Thorne Spring	Franklin	42°09'11"N	111°41'45"W	
Thorne Spring	Owyhee	42°56'08"N	116°57'28"W	
Thoroughbred Pond Spring	Cassia	42°03'04"N	114°13'24"W	
Thoroughbred Spring	Elmore	43°10'09"N	115°11'29"W	
Thoroughbred Spring	Twin Falls	42°09'23"N	114°19'57"W	
Thoroughbred Spring Number 1	Cassia	42°03'20"N	114°13'39"W	
Thoroughbred Spring Number 2	Cassia	42°02'19"N	114°14'28"W	
Thousand Springs	Gooding	42°44'50"N	114°50'33"W	
Three Forks Summit Springs	Owyhee	42°05'14"N	116°03'04"W	
Three Springs	Clark	44°13'34"N	112°31'19"W	
Three Springs	Owyhee	42°26'59"N	116°51'29"W	
Threemile Spring	Owyhee	42°06'37"N	114°38'39"W	
Threemile Spring	Twin Falls	42°06'35"N	114°37'55"W	
Thunder Spring	Clearwater	46°46'12"N	115°47'31"W	
Tiddie Spring	Owyhee	43°06'55"N	116°41'44"W	
Tie Camp Spring	Franklin	42°24'53"N	111°36'55"W	
Tigert Spring	Owyhee	42°38'22"N	116°10'35"W	
Timber Butte Spring	Blaine	43°24'42"N	113°48'24"W	5640
Timber Spring	Twin Falls	42°03'59"N	114°18'09"W	

Spring	County	Latitude	Longitude	Elev.
Time Spring	Bannock	42°28'42"N	111°55'46"W	
Timmerman Spring	Blaine	43°19'13"N	114°16'14"W	
Timmons Field Spring	Elmore	43°17'06"N	115°30'05"W	
Tincan Spring	Lemhi	45°33'13"N	114°08'31"W	
Tincup Spring	Lemhi	45°30'16"N	114°05'56"W	
Tingley Spring	Benewah	47°21'35"N	116°24'43"W	
Tipton Spring	Blaine	43°16'21"N	114°05'34"W	5220
Toe Jam Spring	Washington	44°13'26"N	116°40'55"W	
Tom Gooding Spring	Camas	43°14'39"N	114°25'30"W	
Tommy Field Spring	Elmore	43°27'27"N	115°37'33"W	4300
Toms Canyon Spring	Bannock	42°19'41"N	112°09'37"W	
Tony Seyfreid Spring	Idaho	45°24'37"N	116°25'06"W	
Tony Springs	Owyhee	43°06'38"N	116°53'17"W	
Toolson Spring	Caribou	42°42'16"N	111°56'09"W	
Towsley Spring	Adams	45°09'22"N	116°38'48"W	
Toy Seep	Owyhee	42°45'57"N	116°34'07"W	5794
Trail Canyon Spring	Cassia	42°11'45"N	114°15'57"W	
Trail Canyon Spring	Fremont	44°14'40"N	111°13'24"W	
Trail Creek Spring	Adams	45°14'06"N	116°26'08"W	
Trail Creek Spring	Bannock	42°49'56"N	112°30'47"W	
Trail Spring	Bear Lake	42°33'01"N	111°35'18"W	6945
Trail Spring	Owyhee	43°01'36"N	116°58'49"W	
Tree Spring	Gem	44°20'55"N	116°16'31"W	
Tribe Spring	Washington	44°41'08"N	116°53'51"W	4980
Triple Spring	Cassia	42°04'45"N	114°05'27"W	
Triplet Spring	Owyhee	42°00'18"N	115°42'47"W	
Trouble Spring	Bear Lake	42°32'21"N	111°37'27"W	8180
Trout Creek Pass Spring	Cassia	42°05'23"N	114°10'09"W	
Trout Creek Spring	Caribou	42°27'42"N	111°39'12"W	
Trout Spring	Cassia	42°04'39"N	114°10'30"W	
Trout Spring	Owyhee	42°26'37"N	116°51'54"W	
Tuana Spring	Twin Falls	42°52'31"N	115°00'13"W	
Tub Spring	Cassia	42°16'32"N	114°04'58"W	
Tub Spring	Cassia	42°06'43"N	114°10'55"W	
Tub Spring	Custer	44°15'48"N	114°10'48"W	
Tub Spring	Gooding	43°08'43"N	115°03'19"W	
Tub Spring	Owyhee	42°24'37"N	116°53'04"W	
Tucker Springs	Gooding	42°46'12"N	114°52'22"W	
Tunnel Hill Spring	Twin Falls	42°00'34"N	114°18'47"W	
Turner Spring	Owyhee	43°04'35"N	116°58'51"W	
Turner Spring	Owyhee	42°23'14"N	116°08'14"W	
Twin Deer Springs	Elmore	43°09'08"N	115°16'01"W	5097
Twin Spring	Owyhee	43°11'22"N	116°50'37"W	
Twin Springs	Cassia	42°05'18"N	114°01'44"W	
Twin Springs	Clearwater	46°42'50"N	116°12'37"W	
Twin Springs	Elmore	43°06'04"N	115°13'20"W	
Twin Springs	Oneida	42°15'50"N	112°45'50"W	
Twin Springs	Owyhee	42°29'31"N	116°53'19"W	

Springs

Spring	County	Latitude	Longitude	Elev.
Twin Springs	Owyhee	42°17'37"N	116°41'20"W	
Twin Springs	Owyhee	42°13'52"N	115°54'33"W	
Twin Springs	Owyhee	42°02'54"N	114°38'38"W	
Two Buck Spring	Idaho	45°37'21"N	114°33'34"W	
Two Springs	Owyhee	42°26'20"N	116°58'04"W	
U C Spring	Twin Falls	42°02'38"N	114°18'49"W	
Uphill Water Hole	Lincoln	43°10'05"N	114°34'23"W	
Upper Birch Spring	Owyhee	42°50'08"N	116°22'50"W	
Upper Coal Pit Spring	Cassia	42°14'17"N	114°10'46"W	
Upper Crystal Spring	Clark	44°12'06"N	112°42'30"W	
Upper Fall Creek Spring	Cassia	42°22'32"N	113°04'03"W	6030
Upper Heglar Spring	Cassia	42°24'58"N	113°00'15"W	
Upper Hogpen Spring	Owyhee	42°51'06"N	116°23'34"W	
Upper Lake Fork Spring	Cassia	42°23'33"N	113°02'13"W	6030
Upper Lost Valley Spring	Owyhee	42°23'30"N	116°15'47"W	
Upper Rats Nest Spring	Owyhee	43°23'19"N	116°48'55"W	
Upper Spring	Bannock	42°37'00"N	111°57'37"W	
Upper Spring	Bingham	43°14'48"N	111°42'18"W	6625
Upper Spring	Oneida	42°09'59"N	112°29'40"W	
Upper Squaw Creek Spring	Cassia	42°11'33"N	114°06'43"W	7200
Valieaux Spring	Lemhi	45°27'37"N	114°20'52"W	
Valve House Spring	Bannock	42°41'25"N	112°23'24"W	
Van Eaton Spring	Twin Falls	42°07'27"N	114°21'15"W	
Vulcan Hot Springs	Valley	44°34'02"N	115°41'42"W	
W P A Spring	Cassia	42°00'16"N	114°11'40"W	
Wagon Spring	Twin Falls	42°13'35"N	114°18'47"W	
Wagonbox Spring	Lemhi	44°47'47"N	113°17'44"W	
Wagonhammer Spring	Lemhi	45°23'23"N	113°57'54"W	
Wagonwheel Spring	Twin Falls	42°12'46"N	114°18'23"W	
Walker Spring	Butte	43°43'15"N	113°11'11"W	
Walters Spring	Lemhi	44°29'56"N	113°04'53"W	
Warfield Hot Spring	Blaine	43°38'28"N	114°29'12"W	
Warm River Spring	Fremont	44°12'19"N	111°14'59"W	
Warm Spring	Adams	44°57'58"N	116°12'15"W	
Warm Spring	Bingham	43°02'14"N	112°00'12"W	
Warm Spring	Caribou	42°56'19"N	112°00'04"W	
Warm Spring	Caribou	42°54'39"N	111°33'24"W	
Warm Spring	Cassia	42°13'35"N	113°46'43"W	
Warm Spring	Custer	44°22'55"N	114°05'15"W	
Warm Spring	Elmore	43°00'07"N	115°07'27"W	
Warm Spring	Jerome	42°36'56"N	114°27'54"W	
Warm Spring	Lemhi	44°57'05"N	114°42'19"W	
Warm Spring	Power	42°44'35"N	112°34'35"W	
Warm Springs	Bonneville	43°35'55"N	111°27'55"W	
Warm Springs	Clark	44°15'25"N	112°38'21"W	
Warm Springs	Custer	43°58'36"N	113°49'49"W	
Warm Springs	Idaho	45°51'22"N	114°56'17"W	
Warm Springs	Lemhi	44°36'44"N	113°21'49"W	
Warm Springs	Owyhee	42°04'14"N	116°05'25"W	
Warm Springs	Power	42°32'50"N	112°53'51"W	
Watchabob Springs	Owyhee	42°04'31"N	116°01'07"W	
Water Canyon Spring	Cassia	42°27'51"N	113°34'50"W	
Water Cress Spring	Camas	43°17'54"N	114°26'26"W	
Water Crest Spring	Bonneville	43°33'47"N	111°26'37"W	
Watercress Spring	Blaine	43°23'10"N	113°55'04"W	6060
Watercress Spring	Power	42°43'15"N	112°34'13"W	
Waterhouse Spring	Bannock	42°59'03"N	112°22'15"W	
Wayland Hot Springs	Franklin	42°07'58"N	111°55'38"W	
Waylett Spring	Power	42°34'09"N	112°37'04"W	
Weaner Spring	Owyhee	42°32'22"N	116°46'30"W	
Weasel Spring	Adams	44°56'17"N	116°33'29"W	4800
Weatherby Springs	Elmore	42°59'24"N	115°52'24"W	
Webb Spring	Butte	43°24'47"N	113°01'29"W	
Weir Creek Hot Springs	Idaho	46°27'48"N	115°02'06"W	
Weiser Warm Springs	Washington	44°17'53"N	117°02'59"W	
Welcome Spring	Bannock	42°32'25"N	112°07'15"W	
West Dry Canyon Spring	Cassia	42°09'55"N	113°08'53"W	
West Fork Springs	Power	42°34'41"N	112°37'33"W	
West Spring	Custer	44°11'03"N	113°52'30"W	
West Spring	Owyhee	42°46'17"N	116°32'28"W	
Whir Spring	Twin Falls	42°12'45"N	114°22'49"W	
Whiskey Spring	Cassia	42°19'08"N	114°07'19"W	
Whiskey Spring	Elmore	43°09'35"N	115°12'17"W	
Whiskey Springs	Custer	44°05'30"N	113°50'45"W	
Whisky Spring	Lemhi	45°24'37"N	114°04'01"W	
White Arrow Hot Spring	Idaho	43°02'58"N	114°57'05"W	3310
White Colt Spring	Custer	44°14'48"N	114°14'51"W	
White Rock Spring	Bingham	43°04'42"N	112°12'37"W	
White Rock Springs	Shoshone	46°59'29"N	116°06'38"W	4968
White Spring	Bonneville	43°17'36"N	111°18'57"W	
Whiteside Spring	Owyhee	42°04'08"N	115°05'58"W	5860
Whitey Davis Spring	Cassia	41°59'49"N	114°13'43"W	
Whitson Spring	Elmore	43°25'46"N	115°22'22"W	
Whittle Spring	Twin Falls	42°07'49"N	114°22'47"W	
Wild Horse Spring	Oneida	42°17'14"N	112°54'24"W	
Wild Horse Spring	Owyhee	43°17'20"N	116°59'18"W	
Wildcat Spring	Owyhee	43°27'24"N	116°52'41"W	
Wildcat Spring	Power	42°50'33"N	112°35'10"W	
Wildhorse Spring	Owyhee	42°29'26"N	116°16'50"W	
Wildhorse Springs	Owyhee	42°23'58"N	116°52'25"W	
Williams Creek Spring	Franklin	42°21'25"N	111°39'43"W	
Williams Spring	Twin Falls	42°16'52"N	114°28'13"W	5740
Williams Spring	Twin Falls	42°16'51"N	114°28'13"W	
Willingger Spring	Owyhee	43°04'33"N	116°51'24"W	
Willinicker Spring	Owyhee	43°03'51"N	116°53'34"W	
Willis Spring	Twin Falls	42°02'29"N	114°19'47"W	

Spring	County	Latitude	Longitude	Elev.
Willow Flat Spring	Bear Lake	42°34'24"N	111°36'41"W	
Willow Patch Spring	Bear Lake	42°19'01"N	111°31'29"W	
Willow Spring	Adams	44°56'24"N	116°35'54"W	5260
Willow Spring	Bear Lake	42°19'11"N	111°11'09"W	
Willow Spring	Gooding	43°11'37"N	114°47'22"W	
Willow Spring	Lemhi	44°17'01"N	113°07'54"W	
Willow Spring	Oneida	42°05'05"N	112°10'49"W	
Willow Spring	Owyhee	43°13'39"N	116°36'27"W	
Willow Spring	Owyhee	42°22'15"N	116°24'09"W	
Willow Spring	Owyhee	42°01'53"N	115°33'31"W	
Willow Spring	Owyhee	42°01'40"N	115°49'37"W	
Willow Spring	Twin Falls	42°02'00"N	114°18'09"W	
Willow Springs	Bonneville	43°16'58"N	111°36'27"W	6585
Willow Springs	Butte	44°12'59"N	113°14'57"W	
Wilson Creek Spring	Owyhee	42°29'39"N	116°46'50"W	
Wilson Spring	Cassia	42°06'59"N	113°54'16"W	
Winchell Spring	Caribou	42°34'55"N	111°37'55"W	6760
Windy Point Spring	Owyhee	43°13'44"N	116°43'26"W	
Winecup Spring	Twin Falls	42°06'13"N	114°17'01"W	
Winter Spring	Twin Falls	42°05'56"N	114°30'41"W	5860
Womack Canyon Spring	Cassia	42°05'55"N	113°29'59"W	
Wonder Spring	Owyhee	42°24'43"N	116°48'31"W	
Wood Canyon Spring	Oneida	42°13'41"N	112°30'55"W	
Wood Road Spring	Custer	43°57'20"N	113°51'43"W	
Woodall Spring	Caribou	42°46'42"N	111°32'06"W	
Wooden Shoe Spring	Twin Falls	42°16'07"N	114°17'05"W	
Worm Basin Spring	Franklin	42°10'02"N	111°40'08"W	
Worm Corral Spring	Cassia	42°11'08"N	114°07'30"W	
Worm Creek Spring	Bear Lake	42°09'08"N	111°32'27"W	
Worswick Hot Springs	Camas	43°33'52"N	114°47'49"W	
Worthington Spring	Cassia	42°00'09"N	113°53'02"W	
Wurst Spring	Blaine	43°17'46"N	114°18'54"W	
Wursten Spring	Oneida	42°26'41"N	112°18'14"W	
Y Springs	Lemhi	44°16'50"N	113°18'26"W	
Yandell Springs	Bingham	43°06'52"N	112°10'01"W	
Yarrow Spring	Owyhee	42°29'43"N	116°46'21"W	
Yellow Bull Spring	Idaho	46°05'54"N	116°09'49"W	
Yellow Jacket Spring	Oneida	42°16'40"N	112°54'20"W	
Yellow Jacket Spring	Twin Falls	42°10'05"N	114°21'35"W	
Young Spring	Cassia	42°25'46"N	113°06'25"W	5460
Young Spring	Oneida	42°17'46"N	112°54'42"W	

Spring	County	Latitude	Longitude	Elev.

Personal Waypoints

Feature Name	County	Latitude	Longitude	Elev.

Feature Name	County	Latitude	Longitude	Elev.

Feature Name	County	Latitude	Longitude	Elev.

Feature Name	County	Latitude	Longitude	Elev.

Also Available From Glassford Publishing:

GPS Waypoints: Arizona	**304p**	**$17.95**	**ISBN 1-892182-04-1**
GPS Waypoints: Colorado	**256p**	**$15.95**	**ISBN 1-892182-08-4**
GPS Waypoints: Idaho	**160p**	**$12.95**	**ISBN 1-892182-16-5**
GPS Waypoints: Oregon	**288p**	**$16.95**	**ISBN 1-892182-41-6**
GPS Waypoints: Utah	**208p**	**$14.95**	**ISBN 1-892182-49-1**
GPS Waypoints: Washington	**208p**	**$14.95**	**ISBN 1-892182-53-X**

This series of books provide state-by-state collections of coordinates for summits, lakes, campgrounds, airstrips, highway intersections, and many other categories. They provide GPS users with instant access to thousands of precise waypoints. These books are an essential accessory for anyone who uses GPS for personal navigation, whether on the highways or in the backcountry. Prices range from $12.95 to $18.95, depending on the state. Titles available as of this printing are listed above. Call or write for an updated list.

GPS Land Navigation 272p $19.95 ISBN 0-9652202-5-7

This is the definitive book for land-based users of GPS for personal navigation. Besides its thorough and easy to understand coverage of GPS, this book also provides extensive sections on maps, compasses, altimeters, map-reading tools, and the traditional (pre-GPS) navigation methods. It contains over 150 illustrations and three waypoint appendices. If you want to become an expert GPS land navigator, ***GPS Land Navigation*** is the book for you.

Forthcoming Titles:

GPS: Easy Navigation For Everyone ISBN 0-9652202-6-5

This book is scheduled for release in early 1999. It is a basic primer on how to get the most from your GPS receiver with the least amount of effort. Weighing in at under 100 pages and priced less than $10, it is the "little brother" to ***GPS Land Navigation***, the most complete book available on using GPS for personal navigation. Perfect for field use, anyone who uses GPS will want to keep ***GPS: Easy Navigation For Everyone*** handy. It fills in the many gaps left by your receiver's manual.

Ordering Instructions And Further Information:

U.S. residents may purchase books directly from Glassford Publishing. Be sure to indicate the title, ISBN number, price, and quantity of each book you wish to order. Add $2.00 per order (not per book) for shipping charges. Send a check or money order (no cash, please) to the address below. Write, call, or fax for an updated list of available titles.

Glassford Publishing
P.O. Box 2895
Boise ID 83701-2895

208-343-9205
208-344-2335 (fax)